LICHENOLOGY: PROGRESS AND PROBLEMS

THE SYSTEMATICS ASSOCIATION
SPECIAL VOLUME NO. 8

LICHENOLOGY: PROGRESS AND PROBLEMS

*Proceedings of an International Symposium
held at the University of Bristol*

Edited by

D. H. BROWN

Department of Botany, University of Bristol

D. L. HAWKSWORTH

Commonwealth Mycological Institute, Kew, Surrey

and

R. H. BAILEY

Department of Extra-mural Studies, University of London

1976

Published for the

SYSTEMATICS ASSOCIATION *and the* BRITISH LICHEN SOCIETY

by

ACADEMIC PRESS · LONDON · NEW YORK · SAN FRANCISCO

ACADEMIC PRESS INC. (LONDON) LTD.
24–28 Oval Road
London NW1 7DX

U.S. Edition published by
ACADEMIC PRESS INC.
111 Fifth Avenue
New York, New York 10003

Library of Congress Catalog Card Number: 75-19618
ISBN: 0 12 136750 9

PRINTED IN GREAT BRITAIN BY
T. & A. CONSTABLE LTD., EDINBURGH

Contributors and Participants

Contributors are indicated by an asterisk.

*ARMSTRONG, Dr R. A., *Botany School, University of Oxford, South Parks Road, Oxford OX1 3RA, England.*

ASPERGES, M., *Rijksuniversitair Centrum Antwerpen, Groenenborgerlaan 171, 2020 Antwerpen, Belgium.*

*BADDELEY, Dr M. S., *The Open University, Elden House, Regent's Centre, Gosforth, Newcastle upon Tyne, England.*

*BAILEY, R. H., *Department of Extra-mural Studies, University of London, 7 Ridgmount Street, London WC1E 7AD, England.*

BATES, J. W., *Department of Botany, The University, Bristol BS8 1UG, England.*

BECKETT, P. J., *Department of Botany and Microbiology, University College of Swansea, Singleton Park, Swansea SA2 8PP, Wales.*

BOUSFIELD, Miss J., *School of Pharmacy and Biology, Sunderland Polytechnic, Chester Road, Sunderland SR1 3SD, England*

BRAND, M., *Piet Heinstraat 7, Groningen, Netherlands.*

BRIGHTMAN, F. H., *British Museum (Natural History), Cromwell Road, London SW7 5BD, England.*

*BROWN, Dr D. H., *Department of Botany, The University, Bristol BS8 1UG, England.*

BURNET, Miss A. M., *Appledore, 26 Green Lane, Ford, Salisbury, Wiltshire, England.*

*COPPINS, B. J., *Royal Botanic Garden, Inverleith Row, Edinburgh EH3 5LR, Scotland.*

DILKS, T. J. K., *Department of Biological Sciences, University of Exeter, Exeter EX4 4PS, England.*

DOBBEN, H. van, *Mariaplaats 16, Utrecht, Netherlands.*

DOBBS, Dr G., *University College of North Wales, Bangor, Caernarvonshire, Wales.*

*FARRAR, Dr J. F., *Imperial College Field Station, Silwood Park, Ascot, Berkshire, England.*†

FEARN, Dr G. M., *Department of Chemistry and Biology, Sheffield Polytechnic, Sheffield S1 1WB, England*

*FERRY, Dr B. W., *Botany Department, Bedford College, Regent's Park, London NW1 4NS, England.*

*FLETCHER, Dr A., *Department of Marine Biology, University College of North Wales, Marine Sciences Laboratories, Menai Bridge, Anglesey LL59 5EH, Wales.*

FOX, Dr B. W., *Tryfan, Longlands Road, New Mills, Stockport SK12 3BL, England.*

---

†*Present address: Department of Botany, University of Dar es Salaam, P.O. Box 35060, Dar es Salaam, Tanzania.*

GALLOWAY, Dr D. J., *Botany Division, D.S.I.R., Christchurch, New Zealand.*

GALUN, Professor M., *Tel-Aviv University, Department of Botany, Ramat Aviv, Tel-Aviv, Israel.*

GILBERT, Dr O. L., *Department of Landscape Architecture, The University, Sheffield S10 2TN, England.*

*HALE, Dr M. E., *Department of Botany, National Museum of Natural History, Smithsonian Institution, Washington, D.C. 20560 U.S.A.*

HARDING, D. A., *Juniper Hall Field Centre, Mickleham, Dorking, Surrey, England.*

*HAWKSWORTH, Dr D. L., *Commonwealth Mycological Institute, Ferry Lane, Kew, Richmond, Surrey TW9 3AF, England.*

*HENSSEN, Professor Dr A., *Fachbereich Biologie, Botanik, Universität Marburg, Lahnberge, D-355 Marburg/Lahn, Germany.*

HICKMOTT, Mrs M., *Oakdene, Greyfield, High Littleton, Somerset BS18 SXX, England.*

*HILL, Dr D. J., *Department of Plant Biology, The University, Newcastle upon Tyne NE1 7RU, England.*†

HITCH, C. J. B., *Thackeray House, Hadley Green Road, Barnet, Hertfordshire, England.*

HOLLOWELL, A., *81 Cranbrook Road, Redland, Bristol BS6 7BZ, England.*

*JAMES, P. W., *Department of Botany, British Museum (Natural History), Cromwell Road, London SW7 5BD, England.*

JØRGENSEN, P. M., *Botanical Garden, University of Bergen, Box 12, N-5014 Bergen, Norway.*

KARNEFELT, I., *University of Lund, Department of Plant Taxonomy, Ö. Vallgatan 18, 223 61 Lund, Sweden.*

KEUCK, Mrs A., *355 Marbach, Brunnenstrasse 3, Germany*

KEUCK, G., *355 Marbach, Brunnenstrasse 3, Germany.*

KRISTINSSON Dr H., *Náttúrugripasafnið á Akureyri, P.O. Box 580, Akureyri, Iceland.*

KROG, Dr H., *Botanical Museum, Oslo 5, Norway.*

LALLEMANT, R., *Université de Paris VI, Laboratoire de Cryptogamie, 9 Quai Saint Bernard, 75005 - Paris, France.*

LAUNDON, J. R., *Department of Botany, British Museum (Natural History) Cromwell Road, London SW7 5BD, England.*

LLOYD, A. O., *90 Handside Lane, Welwyn Garden City, Hertfordshire, AL8 6SN, England.*

MANGAN, Miss A., *4 Waltham Terrace, Blackrock, Co. Dublin, Ireland.*

MARGOT, Dr J., *Laboratorie de Botanique, Facultes Universitaires Notre-Dame de la Paix, Rue de Bruxelles, 61, B5000 Namur, Belgium.*

*MILLBANK, Dr J. W., *Botany Department, Imperial College of Science and Technology, London SW7, England.*

MOORBATH, P., *37 Southwood Gardens, Newcastle upon Tyne NE3 3BY, England.*

---

†*Present address: Department of Extra-mural Studies, The University, Bristol BS8 1HR, England.*

MOORE, C., 111 North Circular Road, Dublin 7, Ireland.

MOULT, Miss P. H., Botany Department, Imperial College of Science and Technology, London SW7, England.

NETHERCOTT, P. J. M., 6 Hazlewood Road, Bristol BS9 1PU, England

*NOURISH, R., Department of Chemistry and Applied Chemistry, University of Salford, Salford M5 4WT, England.

*OLIVER, Dr R. W. A., Department of Chemistry and Applied Chemistry, University of Salford, Salford M5 4WT, England.

OUTRED, Dr H. A., Department of Agricultural Science, Parks Road, Oxford OX1 3PF, England.

PEAT, A., Department of Biology, Sunderland Polytechnic, Sunderland SR1 3SD, England.

*PEVELING, Professor Dr E., Botanisches Institut der Universität, Schlossgarten 3, 44 Münster (Westf.), Germany.

*POELT, Professor Dr J., Institut für systematische Botanik der Universität Graz, Holteigasse 6, A-8010 Graz, Austria.

PROCTOR, Dr M. C. F., Department of Biological Sciences, University of Exeter, Exeter EX4 4PS, England.

RICHARDSON, Dr D. H. S., Department of Biology, Laurentian University, Sudbury, Ontario, Canada.

RODGERS, G. A., Department of Biological Sciences, The University of Dundee, Dundee DD1 4HN, Scotland.

*ROSE, Dr F., Department of Geography, Kings College, University of London, Strand, London WC2R 2LS, England.

SANTESSON, Professor Dr R., Swedish Museum of Natural History, Section of Botany, S-104 05 Stockholm 50, Sweden.

*SEAWARD, Dr M. R. D., Trinity and All Saints' Colleges, Horsforth, Nr. Leeds, England.†

SIPMAN, Dr H., Instituut voor Systematische Plantkunde von de Rijksuniversiteit te Utrecht, Botanisch Museum en Herbarium, Transitorium 11, Heidelberglaan 2, De Uithof, Utrecht, The Netherlands.

SKINNER, J., Department of Botany, University of Durham, Science Laboratories, South Road, Durham D1, England.

*SMITH, Professor D. C., Department of Agricultural Science, Parks Road, Oxford OX1 3PF, England.‡

STARKEY, B. T., 3A Loxton Road, Forest Hill, London SE23 2ET, England.

TALLOWIN, S. N., The Moat, Llandyry, Nr. Kidwelly, Caernarvonshire SA17 4EO, Wales.

TAPPER, R. C., Wadham College, Oxford, England.

TIBELL, L., Institute of Systematic Botany, University of Uppsala, P.O. Box 541, S-751 21 Uppsala, Sweden.

*TSCHERMAK-WOESS, Professor Dr E., Botanisches Institut und Botanischer Garten der Universität Wien, 111/40 Rennweg 14, Wien, Austria.

---

†Present address: Postgraduate School of Studies in Environmental Science, University of Bradford, Bradford BD7 1DP, England.

‡Present address: Department of Botany, The University, Bristol BS8 1UG, England.

VÄNSKÄ, H., *Department of Botany, University of Helsinki, Unioninkatu 44, SF-00170 Helsinki 17, Finland.*

VITIKAINEN, O., *Department of Botany, University of Helsinki, Unioninkatu 44, SF-00170 Helsinki 17, Finland.*

VOBIS, G., *Fachbereich Biologie, Botanik, Universität Marburg, Lahnberge, D–355 Marburg/Lahn, Germany.*

WALL, Miss C., *Department of Plant Sciences, King's College, 68 Half Moon Lane, Herne Hill, London SE24, England.*

WARREN, D. H. E., *145 Parkstone Avenue, Parkstone, Poole, Dorset BH14 9LP, England.*

Preface

The Systematics Association, founded in 1937, has as one of its objectives to encourage and assist taxonomic research. In 1940, when the Association published "The New Systematics" (J. Huxley, ed.), systematics was a subject which was undergoing many changes; basic premises of earlier periods were being questioned and new concepts were beginning to be formulated. Lichenology has, like the subjects of its study, grown slowly compared with most scientific disciplines. However, over the last decade in particular, lichenology has become a much more dynamic area of research so that lichenologists of the 1970s find themselves in a comparable position to the founders of the Systematics Association in the 1930s. It was very fitting, therefore, for the Systematics Association and the British Lichen Society to join together at such a time to enable the progress that has been made in the study of lichens to be examined critically.

The subjects discussed at this Symposium, held at the University of Bristol on 8-10 April 1974, were selected to reflect, without encyclopaedic coverage, the current style of lichenological research. Certain topics were chosen because of the rapid progress which had been achieved in them, others because of the application of the new techniques to old areas of study and others because they had not been reviewed in recent years. Such diverse reasons for selecting particular topics is reflected in the different approaches taken by the authors. The brief given to authors was to emphasize, as far as possible, the progress that had been achieved but also to indicate the areas requiring further study. Where the topic had been recently reviewed, further repetition was considered unnecessary. The result has been that authors frequently refer to Ferry, Baddeley and Hawksworth (1973) "Air Pollution and Lichens", Henssen and Jahns (1973★) "Lichenes", and Ahmadjian and Hale (1974★) "The Lichens".

In combination with the publications mentioned above, the present volume attests the vitality of lichenology at the present time. Similarly, it is doubtful whether five years ago such a symposium could have been organised or achieve such a high attendance; over 70 participants representing 15 nationalities. Lichen research is at the stimulating stage where, because one is dealing with a somewhat overlooked group of plants, significant advances can be made in many fields.

★ These works were published on 6 December 1973 and 25 March 1974, respectively, and are cited throughout this volume with these dates, which differ from those on their title pages. In the reference lists in the case of these and other works where the date of publication is incorrectly stated on the volume the date on the work is indicated in square brackets, e.g. Henssen and Jahns (1973) ["1974"].

Undoubtedly one reason for this resurgence of interest in lichens is the result of the substantial progress which has been made, demonstrating that these plants have much to recommend them for basic research on the metabolism of composite, symbiotic, organisms, and as aids in the study of pollution and environmental change.

The editors of this Symposium volume are very grateful to Professor D. C. Smith, whose Chairmanship of the organising committee undoubtedly contributed to the successful formulation of the programme. The British Lichen Society organized a pre-Symposium field meeting, which had a multi-national attendance, led by Mr R. H. Bailey and Dr D. H. Brown. This, and the skilful introduction to the Symposium given by Dr M. E. Hale, set the tone for a very friendly meeting. Thanks are also due to Dr D. H. Brown, Dr O. L. Gilbert, Dr M. E. Hale, Mr J. R. Laundon and Professor D. C. Smith for chairing various sessions, and to Professor Dr R. Santesson for his humorous concluding remarks. The expert assistance of Academic Press in the preparation of this volume is gratefully acknowledged. Our final thanks go, on behalf of the British Lichen Society and the Symposium participants, to the Systematics Association, for without their financial support and encouragement the meeting would have been little more than a pipe-dream.

August 1975   D. H. BROWN
                                                D. L. HAWKSWORTH
                                                R. H. BAILEY

Contents

1 | Lichen Structure viewed with the Scanning Electron Microscope

M. E. HALE

Department of Botany, National Museum of Natural History, Smithsonian Institution, Washington, D.C., U.S.A.

Abstract: The scanning electron microscope (SEM) has permitted viewing of opaque lichen structures with greatly increased depth of field clarity and resolution, and at a much higher magnification than possible by light microscopy. Internal tissues, as seen in cross sections, are also more easily and quickly studied with SEM than with the usual slide preparations. The main result of research so far has been a better comprehension of the structure of rhizines and cilia, cephalodial development and internal organization of the thallus.

The SEM has also uncovered new, previously unsuspected, structures, the most significant being a pored epicortical layer in some of the genera of the Parmeliaceae and Coccocarpiaceae, which potentially functions in gas exchange. Aculeate hyphae coming from the upper cortex have been detected in the Graphidaceae and Thelotremataceae. For the first time crystals of lichen substances and their manner of deposition have also been clearly seen. Much work remains to be done on reproductive structures and spores.

INTRODUCTION

The study of lichen structures began well over a century ago when the light microscope became readily available. By 1860, for example, Schwendener (1860) was able to present an accurate account of the internal structure of several fruticose lichens. While cells can be magnified hundreds of diameters with a light microscope, the specimens must first be sectioned, stained, and viewed in only two dimensions. Translucent and transparent structures are essentially invisible and opaque objects cannot be examined at all. The transmission electron microscope (EM), while greatly extending the range of magnification, suffers from the same problems of specimen preparation as light transmission

Systematics Association Special Volume No. 8, "Lichenology: Progress and Problems", edited by D. H. Brown, D. L. Hawksworth and R. H. Bailey, 1976, pp. 1-15, Academic Press, London and New York.

microscopy. The newest stereoscopic binocular microscopes have permitted three-dimensional viewing of opaque objects at high working magnifications ($\times 100$–$\times 200$) but still below the limit of good visibility of most individual cells.

Since 1965 we have had available an entirely new tool, the scanning electron microscope (SEM), which combines the advantages of the high magnification of an EM and the depth of field of a low power dissecting microscope, revealing objects in lifelike three dimensions (Plate Ia). It is even possible to take pairs of photographs which can be viewed under a stereoscope (cf. Hale, 1973). The effective range of magnification is from $\times 20$ to more than $\times 20\ 000$. We are able to see well known structures with incredible clarity and gain a new appreciation of them, as well as discover others difficult or impossible to detect with past methods. High resolution photographs can be obtained and are far superior to drawings.

There are, of course, certain disadvantages inherent in SEM work. Specimens are placed in a vacuum both during coating and in the machine itself. Thin-walled cells and spores may collapse, although critical-point drying, not so far used for lichens, overcomes this (see p. 3). Only surface features are seen and they all have the same colour, losing the effect of differential staining possible with slide preparation for an ordinary microscope. But by far the most serious drawback is the cost of an SEM machine and its consequent relative inaccessibility to researchers. As more institutions acquire SEM instruments, however, we can expect a great increase in published results.

This paper presents a summary of lichen structures as viewed with the SEM, based on the small but rapidly growing body of literature now available and on the author's own results obtained over the last 3 years from 1300 photographs of about 50 genera and 250 species. At this time SEM results have been published for approximately 25 lichen genera and 130 species (out of a total of about 500 genera and 18 000 species). Many unpublished data are obviously in the hands of lichenologists who have been using the SEM but have not had yet an opportunity to publish their results. It is probable that atlases of SEM pictures of lichens will gradually be assembled, a significant step in a field where good illustrations have been notoriously few and poor. These should provide new insights into the structure of lichens and illustrate this new method for studying the ontogeny of lichen organs.

SPECIMEN PREPARATION

Almost all workers so far have used air-dried specimens. The typically heavily gelatinized lichen cells withstand desiccation and vacuum well with little apparent collapse. Only the algal cells seem to shrivel and this probably happens prior to SEM preparation (see Plates Ic, IVв). Peveling and Vahl (1968) employed a freeze drying technique without demonstrating any superiority over air drying. Another more recent technique, critical point drying, is being used for the more delicate nonlichenized fungi and should also be useful in lichens when one is attempting to preserve the cell contents or prevent collapse. In this process the intracellular water of the living specimen is replaced with a solvent such as acetone and this solvent is then removed. The only drawback here is that phenolic lichen substances would be dissolved away in organic solvents.

Surface details may be observed on small pieces of lichen, only a few millimetres across, since the final area actually observed at × 500 or more is considerably less than 0·5 mm in diameter. Cross sections are prepared by hand with a sharp razor blade, slicing off cross or longitudinal sections 1–2 mm thick from dry specimens. Various common glues or double-sided "Sellotape" are used to mount the specimens on cover slips about 9 mm in diameter or directly on the metal stubs. Poor or loose adhesion will result in electrical discharge of the specimen when pictures are made and give streaked photographs. Coating is normally done with gold, gold–palladium or aluminium to form an ion-emitting layer several hundred angstroms thick. If the coating is too thin, the photographs will have white areas.

Although much information can be derived from SEM photography alone, ideally data from the EM as well as the light microscope should be used to arrive at final interpretations of a particular structure. With SEM treatment, for example, cell contents are destroyed and the hyphae appear as empty tubes. An accompanying EM picture will show protoplasm to be present even in the cortical cells (see Peveling, 1970, for *Parmelia caperata*).

TERMINOLOGY

Lichens have a surprising variety of structures and it is no wonder that a large and sometimes confusing terminology has been built up over the last 80 years to describe them. It is impossible to review the extensive literature in this paper; summaries in various texts should be consulted (Smith, 1921; des Abbayes, 1951; Ozenda, 1963; Henssen and Jahns, 1973). The attempt here at standardizing the terms is drawn entirely from the work of others, modified only in the light of how the tissues appear under SEM.

The terminology used for vascular plants will be reviewed first since this has strongly influenced that used by lichenologists. One must remember that higher plants consist of separate cellular tissues differing in ontogeny, histology and physiology. Lichens consist primarily of fungal hyphae, and, as so aptly stated by Smith (1921), lichen tissues are formed or determined by the prevailing direction of growth of hyphae, branching and crowding of filaments, frequency of septation and thickening of cell walls. This fact was perhaps first recognized by Lindau (1913), who referred to lichen and fungal tissues as plectenchyma rather than parenchyma or prosenchyma, although they mimic higher plant tissues very closely. More specifically, Esau (1965) defines parenchyma as a tissue composed of individual polyhedral cells. Comparable cell-like lichen tissue (cf. Plate IB–D) is more accurately called paraplectenchyma rather than pseudoparenchyma. Prosenchyma consists of elongate tapering cells running in parallel and would include in higher plants collenchyma and xylem. Similarly oriented tissue in lichens is best called prosoplectenchyma (cf. Plate III).

Many lichenologists have adopted and consistently used these terms (e.g. Frey, 1936; Degelius, 1954). Others, however, have been less exact in defining prosoplectenchyma. Ozenda (1963), in particular, referred to the cortical layers of *Cetraria* and *Parmelia* as prosoplectenchymatous (see also Dahl, 1952) (cf. Plate IIB). The cortex in these genera have in fact irregularly arranged (not parallel) hyphae with very thick walls. While it is true that almost all periclinally oriented prosoplectenchyma is thick walled, for the sake of consistency hyphal orientation should be the overriding consideration in defining lichen tissues. Strictly speaking, then, the cortical layers in the Parmeliaceae are composed of pachydermatous (thick-walled) paraplectenchyma.

A third kind of conglutinated tissue that occurs in cortical layers is called palisade plectenchyma (Frey, 1936) (cf. Plate VA). The short hyphae are oriented anticlinally in contrast to the periclinal direction of prosoplectenchyma. It obviously resembles palisade parenchyma in leaves.

Nonconglutinated (free) lichen tissue is confined for the most part to the medulla and is conveniently called medullary plectenchyma (Plate VC). It is also characteristic of tomentum (Plate VD).

Individual hyphae can be classified as short celled (with frequent septations as presumably occurs in paraplectenchyma and palisade plectenchyma) or long celled (less frequent septation as in prosoplectenchyma and medullary plectenchyma). Neither of these characters can be seen well under the SEM. Wall thickness varies considerably; Frey (1936) proposed a useful terminology: leptodermatous (walls thin relative to the diameter of the lumina) (Plate IB); mesodermatous (walls about as thick as the diameter of the lumina (Plate VC); and pachydermatous (walls much thicker than diameter of the lumina) (Plate VD). Lichen hyphae very frequently have these thick cell walls because of secretion of a microfibrillar polysaccharide layer (Jacobs and Ahmadjian, 1969; Peveling, 1974).

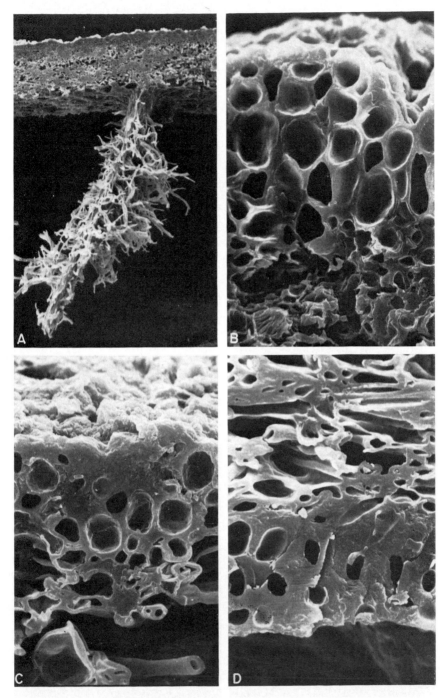

Plate I

A, Cross section of *Physconia venusta* showing the main thallus and a squarrose rhizine (*Rondon* in Vězda, *Lich. Sel. Exs.* 872) (×90). **B,** Upper cortex (long section) of *Pannaria rubiginosa* (*Schauer* in Vězda, *Lich. Sel. Exs.* 281) (×1200). **C,** Upper cortex (long section) of *Physcia vitii* (*Schröppel* in Poelt, *Lich. Alp.* 200) (×1500). **D,** Lower cortex (cross section) of *Physcia pusilloides* (*Mereschkovsky* s.n.) (×2000).

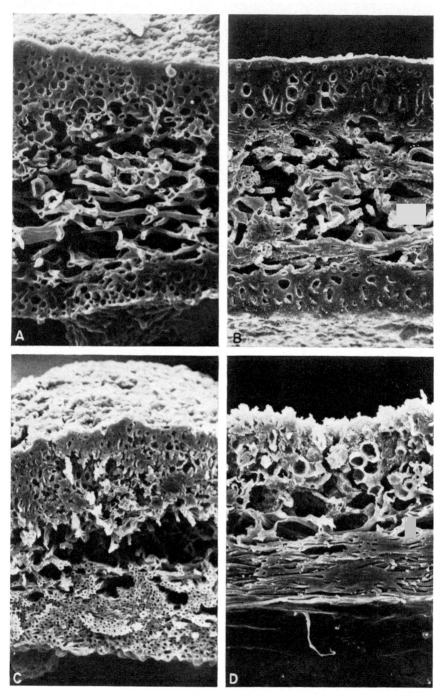

PLATE II

A, Long section of *Xanthoria fallax* (*Hale* 22966) (× 500). **B**, Long section of *Cetraria andrejevii* (*Kärnefelt* 20) (× 600). **C**, Cross section of *Physcia stellaris* (*Huskonen* s.n.) (× 400). **D**, Long section of *Thamnolia vermicularis* (*Scholander* s.n.) (× 500).

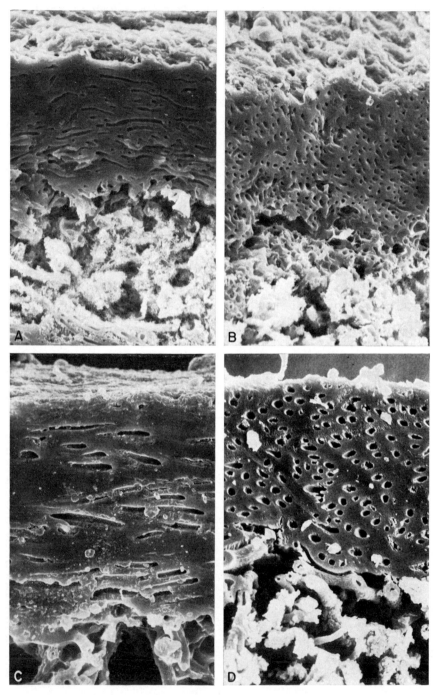

PLATE III

A, Long section of *Heterodermia hypoleuca* (*Hale* 37132) (× 700). **B,** Cross section of *H. hypoleuca* (*Hale* 37132) (× 700). **C,** Long section of *Alectoria cornicularioides* (*Gibaldi* s.n.) (× 1000). **D,** Cross section of *A. cornicularioides* (*Gibaldi* s.n.) (× 1000).

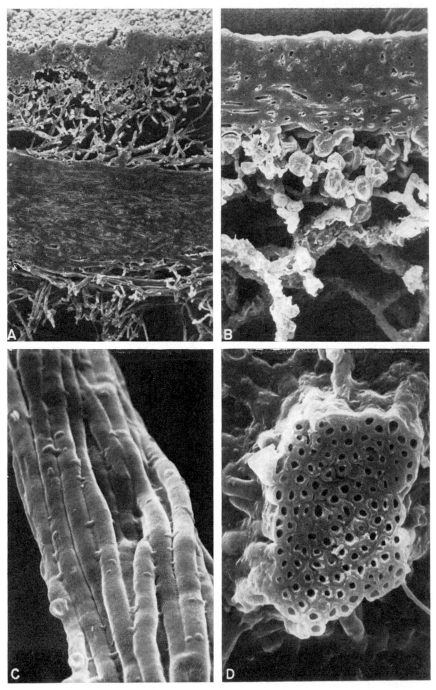

PLATE IV

A, Long section of *Letharia columbiana* (*Wiggins* s.n.) showing upper cortex and medullary strand (×200). **B,** Long section of *Cornicularia aculeata* showing cortex and algal layer (*Pišut* 74) (×625). **C,** Rhizine of *Physcia orbicularis* f. *rubropulchra* (*Willey* s.n.) (×2000). **D,** Cross section of rhizine of *P. orbicularis* f. *rubropulchra* (*Willey* s.n.) (×1500).

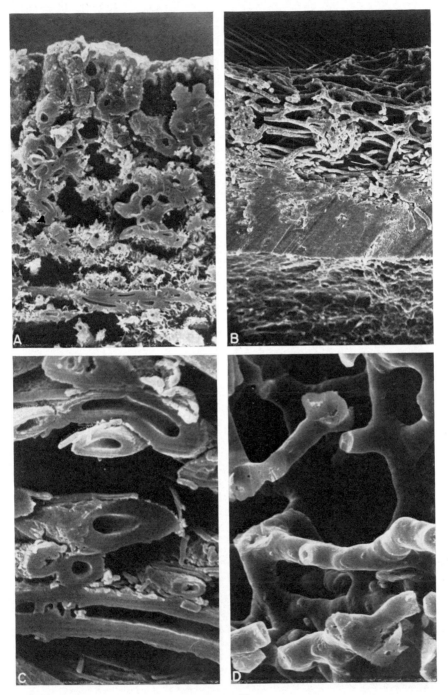

Plate V

A, Long section of upper cortex of *Pseudevernia consocians* (*Hale* 20908) (× 1000). **B**, Long section of podetium of *Cladonia rangiferina* (outer surface at top) (*Plitt* 6) (× 1000). **C**, Medullary hyphae of *P. olivetorum* (*Plitt* s.n.) (× 3000). **D**, Tomental mat of *Pannoparmelia anzioides* (*Imshaug* 45476) (× 1000).

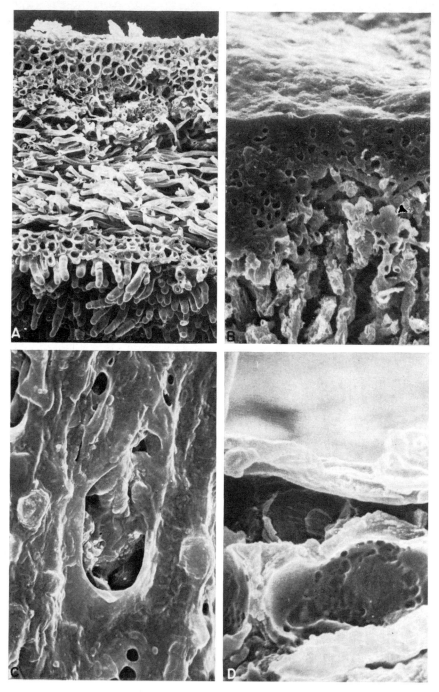

A, Cross section of *Sticta weigelii* (*Green* s.n.) (× 400). **B,** Cross section of upper cortex of *Parmelia bolliana* (*Adams* 34) (× 1000). **C,** Surface of *Coccocarpia granulosa* (*Hale* 35665) (× 2000). **D,** Cross section of cortex and epicortex of *C. parmelioides* (*Ravenel* 59) (× 3000).

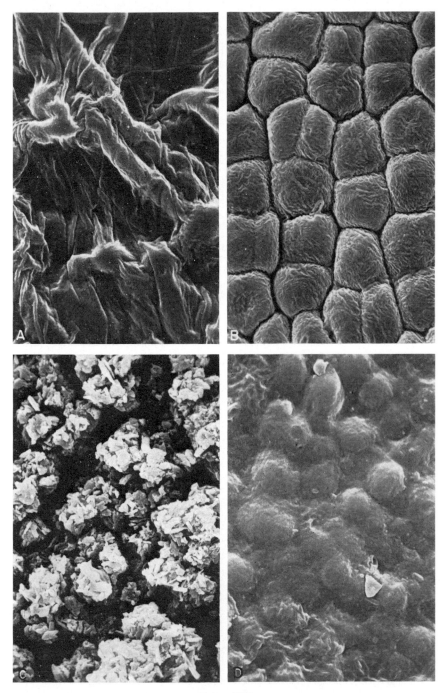

PLATE VII

A, Surface of *Collema nigrescens* (*Schröppel* 23) (× 2000). **B,** Surface of *Leptogium phyllo-carpum* (*Maxon* 5115) (× 2000). **C,** Surface of *Pseudevernia consocians* (*Hale* 20908) (× 2000). **D,** Surface of *Parmelia crozalsiana* (*Skorepa* 1757) (× 2000).

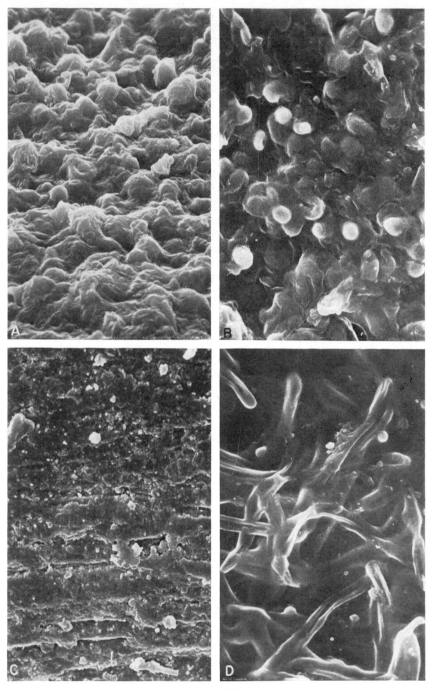

PLATE VIII

A, Lower surface of *Cetraria richardsonii* (*Weber* 7071) (× 2000). **B**, Upper surface of *Physcia setosa* (*Hale* 22273) (× 2000). **C**, Surface of *Alectoria cornicularioides* (*Gibaldi* s.n.) (× 2000). **D**, Surface of *Ocellularia olivacea* (*Hale* 38054) (× 2000).

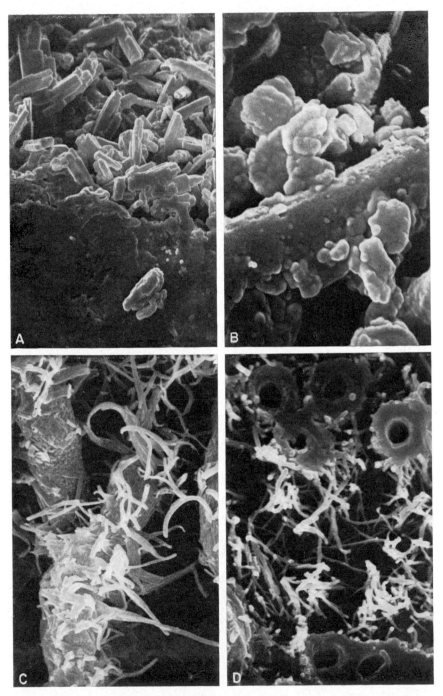

PLATE IX

A, Crystals of lichexanthone on the surface of *Graphina confluens* (Type specimen in P) (× 2000). **B,** Crystals of alectoronic acid on medullary hyphae of *Parmelia rampoddensis Hubricht* B2184) (× 5000). **C,** Crystals of lecanoric acid on medullary hyphae of *Pseudevernia consocians* (*Hale* 20908) (× 5000). **D,** Crystals of divaricatic acid on medullary hyphae of *Anzia colpodes* (*Hale* 10075) (× 5000).

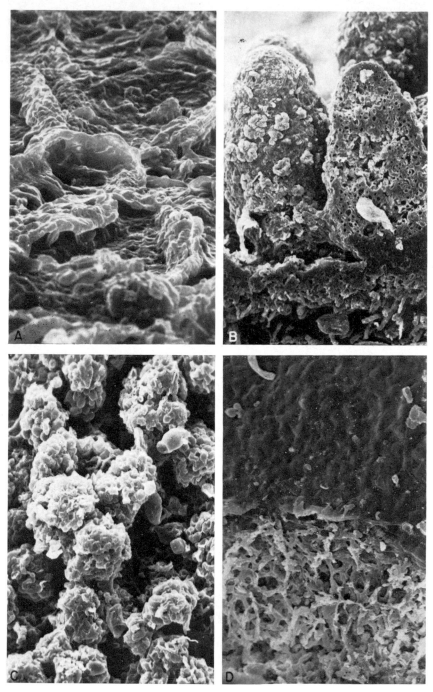

PLATE X

A, Surface of *Lecidea erythrophaea* (outlines of bark cells visible) (*Sundell* in Vězda, *Lich. Sel. Exs.* 1133) (× 500). **B,** Isidia of *Parmelia awasthii* (*Awasthi* 4041) (× 350). **C,** Soredia of *Physcia pusilloides* (*Mereschkovsky* s.n.) (× 500). **D,** Portion of pseudocyphella on upper cortex of *Parmelia rudecta* (*Hale* 14796) (× 500).

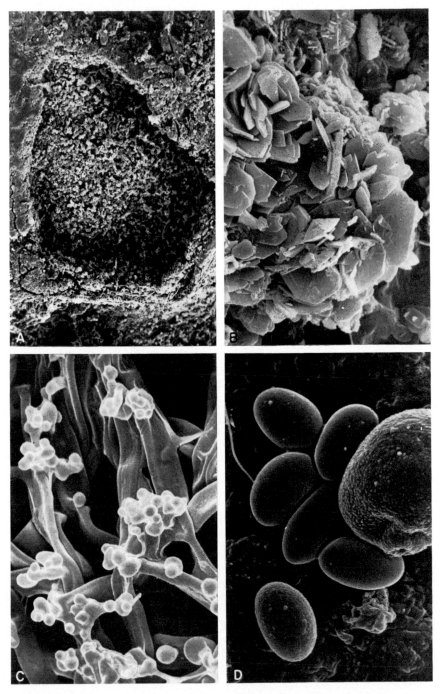

A, Cyphella of *Sticta limbata* (*Tavares* 63) (× 100). **B**, Pruina of *Pyxine caesiopruinosa* (*Hale* 13437) (× 2000). **C**, Lower surface of *Cora pavonia* (*Steyermark* 95120) (× 2000). **D**, Spores on surface of apothecium of *Lecanora subfuscata* (*Santesson* 16374) (× 2000).

A final word should be made about Starbäck's (1895) classification of fungal tissues which has been adopted by Korf (1951, 1958) for the discomycetes. Latin terms, none yet satisfactorily or conveniently translated into English, form the basis of the system. With reference to the terms defined above, textura oblita and textura porrecta appear to be variations of prosoplectenchyma, textura intricata is close to medullary plectenchyma, textura epidermoidea probably does not occur in lichens, and the other terms (textura angularis, textura prismatica and textura globulosa) are variations of paraplectenchyma. There is no reason why Starbäck's terms cannot be used to describe lichen tissues. However, a broader terminology based firstly on hyphal orientation and secondly on cell configuration is preferred here.

The following sections will discuss, in more detail, various thallus tissues (but omitting excipular tissues) in the light of SEM work and suggest new interpretations that can be made. All species listed have been examined, but because of space limitations only a selection of photographs is included.

1. Tissues of Foliose and Fruticose Lichens

(*a*) *Paraplectenchyma*. This tissue forms a supporting and protective layer in lichens and is mostly confined to the cortex. The cells are short, randomly oriented, and closely packed so that in both cross and long sections of a thallus the cellular appearance is the same (Plate II). The simplest example is *Leptogium* which usually has cortical layers only one cell thick, clearly revealed in sections and in a surface view by the regularly arranged cells (Plate VIIB). This aspect is virtually identical in all species studied, including *L. azureum*, *L. corticola*, *L. hirsutum* and *L. phyllocarpum*.

All other genera yet studied (excepting, of course, *Collema* and related groups) have multistratified cortical layers varying in thickness and degree of conglutination of the hyphae. *Lecanora muralis* (Peveling, 1970) and *L. chrysoleuca* have a rather loosely organized, almost porous, upper cortex. A thick, much denser, paraplectenchymatous cortex is typical of all taxa so far examined in the Peltigeraceae (*Peltigera canina*, *P. horizontalis*, *P. polydactyla*, *Solorina octospora* (with vertically elongate cells) and *S. saccata*), Stictaceae (*Lobaria peltigera*, *L. pulmonaria*, *Pseudocyphellaria aurata* and *Sticta weigelii* (Plate VIA), Nephromataceae (*Nephroma arcticum*), Pannariaceae (*Pannaria lurida* (with a double upper cortex), *P. rubiginosa* (Plate IB), *Parmeliella pannosa*, *P. plumbea* and *P. pycnophora*), Verrucariaceae (*Dermatocarpon fluviatile* and *D. miniatum*), Candelariaceae (*Candelaria concolor* and *C. fibrosa*), Physciaceae (*Dirinaria aspera*, *D. confluens*, and *D. picta*; *Physcia albata*, *P. callosa*, *P. ciliata*, *P. constipata*, *P. endococcinea*, *P. nadvornikiana*, *P. orbicularis*, *P. pseudospeciosa*, *P. pusilloides* (Plate ID), *P. setosa* and

P. vitii (Plate Ic); *Physciopsis adglutinata*), Siphulaceae (outer cortex of *Thamnolia vermicularis* (Plate IID), and Teloschistaceae (*Xanthoria fallax* (Plate IIA) and *X. parietina*). All of these genera have an irregularly thickened upper cortex 6–8 cells thick and a similar (or slightly thinner) lower cortex or no lower cortex at all. Among lichens these groups might be considered less advanced since so much of the thallus mass is, on average, occupied by apparently nonassimilative paraplectenchymatous tissue. They have pores, pseudocyphellae or cyphellae, for gas exchange, completely lack a lower cortex, or have comparatively small thalli. It may also be significant that these groups have the poorest and least varied chemistry (Culberson, 1969).

In other families and genera the cell walls of cortical hyphae are much thicker but the cortex correspondingly thinner, 2–4 cells thick. Ozenda (1963) called these prosoplectenchymatous, and Dahl (1952) used the term "platysmoid". It is characteristic of the Anziaceae (Hale, 1973), Parmeliaceae (*Cetraria andrejevii* (Plate IIB), *C. delisei*, *C. islandica*, *C. laevigata* and *C. simmonsii*, and other species in *Asahinea*, *Cetraria*, *Cetrelia*, *Parmelia*, *Parmeliopsis* and *Pannoparmelia*; see Hale, 1973), Hypogymniaceae (Hale, 1973; Hawksworth, 1973), and among the fruticose lichens Alectoriaceae (*Cornicularia californica* and *C. epiphorella*), Ramalinaceae (*Ramalina fastigiata* and *Ramalinopsis mannii*), Usneaceae (lower cortex of *Evernia mesomorpha*, *E. prunastri*, *Everniopsis trulla*, *Letharia columbiana* (Plate IVA) and *L. vulpina*; outer cortex of *Dactylina arctica*, *D. madreporiformis*, *D. ramulosa* and *Usnea strigosa*), and Siphulaceae (*Siphula ceratites*). The stronger mesodermatous cortical cells obviously allow for a reduction in the total mass of supporting parenchymatous tissue and a greater volume of medullary plectenchyma. All groups here also contain *Trebouxia* and are chemically rich.

(*b*) *Prosoplectenchyma.* This tissue also provides a basic structural support for lichens. It occurs in cortical layers as well as in strands in the medulla and in rhizines. The parallel periclinal hyphal orientation is instantly recognizable by comparing cross and long sections of a thallus (Plate III). The cell walls are generally much thicker than in paraplectenchyma and so heavily conglutinated that walls of adjoining cells are indistinguishable, only the small lumina remaining. Other terms used for this tissue are scleroplectenchyma (Degelius, 1954) and "Strang" plectenchyma (Frey, 1936).

Relatively few foliose lichens have periclinal hyphal orientation in the cortex, the most conspicuous exceptions being in the Physciaceae (e.g. *Anaptychia fusca*, *A. palmatula*, *Heterodermia hypoleuca* (Plate IIIA and B), *H. obscurata*, *H. ravenelii*, *H. speciosa* and *Tornabenia atlantica*; and the lower cortex in

Physcia aipolia, P. caesia, P. phaea, P. stellaris (Plate IIc), *P. subdimidiata, P. subobscura* and *P. vainioi; Physconia detersa, P. enteroxantha, P. grisea, P. pulverulenta* and *P. venusta; Pyxine berteriana, P. cocoes, P. eschweileri, P. fallax, P. heterospora, P. isidiophora, P. marginata, P. sorediata* and *P. vulnerata*). The upper cortex in the species of *Physcia, Physconia* and *Pyxine* listed is paraplectenchymatous.

On the other hand, prosoplectenchyma is characteristic of many fruticose lichens: Alectoriaceae (*Alectoria cornicularioides* (Plate IIIc and D), *A. motykae* (Hawksworth, 1972), *A. nitidula, A. ochroleuca* and *A. sarmentosa; Cornicularia divergens* and *C. normoerica*), medullary cords or strands in the Usneaceae (*Everniopsis, Letharia* (Plate IVA) and *Usnea* (Peveling, 1974)), Siphulaceae (inner cortex of *Thamnolia vermicularis* (Plate IID)), Teloschistaceae (*Teloschistes chrysophthalmus, T. flavicans* and *T. nodulifera*), and podetia of *Cladonia* (*C. cristatella, C. furcata, C. gracilis* and *C. rangiferina* (Plate VB)).

Some *Cornicularia* species have an unusual combination of tissues in the cortex, the upper part being paraplectenchymatous grading rather abruptly into a prosoplectenchymatous layer, as in *C. aculeata* (Plate IVB).

Both rhizines and cilia in all families appear to be prosoplectenchymatous (Plate IVc and D). In one of the first SEM studies ever done, Reznik *et al.* (1968) examined in detail the structure of rhizines in *Parmelia caperata* and *Lobaria pulmonaria*, the latter genus having looser, nonconglutinated bundles of hyphae in contrast to the strongly conglutinated strands in *Parmelia*.

(*c*) *Palisade plectenchyma.* This is the third type of cortical tissue of common occurrence in macrolichens. First recognized separately by Zukal (1895), it occurs only as an outer or upper cortical layer and consists of short hyphae emerging from the algal layer, becoming erect, and packing together to form a more or less anticlinally oriented layer with sizeable interstices between the cells. A good example is *Pseudevernia* (Plate VA; see Plate VIIc for a vertical view) (Hale, 1973). A number of sections in the large genus *Parmelia* are similar (Hale, 1973). The lower cortex is always typically rather thin and paraplectenchymatous. The cells are much more tightly packed and conglutinated in fruticose *Roccella* (*R. babingtonii, R. montagnei* and *R. phycopsis*) as well as in other genera of the Roccellaceae (*Darbishirella*) and in *Sphaerophorus* (Sphaerophoraceae).

(*d*) *Medullary plectenchyma.* This is not a supporting tissue but may act as the main site of carbohydrate transfer, metabolism and storage. The hyphae are free and loosely interwoven with large interstices (Plates IIA and B, Vc, VIA;

Hale, 1973; Hawksworth, 1972; Peveling, 1974). The only exception is *Cornicularia*, especially *C. epiphorella* and *C. normoerica*, where the medullary hyphae form cohesive, almost flattened strands. Orientation of hyphae in medullary plectenchyma is usually broadly periclinal but varies to quite random in *Cladonia* and *Cladina* (= *textura intricata* of Starbäck, 1895) (Plate VB; Ahti, 1973; Asperges, 1973). While not strictly medullary in nature, the tomentum on the surface of *Peltigera* (*P. canina*) and *Erioderma* and the lower surface of members of the Stictaceae (Plate VIA), Nephromataceae, etc. are virtually identical to medullary plectenchyma. We could also include here the carbonized anastomosing hyphae that comprise the hypothallus in *Anzia* and *Pannoparmelia* (Plate VD).

The symbiotic algae occupy the upper part of the medulla just below the cortex (Plate IVB). They are usually collapsed in SEM pictures of air-dried material and surprisingly sparse, much sparser than we had been led to believe from pictures of stained and sectioned thalli (cf. Asperges, 1973). The hyphae surrounding the algae may be either indistinguishable from medullary hyphae or somewhat thinner and less encrusted with lichen substances. Jahns (1972) has excellent pictures of free algae (*Stigonema*) associated with cephalodia growing on the surface of *Pilophorus strumaticus*.

(*e*) *Epicortical structures.* The outer parenchymatous cells of vascular plants are protected by a cuticle, a noncellular layer composed of the fatty substance cutin. Cutin is not produced by lichens and most species have not evolved any means of protection other than that provided by the outer layer of closely packed cortical cells (as, for example, in *Candelaria*, Physciaceae, *Pseudevernia*, etc.).

Other lichens have a very thick polysaccharide layer covering the outer surface and which may obscure the orientation of cortical hyphae below. It is best developed in the Parmeliaceae (Plate VIB; Hale, 1973). A further refinement of this is the pored epicortex: a separable, uniform polysaccharide sheet about 0·6 μm thick stretched over the palisade plectenchyma in *Parmelia* (see Hale, 1973, for numerous illustrations). The pores are 10–30 μm in diameter and provide passage into the algal layer and medulla between gaps in the palisade tissue. They function for aeration and could be compared with stomates of higher plants except that they remain permanently open.

A pored epicortex has also been discovered recently in *Coccocarpia* as well (Plate VIc and D). This is quite unexpected since this genus has a paraplectenchymatous cortex. *Physma*, a related tropical genus, appears to have a similar structure. It is perhaps significant that the pored epicortex is characteristic of groups that have evolved or are centred in the tropics, such as *Coccocarpia* or *Parmelia* subgenus *Parmelia* section *Hypotrachyna* (Hale, 1973). Where other groups of the same genus have a boreal origin (as section *Parmelia*, the *P. saxatilis* group), we find no pored epicortex, only a thick polysaccharide layer over the cortex and frequently pseudocyphellae.

The previous interpretation of the nonpored epicortex in the Parmeliaceae (Hale, 1973) is probably not correct. The thick adherent presumably poly-saccharide layer over the cortex may not be comparable in origin to the pored epicortex. In other words, it is probably a continuous layer secreted by the outermost cortical cells, not simply a pored epicortex that has lost its pores and become conglutinated with underlying hyphae.

(*f*) *Surface features*. The earlier studies with the SEM tended to concentrate on surface features since these are observed very easily. Hawksworth (1969) placed much emphasis on ridge and plaque formation in the Alectoriaceae, and differences in surface texture have been noted in the Parmeliaceae (Hale, 1973). It is obvious that these features are for the most part a faithful replica of the shape and orientation of underlying hyphae. For this reason no study of surface texture is complete without a third-dimensional view of a cortical section (Hawksworth, 1972).

Collema has no cortex and individual hyphae at the surface are embedded randomly in the gelatinous matrix (Plate VIIA; Peveling, 1974). *Leptogium* has the simplest cortex, one cell thick, and as previously mentioned the surface view shows regularly arranged cells (Plate VIIB). Species with a nonepicorticate cortex have a knobby surface, the tips of vertical hyphae loosely packed together, as in *Pseudevernia* (Plate VIIc). Multistratified paraplectenchymatous cortices are more irregular but still show outlines of individual hyphae, very clearly in species without the thick epicortex (Plate VIIIA and B), less distinctly when the epicortex is thick (Plate VIID). Species with a prosoplectenchymatous cortex show more or less distinct parallel ridging, as in *Alectoria* (Plate VIIIc; Hawksworth, 1969) and *Heterodermia*.

The configuration of the pored epicortex (Plate VIc) has been discussed in detail elsewhere (Hale, 1973). There is still no evidence concerning the origin of these pores and the mechanism by which they are formed so regularly.

There are many other types of variation in the surface texture of other lichen groups, in particular the Peltigeraceae, which deserve much more detailed study. Another trait, sculpturing, is very pronounced in *Heterodea* (Hale, 1973) but has also been found in *Anzia* and *Hypogymnia*.

(g) *Appearance of lichen substances.* Deposition of cortical substances is easily demonstrated with the SEM. Hale (1973) illustrated many species of *Parmelia* where atranorin, apparently synthesized near the algal layer, seems to migrate into and through the cortex and eventually become deposited on the surface of the epicortex or remain immersed in it. Lichexanthone (Plate IXA) and usnic acid behave similarly. Parietin in *Xanthoria* (Peveling, 1970) and calycin (as in *Candelaria*) also tend to accumulate as distinct crystals on the thallus surface. Long needlelike crystals, suspected to be a diterpene, occur on the surface of *Cladonia boryi* (Ahti, 1973).

Lichen substances produced on the medullary hyphae develop more freely as long ribbonlike crystals or as aggregations of platelets (Plate IXB–D). It does not seem possible at this time to distinguish the individual lichen substances by their shape in the medulla.

2. Tissues of Crustose Lichens

Crustose lichens grow on, and in part within, their substratum and, perhaps for this reason, lack the complex supporting tissues found in foliose and fruticose species. Older workers often refer to the cortex of crustose lichens as decomposed; under SEM the details of structure are indeed obscure. There is usually no clear differentiation into layers. A more or less amorphous unstratified thallus merges into the cork cells of the bark or crystals of rock below. Some species with unusually thick thalli seem to have a distinct paraplectenchymalike cortex, as in *Ocellularia olivacea* (Hale, 1974). Surface features are unpromising (Plate XA) with only one exception so far: several species in the Graphidaceae and Thelotremataceae produce short, semi-erect aculeate hyphae from the surface of the cortex (Plate VIIID). These had not previously been detected with light microscopy.

3. Other Lichen Structures

Lichens have a number of well known and sometimes unique structures that have been described in detail from microscopic studies. They are unfortunately not always well suited to SEM study because tissue organization is weak at best and, without differential staining, little can be seen.

(a) *Vegetative structures.* Soredia have no distinctive features other than irregularly arranged surface hyphae and vary little from genus to genus except in size (Plate XC; see also Asperges, 1973, for *Cladonia* and Hawksworth, 1972, for *Alectoria*). They are virtually identical in families as distant as the Physciaceae (Plate XC) and the Parmeliaceae (*Cetraria*, see Peveling, 1970). Wetmore (1974) illustrates with SEM photographs the peculiar soredia of the Heppiaceae which consist of algae on the outside and fungal hyphae inside. The corticate nature of isidia is clearly seen in sections (Plate XB).

Pseudocyphellae appear as large gaps in the paraplectenchymatous cortex, clearly delimited on the upper cortex (Plate XD) but less so on the lower surface, as in *Pseudocyphellaria*. They are filled with loose, strangely shrivelled medullary plectenchyma. Cyphellae in *Sticta* have a similar kind of hyphae but differ in being produced on the lower cortex within craters that are rimmed by ruptured cortical tissues (Plate XIA).

Cephalodia have been studied successfully by Jahns (1972), who found that in *Pilophorus strumaticus* the superficial colonies of *Stigonema* are gradually enveloped by fungal hyphae. The SEM photographs show the stages of formation with exceptional clarity.

Pruina, the whitish covering on the cortex and apothecial discs of many lichens, is either an amorphous granular layer as in *Cetraria islandica* or crystalline as in many species of the Physciaceae (Plate XIB; Peveling, 1974), in the latter instance appearing to consist of calcium oxalate.

Hairs, which are difficult to study with light microscopy, are ideally suited for SEM examination. In *Anaptychia ulothricodes* they are minute multicellular strands (Peveling, 1974), much thinner of course than cilia; they probably have a similar structure in *Physcia ciliata*.

(*b*) *Reproductive structures*. Reproductive structures in lichens are best examined with staining and light or EM microscopy because cytoplasmic details are so important. Tibell (1971), however, showed how valuable the SEM is in detecting surface sculpturing of spores in the Caliciaceae, using magnifications up to × 10 000. One can also mount apothecia and look for spores ejected on the surface, presuming, of course, that these spores came from that apothecium (Plate XID). Cross sections of apothecia are not particularly successful since the closely packed paraphyses hide much of the detail of asci and spores.

Jahns (1973) has examined trichogynes protruding from the surface of apothecia in *Pilophorus*. The abundance of trichogynes is greater than expected from light microscopy and the development and eventual degeneration of these delicate structures can be easily followed. Further searches with the SEM should produce many more data on the frequency and seasonality of these reproductive structures.

Basidial formation is very clearly seen in *Cora pavonia* (Plate XIc) and the SEM may provide a new means of studying ontogeny of these structures in lichens.

USE OF FINE STRUCTURE IN SYSTEMATICS

There has been a gradual diminution in the number of lichen families based on gross life form and a greater emphasis placed on the ontogeny of the ascocarps (Henssen and Jahns, 1973). It might be profitable, none the less, to see how a better appreciation of anatomical characters afforded by the SEM might provide useful guidelines for redefining the limits of both genera and families.

As we might expect, families with diverse anatomical structure offer the most fertile ground for speculation. For example in the Physciaceae, Poelt (1966) has used certain differences in cortical structure, in part, to separate *Physconia* from *Physcia*. Within *Physcia* itself as redefined one could recognize two genera, one (*Physcia*) clustered around *P. stellaris* which has a paraplectenchymatous upper cortex and a prosoplectenchymatous lower cortex (as well as a pale lower surface and abundant production of atranorin), the other (unnamed), centered on *P. orbicularis* which has a paraplectenchymatous upper and lower (and usually a black lower surface and no atranorin).

Within the Parmeliaceae we find several lines of cortical development. One of these, the nonepicorticate genus *Pseudevernia*, is now accepted as a distinct genus (Hale, 1973). Poelt (1974) recently placed it in a new family Hypogymniaceae along with *Hypogymnia*, but these two genera have little in common in terms of either anatomy or chemistry. Perhaps we should follow older authors and realign *Pseudevernia* with *Evernia*, *Everniopsis* and *Letharia*, similar subfruticose genera which have identical palisade cortical structure. *Candelaria*, which most lichenologists now place in a separate family, Candelariaceae, has a cortex similar to that in *Xanthoria* and assuredly does not belong in the Parmeliaceae, where it had been classified.

Parmelia, which one might think has reached an irreducible state with the removal of *Anzia*, *Hypogymnia* and *Pseudevernia*, might eventually be split up according to type of epicortex, segregating out, for example, *Melanoparmelia* as a separate genus, since it lacks a pored epicortex. *Cetraria* represents a large rather heterogeneous genus. At the least one could segregate *Cetraria richardsonii* as a distinct genus because its very thick, apparently prosoplectenchymatous upper cortex and chemistry (alectoronic acid) set it apart from all other species.

Further study is urgently needed in the genera now placed in the Coccocarpiaceae and Pannariaceae. They have varied cortical development which lends itself well to SEM study.

The Usneaceae present many problems. Henssen and Jahns (1973) recently transferred all of the genera in this family (except for *Ramalina*) to the Parmeliaceae, a solution justifiable perhaps on the basis of similar ascocarp structure but impractical in view of the tremendous morphological and anatomical diversity in the family. *Alectoria* is a homogeneous group because of its distinctive and uniform prosoplectenchymatous cortex (Hawksworth, 1969). *Cornicularia*, however, is quite diverse and has significantly different cortical structures as illustrated by *C. californica* (a thin paraplectenchymatous cortex) and *C. aculeata* (Plate IVB, mixed paraplectenchymatous and prosoplectenchymatous), as well as the peculiar medullary strands mentioned above for *C. epiphorella* and *C. normoerica*.

Discrimination at the species level with SEM is less successful as a rule since the kinds of characters we are discussing are more or less uniform from species to species within genera or subgenera. Hale (1973) in fact was able to reassign several species of *Parmelia* to a more appropriate subgenus by looking for the presence or absence of the pored epicortex. Pore size alone (Hale, 1972) can be used to separate *Parmelia formosana* and *P. pustulifera*. Krog (1973) found very obvious differences between *Umbilicaria papulosa* and *U. pustulata* in terms of smoothness and size of cortical papillae at × 500. These cases, however, are rather rare when one considers the great amount of homogeneity within genera.

In summary, the surface structure of lichens as well as the internal organization, both seen so vividly with the SEM, have great potential in lichen systematics. SEM studies will undoubtedly revitalize anatomical research in lichens and offer additional characters for revising systematic concepts at the generic and family level.

ACKNOWLEDGEMENTS

I have had profitable discussions with Dr D. L. Hawksworth, Prof. Dr A. Henssen, Canad-real. P.M. Jørgensen, Mr I. Kärnefelt, Dr H. Robinson, Dr R. Santesson, and Dr C. Wetmore on interpretation of photographs presented here. I am especially thankful to Mr W. Brown of the Smithsonian Scanning Electron Microscope Laboratory for his patient help over many hours in taking the SEM pictures. The photographs were printed by the Smithsonian Photographic Laboratory.

REFERENCES

DES ABBAYES, H. (1951). "Traité de Lichenologie." [*Encycl. Biol.* **41**, i–x, 1–217.] Lechevalier, Paris.

AHTI, T. (1973). Taxonomic notes on some species of *Cladonia*, subsect. *Unciales. Annls bot. fenn.* **10**, 163–184.

ASPERGES, M. (1973). Preparation and study of *Cladonia floerkeana* (Fr.) Sommerf. for scanning electron microscopy. *Bull. Jard. bot. natn. Belg.* **43**, 353–356.

CULBERSON, C. F. (1969). "Chemical and Botanical Guide to Lichen products." University of North Carolina Press, Chapel Hill, N.C.

DAHL, E. (1952). On the use of lichen chemistry in lichen systematics. *Revue bryol. lichén.* **21**, 119–134.

DEGELIUS, G. (1954). The lichen genus *Collema* in Europe. *Symb. bot. upsal.* **13** (2), 1–499.

ESAU, K. (1965). "Plant Anatomy." Wiley, New York.

FREY, E. (1936). Vorarbeiten zu einer Monographie der Umbilicariaceen. *Ber. schweiz. bot. Ges.* **45**, 198–230.

HALE, M. E. (1972). *Parmelia pustulifera*, a new lichen from southeastern United States. *Brittonia* **24**, 22–27.

HALE, M. E. (1973). Fine structure of the cortex in the lichen family Parmeliaceae viewed with the scanning-electron microscope. *Smithson. contr. bot.* **10**, 1–92.

HALE, M. E. (1974). Morden–Smithsonian expedition to Dominica: The lichens (Thelotremataceae). *Smithson. contr. bot.* **16**, 1–46.

HAWKSWORTH, D. L. (1969). The scanning electron microscope. An aid to the study of cortical hyphal orientation in the lichen genera *Alectoria* and *Cornicularia*. *J. Microscopie* **8**, 753–760.

HAWKSWORTH, D. L. (1972). Regional studies in *Alectoria* (Lichenes) II. The British species. *Lichenologist* **5**, 181–261.

HAWKSWORTH, D. L. (1973). Two new species of *Hypogymnia* (Nyl.) Nyl. *Lichenologist* **5**, 452–456.

HENSSEN, A. and JAHNS, H. M. (1973) ["1974"]. "Lichenes. Eine Einführung in die Flechtenkunde". Georg Thieme Verlag, Stuttgart.

JACOBS, J. B. and AHMADJIAN, V. (1969). The ultrastructure of lichens. I. A general survey. *J. Phycol.* **5**, 227–240.

JAHNS, H. M. (1972). Die Entwicklung von Flechten-Cephalodien aus *Stigonema*-Algen. *Ber. dt. bot. Ges.* **85**, 615–622.

JAHNS, H. M. (1973). The trichogynes of *Pilophorus strumaticus*. *Bryologist* **76**, 414–418.

KORF, R. P. (1951). A monograph of the Arachnopezizeae. *Lloydia* **14**, 129–180.

KORF, R. P. (1958) Japanese discomycetes. Notes I–VIII. *Sci. Rep. Yokohama natn. Univ.* **7**, 7–35.

KROG, H. (1973). On *Umbilicaria pertusa* Rass. and some related lichen species. *Bryologist* **76**, 550–554.

LINDAU, G. (1913). "Kryptogamenflora für Anfänger. 3. Die Flechten". Springer, Berlin.

OZENDA, P. (1963). Lichens. *In* "Handbuch der Pflanzenanatomie", second edition (W. Zimmermann and P. Ozenda, eds), **6** (9), i–x, 1–199. Borntraeger, Berlin.

PEVELING, E. (1970). Die Darstellung der Oberflächenstrukturen von Flechten mit dem Raster-Elektronenmikroskop. *Vortr. bot. Ges.* [*Dtsch. bot. Ges.*] *n.f.* **4**, 89–101.

PEVELING, E. (1974) ["1973"]. Fine Structure. *In* "The Lichens" (V. Ahmadjian and M. E. Hale, eds), pp. 147–182. Academic Press, New York and London.

PEVELING, E. and VAHL, J. (1968). Die Anwendung der Gefrierätzmethode für Untersuchungen mit dem Raster-Elektronenmikroskop im Vergleich mit herkömmlichen Untersuchungsmethoden. *Beitr. elektronenmikroskop Direktabb Oberfl.* **1**, 205–212.

POELT, J. (1966). Zur Kenntnis zur Flechtengattung *Physconia* in der Alten Welt und ihrer Beziehungen zur Gattung *Anaptychia*. *Nova Hedwigia* **12**, 107–136.

POELT, J. (1974) ["1973"]. Classification. *In* "The Lichens" (V. Ahmadjian and M. E. Hale, eds), pp. 599–632, Academic Press, New York and London.

REZNIK, H., PEVELING, E. and VAHL, J. (1968). Die Verschiedenartigkeit von Haftorganen einiger Flechten. Untersuchungen mit der Overflächen-Raster-elektronenmikroskop. *Planta* **78**, 287–292.

SCHWENDENER, A. (1860). Untersuchungen über den Flechtenthallus. *Naegeli's Beitr. wiss. bot.* **4**, 109–186.

SMITH, A. L. (1921). "Lichens." Cambridge University Press, Cambridge.

STARBÄCK, K. (1895). Discomyceten-Studien. *Bih. K. svenska VetenskAkad. Handl.* **21** (5), 1–42.

TIBELL, L. (1971). The genus *Cyphelium* in Europe. *Svensk bot. Tidskr.* **65**, 138–164.

WETMORE, C. M. (1974). New type of soredium in the lichen family Heppiaceae. *Bryologist*, **77**, 208–215.

ZUKAL, H. (1895). Morphologische und biologische Untersuchungen über die Flechten. II. *Sber. Akad. Wiss. Wien, II*, **104**, 1303–1395.

2 | Investigations into the Ultrastructure of Lichens

E. PEVELING

Botanical Institute, University of Münster, West Germany

Abstract: A description is given of the osmiophilic globules of algal pyrenoids, vesicle complexes, mesosomes and lomasomes within the phycobionts and mycobionts of a variety of lichens. Consideration is given to the significance of such structures in the metabolism of lichen thalli. The mobility of storage products and transfer of material within the thallus is discussed.

INTRODUCTION

During the last 5 years studies on lichens with the transmission electron microscope have provided new insight into the micromorphology of these cells. In the lichen symbionts protoplasmic structures have been detected that were either previously unknown or occurred with some similarities to those of their free-living counterparts.

Lichens with blue-green and green algal phycobionts have been investigated. The mycobionts studied have been ascomycetes with the exception of the basidiolichen *Cora pavonia*. The characteristic intracellular difference between blue-green phycobionts and their free-living counterparts is the occurrence of a large number of osmiophilic globules between the thylakoids (Peat, 1968; Peveling, 1969a; Paran *et al.*, 1971; Jacobs and Ahmadjian, 1973). Green phycobionts are also characterized by an abundance of such osmiophilic globules. In *Trebouxia* and *Stichococcus* these are aligned along the thylakoids which cross the pyrenoid (Brown and Wilson, 1968; Ben-Shaul *et al.*, 1969; Peveling, 1969b; Ahmadjian and Jacobs, 1970; Fisher and Lang, 1971), while in *Coccomyxa* they are scattered between the thylakoids within the chloroplast. Other common

Systematics Association Special Volume No. 8, "Lichenology: Progress and Problems", edited by D. H. Brown, D. L. Hawksworth and R. H. Bailey, 1976, pp. 17–26. Academic Press, London and New York.

structures of green phycobionts include storage bodies located peripherally in the protoplast which may be of different electron densities.

The most characteristic structures in mycobiont protoplasts are concentric bodies (Griffiths and Greenwood, 1972). These consist of an electron transparent core with two surrounding shells; finger-like projections are associated with the outer shell. Many mycobiont cells have an invaginated plasmalemma (Brown and Wilson, 1968; Jacobs and Ahmadjian, 1969; Peveling, 1969c), resulting in the protoplast having an enlarged surface. Protoplasts in the medullary hyphae are sometimes filled with large storage bodies, while those of the cortical layers are more or less vacuolated.

Peveling (1974) has reviewed lichen ultrastructure in more detail. The present paper is concerned with vesicular and membranous inclusions which have been observed recently, and the metabolic significance of the various kinds of intracellular structures.

ULTRASTRUCTURE OF PHYCOBIONTS

1. Osmiophilic Globules

The chlorococcalean alga *Trebouxia*, probably the most frequent green phycobiont of lichens, has a very large number of osmiophilic globules in the pyrenoid, which is located in the centre of the chloroplast. Such osmiophilic globules are found in the chloroplasts of most algal classes, but do not represent a dominant component. In the free-living algae these globules are scattered in the stroma between the thylakoids and no osmiophilic globules are reported to occur in the pyrenoids themselves (Dodge, 1973). In contrast, the pyrenoid in *Trebouxia* (investigated in about 50 lichens) is characterized by numerous osmiophilic globules, each 40–100 μm in diameter. Only a few globules are located outside the pyrenoid. Because of their distribution these globules have been termed plastoglobuli (Jacobs and Ahmadjian, 1969). The globules are aligned along the thylakoids which cross the pyrenoid either as parallel lines (Galun *et al.*, 1970) or as long arcs (Plate IA). Another location for thylakoids in the pyrenoids is in the widened channels or vesicles, where again they are bordered by globules (Plate IB). It may be of special interest that the typical starch shells which surround the pyrenoids in many green algae (Griffiths, 1970) are absent in the *Trebouxia* chloroplasts. Depending on the environmental conditions, starch appears only in the stroma of the chloroplast.

The osmiophilic globules which exist generally in the stroma of plastids consist of lipids, mainly plastoquinones. In degenerating chloroplasts they also contain carotenoids (Lichtenthaler, 1968). Pyrenoglobuli have not, however, yet been isolated from the phycobionts to enable their components to be analysed. As a result of their high osmiophily and positive staining after treatment with Sudan IV, Jacobs and Ahmadjian (1969) concluded that the globules in phycobionts also contain lipids. After incubation for 24 h with $^{14}CO_2$ and a study of these cells by high-resolution autoradiography, no label was discovered in the pyrenoid or its globules (Jacobs and Ahmadjian, 1971a). From this experiment it is evident that the globules are later products of algal metabolism.

Studies on the number and location of globules in the phycobionts of hydrated and desiccated lichens suggests a relationship to physiological changes in the thalli themselves. However, there is no agreement between the experiments on this topic so far carried out (Brown and Wilson, 1968; Jacobs and Ahmadjian, 1971b). Further investigations are needed before a more definitive assessment as to their significance can be made.

2. Vesicle Complexes as Lysosomes?

In addition to the well-known organelles, there are vesicle complexes in the phycobionts. In the phycobiont *Coccomyxa* these vesicle complexes are very obvious and they may occupy almost half of the cell (Plate II). In *Trebouxia* small vesicle complexes occur in the deep inlets of the irregular chloroplast (Plate VA and c). These structures are termed "vesicle complexes" following Crawford (1973) who described similar organelles in the centric diatom *Melosira varians*.

The vesicle complexes observed in *Coccomyxa* and *Trebouxia* are separated from the ground cytoplasm by a single membrane. Each consists of a variable number of vesicles with different appearances (Plates III and IV). Most vesicles are formed by a single membrane and have a diameter of 40–80 nm and in their centre an electron dense core may be visible (Plate IIIc and e). Some of the vesicles have two membranes concentrically arranged with diameters of 100–140 nm. Moreover, there are aggregates of three to seven separate small vesicles which are surrounded by a single membrane (Plate IIIB). In most cases the vesicles of the complexes are tightly packed (Plate IIIA and c), yet sometimes they are dispersed (Plate IIID and e). The ground substance between the vesicles is electron transparent (Plate IIID) or lightly electron dense (Plate IIIE). In addition, it is noteworthy that tubular elements are present in the complex (Plate VB). Between vesicles within the complex, two differently staining inclusions can be found (Plate IIIc), one showing a strong reaction with osmium tetroxide, which looks almost black, and one where the staining seems to be much less, leading to a lighter appearance. Both types of inclusions lack a visible surrounding membrane and occur as spherical, ovoid or droplet-shaped particles.

In thin sections some of the complexes contain either both types of inclusions (Plate IIIc) or the dark and light inclusions are to be found in different complexes (Plate IIID and E). Finally, the vesicles are occasionally concentrated around the inclusions (Plate IIID).

Outside the vesicle complexes there are similar bodies to those which occur as inclusions (Plates IIID and E, IVB). In many cases they are located so close to the vesicle complexes that it is difficult to discover the limiting membrane of the vesicle complexes themselves (Plate IVB). These bodies appear either to migrate into a vesicle complex or to be absorbed. The dark and light bodies closely associated with the vesicle complexes look very similar to those of similar contrast scattered in the cytoplasm and which are generally considered to be storage bodies.

In addition to vesicle complexes within the cytoplasm there are very often vesicles of almost the same size located in the area between the plasmalemma and the cell wall (Plate IVA and c). In such a case the plasmalemma has a very irregular coarse appearance and is more or less separated from the cell wall. Deep invaginations formed by the plasmalemma can be filled with vesicles (Plate IVB). Not only vesicles are visible outside the plasmalemma. The zone between the cell wall and the plasmalemma can have the same appearance as the ground-substance in the protoplast (Plate IIIA).

The two types of inclusions described above distinguish the vesicle complexes of phycobionts from those found in diatoms (Drum, 1964; Stoermer et al., 1965; Crawford, 1973) and in a silicoflagellate (van Valkenburg, 1971). The vesicle complexes with their different inclusions, might, however, be compared with the lysosomes in animal cells (Weissenfels, 1973). In particular, cells with a high lytic activity have lysosomes with varying tubular and vesicular structures as well as inclusions of substrate particles. The substances which are hydrolysed by the enzymes of the lysosomes may be exogenous or endogenous in origin.

When considering lysosomes in plant cells, investigators have mainly been concerned with the analysis of their enzymes and function (summarized by Frey-Wyssling, 1973). Sievers (1966) reported about lysosome-like complexes in ageing cells of Chara foetida with a structure very similar to those described in lichen phycobionts. In some electron micrographs Galun et al. (1970, 1971) pointed out that there are some lysosomes in Trebouxia but did not discuss them in detail.

From all the published data on lysosomes and our knowledge of the fine structure of lichen phycobionts, it appears reasonable to conclude that the vesicle complexes in lichens function as lysosomes. It seems possible that sub-

General key to plates
Abbreviations: A = alga; F = fungus; Ch = chloroplast; Pl = plasmalemma; I = invaginations; V = vesicles; VC = vesicle complex; SB = osmiophilic storage body; LB = lightly stained storage body; OI = osmiophilic inclusions in a vesicle complex; LI = lightly stained inclusions in a vesicle complex.

All material was fixed with glutaraldehyde, postfixed with osmium tetroxide and stained with lead citrate. Material in Plates II and IV was stained with ruthenium red in addition. Scale = 1 μm unless marked.

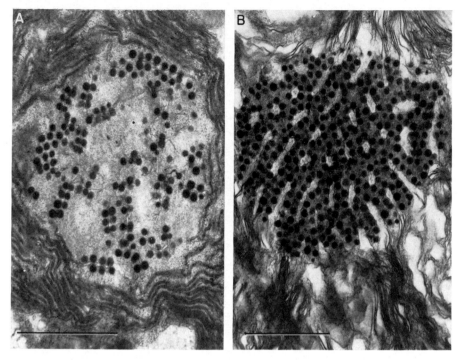

PLATE I

Osmiophilic globules aligned along the thylakoids which cross the pyrenoid in the centre of the chloroplast in the phycobiont *Trebouxia*. **A**, Pyrenoid of *Trebouxia* in *Cladonia fimbriata*. **B**, Pyrenoid of *Trebouxia* in *Physcia aipolia*.

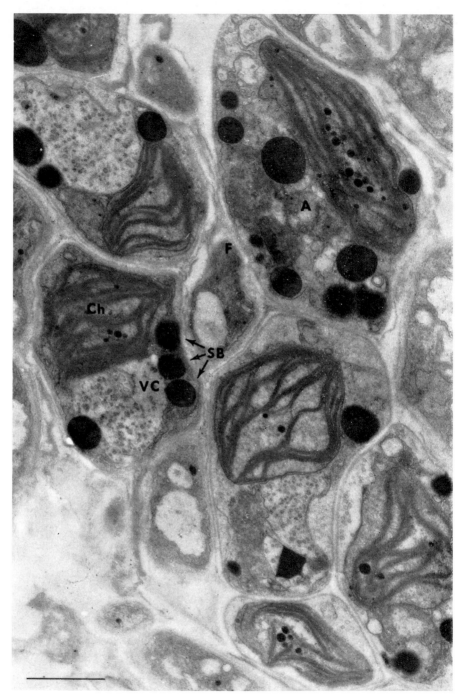

Part of the algal layer of *Icmadophila ericetorum* showing close contacts between the algae and fungi. In the algal cells (*Coccomyxa*), in addition to the chloroplast, large vesicle complexes and osmiophilic storage bodies occur.

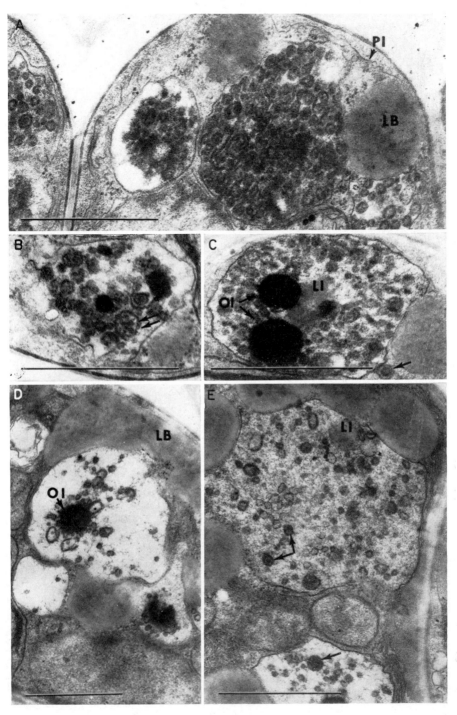

PLATE III

Different aspects of vesicle complexes in the *Coccomyxa* phycobiont of *Icmadophila ericetorum*. Arrows indicate vesicles with a dense core; the double arrow indicates an aggregate of several small vesicles in a larger one. See text for further explanation.

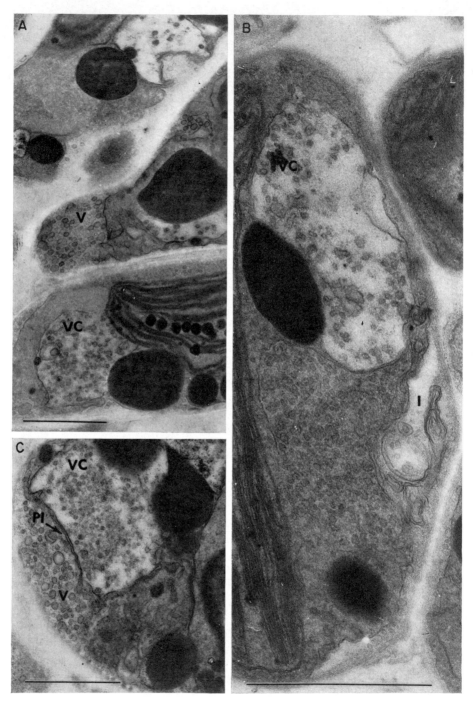

PLATE IV

Vesicles outside the protoplast and in the vesicle complexes within the cytoplasm of the *Coccomyxa* phycobiont of *Icmadophila ericetorum*; osmiophilic inclusions as well as osmiophilic storage bodies of different sizes are present.

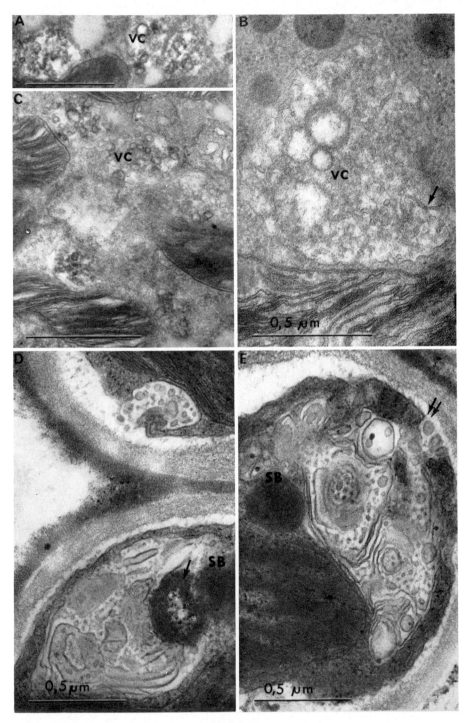

A–C, Small vesicle complexes between the chloroplast tips in the *Trebouxia* phycobiont of *Ramalina maciformis*; in the complex shown in **B** there are additional tubular elements (arrow). **D** and **E** show lomasomes in the *Trentepohlia* phycobiont of *Dimerella diluta* with neighbouring osmiophilic storage bodies which in **D** are partly dissolved (arrows). The double arrow points to the connection of a large lomasome to the space between the plasmalemma and cell wall.

stances stored in the autotrophic phycobiont are hydrolysed by lysosomes into transportable metabolities making them available to the heterotrophic myco-biont.

Many problems are still to be solved: for example what are the active enzymes and which are the main metabolites in the phycobiont protoplasts? Also where does the transfer to the mycobiont take place? In the *Coccomyxa* phycobiont additional vesicles can be found between the plasmalemma and the cell wall. Without more experimental evidence, however, it is impossible to conclude that those are transfer vesicles. It might also be possible that those vesicles are only parts of the lysosome complex which have passed through the plasmalemma.

3. Mesosomes, Lomasomes or Plasmalemmasomes?

Structures associated with the plasmalemma have been described in bacteria as mesosomes (Fitz-James, 1960) and in fungi as lomasomes (Moore and McAlear, 1961). In some algae and higher plants such structures have also been observed (see the review by Marchant and Robards, 1968). By freeze etching it was shown that the tubules and (or) vesicles in the lomasomes are not artefacts (Marchant and Moore, 1973).

The *Trentepohlia* phycobiont of *Dimerella diluta* shows very extended loma-somes (Plate VD and E) which appear as aggregations of membranes and vesicles located close to the cell wall. Sometimes a direct connection with the space between the plasmalemma and the cell wall is found (Plate VE). Besides the large lomasomes there are small zones filled with vesicles and bordered by the plasmalemma (PLATE VD). Between the small vesicle areas and the large loma-somes there are intermediates and it is therefore almost impossible to term them either lomasomes or plasmalemmasomes according to the criteria of Marchant and Robards (1968). Furthermore, they also show similarities to bacterial mesosomes.

Storage bodies occur very close to almost all lomasomes, as in all other phycobionts where they are adjacent to the vesicle complexes. The storage bodies are always located on the side of the lomasome away from the cell wall. The storage bodies can partly be surrounded by membranes of the lomasome. Some of the storage bodies seem loosened as if digestion has taken place (Plate VD).

The close connection between lomasomes and storage bodies may point to significance of the lomasomes in the transfer of metabolites. A similar function of lomasomes had been discussed for such differentiations in the host–parasite interface of *Peronospora* (Peyton and Bowen, 1963). Otherwise the main func-

B

tion of lomasomes and of mesosomes is regarded as the synthesis of cell walls (Marchant and Robards, 1968; Burdett and Rogers, 1972). In the mycobiont of *Peltigera canina* the participation of lomasomes in cell wall synthesis can be demonstrated by a specific histochemical technique (Boissière, 1972).

After discussing such different possible functions of lomasomes and the significance of mesosomes the question arises as to whether they have a specific function at all or they represent a structure which is necessary to enhance the activity of plasmalemma-bound enzymes. From this point of view the great variety of lomasomes and mesosomes on the one hand and the obvious similarity in their appearance on the other would be understandable (Hashimoto and Yoshida, 1966; Zachariah and Fitz-James, 1967).

TRANSPORT OF METABOLIC PRODUCTS

From physiological studies it has been established that phycobionts assimilate as autotrophic organisms and provide metabolites for the heterotrophic mycobiont. This concept can be further confirmed by investigations into the ultrastructure of phycobionts and mycobionts.

1. Presence of Storage Bodies in the Phycobionts

The fact that the algae take up carbon dioxide has been demonstrated autoradiographically (Jacobs and Ahmadjian, 1971a; Peveling and Hill, 1974), but so far there has been no success in locating radioactive label in specific structures, for example those in the phycobionts interpreted as storage bodies. The osmiophilic bodies and those with a lighter appearance are considered to be storage products as a result of comparisons with known storage bodies (Dodge, 1973).

2. Activation of Storage Products for Transfer

If the mycobiont is to benefit from the photosynthetic products of the phycobionts, the storage bodies in phycobiont cells should be degraded to facilitate transfer to the mycobionts. The presence of lysosomal vesicle complexes adjacent to storage bodies, as well as inclusions within the complexes, may suggest a degradation process. At least the vesicle complexes, as lysosomes, could partly hydrolyse the storage bodies and this could conceivably make some stored products transferable.

Enzymatically active lomasomes are another possibility for transforming metabolic products. The appearance of storage bodies close to the lomasomes and the open connection of the lomasomes to the cell wall in particular may be suggestive of preparation for the transfer of metabolites.

3. Type of Transfer

All substances observed between the plasmalemma and cell wall may be considered as transferred metabolites. In the phycobionts described above a substance similar to the ground-substance and vesicles was found outside the protoplast. Moreover, vesicles were detected in the sheath of the *Calothrix* phycobiont of *Lichina pygmaea* (Peveling, 1973). In addition to these visible components, it is necessary to consider the possibility that other substances affecting transfer might exist undetected either because they are completely electron transparent or because they are washed out during preparative procedures.

4. Movement of Metabolites into Mycobionts

By different techniques it has been shown that photosynthetic products from the algae reach the mycobionts. The amounts and rates of movement of the transferred carbohydrates have so far been studied mainly from the physiological standpoint (see review by Richardson, 1974).

In active living thalli the mycobiont hyphae close to phycobiont cells are characterized by numerous plasmalemma invaginations. These invaginations occur all around the protoplast and provide an increased protoplast surface. Such an increase is a clear indication of high metabolic activity. In the cytoplasm of such hyphae a large number of ribosomes can be found simultaneously. These observations may suggest an uptake of metabolic products.

Extensive membranous structures have been observed in the mycobiont of *Lichina pygmaea* (Peveling, 1972) which start at the plasmalemma and extend far into the cytoplasm. With their tubular and membranous structures and their vesicles they resemble the large lomasomes of the *Trentepohlia* phycobiont but otherwise they show similarities to the mesosomes of bacteria. The function of these structures, termed mesosome-like structures, was discussed in connection with the deposition of glycogen and another storage product at their border in the cytoplasm. In addition to the storage areas of the mycobiont in *Lichina pygmaea* storage bodies have been found in all other lichens investigated. These appear as dark- or light-stained bodies, very similar to those in the cytoplasm of the phycobionts but they are usually more numerous in the mycobionts.

5. Transfer between the Fungal Cells

A question that should also be considered is whether each fungal cell must take up its own substances or whether transfer within hyphae is possible. In free-living fungi, as well as in mycobionts, several round or oblong bodies close to the septal pores have been described (Reichle and Alexander, 1965; Wetmore,

1973). These have been interpreted both as lipid bodies and as bodies of an unknown nature associated with pores and plugs. Close to the septal pores these bodies (termed Woronin bodies) which are considered part of the storage material, become smaller and their limits increasingly diffuse, giving the impression that they are flowing through the pore. Since, particularly in lichens, multiperforate septa are found (Strasburger, 1901; Meyer, 1902; Kienitz-Gerloff, 1902) it may be concluded that metabolic substances are transferred through these pores within the hyphae.

CONCLUSIONS

The observations reported here and the interpretations suggested together with data about lichen ultrastructure already published, provide some indication as to how all the observed structures take part in the metabolism of the symbiotic algae and fungi. If the assumptions made in this paper seem obvious and acceptable it should be kept in mind that almost all results were obtained from fixed material. New techniques need to be developed to establish the precise connection between structure and function. The first steps are being made by high resolution autoradiography and by refinements of this technique we hope we shall eventually be able to obtain more precise information as to the metabolic activities of lichens.

ACKNOWLEDGEMENT

Acknowledgement is made to the Deutsche Forschungsgemeinschaft for the support of this work.

REFERENCES

AHMADJIAN, V. and JACOBS, J. B. (1970). The ultrastructure of lichens. III. *Endocarpon pusillum*. *Lichenologist* **4**, 268–270.

BEN-SHAUL, Y., PARAN, N. and GALUN, M. (1969). The ultrastructure of the association between phycobiont and mycobiont in three ecotypes of the lichen *Caloplaca aurantia* var. *aurantia*. *J. Microscopie* **8**, 415–422.

BOISSIÈRE, J.-C. (1972). Étude ultrastructurale de l'édification des parois des hyphes du *Peltigera canina* (L.) Willd. *Rev. Cytol. Biol. vég.* **36**, 1–6.

BROWN, R. M. and WILSON, R. (1968). Electron microscopy of the lichen *Physcia aipolia* (Ehrh.) Nyl. *J. Phycol.* **4**, 230–240.

BURDETT, J. D. J. and ROGERS, H. J. (1972). The structure and development of mesosomes studied in *Bacillus licheniformis* Strain 6346. *J. Ultrastruc. Res.* **38**, 113–133.

CRAWFORD, R. M. (1973). The protoplasmic ultrastructure of the vegetative cell of *Melosira varians* c. a. Agardh. *J. Phycol.* **9**, 50–61.

DODGE, J. D. (1973). "The Fine Structure of Algal Cells." Academic Press, London and New York.

DRUM, R. W. (1964). Post mitotic fine structure of *Gomphonema parvulum*. *J. Ultrastruc. Res.* **10**, 217–223.

FISHER, K. A. and LANG, N. J. (1971). Ultrastructure of the pyrenoid of *Trebouxia* in *Ramalina menziesii* Tuck. *J. Phycol.* **7**, 25–37.

FITZ-JAMES, P. (1960). Participation of the cytoplasmic membrane in the growth and spore formation of Bacilli. *J. biophys. biochem. Cytol.* **8**, 507–528.

FREY-WYSSLING, A. (1973). Comparative Organellography of the Cytoplasm. [Protoplasmatologia III/G] Wien–New York.

GALUN, M., PARAN, N. and BEN-SHAUL, Y. (1970). Structural modifications of the phycobiont in the lichen thallus. *Protoplasma* **69**, 85–96.

GALUN, M., PARAN, N. and BEN-SHAUL, Y. (1971). Electron microscopic study of the lichen *Dermatocarpon hepaticum* (Ach.) Th. Fr. *Protoplasma* **73**, 457–468.

GRIFFITHS, D. J. (1970). The pyrenoid. *Bot. Rev.* **36**, 29–58.

GRIFFITHS, H. B. and GREENWOOD, A. D. (1972). The concentric bodies of lichenized fungi. *Arch. Mikrobiol.* **87**, 285–302.

HASHIMOTO, T. and YOSHIDA, N. (1966). Unique membranous system associated with glycogen synthesis in an imperfect fungus *Geotrichum candidum*. *Proc. 6th Intern. Congr. Electron Microscopy, Kyoto* **2**, 305–306.

JACOBS, J. B. and AHMADJIAN, V. (1969). The ultrastructure of lichens. I. A general survey. *J. Phycol.* **5**, 227–240.

JACOBS, J. B. and AHMADJIAN, V. (1971a). The ultrastructure of lichens. IV. Movement of carbon products from alga to fungus as demonstrated by high resolution autoradiography. *New Phytol.* **70**, 47–50.

JACOBS, J. B. and AHMADJIAN, V. (1971b). The ultrastructure of lichens. II. *Cladonia cristatella*: The lichen and its isolated symbionts. *J. Phycol.* **7**, 71–81.

JACOBS, J. B. and AHMADJIAN, V. (1973). The ultrastructure of lichens. V. *Hydrothyria venosa*, a freshwater lichen. *New Phytol.* **72**, 155–160.

KIENITZ-GERLOFF, F. (1902). Neue Studien über Plasmodesmen. *Ber. dt. bot. Ges.* **20**, 93–117.

LICHTENTHALER, H. K. (1968). Plastoglobuli and the fine structure of plastids. *Endeavour* **27**, 144–149.

MARCHANT, R. and MOORE, R. T. (1973). Lomasomes and plasmalemma in fungi. *Protoplasma* **76**, 235–247.

MARCHANT, R. and ROBARDS, A. W. (1968). Membrane systems associated with the plasmalemma of plant cells. *Ann. Bot.* **32**, 457–471.

MEYER, A. (1902). Die Plasmaverbindungen und die Fusionen der Pilze der Florideenreihe. *Bot. Ztg* **60**, 139–178.

MOORE, R. T. and McALEAR, J. H. (1961). Fine structure of mycota. 5. Lomasomes—previously uncharacterized hyphal structures. *Mycologia* **53**, 194–200.

PARAN, N., BEN-SHAUL, Y. and GALUN, M. (1971). Fine structure of the blue-green phycobiont and its relation to the mycobiont in two *Gonohymenia* lichens. *Arch. Mikrobiol.* **76**, 103–113.

PEAT, A. (1968). Fine structure of the vegetative thallus of the lichen *Peltigera polydactyla*. *Arch. Mikrobiol.* **61**, 212–222.

PEVELING, E. (1969a). Elektronenoptische Untersuchungen an Flechten. IV. Die Feinstruktur einiger Flechten mit Cyanophyceen-Phycobionten. *Protoplasma* **68**, 209–222.

PEVELING, E. (1969b). Elektronenoptische Untersuchungen an Flechten. II. Die Feinstruktur von *Trebouxia*-Phycobionten. *Planta* **87**, 69–85.

PEVELING, E. (1969c). Elektronenoptische Untersuchungen an Flechten. III. Cytologische Differenzierungen der Pilzzellen im Zusammenhang mit ihrer symbiontischen Lebensweise. *Z. Pflanzenphysiol.* **61**, 151-164.

PEVELING, E. (1972). Mesosomen-ähnliche Strukturen im Mycobionten von *Lichina pygmaea. Z. Naturf.* **27b**, 1388–1392.

PEVELING, E. (1973). Vesicles in the phycobiont sheath as possible transfer structures between the symbionts in the lichen *Lichina pygmaea. New Phytol.* **72**, 343–345.

PEVELING, E. (1974) ["1973"]. Fine Structure. *In* "The Lichens" (V. Ahmadjian and M. E. Hale, eds), pp. 147–184. Academic Press, New York and London.

PEVELING, E. and HILL, D. J. (1974). The localisation of an insoluble metabolite in the lichen *Peltigera polydactyla. New Phytol.* **73**, 767–769.

PEYTON, G. A. and BOWEN, C. C. (1963). The host-parasite interface of *Peronospora manshurica* on *Glycine max. Am. J. Bot.* **50**, 787–797.

REICHLE, R. E. and ALEXANDER, J. V. (1965). Multiperforate septations Woronin bodies, and septal plugs in *Fusarium. J. Cell Biol.* **24**, 489–998.

RICHARDSON, D. H. S. (1974) ["1973"]. Photosynthesis and carbohydrate movement. *In* "The Lichens" (V. Ahmadjian and M. E. Hale, eds), pp. 249–288. Academic Press, New York and London.

SIEVERS, A. (1966). Lysosomen-ähnliche Kompartimente in Pflanzenzellen. *Naturwissenschaften* **53**, 334–335.

STOERMER, E. F., PANKRATZ, H. S. and BOWEN, C. C. (1965). Fine structure of the diatom *Amphipleura pellicuda.* II. Cytoplasmic fine structure and frustule formation. *Am. J. Bot.* **52**, 1067–1078.

STRASBURGER, E. (1901). Über Plasmaverbindungen pflanzlicher Zellen. *Jb. wiss. Bot.* **36**, 493–610.

VAN VALKENBURG, S. D. (1971). Observations on the fine structure of *Dictyocta fibula* Ehrenberg. II. The protoplast. *J. Phycol.* **7**, 118–132.

WEISSENFELS, N. (1973). Lysosomen. *In* "Grundlagen der Cytologie" (G. C. Hirsch, H. Ruska and P. Sitte, eds), pp. 297–304. Stuttgart.

WETMORE, C. M. (1973). Multiperforate septa in lichens. *New Phytol.* **72**, 535–538.

ZACHARIAH, K. and FITZ-JAMES, P. (1967). The structure of *phialides* in *Penicillium claviforme. Can. J. Microbiol.* **13**, 249–256.

3 | The Morphological and Taxonomic Significance of Cephalodia

P. W. JAMES

Department of Botany, British Museum (Natural History), London, England

and

A. HENSSEN

Department of Plant Biology, University of Marburg, West Germany

Abstract: The occurrence of cephalodia in lichens is reviewed in relation to the recent discovery of several instances where different morphotypes in the Peltigerineae and Pannariaceae are formed by the same mycobiont but with different genera of phycobionts. In some cases composite (dimorphic) plants are produced and the result of a detailed anatomical investigation of these structures is presented. The phycobiont is shown to play an important role in the determination of thallus form. Affinities are established or suggested between species with blue-green or green algae in the same and different genera, and particular attention is drawn to the status of *"Dendriscocaulon"*. Environmental factors appear to play an important rôle in determining which phycobiont is selected by the mycobiont. The significance of this phenomenon is discussed in relation to lichen chemistry, morphogenesis and nomenclature.

The new combination *Dendriscocaulon intricatulum* (Nyl.)Henss. **comb. nov.** (syn. *Leptogidium intricatulum* Nyl.) is made.

INTRODUCTION

It is generally accepted that each fungal partner (mycobiont) of most lichens forms an association with a single algal partner (phycobiont) which belongs either to the Chlorophyceae (green algae) or the Cyanophyceae (blue-green algae). However, in a small but significant number of genera (Table I), thalli

---

Systematics Association Special Volume No. 8, "Lichenology: Progress and Problems", edited by D. H. Brown, D. L. Hawksworth and R. H. Bailey, 1976, pp. 27–77. Academic Press, London and New York.

incorporating green algae as their main phycobiont may also form subsidiary associations with blue-green algae, in particular *Nostoc*, but also *Calothrix*, *Gloeocapsa*, *Scytonema* and *Stigonema*. The areas circumscribed or the special

TABLE I. Synopsis of the occurrence and types of cephalodia

Family and genus	Position and alga	Remarks
CALICIACEAE		
Chaenotheca	External, internal when immature, *Nostoc*.	One species, *C. trichialis*, possibly a few others.
Sphaerophorus	External, genus unknown, probably *Nostoc*.	One species, *S. stereocaulonoides*. Originally placed in the monotypic genus *Thysanophoron*.
LECIDEACEAE		
Lecidea	External, *Stigonema*; *Gloeocapsa* in *L. pelobotryon* (Henssen and Jahns, 1973).	Few species, e.g. *L. pelobotryon*.
Lopadium	External, ?*Scytonema* (Santesson, 1952).	Two foliicolous species, *L. vermiculiferum*, *L. tayabasense* (Santesson, 1952).
Lasioloma	External, ?*Scytonema* (Santesson, 1952).	In all species (3); foliicolous. Closely related to *Lopadium*.
PERTUSARIACEAE		
Placopsis	External, *Nostoc*, *Stigonema*, *Scytonema* (Lamb, 1947).	In all species (*c.* 40).
Aspiciliopsis	External, *Nostoc* (Lamb, 1947).	Monotypic genus, probably to be subsumed with *Placopsis*.
COCCOTREMATACEAE		
Coccotrema	External, *Calothrix* (Brodo, 1973).	In all species (*c.* 12). (See Ch. 6).
Lepolichen	External, *Calothrix*.	Monotypic genus.
STEREOCAULACEAE		
Stereocaulon	External, *Nostoc*, *Stigonema*, *Scytonema*.	All species (*c.* 200).
Argopsis	External, *Nostoc*, *Stigonema* (Lamb, 1974).	Monotypic genus.

Family and genus	Position and alga	Remarks
Pilophorus	External, *Nostoc, Stigonema* (Jahns, 1970).	All species (*c.* 25).
Compsocladium	Internal, *Scytonema* (Lamb, 1956).	Monotypic genus.
PANNARIACEAE		
Psoroma	External, rarely internal, *Nostoc, ?Scytonema.* Extra-generic relationship with with *Pannaria* suggested (Forssell, 1884c). One (?or two) species with sorediate cepha-lodia (James and Henssen, 1975).	Nearly all species (*c.* 45). Internal or on underside of squamules or as complete squam-ules in *P. hypnorum* aggregate; mostly laminal in rosette-forming species, e.g. *P. sphinctrinum.*
Psoromaria	External, *Nostoc.*	All species (2 ?or 3). Closely related to subsquamulose species of *Psoroma.*
PELTIGERACEAE		
Peltigera	External, very rarely internal, *Nostoc.* Intrageneric relation-ships possible, perhaps certain.	In all species with green phycobionts (*c.* 12).
Nephroma	Internal, *Nostoc.* Intrageneric relationships possible but no evidence of occurrence.	In all species with green phycobionts (*c.* 10).
Solorina	Internal (*S. crocea*) or external (*S. spongiosa*), *Nostoc*; can lead independent existence in some species. Intrageneric relation-ships possible but not observed.	In all species (*c.* 20), sometimes dominant; in one undescribed species green phycobiont absent. Relationship between ascocarp formation and green morphotype very noticeable in the genus.
STICTACEAE		
Sticta	Internal or external, *Nostoc,* either as foliose blue-green phycobiont counterpart or *Dendriscocaulon.* Intrageneric relationships proven. Extra-generic relationships with *Dendriscocaulon* proven.	All species with green phycobiont (*c.* 48).

TABLE I.—*continued*

Family and genus	Position and alga	Remarks
Pseudocyphellaria	Internal, *Nostoc*. Number and (or) extent depends on environmental conditions. Intrageneric relationships possible but not proven.	All species with green phycobiont (*c.* 40).
Lobaria	Internal or external, *Nostoc*, as *Dendriscocaulon*. Intrageneric relationships probable but not conclusively proven. Extra-generic relationships with *Dendriscocaulon* proven.	All species with green phycobiont (*c.* 50).

structures formed incorporating these subsidiary algae are broadly termed cephalodia. When developed within the thallus as internal cephalodia (*cephalodia thalloidea*), the blue-green algae may be arranged in a more or less interrupted layer below the green phycobiont (as in *Solorina crocea*), but generally they are more frequently restricted to delimited areas and are chiefly located in the medulla of the thallus. The position, shape and size of external (erumpent) cephalodia (*cephalodia epigaea*) are characteristic for each of the species in which they occur and their constancy in these aspects has led to their use as a reliable taxonomic criterion as, for example, in the species of *Placopsis* (Lamb, 1947) and *Peltigera* (Thomson, 1950). The reversal of this interrelationship, one in which the main thallus incorporates blue-green algae but also develops cephalodia containing green algae, has not been recorded. In the sub-order Peltigerineae (see Henssen and Jahns, 1973) some genera, *Nephroma*, *Solorina* (rare), *Lobaria*, *Pseudocyphellaria* and *Sticta*, have a range of species which, although having immediately recognizable generic affinities in their morphology and anatomy, contain either blue-green algae or green algae and cephalodia. Such a parallel morphology is well illustrated by the British species of *Peltigera* where *P. canina*, *P. horizontalis*, *P. malacea*, *P. polydactyla* and others containing blue-green algae, are similar and obviously related to *P. aphthosa*, *P. leucophlebia* and *P. venosa* with green phycobionts and external cephalodia. Conversely, in other genera of this suborder, namely some species of *Lobaria* and *Sticta*, the cephalodia assume a coralloid form capable of either an independent existence or remaining attached to an associated thallus containing a green phycobiont. The free-living blue-green structures were first assigned to the genus *Dendris-*

cocaulon by Nylander (1888), since which time a considerable controversy has arisen as to the interpretation of their true nature, either as independent and unrelated lichens, or as attached or liberated cephalodia of *Lobaria amplissima*; some authors (Degelius, 1935, 1941; Henssen, 1963; but see Henssen and Jahns, 1973) came to different conclusions at different periods of study of this problem[*]. However, the correct interpretation of the nature and relationships of these heteromorphic structures containing different phycobionts, hinted at by Wilson (1891) as a result of his studies on the Australian species *Sticta stipitata*, was contained in two papers by Dughi (1944, 1954). As a result of his re-examination of Wilson's collections and his own work on the origin and development of the cephalodia of *Lobaria amplissima*, Dughi concluded that the *Dendriscocaulon* growth on these two species was derived from the same fungus as the foliose part but under the influence of a different alga. Our investigations based on a detailed anatomical analysis of these and additional species confirm the statements of Dughi.

---

[*] The type material of *Dendriscocaulon umhausense* (Auersw.) Degel. (syn. *Polychidium umhausense* (Auersw.) Henss.) was from moist rocks in the Alps (Plate VIIA) and cannot be distinguished from the cephalodia of *Lobaria amplissima* on the basis of internal and external morphological characters; such plants have consequently often been considered as the "escaped cephalodia" of *L. amplissima*. However, although the similarity is striking, no experimental work has been undertaken to prove that the mycobionts of all free-living specimens of *D. umhausense* in Europe are the same as that of *L. amplissima*. In the case of the *Dendriscocaulon* growing in the vicinity of *L. amplissima* in Portugal this relationship seems very probable, but in quite different biotopes (e.g. the Lapland tundra) it is conceivable that mycobionts of other *Lobaria* species with green phycobionts might produce similar *Dendriscocaulon* morphotypes under certain environmental conditions.

In North America *D. umhausense* is replaced by **Dendriscocaulon intricatulum** (Nyl.) Henss. **comb. nov.** (basionym: *Leptogidium intricatulum* Nyl., *Syn. Lich.* **1**, 135, 1858; synonym: *Polychidium intricatulum* (Nyl.) Henss., *Symb. bot. upsal.* **18** (1), 106, 1963). Type material of the latter (H–Nyl. no. 41299) is very poor (see Henssen, 1963) but abundant material collected by one of the authors (A.H.) in different parts of Canada and the U.S.A. revealed that this lichen is very closely related to *D. umhausense*. We have only seen *D. intricatulum* in an independant state, although in one case growing beneath a lobe of *Lobaria erosa*. According to Moore (1969, p. 406), cephalodia should generally occur on thalli of both *L. erosa* and *L. lobulifera*; thus, *D. intricatulum* might be the blue-green algal morphotype of either of these two species or of *L. quercizans* (a species very similar in habit to *L. amplissima*) (see p. 42).

As long as the identity of *D. umhausense* and *D. intricatulum* mycobionts with mycobionts of particular *Lobaria* species remains unclear, it seems most appropriate to adopt these names for the free-living blue green algal morphotypes.

The type specimen of *D. dendroides* Nyl. (H–Nyl. no. 41030), described from New Zealand, is extremely small, making it impossible to relate it with certainty to any of the *Dendriscocaulon*-like morphotypes of the Peltigerineae in New Zealand discussed elsewhere in this paper (p. 42).

The biology of the various types of cephalodia has been studied from several different aspects. Their initiation and comparative accounts of their morphology and anatomy have, in particular, received detailed examinations by Forssell (1883, 1884a–c), Moreau (1928, 1956), Kaule (1931, 1934) and others. In an important but largely neglected series of papers, Dughi (1936, 1937, 1944, 1945, 1954) discussed in detail their origin and compared both free-living and attached forms of cephalodial development with other related aspects of thallus development and morphogenesis. He noted that one mycobiont could be shared between some species of *Dendriscocaulon* and members of the genera *Sticta* and *Lobaria* and also drew attention to a possible similar relationship between *Psoroma hypnorum* and *Pannaria pezizoides* previously discussed by Forssell (1884c). As a result of his researches he proposed significant and immediately controversial hypotheses on the interpretation and value of the lichen thallus in the classification of the group and proposed a scheme for dealing with those nomenclatural problems that naturally arise from the interpretation of pairs of morphologically different thalli having the same mycobiont. Theories as to the morphogenetic processes in lichens are few: Beyerinck (1877) and Moreau (1956) considered the development of cephalodia as a form of cecidization (gall formation) (see also Mani, 1964). Moreau suggested a broad analogy between the processes of lichenization and gall formation as an explanation of the wide diversity in lichen form and behaviour. In his view the interrelationship between the bionts in lichens was broadly analogous to that of the development of certain galls. The physiological differences between the cephalodia and the main thallus have received little attention: Richardson *et al.* (1968) have found fundamental differences between the carbohydrate pathways from the phycobiont to the mycobiont in the "cephalodia" and main thallus of *Lobaria amplissima*, and Kershaw and Millbank (1970) and Millbank (Chapter 18) have examined nitrogen fixation in the cephalodia of *Peltigera aphthosa* and the transport of the products of nitrogen metabolism to the main thallus. Culberson (1967) reported on the different chemistry of the main thallus and "cephalodia" of *Lobaria amplissima*, results which seem to contradict those reported by Jordan (1972) for *Lobaria* cf. *erosa*.

A detailed analysis of the morphology, anatomy, ecology and chemistry of several recently discovered, as well as relatively familiar, pairs of intimately associated but disparate conspecific morphotypes forms the main topic of this paper. The biological and taxonomic implications of such relationships have, we believe, far-reaching implications concerning the interpretation of some of the fundamental problems of lichen ecology and taxonomy.

MORPHOTYPES KNOWN

Figure 1 illustrates the range of form and arrangement of the composite thalli discussed in this paper. An asterisk (*) indicates species not studied in detail and a

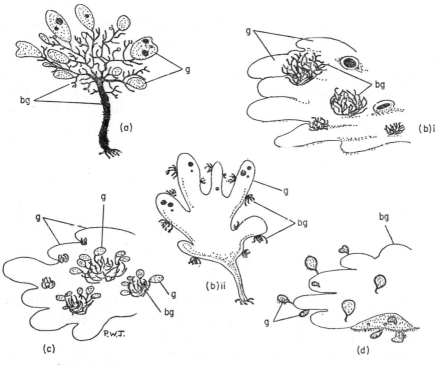

Fig. 1. Relationship between green (g) and blue-green (bg) algal morphotypes. (a): Caulescent blue-green algal morphotype (*Dendriscocaulon* spp.) with fertile leaflets of the green algal morphotype (*Sticta filix, S. latifrons*). (b)i: Foliose thallus of the green algal morphotype (e.g. *S. glomuligera, Lobaria amplissima, L. ornata*) with coralloid excrescences of the blue-green algal morphotype (*D.* spp.). (b)ii: Caulescent and fertile green algal morphotype (*S. dichotomoides*) with marginal coralloid clusters of the blue-green algal morphotype (*D.* spp). (c): as (b)i, but with the secondary development of the green algal morphotype as leaflets on the clusters of the blue-green algal morphotype (e.g. *L. amplissima, L.* cf. *erosa*). (d): Foliose thallus of the blue-green algal morphotype (*Peltigera* sp., *Sticta dufourii*) with developing marginal and laminal leaflets of the green algal morphotype (*P.* sp., *S. canariensis*). See text for further information.

dagger (†) denotes those on which the data are based on information in the literature; all other species cited have been examined in detail.

The following combined morphotypes are known:

A. Primary part *Dendriscocaulon*, incorporating blue-green algae, fruticose and attached by a conspicuous rooting stalk. Secondary part with green algae, dorsiventral leaflets of either *Sticta filix* (Räusch.) Nyl., *S. latifrons* Rich., or *S. stipitata* C. Kn. ex Wils. (syn. *S. shirleyana* Müll. Arg.). The species of *Dendriscocaulon* seem to differ for each green algal morphotype; that for *S. stipitata* is *Dendriscocaulon dendro-thamnodes* Dughi (nom. inval.); the others are undescribed. Composite morphotypes are frequent in *S. stipitata* (80%); only seen once or twice in the other species.

B. (i). Primary thallus foliose, prostrate, dorsiventral with green phycobiont: *Lobaria amplissima* (Scop.) Forss., *L. dendrophora* Zahlbr.† (Yoshimura, 1971), *L. ornata* Malme* and *Sticta glomuligera* Nyl.*. Secondary development of *Dendriscocaulon* spp., ?*D. umhausense* (see footnote on p. 31) in *L. amplissima*, undescribed in other taxa. Composite morphotypes usual in *L. amplissima* (95%); very rare in other species.

(ii). Primary thallus foliose, erect or pendent, more or less palmate or entire, dorsiventral, caulescent: *Sticta boschiana* Mont.*, *S. dichotomoides* Nyl., *S. pendunculata* Kremp.*. Secondary development of *Dendriscocaulon* on the upper surface and margins (*S. dichotomoides*), species all different and undescribed, present in all individuals of *S. dichotomoides*, known from two specimens of *S. pedunculata*, and one of *S. boschiana*.

C. As in **B**(i) but with the additional development of dorsiventral leaflets of the green algal morphotype on the *Dendriscocaulon* morphotype as in **A**: *Lobaria amplissima* (Scop.) Forss. †(Dughi, 1936) and *L.* cf. *erosa* (Eschw.) Trev. †(Jordan, 1972). Both are only known from a single gathering.

D. Primary thallus foliose with blue-green algae, dorsiventral like the secondary thallus which has a green phycobiont: *Sticta dufourii* Del. (blue-green algal morphotype)—*S. canariensis* Bory ex Del. (green algae morphotype); *P. aphthosa* (L.) Willd. or *P. leucophlebia* (Nyl.) Gyeln. (green algal morphotype) upon specimens resembling *P. malacea* (Ach.) Funck or the *P. polydactyla* (Neck.) Hoffm. aggregate (blue-green algal morphotype)*.

In addition, *Sticta marginifera* var. *coralloidea* Müll. Arg. and many caulescent species of *Sticta* with green or blue-green algae were examined. Other cephalodiate genera studied were *Solorina*, *Stereocaulon* and *Placopsis*. A recently described species of *Psoroma* from New Zealand and Tasmania (*P. durietzii* P. James & Henss.) with sorediate cephalodia is noted which, with the possible exception of *P. dimorphum* Malme, is unique among cephalodiate species.

MORPHOLOGY AND ANATOMY

A characteristic feature for all types of *Dendriscocaulon*, whether attached or free living, is the unequal differentiation of the thallus. A great variability of the cortex is typical; increasing considerably towards the older parts at the base. In the northern group, allied to *Lobaria*, the tips of the branchlets are formed by the medullary hyphae pushing forwards (Plate VIIID), while in the southern group, allied to *Sticta*, the tips remain delimited by a single celled cortex (see pp. 42–44. The algae fill the small branchlets at the tip uniformly, but in older parts a distinct algal zone is formed below the cortex which extends all round or is mainly restricted to the upper part (Plate VIIIE). Flattened lobes do not always have a dorsiventral structure but may also be isolateral. Towards the base the algal cells diminish in quantity becoming accumulated in groups or chains which are easily distinguished when the lobes are mounted in water and examined under low magnification. The base of the stalk is usually completely free of algal cells. This general pattern of differentiation of the *Dendriscocaulon* type should be kept in mind when reading the following descriptions.

1. Sticta filix–Dendriscocaulon *sp.*

Our particular interest in the nature of composite morphotypes and their significance originates from the discovery by one of the authors (P.W.J.) in 1962, near Lake Te Anau, on South Island, New Zealand, of an apparently unique clone of thalli (see p. 48), some of which are typical of *Sticta filix*, some of an undescribed species of *Dendriscocaulon*, and the remainder, strange and seemingly anomalous plants compounded of both species. Plates IA and IIA illustrate the diversity of form within the clone. Each composite thallus is up to 6 cm tall and has a lower primary part identical with the nearby free-living *Dendriscocaulon* sp., with a stout, terete, red-brown rooting stalk surmounted by a dense crown of richly branched branchlets containing blue-green algae (*Nostoc*). One to twenty leaflets, up to 1·5 cm in length, occur on each composite plant, apparently derived from small nodules of green algae (?*Myrmecia*) probably captured by the blue-green algal morphotype. The leaflets are dorsiventral and have a smooth, naked, pale grey-green upper surface and a pale brown, sparingly tomentose lower surface with numerous small and evenly spaced cyphellae. A few of the larger leaflets bear red-brown ascocarps; comparison of the structure and anatomy of these and the vegetative parts of the thalli confirmed that the leaflets were identical with the independent plants of *S. filix* close by. The latter differ only in their development of large, up to 10 cm tall, deeply lobed, palmate thalli which often develop secondary laciniae along the margins of the main lobes (Plate IA). Attachment is by means of a short, but well developed, more or less terete, rooting stalk.

A detailed examination of the transitional area between the two entities in the composite plants clearly demonstrated that there is an unbroken continuity between the hyphae in the leaflets and the coralloid branchlets of the *Dendris-cocaulon* (Plate IIIA). This convincingly indicates that not only is a single myco-biont involved but, more importantly, that its interaction with different genera of phycobionts results in strikingly different morphotypes. With *Nostoc* it forms a fruticose, coralloid, non-fruiting thallus and with ?*Myrmecia* a foliose, dorsi-ventral, fruiting morphotype. On this evidence, the green algal morphotype is in no sense to be considered as epiphytic or parasitic on the blue-green algal morphotype, but is an integral part of the same thallus. These observations agree with and confirm those of Dughi (1944) for *Sticta stipitata* and *Dendris-cocaulon dendro-thamnodes*, a similar composite plant first reported from south-eastern Australia by Wilson (1891) (Plate VIII).

The transitional area between the two morphotypes (Plate IIIA) is entirely devoid of algal cells, at least in the sections studied, and is notable for the abrupt change from the radial organization of the blue-green algal morphotype to the more complex dorsiventral structure of the green algal morphotype.

The tips of the blue-green algal morphotype, like other species of *Dendrisco-caulon* examined, have a homoiomerous organization without tissue differen-tiation. However, older branchlets soon become radially organized with a well developed central core of closely compacted, longitudinally arranged hyphae surrounded by a well defined layer containing the blue-green phycobiont. External to this there is a well defined cortex, comprising a layer three to six cells in thickness, arranged in anticlinal fashion (Plate IIIc). The outermost layer of the cortex remains in a not strongly gelatinized or relatively thin-walled condition; a few cells may develop simple to short-celled tomental hairs. In the region of the attachment stalk (Plate IIIB) and main branches, the phycobiont zone is partially occluded, mostly devoid of algae and is compressed between a much thickened cortex and a very stout, chondroid-like axial strand. The cortex here is 20–40 or more cells thick and of a double structure. The upper cells are radially stretched and thus relatively narrow, whilst below them groups of very small cells form a distinct pattern.

In the green algal morphotype, a dorsiventral arrangement of the tissues predominates, producing a very marked difference between the upper and lower cortices (Plate IIID). The upper cortex has a markedly gelatinized outer layer beneath which are three to four layers of thinner walled, isodiametric cells in a more or less periclinal arrangement. The phycobiont layer is strictly con-fined to a region between the upper cortex and a very broad layer of loosely interwoven hyphae which constitute the medulla; large internal cephalodia containing *Nostoc* are also frequent in the medulla. The lower cortex comprises a single layer of cells with reddish-brown walls, the outer, external side of

PLATE I
A, *Sticta filix*: abundantly fertile, stalked, green algal morphotype, habit ($\times$ 0·6). **B**, *S. latifrons*: fertile green algal morphotype showing the fusion of several stalks near the point of attachment, habit ($\times$ 0·6). **C**, *S. hypochroa*: a stalked species containing blue-green algae; note the presence of numerous marginal and few laminal isidia; a few ascocarps are also present ($\times$ 0·6). **D–G**, *Psoroma durietzii*: **D**, Primary thallus of the blue-green algal morphotype with two developing lobes of the green algal morphotype ($\times$ 40). **E**, Later stage in development of the green-algal morphotype showing the characteristic spreading marginal lobes containing green algae; the precursor lobe with blue-green algae is indicated by an arrow ($\times$ 24). **F**, Mature thallus with lobed cephalodia each with sorediate margins and underside ($\times$ 5·5). **G**, Habit, showing the completely sorediate cephalodia as well as numerous ascocarps with strongly convoluted margins and a sterile "plug" of thalline tissue ($\times$ 1).

PLATE II
Sticta filix and *S. latifrons*, habit. **A**, *S. filix*: collection of morphotypes showing independent *Dendriscocaulon* phase and combined morphotypes. The presence of ascocarps on the leaflets of the green algal morphotype should be noted ($\times$ 1). **B–E**, *Sticta latifrons*: **B**, Compound plant with basal *Dendriscocaulon* with leaflets of *S. latifrons* ($\times$ 2). **C**, Internal cephalodia as seen between the cyphellae on the underside of the green algal morphotype ($\times$ 15). **D**, Upper part of stalk of *Dendriscocaulon* with developing lobes of the green algal morphotype and spuriously lateral fruticose branchlets of *Dendriscocaulon*; the presence of an internal cephalodium at the summit of the unbranched part of the stalk is indicated by an arrow ($\times$ 20). **E**, Underside of juvenile thallus of *S. latifrons* showing remains of external *Dendriscocaulon* branchlets on the stalk and internal cephalodia in the flattened part ($\times$ 5·5).

PLATE III
Sticta filix and *S. marginifera*, var. *coralloidea*, anatomy. **A–C**, *S. filix*: **A**, L.S. through transition zone showing continuum of the mycobiont in association with blue-green algae on the left-hand side and green on the right; note the presence of tomental hairs on both sides of the middle part of the section and their restriction to the lower side in the vicinity of the green alga. The cortices are uniformly composed of a pseudoparenchyma of isodiametric cells ($\times$ 130). **B**, L.S. of lower part of stalk of *Dendriscocaulon* showing thickening of cortical region and the obliteration of the phycobiont zone ($\times$ 520). **C**, L.S. thallus containing green algae; note the dorsiventral stratification and the well developed cortex with the cells arranged periclinally ($\times$ 650). **D**, L.S. upper part of a main branchlet with a radial organization of the tissues, a cortical layer one cell thick, and tomental hairs on both sides (i.e. all round the branchlet) ($\times$ 130). **E–G**, *S. marginifera* var. *coralloidea*: **E**, L.S. of apex of a terete lobe with a continuously corticate apex and radial organization of the tissues ($\times$ 260). **F**, L.S. of a more flattened lobe still retaining a radial organization of the tissues ($\times$ 325). **G**, T.S. of main dorsiventral thallus; note that the blue-green algae tend to accumulate under the upper cortex and the general tissue organization is more in keeping with the dorsiventral nature of the thallus. The section is not quite typical, having abnormally enlarged cells in the lower cortex ($\times$ 260).

PLATE IV
Sticta latifrons, anatomy. **A**, L.S. green algal morphotype with internal cephalodium ($\times$ 200). **B**, Margin of apex of the lobe of the green morphotype showing the continuity

of the cortical layer between the upper and lower sides ($\times$ 370). **C**, L.S. upper side of lobe of green algal morphotype with proliferating cortical cells ($\times$ 500). **D**, L.S. upper cortex of green algal morphotype with the periclinal arrangement of the cells ($\times$ 200). **E**, L.S. stalk of *Dendriscocaulon* showing the thickened cortex and the obliteration of the phycobiont zone ($\times$ 370). **F**, Section of an abnormal cyphella with elongated cells on the left-hand side and normal isodiametric cells on the right, some of which are characteristically loose and detached ($\times$ 500). **G**, L.S. just below transition zone showing the relationship between internal cephalodia and the external branchlets of the *Dendriscocaulon* ($\times$ 100). **H**, L.S. transition zone; note the presence of tomental hairs on both sides of the blue-green algal zone and their restriction to the lower surface at the onset of influence by the green algal biont. The organization of the *Nostoc* cells into internal cephalodia and the zonation of the green algae, and the uniform nature of the cortices are other important features ($\times$ 60). **I**, L.S. basal part of *Dendriscocaulon* stalk; the absence of any algal biont should be noted ($\times$ 60). **J**, L.S. young branchlet of *Dendriscocaulon* showing the homoiomerous arrangement of the bionts ($\times$ 500).

PLATE V

Sticta canariensis–S. dufourii and *Solorina spongiosa*, habit. **A–E**, *S. canariensis–S. dufourii*: **A**, *S. canariensis* with laminal ascocarps ($\times$ 1·5). **B**, *S. dufourii* with folioles acting as vegetative diaspores ($\times$ 1·5). **C**, Combined morphotypes; paler leaflets of the green algal morphotype, corresponding to *S. canariensis*, on darker thallus of blue-green algal morphotype, corresponding to *S. dufourii* ($\times$ 2). **D**, Development of a lobule of *S. canariensis* (paler, right) on upper surface of *S. dufourii* (darker, left, with folioles) ($\times$ 4). **E**, Two small lobes of *S. canariensis* developing from the under surface of *S. dufourii* ($\times$ 5). **F**, *Solorina spongiosa* showing a more or less continuous thallus containing *Nostoc*; the green algal morphotype is restricted to the margins of the ascocarps ($\times$ 1).

PLATE VI

Sticta canariensis–S. dufourii, anatomy. **A**, L.S. showing capture of green algae by the cortical cells of *S. dufourii* ($\times$ 600), **B**, L.S. development of two small lobes of *S. canariensis* on the underside of the thallus of *S. dufourii*; note the orientation of the green algae in the larger lobe ($\times$ 100). **C**, A small developing nodule containing green algae on the underside of *S. dufourii* ($\times$ 200). **D–F**, L.S. of connections between *S. canariensis* and *S. dufourii*: **D**, A normal connection between the two morphotypes showing a continuity of fungal hyphae ($\times$ 100). **E**, An abnormal secondary attachment of the green algal morphotype to *S. dufourii* showing the disorganization of the cortical cells of the latter ($\times$ 75). **F**, Shows the relationship between **D** and **E**; the central tongue of tissue contains green algae, the connection to the blue-green algal morphotype is normal on the left-hand side and is a secondary attachment to a different lobe of the same thallus on the right-hand side. **D** and **E** have been turned through 90° for comparative purposes ($\times$ 30). **G**, L.S. *Sticta dufourii* ($\times$ 250). **H**, L.S. of a young lobe, near the transition zone between the two morphotypes, containing green algae; note the similarity of the cortices to those in *S. dufourii* ($\times$ 200). **I**, T.S. of older thallus of *S. canariensis* showing the dissimilarity of the cortices to those of *S. dufourii* ($\times$ 200).

PLATE VII

Dendriscocaulon and associations: **A**, *Dendriscocaulon umhausense* ($\times$ 1·5). **B**, *Sticta dichotomoides*; enlargement of attached clusters of granular sorediate *Dendriscocaulon* ($\times$ 4). **C**, *Sticta marginifera* var. *coralloidea*; dorsiventral thallus bearing numerous finger-like terete foḷioles ($\times 2.5$). **D**, *Sticta filicinella* with conspicuous brown veins on the underside and small marginal lobules ($\times 1$). **E**, *Sticta dichotomoides* ($\times 0.5$). **F**, *Sticta marginifera* var. *coralloidea* with terete appendages, one of which has reverted to the dorsiventral form of the main thallus ($\times 4$). **G**, *Dendriscocaulon* sp., *Galloway* 3291, with the flattened main part of the thallus and developing cyphellae ($\times 1.5$). **H**, *Dendriscocaulon* sp., *Galloway* 3291, enlargement showing flattened main stems with prominent central ridges and the development of shallow, pale cyphellae ($\times 4$). **I**, *Sticta stipitata*, a basal part of *Dendriscocaulon* with attached lobes of *S. stipitata* containing green alagae ($\times 4$).

PLATE VIII

A, L.S. of granular sorediate branchlets of *Dendriscocaulon* on *S. dichotomoides* ($\times 120$). **B**, L.S. branchlet of *Dendriscocaulon umhausense* ($\times 240$). **C**, L.S. thallus of *Lobaria amplissima*, showing relationship between thallus containing green algae and associated *Dendriscocaulon* ($\times 90$). **D**, L.S. apices of *Dendriscocaulon* associated with *Lobaria amplissima* ($\times 440$). **E**, L.S. branchlet of *Dendriscocaulon* sp. (cf. *Sticta calithamnia*) *Henssen 24276a* ($\times 440$). **F**, L.S. of a cyphellum of *Dendriscocaulon* sp., *Galloway* 3291; note also the almost complete concentration of the algal cells to a zone near the non-tomentose upper surface of the thallus ($\times 240$).

PLATE IX

Psoroma durietzii: **A–F**, L.S. through cephalodium and associated thallus: **A**, Section through a cephalodium and thallus, the main thallus with green algae on the right-hand side. Note that the cephalodium, like the main thallus, is anchored to the substrate by rhizines ($\times 75$). **B**, Section through a young thallus of a recently liberated soralium, showing the development of the blue-green algal morphotype from the cluster of soredia ($\times 200$). **C**, Section of mature thallus of *Psoroma durietzii* containing green algae; the medulla becomes opaque due to the abundance of lichen substances ($\times 200$). **D**, Part of a cephalodium, showing the origin and liberation of the soredia containing blue-green algae ($\times 200$). **E, F**, Transition zone between a developing blue-green morphotype, derived from a small cluster of soredia liberated from a cephalodium, and the subsequent development of spreading lobes, containing only green algae; the green algal morphotype is on the right-hand side, the blue-green counterpart, on the left ($\times 200$). **E**, L.S. green algal morphotype at higher magnification ($\times 350$). **G**, *Sticta filix*: L.S. margin near apex of a lobe of the green algal morphotype showing a continuity of cortex between the upper and lower surfaces; the hyphae are periclinally arranged ($\times 150$).

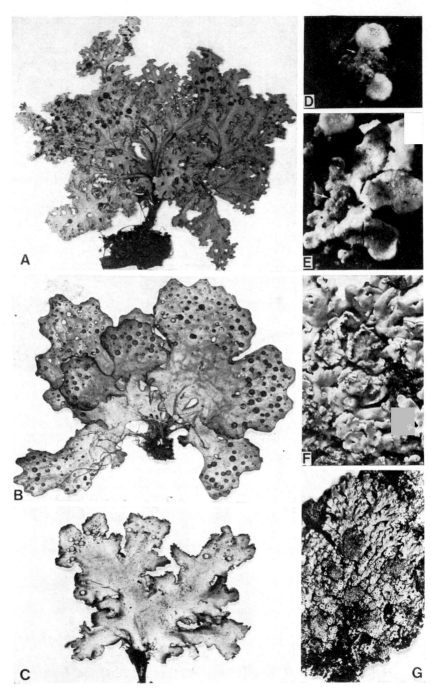

Plate I

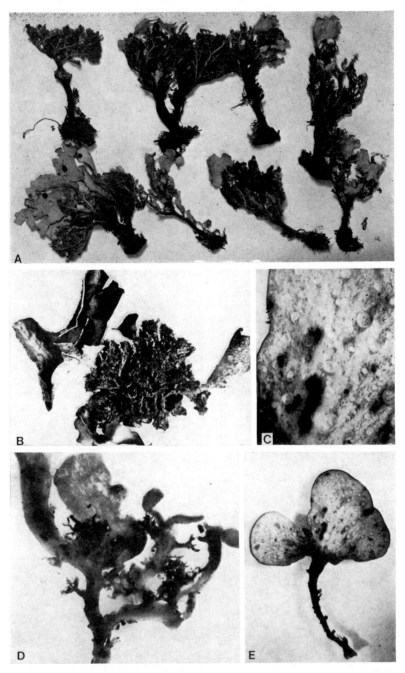

PLATE II

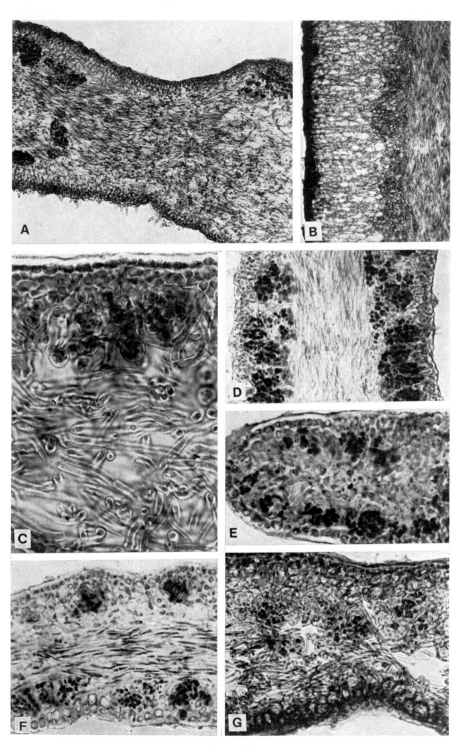

PLATE III

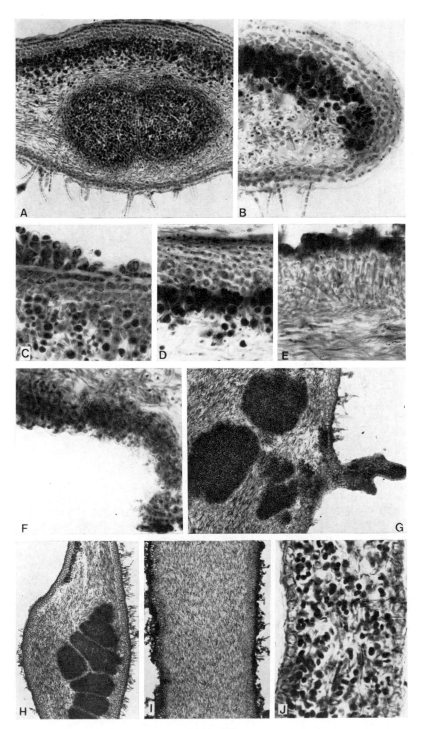

PLATE IV

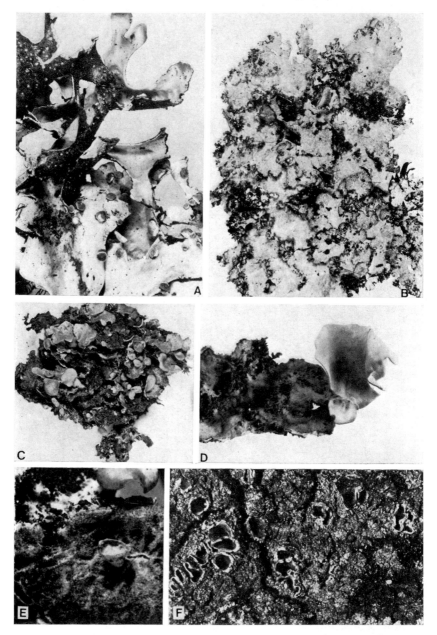

Plate V

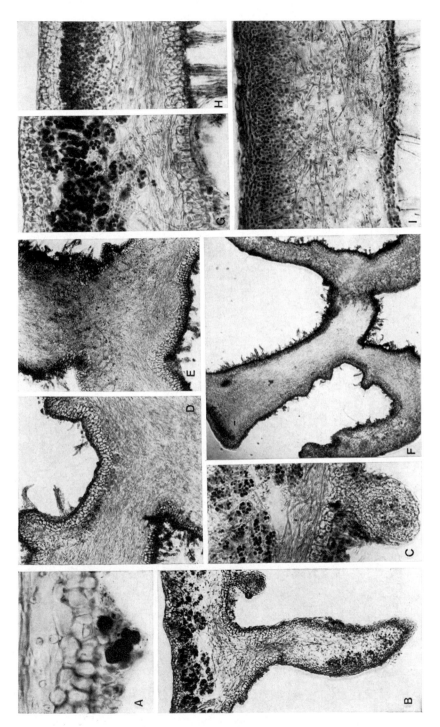

PLATE VI

PLATE VII

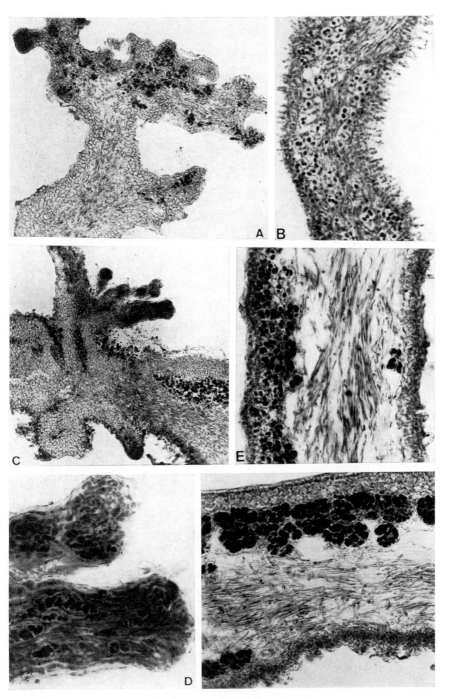

PLATE VIII

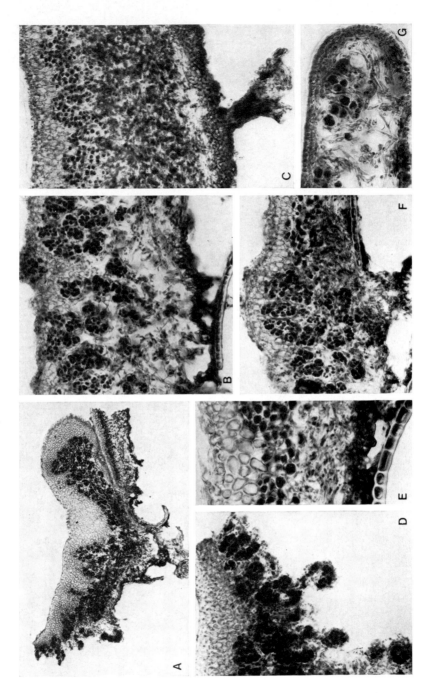

Plate IX

which has a common gelatinized layer. Simple, unbranched, septate, tomental hairs may develop from these cells. There is a well defined, corticate marginal and apical differentiation into an upper and lower cortex and a distinct algal zone in contrast to the homoiomerous organization observed in the tips of the blue-algal morphotype. The structure of the leaflets of the green algal morphotype is identical with that of the independent thalli of *Sticta filix* (Plate IA); it is worth noting that the stalk of the latter has an identical anatomy to that of the blue-green algal morphotype of the composite plant (Plate IIIB) and juvenile specimens may sometimes contain a few blue-green algal cells between the thickened double cortical layer and the stout chondroid axis.

2. Sticta latifrons–Dendriscocaulon *sp.*

Additional data concerning the relationship between the three bionts in composite morphotypes came from the discovery of a compound thallus similar in appearance to *Sticta filix*. This material, also from the South Island of New Zealand, consists of three thalli, the largest of which measures about 3 cm in height (Plates IIB–E). Each thallus comprises one to three small dorsiventral lobes, typical in form and anatomy to those of free-living *S. latifrons* (Plate IB), connected to the branchlets of a fruticose *Dendriscocaulon*-like blue-green algal morphotype. The latter represents an undescribed entity which differs from the equivalent morphotype in *Sticta filix* in possessing a markedly more richly branched–coralloid thallus with a tendency for the main stem to separate into three or four main branches originating near to the rooting holdfast. This early separation is probably the result of a noticeable thickening throughout their entire length of those branches which support the green algal morphotype and connect it to the blue-green algal partner. As a result of this development, the coralloid branchlets of the *Dendriscocaulon*, which are not associated with leaflets of *S. latifrons*, appear spuriously as lateral appendages although most are probably well developed before inoculation by the green alga (?*Myrmecia*) and the development of the green algal morphotype had taken place. As in *Sticta filix*, there is a marked contrast between the dorsiventral arrangement of the green algal morphotype and the terete, radial organization of the blue-green morphotype.

The transitional zone connecting the two morphotypes (Plate IVH) is less well defined than in *S. filix*. In longitudinal section both phycobionts are present, the green algae disposed towards one side of the still almost terete upper part of the zone. In the lower part the blue-green algae are largely present as closely compacted, relatively large, internal cephalodia, similar to those found in the medullary region of older parts of the green algal morphotype (Plate IVA). The origin of the transitional zone from the internal cephalodia is shown in Plate IVG. They arise from algal cells migrating in from small branchlets and are not freshly captured *Nostoc* cells like those encountered in the origins of cephalodia on the green algal morphotype. It is important to note the influence of the *combined* mycobiont and green phycobiont on that of the same mycobiont with blue-green algae; the joint influence of the former causes the latter to revert from an anatomically complex, external coralloid form to an incorporated, largely structureless inclusion within the thallus of the green algal morphotype. The incorporated cephalodia can easily be observed as oval to elongate areas on the underside of the thallus of the green algal morphotype (Plate IIc). The central part of the transitional zone is terete and tomentose all round; in the upper part, towards the attached leaflets, the tomentum has already become restricted to one side, corresponding to the underside of the free-living thalli of *S. latifrons* (Plate IVн). In section the cortex of the transitional zone is six to eight cells thick; these are arranged anticlinally and are uniformly thin-walled. This arrangement persists some distance within the leaflets of the green algal morphotype from their point of attachment to the transitional zone. The rest of the transitional zone comprises a broad area of closely compacted, largely disorientated hyphae and large cerebriform internal cephalodia intersected by thin layers of closely compacted hyphae, the latter derived from the medullary region (Plate IVн).

Details of the anatomy and morphology of the connected morphotypes emphasize the great differences in structure of the two entities, even though the same fungus is present in both. The markedly coriaceous thallus of the green algal morphotype (Plate Iв) is dorsiventral and palmate with rather shallow lobes. The upper surface is smooth, slightly foveolate, with a few minute raised spots. The underside is pale brown with a paler tomentum and scattered shallow, irregularly dispersed, cyphellae. The lower surface incorporates a few low, almost imperceptible, ridges. Independent thalli of *S. latifrons* are similar, but reach a size of 15 cm tall and 8 cm wide. They are attached to the substrate by a stout, rather irregularly ridged, basal stalk. Ascocarps are frequently present, although these were not seen on leaflets of the composite material. Internally the cortex on both surfaces is conspicuously gelatinized (Plate IVв and D); the upper cortex comprises 11–13 rows of isodiametric cells with very thick walls, more or less periclinal in their arrangement (Plate IVD). The lumina of the cells decrease in size towards the surface. A peculiar feature is the presence of small circumscribed areas of meristematic cells involving a minor proliferation of the upper cells of the upper cortex (Plate IVc). These cell groups are seen as pale dots on the upper surface.

Similar proliferating cells were also observed on the lower surface associated with abortive cyphellae (Plate IVF). The lower cortex is three to four cells thick and of similar organization to the upper although the cell lumina are more uniform in size; both surfaces are invested in a thick, semi-transparent gelatinous outer layer derived from the walls of those cortical cells nearest to the surface. The green algal cells (?*Myrmecia*) are strictly confined to a layer immediately below the upper cortex. The upper and lower cortices are continuous at the margins of the thallus (Plate IVB).

The branchlet tips of the blue-green algal morphotype are entirely homoiomerous (Plate IVJ) with markedly encapsulated algae (*Nostoc*). The cortex appears to remain only one cell thick for much of its length. The cortical cells are isodiametric and thin walled except towards the side facing the exterior where they are covered with a conspicuous, gelatinized transparent layer. The older main branches are similar in structure to those of the equivalent morphotype of *Sticta filix* (Plate IVI). As in this morphotype, numerous hairs make up a dense pubescent tomentum; internally there is a great increase in cortical cells until, in the main attachment stems, a layer of about ten cells thick is formed (Plate IVE), but the cortex has no double structure. Again algal cells are largely absent from the main stalk.

3. *Other Morphotypes involving* Dendriscocaulon *spp.*

(*a*) *Lobaria amplissima–Dendriscocaulon* "*umhausense*". Details of the initiation, development and anatomy of the two morphotypes of this species, the oldest known example, are well documented (Forssell, 1883; Moreau, 1928; Kaule, 1931; Dughi, 1936). The blue-green algal morphotype, which is derived from colonies of *Nostoc* captured by the tomentum of the underside of the green algal morphotype, develops into internal cephalodia. Some of the internal cephalodia may burst through the upper cortex forming characteristic *Dendriscocaulon*-like clusters containing blue-green algae (Plate VIIIc). The branchlets of the blue-green algal morphotype, often named as *D.* '*umhausense*', have homoiomerous, ecorticate tips (Plate VIIID); older parts are heteromerous, with a well defined cortex, one to four cells thick, beneath which lies the phycobiont layer, the lower part of which abuts directly onto a rather loosely organized medulla of interwoven, longitudinally arranged, hyphae. In common with other species of *Dendriscocaulon*, the anatomy of the basal stalk comprises a much thickened cortical layer, a compressed phycobiont zone, in which most of the algae have been lost, and a central axis of densely compacted, longitudinally orientated, pachydermatous hyphae. The outermost layer of cortical cells may give rise to a few short tomental hairs consisting of one to three cells with constricted walls, typical for the northern group of *Dendriscocaulon* taxa (cf. Plate VIIIB, from a free-living specimen).

The arrangement of the tissues of the green algal morphotype, unlike that of the radially organized, blue-green algal counterpart, is strictly dorsiventral. The upper and lower cortex are very different, the lower bearing a short tomentum. The medulla is more or less lacunose, the hyphae leptodermatous and with thickened walls; numerous crystals of calcium oxalate are present throughout the thalli of some specimens. Ascocarps are frequently developed on the green algal morphotype but are not correctly recorded from the blue-green algal counterpart.

The blue-green algal morphotype is usually present. Exceptionally, such as in areas with moderate levels of air pollution (mean winter sulphur dioxide levels up to 40 $\mu g/m^3$), they are not developed, although internal cephalodia are still abundantly produced (see Rose and James, 1974).

Conversely, independent blue-green algal morphotypes are rare, tending to occur where the green algal morphotype (i.e. *Lobaria amplissima*) has been killed, such as by browsing snails or other factors. However, given relatively damp, sheltered conditions, these isolated thalli can survive for periods of at least 5 years (James, unpublished) without the presence of an associated green phycobiont. As mentioned above (p. 31), the free-living *Dendriscocaulon umhausense* cannot be distinguished from the cephalodia of *L. amplissima* on the basis of its morphological characters.

The secondary development of lobules of the green algal morphotype on the blue-green algal counterpart has been reported once and is illustrated by Dughi (1936). This development is exactly parallel to that reported and described by Jordan (1972) for *Lobaria* cf. *erosa*—also from a single gathering. The present authors have unfortunately been unable to locate and examine Dughi's material.

(*b*) *Lobaria* cf. *erosa–Dendriscocaulon* sp.*–Lobaria* cf. *erosa* morphotypes. The morphology, anatomy, and interrelationships of the two major morphotypes of this combination have been discussed in detail by Jordan (1972). The origin and structure of the blue-green morphotype is similar to that found in *Lobaria amplissima*, *L. ornata* and *Sticta glomuligera*. Like the two latter species, the development of the blue-green algal morphotype as external *Dendriscocaulon*-like excrescences in *Lobaria* cf. *erosa*, perhaps appears to be an aberrant state; in most material examined by us only numerous internal cephalodia seem to be normally developed although, according to Moore (1969), cephalodia (referred to as

---

* Possibly *D. intricatulum* (see footnote to p. 31).

"*Polychidium*" by her) are generally found on *L. erosa* and *L. lobulifera* (see footnote to p. 31), an observation which was not confirmed by our studies.

The secondary development of the green algal morphotype in the form of numerous, small, dorsiventral leaflets and primordial nodules attached on or near the extremities of the coralloid branchlets of the blue-green algal morphotype, is of particular interest in this species. The organization of these leaflets bears a close similarity to that of the green algal morphotype of the main and parent thallus. Jordan (1972) noted several modifications of the lichenized fungal tissue that may be correlated with the type of alga present:

"The medulla associated with the nearby algae appears so as a lax network, the medullary tissue in the vicinity of the blue-green phycobiont is more compactly arranged. The individual hyphae are indistinguishable from this in the *Lobaria* thallus proper. The dorsiventral orientation of the lobules is reflected in the structure of the cortices. The upper cortex of the lobules is a compact layer, 4 to 5 cells thick, of paraplectenchymatous tissue. The lower cortex is a loosely arranged layer of only 2 or 3 cells thick. The cortex of the erect fruticose tissue is homogeneous, the cellular arrangement being similar to the upper cortex of the foliose tissue."

The secondary recapture of green algae by the blue-green phycobiont phase and its organization into dorsiventral leaflets with all the internal characters of the parent green algal morphotype, is of considerable morphogenetical interest.

(c) *Sticta dichotomoides–Dendriscocaulon* sp. The development of the coralloid outgrowths of the blue-green algal morphotype is of rare occurrence in the genus *Sticta*. The two examples studied were *S. dichotomoides* and *S. glomuligera*. In the former, the capture and organization of the blue-green algae, terminating in the coralloid morphotype, is similar to that in those species of *Lobaria* cited above. In structure, the blue-green algal morphotype in *S. glomuligera* resembles more closely that of *Polychidium* species (e.g. *P. muscicola*; see Henssen, 1963, pl. 27c–D) than that of cephalodia of *Lobaria amplissima*. *S. glomuligera* probably represents an aberrant phase of the entities contained within the aggregate of species included in the variable *S. sinuosa–S. damaecornis* complex, where internal cephalodia are always abundantly produced.

Twelve specimens of *S. dichotomoides*, a species so far known only from Tahiti, have been examined. All bear relatively coarse clusters of the blue-green algal morphotype (Plate VIIB and E). In this species the coralloid excrescences of this phase are less apparent than usual and the main supporting branches are, in consequence, more conspicuous. In section, the blue-green algae (*Nostoc*) are restricted to the terminal branchlets (Plate VIIIA) and appear to become naturally abraded, possibly acting as a type of vegetative diaspore for this species. The main thalli are attached by a short, but definite, canaliculate stalk without any algal cells. Ascocarps are only present on the holotype (BM) and one other sample. A noteworthy feature of this species is the restriction of the blue-green algal morphotype to the margins of the dorsiventral, erect or weakly decumbent lobes of the green algal morphotype. The explanation for this very specific distribution remains unknown.

(*d*) Free-living *Dendriscocaulon* spp. and some caulescent *Sticta* spp. A detailed study follows of both free-living *Dendriscocaulon* taxa and those constituting blue-green algal morphotypes of certain species of *Lobaria* and *Sticta*. In the free-living forms two distinct groups can be recognized. First, there is a northern group allied to *Lobaria* with a densely fruticose thallus the terete branchlets of which bear numerous bead-like hairs the cells of which are constricted at the septa (Plate VIIIB; Henssen, 1963, *pl.* 28c); two species in this group were studied in detail, *D. umhausense* (European) and *D. intricatulum* (North American but extending into Brazil)★. Secondly, there is a southern group (known from Australia, New Zealand and South America) containing taxa allied to *Sticta* blue-green algal morphotypes, characterized by a tendency to form flattened branches with a dorsiventral structure (Plate VIIIE) and in having straight tomental hairs lacking bead-like constrictions. In this latter group the morphology is very variable and specimens often offer considerable difficulties in identification★. The southern group also includes types intermediate with respect to the genus *Sticta* in having not only extremely flattened lobes but also precursors of rudimentary cyphellae (e.g. *Dendriscocaulon* sp., *Galloway* 3291; *Sticta calithamnia*; see p. 43). Though those of the southern group are mostly very similar in their morphology, anatomically they exhibit small but constant differences which would enable their segregation into a genus encompassing perhaps as many as 30 putative species.

Comparable, often subtle, but constant differences are also a feature of those *Dendriscocaulon*-like morphotypes occurring as composite plants with green alga-containing *Sticta* species (e.g. *S. filix*, *S. latifrons*). The characters of each kind of *Dendriscocaulon* are constant, whether the particular blue-green algal morphotype concerned is free-living, or is supported by or gives rise to a green algal morphotype. Plates VIIIB and E represent two typical variants of the terete, coralloid organization of the branchlets near their apices in two different free-living species of *Dendriscocaulon*; the disposition of the cortical cells, algae, and the organization of the cortex and axis in older parts of the thallus are all

---

★ See footnote to p. 31.

constant for particular entities. The nature and distribution of the tomentum is also an important distinguishing feature between species (e.g. *D. dendrothamnodes* and the unnamed entity associated with *S. filix*). Amongst this assemblage there are also a few important species showing a tendency towards the development of a flattened thallus with a correspondingly dorsiventral arrangement of tissues.

This trend in development among the Dendriscocaula is particularly interesting in that it is also accompanied by the development of a tomentose undersurface in which shallow, rudimentary to well-defined cyphellae are developed. A particularly good example of this tendency is afforded by a specimen (*Galloway* 3291) from the South Island of New Zealand (Plate VIIg). In this specimen, although the terminal branchlets are terete with an anatomy corresponding to most other *Dendriscocaulon* entities examined, the main branches are more or less flattened and dorsiventral in structure. The upper surface is blue-grey and smooth, the lower has prominent ridges and is wholly finely pubescent-tomentose. Between the ridges shallow but immediately characteristic and recognizable cyphellae are developed (Plate VIIн). In section (Plate VIIIf) the older dorsiventral parts of the thallus have a well developed upper cortex three to four cells thick, the outer surface of which has a common gelatinous layer. The phycobiont (*Nostoc*) is largely aggregated in a zone beneath the cortex, although a few isolated clusters of algae are also present in close proximity to the lower cortex. The lower cortex is mainly one cell thick with numerous, one to five-celled, uniseriate, tomental hairs. The medulla is lax, consisting of distinct, thin-walled hyphae. The stalk region is devoid of algae and has a well developed cortex and axial strand as is usually present in all other species of *Dendriscocaulon*.

Another morphological divergence from the generally terete and internally radial organization of most Dendriscocaula, is seen in *Sticta calithamnia*, a species possibly endemic to the island of Juan Fernández. This taxon is characterized by a dorsiventral flattening of all parts, including the apices of the smallest branchlets. The attachment stalk is, however, usually more or less terete. The median part of the thallus is noticeably monophyllous and shallowly lobate. The upper surface is smooth, blue-grey; the lower pale- to red-brown, with scattered, often inconspicuous, ribs. A tomentum is generally more or less present and is best developed on or near the ribs on the under surface. Cyphellae are mostly absent; occasionally, in a few samples, shallow breaks in the lower cortex correspond to rudimentary cavities, often filled with rounded, rather loosely packed cells. These structures probably represent abortive cyphellae, but more detailed studies are required to ascertain their true origin and nature;

recent gatherings from Chile of this, or possibly a related, species by one of the authors (A. H.) may prove valuable in these studies. In shape and structure, *Sticta calithamnia* lies midway between a distinct *Dendriscocaulon* (*Galloway* 3291), and the blue-green algal morphotype of *Sticta canariensis* (i.e. *S. dufourii*).

The discovery of flattened forms of *Dendriscocaulon* with *Sticta*-like features, such as cyphellae, suggests that there is now a much closer affinity between this genus and the caulescent species of *Sticta* with blue-green phycobionts. These affinities are discussed later in this paper. Some species of this particular assemblage of *Sticta* species, such as *S. filicinella*, have well developed midribs (Plate VIID); most species are notably sterile, reproducing vegetatively by means of elongate, often rather *Dendriscocaulon*-like, marginal (rarely laminal), folioles or isidia. *Sticta elmeri* and *S. cyphellulata* are good examples of this type of morphology.

A particularly interesting case is *Sticta marginifera* var. *coralloidea*, first described from New Caledonia (Plate VIIc). This taxon is stalked, has a dorsiventral, almost monophyllous, blue-green algal morphotype, from the margins of which elongated, rather irregular, more or less terete *Dendriscocaulon*-like outgrowths develop. The appendages are radial in structure (Plate IIIF), and have corticate apices (Plate IIIE) similar to the apices of the southern group of *Dendriscocaulon* taxa (see p. 42). An interesting feature of this species is that these outgrowths appear to be capable of reverting to the flattened aspect of the main thallus (Plate VIIG). In this case there is a transformation from a radial to a dorsiventral organization in the presence of the *same* alga. The same type of small lobes may be seen occasionally at the tips of isidia produced on thalli of *Lobaria pulmonaria*, but in this species they are shorter than those of *S. marginifera* var. *coralloidea*.

4. Sticta canariensis–S. dufourii *morphotypes*

The discovery of composite thalli of two morphotypes, hitherto classified as two entirely unrelated species in the same genus, is of particular significance and interest, especially in relation to this and other genera which include species with either blue-green or green algae as their main phycobiont (e.g. *Lobaria*, *Peltigera*, *Pseudocyphellaria*). Furthermore, unlike the composite thalli discussed earlier in this paper, such as *Sticta filix* and *S. latifrons*, the blue-green morphotype (i.e. *S. dufourii*; Plate VB), like the green (i.e. *S. canariensis*; Plate VA), is foliose in shape and dorsiventral in structure. The form and nature of this pair of morphotypes suggests that other, as yet unrelated, pairs of species may occur in the same genus which are, in reality, undiscovered morphotypes of the same mycobiont.

Anatomical studies of the composite thalli have revealed all stages involving the capture of free-living algae and their subsequent development into attached, small, but characteristic, lobes of *Sticta canariensis*. The green algal morphotype in this species may originate from both surfaces, as well as along the margins, of the lobes of the parent blue-green algal morphotype (Plates VC, D, E). Initial stages of inoculation involve the capture of one or a few green algal cells and their encapsulation by the active outgrowth of the fungal hyphae of the cortex (Plate VIA). At first the dividing algal cells are amassed in the centre of an expanding nodule of fungal tissue (Plate VIc), but, as the latter assumes a more foliose shape, the algae come to lie nearer the surface which is destined to become the upper cortex (Plate VIB). At this stage the anatomical structure of the developing leaflet closely corresponds, except for its algal content, to that of the blue-green algal morphotype which supports it (Plate VIG). In both there is a similar upper and lower cortex, consisting of about four layers of anticlinally arranged cells with slightly thickened walls; the rather lax medulla has loosely organized hyphae dispersed in all directions. In both the blue-green algal morphotype, and the developing propagules with green algae, the algal cells, although mainly concentrated in a well defined layer below the upper cortex also tend to penetrate into the middle and upper part of the medulla (Plate VIG). With increasing size and age, the attached leaflets with green algae come to assume the structure of mature, independent thalli characteristic of *Sticta canariensis* (Plate VIi). In this morphotype, the upper and lower cortices have a similar cellular structure, consisting of three or four rows of very thick-walled cells, anticlinally arranged and with noticeably thickened walls and much reduced lumina. The surface layer of cells of the upper cortex are slightly enlarged and are incorporated in a thick, common gelatinous layer which covers both the upper and lower surfaces of the thallus. Thick walled tomental hairs, consisting of uniseriate chains of cells, are also frequently developed from the lower cortex. The green phycobiont is very closely compacted below the transparent upper cortex; the algal cells (?*Myrmecia*) are notable for their relatively small size. The medulla is similar in structure to that of the blue-green algal morphotype, but the individual hyphae have thicker walls and are more closely compacted.

A particularly interesting discovery was made during the examination of the transitional zone between the two morphotypes of this species (Plates VID–F). In one longitudinal section it was observed that a leaflet of the green algal morphotype had produced a foot-like outgrowth which had become secondarily attached to the parent thallus containing blue-green algae (Plate VIe). Since the original attachment was still also intact (Plate VID), it was possible to distinguish between the two types of connections (Plate VIF); that of a secondary nature shows a clear displacement and partial dispersal of the original cortical cells of the blue-green morphotype whereas no such disorganization is apparent in the green algal morphotypes which have originated and developed from the capture of a single algal cell. This discovery is a conclusive proof that the combinations of this and other attached morphotypes discussed in this study are the result of a continuity of development between two contrasted morphotypes sharing the same fungus and not a secondary or epiphytic attachment of two otherwise unrelated entities.

5. Other Pairs of Related Morphotypes

Psoroma durietzii, a recently described taxon known from New Zealand, the Auckland Islands, Campbell Island and Tasmania (James and Henssen, 1975), is unusual in that the cephalodia become sorediate; subsequent development of the liberated soredia, containing blue-green algae, is of particular interest. In this species, the dominant thallus is rosette-shaped, up to 5 cm in diameter, comprising radiating lobes with largely contiguous margins (Plate IG). The upper surface is grey to pale yellow-grey and smooth, while the lower is pale with a dense felt of blue-black rhizinae. The cephalodia appear laminal although most originate along the margins of the inner lobes. They are flattened, more or less peltate, with shallowly lobate margins and are concolorous with the main thallus or flesh-coloured, or more rarely bluish-grey (Plate IF). The granular soredia, which originate from a breakdown of the cortex on the underside of a cephalodium, come to involve the whole of the cephalodium, mainly by an upward turning and peeling back of their upper surface which reveals more of the sorediate surface on the underside. The sorediate areas are bright glaucous blue in colour. Ascocarps are rarely developed and are similar to those described for *Psoroma contortum*, *P. pulchrum* and *P. xanthomelanum*; these have a central plug of sterile tissue which seems, in these species, to arise either from an invagination of the thalline margin surrounding the ascocarp or by an infilling of a fissural division of the disc of the developing ascocarp.

Anatomically, the main thallus, which incorporates the green phycobiont, has a well developed pseudoparenchymatous upper cortex up to six cells in thickness, the upper layer of which is covered by a rather irregular gelatinous layer, corresponding to the smooth upper surface as seen under low magnification (Plate IXc). The phycobiont zone lies below the cortex in a more or less well compacted layer under which there is a rather dense medulla which remains opaque in thin section due to the aggregation of numerous crystals of an unknown lichen substance. A well developed lower cortex is present, composed of cells with brown pigmented walls. Simple rhizinae, four to eight cells thick, anchor the thallus to the substrate.

In vertical sections of the cephalodia there is a thick, irregular cortex six to 15 cells thick, those cells nearest the surface having thickened gelatinous walls. The alga (*Nostoc*) is irregularly dispersed and compacted below the cortex, eventually rounding up into coarse clusters surrounded by hyphae to become liberated as soredia by a breakdown of the cortex on the lower side of the cephalodium (Plate IXA and D). Young cephalodia have a compacted, non-opaque medulla without crystalline inclusions and a disrupted lower cortex which develops a few simple rhizinae that anchor the structure to the substratum, although appearing as a superficial development on the main thallus (containing green algae).

Subsequent development of a liberated cluster of soredia can be seen in Plate ID and E. This development takes the form of a small pad of tissue a few millimetres across. Anatomically, this structure is corticate with a rather disorganized cortex of thin-walled, isodiametric cells, partly disrupted by clusters of the blue-green phycobiont (Plate IXB). A rather lax medulla and narrow lower cortex are also displayed, and the whole is anchored by small simple rhizinae, one to four cells thick. Subsequent development beyond this stage appears only to take place after inoculation by the appropriate green alga, whereupon the thallus is organized into a structure approaching that of the lobes of the fully mature green algal morphotype. Concomitant with this development of the fungal partner is the deposition of lichen substances on the walls of the medullary hyphae in the vicinity of the green algae (Plate IXc). As development of the green algal morphotype proceeds there is a thickening of the upper cortex, a marked increase in the number of algal cells, and the development of a well defined lower cortex and thicker rhizinae (Plate IXE and F).

This development is interesting in that the liberated soredia form a small rudimentary, blue-green algal pabulum which acts as a precursor to the green algal morphotype. There is no evidence that this blue-green algal thallus can increase in size to form a larger and identifiable species under different environmental conditions. However, this may indeed be the case in another pair of species based on a study of their behaviour made last century. Forssell (1884c), in his studies on *Psoroma hypnorum* and *Pannaria brunnea* (syn. *P. pezizoides*), observed that the undersides of the green squamules of the former were often associated with blue-green algae (*Nostoc*) which became compacted into what he considered a cephalodium. More frequently, in this species, he also noted the occurrence of actual squamules containing blue-green algae intermixed with those with a green phycobiont. Furthermore, hyphae from the germinating spores of *Psoroma hypnorum* were seen by him to encircle blue-green algae on mosses close by; subsequent development of this preliminary association is less clear and Forssell could not decide whether the green or blue-green algal associations preceded each other or were simultaneous. If it can be proved that *Psoroma hypnorum* and *Pannaria pezizoides* have the same mycobiont, this will provide the first instance of an example in which both morphotypes can occur with ascocarps in an independent condition. Forssell's studies are urgently in need of verification.

A recent and very important discovery of a series of joined morphotypes of *Peltigera aphthosa* or *P. leucophlebia* (green algal morphotypes) developing from and on *P. malacea–P. polydactyla–P. degenii* or intermediates between these species (blue-green algal morphotypes) (kindly submitted to us by Mr O. Vitikainen) not only indicates that several species in the genus may have the same mycobiont but are different because of a different phycobiont, but also may be another case where both partners are capable of sexual reproduction. Investigations on this particular example are continuing and will be published elsewhere.

ECOLOGY AND DISTRIBUTION

1. Sticta filix *and* S. stipitata

The green algal morphotypes of *Sticta filix* and *S. latifrons* are widespread in most well wooded areas in New Zealand and are particularly frequent in sunny, sheltered creeks on the South Island, especially in proximity to freshwater streams and lakes. The large palmate thalli of both species often form extensive conspicuous clones on mossy rocks and tree boles, there being no preference for either substrate. Unfortunately there is little information concerning the distribution of the corresponding blue-green algal morphotype due to a failure in the past to adequately distinguish between several closely related Dendriscocaula which have a very similar morphology, structure and ecological requirements. However, the blue-green algal morphotype of *S. filix* appears to share, along with these other Dendriscocaula, a preference for very sheltered habitats of unvarying humidity and low illumination. In contrast, the green algal morphotype requires a more variable regime of greater illumination and more fluctuating humidity.

Figure 2 illustrates the composition and zonation of 93 morphotypes of *Sticta filix* independently blue-green, green and also united (dimorphic), at a site in New Zealand, examined by one of the authors (P.W.J.). The site consists of two very large boulders, *c.* 6 m in height, strongly influenced by the spray from a small waterfall, especially when the nearby river is in full spate. The upper part of the boulders, though partially shaded by overhead branches of *Nothofagus menziesii*, received a potential of 5 h of sunshine per day between the months of October and May; the lower aspect of the boulders was in permanent shade.

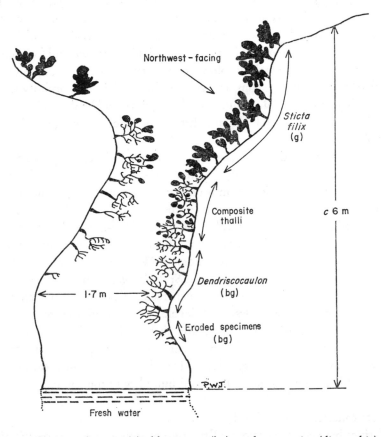

FIG. 2. Distribution of green (g), blue-green (bg), and composite (dimorphic) algal morphotypes of *Sticta filix*, in relation to aspect and associated environmental factors, at a site in New Zealand (South Island, Lake Te Anau, Middle Fiord, above Lake Hankinson, 1962, *James*). See text for further information.

The composition of the community was as follows: 24 green algal; 38 blue-green algal; 31 composite morphotypes; each entity zoning in the order illustrated. There was also evidence of drift in algal preference in the zone containing most composite plants; those nearest the top of the zone having many leaflets of the green algal morphotype some of which were fertile, alternatively, those nearest the blue-green algal morphotype zone had only a few leaflets and a few scattered nodules containing green algae. The blue-green algal morphotype persists in sites with a low, even light intensity and a relatively high non-fluctuating humidity. Such observations are important since they suggest (1) that the mycobiont, by virtue of its association with two bionts, can considerably increase the range of its ecological tolerance, and (2) that the type and degree of variation in such factors as light and humidity determine whether the mycobiont of this species is expressed and (or) persists as a blue-green or green algal morphotype or, very rarely, as a composite thallus in any given locality. The comparative abundance of either morphotype within the overall geographical distribution of the species could be due to isolated local conditions or to the overall climatic drift within the total range of the species (see pp. 51-52).

In *Sticta stipitata*, the difference in the ecological requirements for both morphotypes is not so noticeably apparent; this may be due to the fact that the green algal morphotype is seldom found without a few subtending blue-green coralloid excrescences, particularly near the apex of the stalk. In this species, initiation of the green algal morphotype on the blue-green algal phase can be delayed so that well developed specimens of the *Dendriscocaulon* may have only a few leaflets of the green algal morphotype. Well developed and fertile specimens of the green algal morphotype (i.e. *S. shirleyana*) with very few branchlets of the blue-green algal morphotype are characteristic of sunny branches in tree and shrub crowns, whereas less well developed thalli with correspondingly a more *Dendriscocaulon*-like structure, are characteristic of more shaded tree trunks and mossy rocks. In this species the composite phase seems to be an almost permanent feature of the species where the development of each morphotype depends on the amount of light and humidity in the habitat.

2. Sticta canariensis–S. dufourii

Similarities between the ecology of the two morphotypes of *S. canariensis* and that described for *S. filix* above, are also apparent from a study of a community containing all three phases occurring on coastal rocks near Campbeltown, Kintyre, Scotland. Swards of *S. canariensis* and *S. dufourii* were abundant on rock shaded by *Pteridium aquilinum* in this locality although the composite plants were rather rare. The position of the composite thalli appears in this site to be very dependent on available herbaceous cover, especially during the summer months; heavy moss cover and closeness to the soil level also appeared to favour the development of the blue-green algal morphotype. As a general rule the green algal morphotype occurred highest up the rock faces, was not usually associated with mosses and was only partly covered by the upper canopy of adjacent herbaceous phanerogams, most of which die back to ground level during the winter months.

Conversely, the blue-green algal morphotype is most frequent in the most shaded sites which also have an even humidity. On the sites where all three morphotypes are present, the paired state is produced in a more or less intermediate position between the two independent phases. Only a few of the leaflets of *S. canariensis* seem able to form a viable independent thallus and these lie within the upper limit of tolerance of the parent blue-green algal morphotype. The actual liberation and consequent survival of the green algal morphotype in

TABLE II. Distribution of independent and combined morphotypes of *Sticta dufourii–S. canariensis*

Morphotype	Country								
	Norway	Britain[a]	Ireland[a]	France[a]	Spain	Portugal[a]	Madeira	Canary Islands[a]	Azores
Blue-green (*S. dufourii*)	1	159	52	43	27	21	5	22	7
Green (*S. canariensis*)	0	60	9	0	0	1	28	39	10
Composite plants (see text)	0	30	8	2	0	0	0	6	1
Total	1	249	69	45	27	22	33	67	18

[a] Countries visited by one of the authors (P.W.J.) where a special study of the distribution was made.

this species is obviously critical. Anatomical examination shows that many initial inoculations of green algae do not advance beyond the development of a small leaflet at which stage the algae die and no further development takes place. Alternatively they are often poorly attached, and when detached die as a result of unsuitable conditions for their further development. Many are consumed by molluscs, especially *Clausilia* spp. which show a marked preference for the green but not the blue-green algae in this species.

Table II summarizes the range and morphotype composition of *Sticta canariensis*. From the data available it appears that the widest diversity (i.e. the greatest numerical variation in morphotypes) occurs in Britain and on the Islands of the Atlantic (Madeira, Canary Islands and Azores), the blue-green algae predominating in the former and the green algal morphotype in the latter. In Europe outside the British Isles, composite morphotypes have only been reported from France (Brittany) where they appear to constitute a minor proportion of the total population of the blue-green algal morphotype. The green algal morphotype is very rare or absent in France, Spain and Portugal—all areas where the blue-green algal morphotype becomes widely distributed and sometimes locally frequent, especially in the west. Conversely, the green algal stage becomes more frequent and widely distributed in the southern and western part of its range where, on the evidence available, the blue-green algal phase is correspondingly rare. The green algal morphotype, however, does not occur outside the range of the blue-green algal morphotype. The reasons for this difference in distribution between the two morphotypes are currently unknown but may reflect a response to a climatic drift where the green algal morphotype is favoured by an increased rainfall, greater humidity fluctuations, increased sunshine and higher mean temperature.

The British distribution of the individual morphotypes and composite thalli is given in Figs 3–5. These indicate that whilst the blue-green algal (*S. dufourii*) phase is locally common and widespread in western areas of the British Isles, especially in sheltered valley and coastal sites, records for independent thalli of *S. canariensis* are correspondingly few and scattered, lying within the distributional range of the blue-green algal morphotype. The green algal morphotype reaches maximum abundance in south-west Ireland; here and in Connemara (western Ireland) are the only stations in Europe where *S. canariensis* is fertile, although the species is often so on the Atlantic Islands.

The united morphotypes are thinly scattered throughout the range of the species. Their abundance in relation to the green algal morphotype emphasizes the difficulty which this stage has in establishing itself as an independent thallus away from the parent blue-green algal morphotype. The marked differences in the ecological parameters of the two morphotypes of a single species such as *S. canariensis*, clearly accounts for the apparent discrepancy in their distributions. In *S. canariensis*, for instance, the blue-green algal morphotype appears to occupy the greater part of the range of the species to the exclusion of the green algal morphotype. This suggests that the occurrence of a particular suspected morphotype does not necessarily indicate that the corresponding morphotype with a different phycobiont is present in the vicinity; this will depend on the ecological tolerance and requirements of each morphotype and the climatic variation in the locality in which it occurs.

3. Pseudocyphellaria punctulata *and* P. junghuhniana

From the above observations on the respective ecology of the two morphotypes of both *Sticta filix* and *S. canariensis*, it appears that certain environmental para-

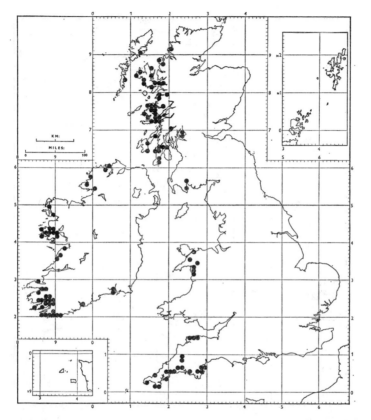

FIG. 3. British distribution of the blue-green algal morphotype (i.e. "*Sticta dufourii*") of *Sticta canariensis*.

meters, particularly light and humidity, are extremely important factors in determining the behaviour of the mycobiont and its subsequent relationship with either (or both) of the two possible phycobiont genera. Replacement of the blue-green by the green algal morphotype usually occurs at a very early stage, with the result that united morphotypes of the kind seen in these two instances are only rarely encountered in nature. However, a different pattern of variation in algal content with relation to environmental factors seems to be established in a few species of *Pseudocyphellaria*. The discovery by one of the authors (P.W.J.) that the ratio of green to blue-green algal cells in a few species

c

of this genus may alter according to the degree of shading and shelter of the habitat indicates an important alternative mode of morphotypic variation in the Peltigerineae. This previously unreported phenomenon is particularly well

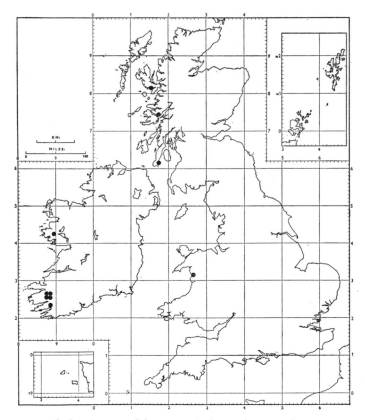

Fig. 4. British distribution of the green algal morphotype of *Sticta canariensis*.

marked in the aggregate of species including both *P. punctulata* (green phyco-biont) and *P. junghuhniana* (blue-green phycobiont), which is common in south-east Asia and Oceania. Although very variable in their algal content and morphology, the unifying feature of this assemblage is the presence of evenly spaced, punctate indentations in the upper surface of the thallus. In two separate collections from the British Solomon Islands (*Hill*, BM) and Sabah (*Lehmann*, BM), the thalli from the upper tree canopy, growing in conditions of high light intensity and low humidity, correspond to *P. punctulata* in having a well developed green algal zone and a few, scattered, clearly delimited, medullary cephalodia. In more sheltered situations on the main boughs and the upper parts of the trunk, whilst a zone of green algal cells is still present within the thalli, the blue-green algae constituting the internal cephalodia are more randomly dispersed throughout the medullary hyphae of the thallus. When the same

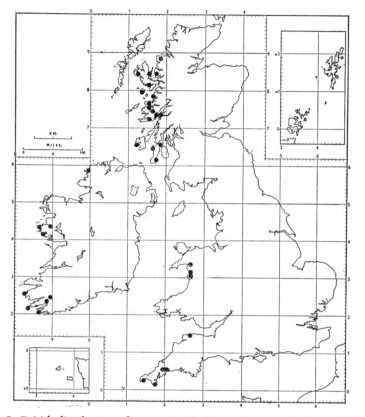

FIG. 5. British distribution of composite algal morphotypes of *Sticta canariensis*.

species grows on the bases of trees and, in one case, on adjacent rocks, the zone which previously incorporated green algae is replaced by an equivalent layer of blue-green algal cells; no internal cephalodia are present. This arrangement and corresponding morphology agrees with *P. junghuhniana* and is indicative of the lowest light intensities and an even, high humidity. A continuous change from a green to a blue-green algal phase according to the environmental conditions is clearly a distinctive feature of the biology and ecology of this species. Both the green and the blue-green algal phases and their intermediates freely develop pycnidia and ascocarps and, significantly, there is relatively little alteration in their respective thalline anatomy and morphology according to which of the algal partners predominates. In this case it is also interesting to note that the observed wide variation in the morphology of the thallus in this species aggregate appears to be entirely independent of those factors which control the proportion of green and blue-green algal cells; individuals with abnormally narrow, linearly elongate lobes appear to occur in all algal phases and are a response to different environmental parameters.

DISCUSSION

1. Some Observations on the Biology of Caulescent Species of Sticta

It is possible to segregate a small, but distinctive, group of taxa with either green or blue-green algae, within the genus *Sticta* which are all characterized by a terete or slightly flattened attachment stalk supporting a more or less palmately lobed frond with a dorsiventral arrangement of tissues. Those species containing green algae are assigned to the section *Caulata*, which, besides *S. dichotomoides*, *S. filix*, *S. latifrons* and *S. stipitata*, includes *S. carpolomoides*, *S. ferax*, *S. hypopsila*, *S. lineariloba*, *S. pluriseptata*, *S. recedens*, *S. sayeri*, *S. seemanii*, *S. subdamaecornis* and *S. tjibodensis*, as well as twelve other (as yet undescribed) entities. A complementary group of species, with a blue-green phycobiont are placed in section *Cyanisticta*, to include such species as *S. boschiana*, *S. caulescens*, *S. cyphellulata*, *S. filicinella*, *S. gyalocarpa*, *S. heppiana*, *S. hypochroa*, *S. marginifera*, *S. pedunculata*, *S. peltigerella*, *S. orbicularis*, *S. suberecta* and approximately forty other taxa whose taxonomic status has yet to be fully investigated. The identification of the species in section *Cyanisticta*, in particular, presents considerable difficulties as many of the "species" are poorly defined and polymorphic, responding markedly to seemingly minor changes in environmental conditions. Furthermore, sample collections of superficially the same species from neighbouring islands or adjacent territories show, on closer examination, small but consistent differences indicative of some degree of localized clone formation and morphological plasticity. The caulescent *Sticta* species are particularly well represented, both in numbers of species and abundance, on the islands of the East Indies; a smaller number of species occur in Japan, the Philippine Islands, New Caledonia, south-eastern Australia (mainly Queensland and New South Wales), New Zealand, southern Chile and the Maldive Islands. Many of the species with a green phycobiont are very local in distribution, those with blue-green algae are more widely dispersed; *S. cyphellulata*, for instance, is widely distributed in the whole Pacific Basin. The geographical range of those species with a green phycobiont mostly overlaps that of blue-green algal taxa, except in southern India, where only two species with *Nostoc* have so far been recorded. Although the relationship between *Dendriscocaulon* and certain caulescent species of *Sticta* is now well established, it is worth noting that although forms of *Dendriscocaulon* are frequent within the geographical range of the caulescent species of *Sticta*; they are also widespread in other, mainly temperate, oceanic areas of the world. It is unfortunately impossible, in the absence of certain evidence from united pairs, to ally most caulescent species of *Sticta* incorporating green algae with a corresponding blue-green *Dendriscocaulon* morphotype. At least some morphologically similar Dendriscocaula within, as well as outside, the range of caulescent *Sticta* species with green algae, can be expected to have as yet undetected affinities with either non-caulescent species of *Sticta* (as in *S. glomuligera*) or *Lobaria* (as in *L. amplissima*).

In those caulescent species of *Sticta* with green algae which have been studied in detail, there is evidence which suggests that the green algal morphotype is preceded by a short-lived fruticose, coralloid growth containing blue-green algae. This germinal phase is presumably derived either from (1) the capture of blue-green algal cells by the germinating spores from mature thalli of the corresponding green algal morphotype, or (2) from the fragmentation and dispersal of branchlets of an older and persisting blue-green algal morphotype of the same species. Subsequent development and persistence of the primary thallus is determined by environmental conditions; higher light intensities and uneven levels of humidity tending to favour the early capture of suitable green algal cells and the establishment of the corresponding green algal morphotype. Under conditions of even, high humidity and low light levels, the blue-green morphotype tends to persist (see p. 48). The conditions for the exchange of algal partners are obviously very critical at an early stage of development; apart from *S. stipitata*, where the delayed incorporation of the green algal morphotype is the rule rather than the exception, there is little visible evidence of an early blue-green algal origin in mature green algal morphotypes. United pairs, as in *S. filix* and *S. latifrons*, are extremely rare and probably represent a response on the part of the mature plant of the blue-green algal morphotype to a sudden change in environmental parameters which favour the establishment of the green algal morphotype. In the case of *S. filix*, the collapse of a nearby large tree (*Nothofagus menziesii*) permitted more light to pass through a normally dense canopy thereby appreciably altering those conditions which normally favoured the development of the blue-green algal morphotype to those necessary for the initiation of the green algal phase.

Some indication of the progressive incorporation and reduction of the blue-green algal cells in maturing green algal morphotypes is deduced from a detailed anatomical study of the transitional zone between the two morphotypes in united pairs of *Sticta latifrons* (Plate IIB–E). In this species, there is a notable transformation of the *Dendriscocaulon* branchlets into internal cephalodia which occurs under the combined influence of the green algal phycobiont and mycobiont. This transformation is also accompanied by a reduction of the blue-green algal cells within the main stalk region as well as a synchronous loss, by shedding, of the *Dendriscocaulon*-like branchlets (Plate IIᴇ). Although most mature specimens of caulescent species with green algal phycobionts examined have no blue-green algae in their stalk region, internal cephalodia are a universal feature of the group. In a few species, for example in all specimens of *S. dichotomoides* and very occasionally a few samples of *S. sayeri*, the internal cephalodia may find re-expression as coralloid *Dendriscocaulon*-like excrescences, these probably resembling their respective, as yet unidentified, free-living blue-green algal counterparts.

Although absent in mature thalli, blue-green algae are often present in the stalk region of the juvenile stage of several species of caulescent *Sticta* which incorporate a green phycobiont as, for instance, *S. ferax*, *S. lacera*, *S. lineariloba*, *S. sayeri*, *S. subdamaecornis* and at least two unidentified taxa. In most cases when a green phycobiont is taken up, the primary phase of the blue-green algal morphotype is probably as ephemeral and diminutive as that observed in *Psoroma durietzii* (Plate Iᴅ) and once incorporated the combined influence of the green phycobiont and mycobiont acts as a curb to any further expression of the blue-green algal morphotype, except as internal, or rarely (as in *Sticta dichotomoides*), external, cephalodia. It will be of considerable significance if all caulescent species of *Sticta* with a green algal partner are found to rely on a primary association based on the same mycobiont in association with a blue-green phycobiont.

From the evidence which has become available as a result of the investigations reported in this paper, it is possible to suggest two tentative hypotheses which may explain the mode of evolution and subsequent diversification of the development and relationships between *Dendriscocaulon*-like plants and the leaf-like caulescent species of *Sticta* which either have a blue-green phycobiont or incorporate a green alga (often associated with the development of internal cephalodia). In the lichens circumstantial evidence from studies on related pairs of species suggests that the manifestation of sexuality (as indicated by ascocarp formation) is a more primitive feature than the secondary formation of asexual propagules (see e.g. Poelt, 1970; Bowler and Rundel, 1976). Likewise, it may be assumed that the completely sterile *Dendriscocaulon* morphotypes, both as united pairs or in the free-living state, indicate that the corresponding, normally fertile, green algal morphotypes are more primitive, or at least primary. It is possible that the mycobionts of one of several caulescent species of

Sticta incorporating green algae may have evolved or retained a capacity to utilize an alternative phycobiont (*Nostoc*). In this process, a particular mycobiont of a *Sticta* with a green algal partner could have rapidly developed such an overall compatibility for a corresponding blue-green alga that respective pairs of free-living morphotypes were rapidly established and independently developed, depending on the particular environmental conditions now known to be required for the satisfactory development of each morphotype (see pp. 48–56). However, it is more likely that the affinity of those mycobionts normally associated with green phycobionts for blue-green algae developed more gradually, beginning with the capture and subsequent incorporation, without ensuing death or rejection, of alien free-living *Nostoc* cells. The blue-green algae thus captured were subsequently restricted and modified within the thallus of the host as internal cephalodia. This behaviour of the encapsulated blue-green algae is apparently determined by the predominating and combined interaction of the mycobiont and green phycobiont, a feature well demonstrated in the formation of cephalodia within the transitional zone of *Sticta latifrons* (Plate IVG). Internal cephalodia of this type are of widespread occurrence in the Peltigerineae and in particular cases in those genera known to develop, commonly or spasmodically, external *Dendriscocaulon*-like growths (i.e. *Lobaria*, *Sticta*). Interestingly, it is only the southern group of *Dendriscocaulon* species (see p. 42) which is associated with the caulescent species of *Sticta*. While the distribution of the Dendriscocaula is poorly known, the available data suggest that the two groups within it are allopatric. Since there are many lichen mycobionts which are associated with blue-green phycobionts (especially *Nostoc*) the potentiality of some mycobionts to assume viable associations with more than one algal genus should not be underestimated. Internal cephalodia in green algal morphotypes may contribute to the nutritional needs of the thallus as an extra source of carbohydrates (glucose; see Chapter 19, p. 457) and, possibly, nitrogenous compounds (but see Kershaw and Millbank, 1970). However, as observed in all specimens of *Sticta dichotomoides* (Plates VIIB and E, VIIIA) and more rarely in *S. sayeri* and other caulescent species of the genus with green phycobionts as well as the non-stalked *Lobaria* cf. *erosa*, *L. ornata* and *S. glomuligera*, the control over the secondarily incorporated *Nostoc* cells appears to be less restricted permitting, under certain conditions, the internal cephalodia to burst through the host cortex and give rise to attached *Dendriscocaulon*-like plants.

At first these partially liberated structures may have been incapable of independent survival but a subsequent process of adaptation may have resulted in the development of one or more pairs of independent blue-green morphotypes requiring sets of environmental parameters quite different from those of the green algal morphotype from which they were originally derived (see p. 48 *et seq.*). Whilst, as in the case of some caulescent species of *Sticta* with green phycobionts, a few mycobionts have retained the capacity to combine with either a green or a blue-green alga, other Dendriscocaula may have become entirely independent, lacking this capacity altogether. The subsequent evolution of this latter group appears to have been towards the realization of their latent *Sticta*-like characters. Such a possible intermediate is the *Dendriscocaulon* sp. (*Galloway* 3291; Plates VIIG and H, VIIIF) which, although retaining a basic *Dendriscocaulon* habit, has smaller branchlets which are noticeably dorsiventrally flattened and with a tomentose, ribbed, and rudimentarily cyphellate under surface, and a smooth upper surface in older or median parts of the thallus. Accordingly, the internal anatomy becomes correspondingly realigned to form the typical horizontal zonation characteristic of the foliose species of *Sticta*. A similar pattern is seen in *Sticta calithamnia* and a possibly related species from Chile (*Henssen* 24276a; Plate VIIIE). In the latter, although the thallus is characteristically *Sticta*-like, the cyphellae are extremely rudimentary or perhaps not developed. The retention of veins, as seen in *S. filicinella* (Plate VIID), might also be interpreted as a primitive feature. It may also be significant that the development of ascocarps is a relatively rare phenomenon in foliose species of *Sticta* which have blue-green phycobionts and occurs principally in those with fully rounded lobes and few or no lateral folioles or isidia (e.g. *S. caulescens*); most such species are narrow lobed and retain numerous flattened or terete, predominantly marginal, folioles or isidia which can act as efficient diaspores by severance or fragmentation (P. W. James, unpublished observations on the blue-green algal morphotype of *S. filix*). These propagules retain anatomical features similar to those found in the terminal branchlets of *Dendriscocaulon* morphotypes known to be associated with caulescent species of *Sticta* (e.g. *S. marginifera* var. *coralloidea*; Plates IIIE–G, VIIc and F).

The possible alternative hypothesis that should be considered is that the caulescent *Sticta* species and Dendriscocaula might have been independent from the first, arising from the ability of one or a few mycobionts to form viable thalli with either *Nostoc* or a particular green algal species according to the environmental conditions. A few of the Dendriscocaula so evolved might have retained the ability to accept green algal partners, hence the implementation of the secondary method of development of at least some of the green morphotypes of those caulescent *Sticta* species known to have *Dendriscocaulon* precursors (see p. 58). In this suggested alternative method of evolution, the nature of the internal and external cephalodia may be interpreted as that of a secondary infection; the behaviour of a receptive mycobiont to *Nostoc* in the presence of a green phycobiont being controlled by different sets of environmental parameters. Some of these secondary blue-green algal associations might conceivably become independent Dendriscocaula in their own right. In this alternative system it is possible to envisage independent evolution of both the green and the blue-green algal species along different lines of development from these initial precursors derived from a mycobiont with an affinity for two kinds of algal partners. Although there is some evidence of interrelationships between the blue-green *Sticta* species, none is available for the corresponding green algal species. These latter show remarkably few species affinities or even much evidence of any active speciation. Although most of the species so evolved may have lost the basic affinity for incorporating different algal partners, a few taxa of either algal affinity may still be able or may have secondarily reinstated this capacity; these species may have given rise independently to corresponding flattened and more or less leafy morphotypes—a development of considerable importance in the evolution of the flattened blue-green algal species. However, there is no evidence of any interrelationship between two dorsiventral morphotypes in the caulescent *Sticta* species at present, although such a development is characteristic of the *S. dufourii* morphotype of *S. canariensis* (see. p. 45).

2. Chemistry

No lichen substances were detected in any of the paired morphotypes of *Sticta* by either thin-layer chromatography or microcrystal tests; this finding is in agreement with data from many other as yet unpaired species of the genus and is contrary to the diversity of lichen substances which occur in many other genera. Other genera of the Stictaceae, such as *Lobaria* and *Pseudocyphellaria*, are rich in triterpenes, depsides, depsidones, terphenylquinones and pulvinic acid derivatives (Culberson, 1969). Yoshimura (1971) gives a detailed account of the chemistry of the eastern Asiatic species of *Lobaria*. These observations suggest that the genus *Pseudocyphellaria* has closer affinities with *Lobaria* than with *Sticta* and lends support to the separation of *Pseudocyphellaria* from *Sticta*, with which it has sometimes previously been subsumed.

However, major differences in the products of nitrogen metabolism are found in the two morphotypes of *Sticta canariensis*. The blue-green algal morphotype (i.e. *S. dufourii*), gives off a strong odour of stale fish when moistened and crushed; this is due to the rapid release of ammonia, methylamine and trimethylamine, substances which are absent in the green algal morphotype.

Using thin-layer chromatography, Jordan (1972) reported the presence of gyrophoric and 4-O-gyrophoric (i.e. congyrophoric) acids in quantity in the green algal morphotype of *Lobaria* cf. *erosa* and, in smaller amounts, in the erumpent blue-green algal morphotype. The presence of atranorin in both morphotypes was revealed by appropriate microcrystal tests. Contrary to the findings of Jordan, although the present authors were able to demonstrate the presence of lichen substances in the green algal morphotype of *Lobaria amplissima* and *L. ornata*, no equivalent or different chemistry was recorded in the corresponding blue-green algal morphotype in either case. Using thin-layer chromatography and microcrystal tests, 60 samples of *L. amplissima* were found to contain scrobiculin and two unidentified substances (possible breakdown products of this meta-depside); in addition, 28 of these samples contained traces of atranorin. No trace of any of these substances was recorded from the attached or isolated blue-green algal morphotypes. More significantly, both gyrophoric and congyrophoric acids were demonstrated in the green algal morphotype of *Lobaria ornata*, but unlike the situation reported in *L.* cf. *erosa*, these substances were absent in the attached blue-green algal morphotype of this species.

On the basis of his findings, Jordan (1972) suggested that the presence of the same lichen substances in both attached morphotypes of *Lobaria* cf. *erosa* could either be attributed to the presence of independent but similar biosynthetic pathways in both morphotypes, or in their exclusive production in the green algal morphotype and subsequent diffusion to the attached blue-green morphotype. Certainly, in the same lichen, substances are found in the respective cephalodia and thalli of *Stereocaulon* and *Placopsis* (P. W. James, unpublished) although it is not known which pathways are adopted in these genera.

Alternatively, the presence of lichen substances in the thallus of the blue-green morphotype of *L.* cf. *erosa* could be due to the inclusion of scattered aggregations of green algal cells within the matrix of the coralloid tips of the blue-green morphotype as reported by Jordan (1972). Such unobtrusive packets of cells were also seen by us in *Sticta filix*, where they form the initials for the leaflets of the green algal morphotype on its parent *Dendriscocaulon*. The presence of only a few green algal cells could account for the lowered concentrations of gyrophoric acid and atranorin reported by Jordan in the blue-green algal morphotype of *Lobaria* cf. *erosa* and the absence of this depside in the corresponding morphotype in *L. ornata*, where such cells are not present. Furthermore, an analysis of 100 free-living *Dendriscocaulon* thalli, representing 15 distinct taxonomic entities, failed to reveal the presence of any lichen substances. Disregarding the chemistry of the blue-green algal morphotype of *L.* cf. *erosa*, which may represent an anomalous situation, there is strong evidence, on the basis of the other pairs of morphotypes examined, that the production of a least some lichen substances is a product of the relationship between the particular phycobiont and mycobiont and suggests that the biosynthesis of at least some, for example, gyrophoric acid and scrobiculin by the mycobiont, is possible only in the presence of a certain phycobiont, and, more significantly, is dependent on the particular genus of algae present. It is possible that the presence of blue-green algae may, in some cases, actively inhibit the production of certain lichen substances. However, scrobiculin is not produced by the blue-green algal morphotype of *Lobaria amplissima* but is present in *L. scrobiculata* which also has a blue-green phycobiont. The biosynthesis of salazinic acid by an isolated mycobiont in the absence of algae has been reported by Komiya and Shibata (1969). It is of interest to note that salazinic acid is allied to stictic acid (see Chapter 7, Fig. 4). Stictic acid, and the biosynthetically related depsides, constictic and norstictic acids, are the only substances of this group commonly found in many species of *Pseudocyphellaria*, *Lobaria* and *Nephroma* with blue-green phycobionts. It is conceivable that the frequent occurrence of this chemical grouping may represent a common biosynthetic pathway which is independent of the influence of an algal partner, as is the case with (+)-usnic acid, zeorin and bellidiflorin (Shibata, 1974), whereas other substances, such as gyrophoric acid (and possibly scrobiculin in some cases; see above), require the presence of certain phycobionts in order to be produced.

These observations have considerable implications with respect to the usage of chemical criteria in lichen taxonomy (see Chapter 7, p. 152).

3. *Morphogenesis*

It is apparent from our studies on lichen morphotypes that in those lichens where there is a complex structure derived from an apparently dominant fungal partner (e.g. *Peltigera*, *Lobaria*, *Sticta*) it is the presence and subsequent interaction of the phycobiont with a particular mycobiont that plays a vital role in the phenotypic expression of the genotype of the fungal partner. This interaction, initiated, stimulated and directed by the presence of the algal partner, effectively determines the final morphology, anatomy, physiology, sexuality and probably many aspects of the chemistry of each mycobiont, thus determining the shape and biology of each lichen species so formed. Some indication of the influence of the phycobiont in the determination of the thallus form is apparent from the united pairs of lichen morphotypes studied where there is a startling diversity in structure and behaviour of the same mycobiont in the presence of two dissimilar phycobionts. In complex lichens, each species is a unique expression of an algal stimulated interreaction with a suitable mycobiont which shows a remarkable stability, is self-reproducing, and behaves for all outward intents and purposes as a single organism. Some idea of the complexity of the interrelationship between bionts is afforded by the interreaction between the three components in the transitional zone between united morphotypes of *Sticta latifrons*. In this case the combined blue-green phycobiont and mycobiont resulting in a *Dendriscocaulon* is transformed into internal cephalodia under the influence of a green algal phycobiont and the same mycobiont (Plate IVG).

The importance of the rôle of the phycobiont in the morphogenesis of lichen thalli has been much underestimated. Part of this neglect is due, in all probability, to the apparent visual dominance of the mycobiont and its complexity of structure in most lichens. In many species the mycobiont constitutes from 80 to 98% of the total volume of the thallus. Furthermore, before the studies on lichen morphotypes reported here, there was little *direct* evidence which gave support to the importance of the phycobiont in morphogenesis. There has also been considerable reluctance to forsake those older theories on the nature of lichenisation which lend support to a more or less balanced mutualism between the two bionts. In effect, all recent evidence is to the contrary and suggests that the algal partner provides all the vital requirements that are needed for the survival of the mycobiont and, secondarily, the necessary stimulus for successful lichenization. In return the phycobiont appears to derive very little, if any, direct benefit from its association in a lichen. Indeed, in some of the more complex lichens, the mycobiont seems to have become specially evolved as a receptacle for the most advantageous display of its captured phycobiont, the resulting thallus organization ensuring the continued survival and propagation of the mycobiont. One unresolved enigma is that there is little direct experimental evidence as to why the mycobionts in lichens cannot exist in a free-living state and why lichenization has become a necessary part of their survival.

When considering the role of the phycobiont in morphogenesis, it is worth recalling the behaviour of both bionts in axenic culture. As a general rule, whereas most of the phycobionts appear to be able to survive in a free-living state, the mycobionts are not able to do so; in axenic cultures the latter seem to be rarely capable of assuming even an approximation of the parent lichen thallus. Usually only an unstructured weft of vegetative hyphae is formed, closer in organization to the normally substrate-interred, vegetative part of non-lichenized ascomycetes. Only in a few species has the cultured mycobiont been induced, in the absence of the phycobiont, to form pseudoparenchymatous tissue, a development generally accepted as a primary indication of successful lichenization (Tobler, 1925, 1934; Werner, 1927, 1931; G. Vobis, unpublished). Ahmadjian (1966, 1967) has successfully demonstrated that the mycobiont of *Cladonia cristatella* can be induced to undergo sexual reproduction and to produce podetia-like structures in the absence of the phycobiont. Notwithstanding these observations, most attempts, although relatively few in number, at altering the ratio and type of nutrients, the use of growth substances, or changing the experimental parameters, have singularly failed to produce equivalent highly structured thalli comparable to those of the parent lichens involved. Further experiments, involving the application of extracts of the phycobiont to axenic cultures of the fungus, are now urgently needed (Quispel, 1945); these could give important data on important aspects of the relationship between the two bionts and the actual processes which cause such spectacular types of morphogenesis to occur in many lichens. All evidence available suggests that the phycobiont plays an important rôle in effecting the necessary stimulus for lichenization and the growth forms associated with this phenomenon; no real evidence to the contrary is yet available to refute this tenet. Another notable feature of the behaviour of the lichen fungus in axenic culture is an almost total loss of sexuality; the formation of ascocarps under experimental conditions is very rare, as are the structures equivalent to biont propagules such as isidia and soredia. Only rarely have macroconidia and blastospores been observed for a small number of species in axenic culture (Winter, 1877, Hale; 1957; G. Vobis, unpublished) and it is of interest to note that these structures are usually absent, or at least not yet observed, in the appropriate species of lichen.

Whilst it is feasible to assume that the ultimate survival of the mycobiont of all lichens depends on the presence of a suitable phycobiont, evidence suggests that the final form of the thallus is an expression of the interaction of the two, or sometimes three, bionts. The subsequent variation that occurs, as seen from the multifarious variety of thallus forms, can be explained in terms of the variety of bionts involved and the polyphyletic origins of the lichens. The taxonomic disparity amongst mycobionts suggests that lichenization (i.e. the appropriation of an alga by a fungal partner) has occurred on several occasions within the evolution of the ascomycete fungi. This polyphyletic origin has resulted in the development of several discernible patterns of association between the bionts. In the Arthopyreniaceae, the association with an algal partner, in those species which are obligately lichenized, appears as no more than a process of annexation; no distinctive thallus form is evolved and often the phycobiont (mainly *Trentepholia* species) is incorporated into the substrate with the mycobiont. Closely related taxa may either be with or without a phycobiont according to the species and a few only cohabit with algae, probably in response to some deficiency in the nutrient supply of the substrate.

In the Collemataceae, however, complex thalli, identifiable by their vegetative morphology, are developed. Each species represents the morphological and

anatomical modification when a free-living algal partner, always *Nostoc*, of well
defined morphology, is penetrated by a particular mycobiont. The invasion
and spread of the mycobiont in the gelatinous matrix of the phycobiont and its
use of the alga as a nutrient pabulum causes an interaction between the two
bionts which gives rise to a growth form characteristic for the combination.
Not only is sexuality achieved by the fungal partner, but the differentiation of
asexual propagules containing both bionts is alternatively achieved in some
species (e.g. *Collema auriculatum*). In this group the algal partner is dominant and
is accordingly modified by its relationship with the invading fungus.

In the majority of lichens the apparent dominance of the relationship is
reversed and the mycobiont is elaborated to enclose the algal cells and to take on
the responsibility for the major bulk and form of the thallus (up to 98%) (see
Henssen and Jahns, 1973, for examples). However, it is the presence of the
phycobiont and the subsequent interaction between the bionts which is responsible for the ultimate stimulation of the phenotype of the mycobiont and its
subsequent organization into the highly structured thalli of relatively stable
organization characteristic of most lichens. Evidence of the vital and different
formative rôle of the algal partner in lichens is clearly observed in our studies on
united morphotypes, particularly in those where a considerable degree of
dimorphism is manifest. For instance, in the respective morphotypes of *Sticta
filix*, an association with a green algal partner is capable of producing an
independent classifiable growth form very distinct from that produced by the
same mycobiont and a blue-green phycobiont. In the latter, the thallus form is
terete with a radial organization of tissues, and growth is apical at the non-
corticate ends of homoiomerous branchlets. In contrast, with a green phycobiont and the same mycobiont the interaction produces a dorsiventral, palmate
shape with a stratified structure, a tomentose, cyphellate lower surface, and a
smooth, ascocarp-bearing upper cortex; growth in this case is along a diffuse front
at the corticate, heteromerous, margin of the morphotype. Such a degree of dimorphism is a spectacular feature of this pair of morphotypes, and other species in
which the blue-green phase is *Dendriscocaulon*-like. In all the examples of paired
morphotypes recorded, some degree of dimorphism has been observed; in
some, such as *Peltigera* and *Pseudocyphellaria*, only a relatively small difference
between the two phases was noted; in others the degree of dimorphism is
greater (e.g. *S. canariensis*) and involves considerable changes in the thallus
structure and morphology.

Whilst the behaviour of the mycobiont appears to be largely determined by
an appropriate phycobiont, the latter can also undergo considerable morphological change as a result of lichenization. For instance, many phycobionts are

not recognizable at the species level and need to be subcultured before satisfactory determination can be attempted (see Chapters 4 and 5). A particularly interesting example of the fungal influence on the algal partner is seen in the genera *Staurothele* and *Endocarpon* where the green phycobiont (*Stichococcus*) assumes an entirely different shape and size within the perithecium contrasting with that adopted within the vegetative thallus (see Ahmadjian and Heikkilä, 1970).

In spite of the apparent dependence of the mycobiont on the algal partner it seems that the morphogenesis of lichens is based on a mutual interaction between the two, or sometimes three, bionts. The morphology of each lichen species achieved by each combination of bionts is an expression of a delicate balance of the presumed physiological interaction that takes place in the thallus. Some evidence that the fungal partner also retains the power to influence the morphology of the thallus is seen in *Sticta marginifera* var. *coralloidea*. In this taxon the main part of the thallus is dorsiventrally flattened but from this, elongate, more or less terete marginal folioles are developed. These folioles still retain the power of reverting to the original thallus form (Plates IIIE–G, VIIF).

The diverse effect of two different algal phycobionts on the same mycobiont and, in consequence, the ultimate morphogenesis of the respective species of lichen has some important implications for our understanding of the behaviour between the two bionts. It suggests that the mycobionts of at least some species in the Peltigerineae and Pannariaceae can effectively form two different physiological and morphological entities in the presence of different phycobionts. Fundamental differences in the carbohydrate pathways of the respective morphotypes of *Lobaria amplissima* have already been elucidated (Richardson *et al.*, 1968; Richardson, 1974) and a similar divergence in nitrogen metabolism for this species and *Peltigera* have been documented (Kershaw and Millbank, 1970; Millbank and Kershaw, 1974). Different chemical pathways are suggested by our observations on the lichen substance in the morphotypes of *Lobaria ornata* and *L. amplissima* (see p. 62). In reality, apart from a common mycobiont, the behaviour of the two morphotypes is entirely different; even in respect to their different ecological requirements. It is not known what processes in the interaction between the two bionts account for the diversity of form, physiology and chemistry. Considering the relationship to be analogous to that found in certain galls (Mani, 1964) the effect of certain proteins, growth substances, hormones, and certain enzymes, can possibly be invoked, though which factor (or factors) are involved and their method of action remains highly speculative and a most important field for future research.

Since the algal partner does play an important rôle in determining the form of the lichen thallus in which it is involved, it is not impossible to speculate that strains of the same phycobiont, just as slight changes in the genetic make-up of the mycobiont, could be effective in bringing about subtle but significant changes in the behaviour of the whole lichen thallus involved. It might be possible, for instance, for such strain formation, now reported in a number of instances (see Chapter 7, p. 154), to influence the biosynthetic pathways of the lichen products as well as a responsibility for the morphological plasticity of closely related groups of species such as occur in the genus *Usnea*.

Examples of associations between a single mycobiont and two phycobionts are fairly widespread throughout the lichens and are always a feature of species in which the green phycobiont predominates (see Table I). In many genera, such as *Coccotrema*, *Pilophorus*, *Placopsis* and *Stereocaulon*, the cephalodia are small in relation to the parent thallus and, although often characteristic in shape for the species on which they occur, show little evidence of being able to survive as identifiable morphotypes capable of an independent existence. All available evidence suggests that the cephalodia in these genera reflect a very localized response, on the part of the main thallus, to the intrusion of particular blue-green algae (commonly species of *Nostoc*). However, in the sub-order Pelti-gerineae, there is now more than circumstantial evidence which suggests that a more harmonious and significant interchange may occur between two different algal partners and the same mycobiont, often resulting in pairs of independent thalli which exhibit a considerable range of dimorphism. The resulting morpho-types of either algal affinity may be capable of an independent existence, with the result that it is possible that pairs of as yet unrelated morphotypes may occur in those genera that include taxa with either blue-green or green algae. Environ-mental conditions, particularly in the early stages of development, may be critical in determining which morphotype of a particular mycobiont is developed.

A preliminary survey of selected genera in the Peltigerineae suggests that there may be several distinct patterns of algal interchange, more than one of which may be represented within a particular genus. In *Peltigera* there is recent evidence which suggests the occurrence of pairs of related morphotypes which only show a relatively small change in their morphology and anatomy, according to which phycobiont is present (see p. 48).

There are few data concerning the development of morphotypes in the genus *Pseudocyphellaria*; unlike *Sticta*, no Dendriscocaula have been recorded. However, one unique feature of this genus is the discovery of a gradual shift in the balance of the green and blue-green algal content of the thallus according to the existing environmental conditions. This phenomenon, which has been described here for the *P. punctulata–P. junghuhniana* complex (pp. 53–56) is probably a feature of several other pairs of species in the genus which are similar except in their algal content. It seems that this particular mode of interchange in the phycobiont is accompanied by a relatively small alteration in the morphology and anatomy of the thallus concerned. Such a change also appears to have little effect on the sexuality of the mycobiont.

In *Sticta*, two types of algal interchange occur, one involving the production of an alternative *Dendriscocaulon* as a blue-green algal morphotype, and the other an interchange between two relatively similar foliose phases. In the former, the *Dendriscocaulon*-like forms may act as precursors for those caulescent species of the genus with a green algal phycobiont. The degree of dimorphism between the two morphotypes of the same species is particularly striking; only the green algal morphotype is capable of producing ascocarps. Most united pairs of morphotypes are of short duration, apart from that formed in *S. stipitata*; they are consequently only very rarely reported. There is some evidence in this genus which suggests that the blue-green algal morphotypes may have been responsible for the evolution of some species of caulescent *Sticta* with blue-green algae (p. 56 *et seq.*). A few non-caulescent species of the genus may also produce *Dendriscocaulon*-like forms, but it is not known whether these can maintain an independent existence. Their origin, unlike that of the caulescent *Sticta* species, is of a secondary nature resulting from the capture of blue-green algae by the green algal morphotype in a manner similar to that described for *Lobaria amplissima* (p. 39 *et seq.*).

The discovery of interchange of foliose morphotypes in *S. canariensis* suggests that the blue-green phase in some species of *Sticta* is not always represented by a species of *Dendriscocaulon*, an observation which opens up possibilities for the occurrence of other, similar foliose pairs. As in the caulescent *Sticta* species, the green algal morphotypes of *S. canariensis* are derived from the blue-green algal phase and united pairs are not uncommon, individual unions persisting for up to 5 years. It is notable that, in this species, the establishment of the green algal morphotype is much more difficult to achieve over much of the geographical range of the species and in consequence both the blue-green algal morphotype and the united pairs are much more frequent throughout a considerable part of its distribution (Table II).

Although many species of the genus *Lobaria* with green algal phycobionts have well developed and often numerous internal cephalodia, the development of these into *Dendriscocaulon* forms seems to be a relatively rare occurrence in the genus. Only in the European species *Lobaria amplissima* are *Dendriscocaulon* outgrowths frequently developed. The formation of the blue-green morphotype in this species, like that in non-caulescent *Sticta* species, is of a secondary nature and results from the capture of free-living blue-green algae by the developing green algal morphotype. On two occasions only (*L. amplissima, L.* cf. *erosa*) have the coralloid blue-green algal excrescences been observed with developing leaflets of the same green algal morphotype as the basal thallus. The blue-green algal morphotype in *L. amplissima* is known to be able to lead an independent existence and this phase is generically indistinguishable★ from the *Dendriscocaulon* forms associated with species of *Sticta*.

The morphological and anatomical similarities, apart from their algal content, of the taxa in the *Lobaria retigera* and *L. pulmonaria* complex is to be noted. This suspected affinity may suggest that a morphotype relationship, akin to that found in *Peltigera* or perhaps, in some cases, to that in *Pseudocyphellaria*, may occur in this genus.

The genus *Solorina* is unique in that there is an overall tendency in the different species towards the liberation of the blue-green algal morphotype; this series ranges from the formation of internal cephalodia, as in *S. saccata*, to a fertile species with blue-green algae which is independent of any association involving green algae. This ascendency in the rôle of the cephalodia is accompanied by a progressive reduction in the development of the green algal morphotype. Such a clearly defined transition of the transference of importance from a green to blue-green algal phase has not been encountered in any other genus in this study.

Apart from *Dendriscocaulon* and *Sticta*, the only other intergeneric relationship that has been suggested on the basis of algal morphotypes is one that possibly occurs between *Psoroma* and *Pannaria*. Such a relationship has been suggested by Forssell (1884c), based on his observations on the early stages of development of the thalli of *Pannaria pezizoides* and *Psoroma hypnorum* (p. 47). It is noteworthy that the ascocarps of the two species in question show some similarities, not least in the characteristic ornamentation of their respective spores. Those terricolous species of *Psoroma* having affinities with *P. hypnorum* are mainly restricted to sub-alpine or cold temperate areas especially in the southern hemisphere. They form a distinctive assemblage, probably worthy of generic separation, from the saxicolous and (or) corticolous squamulose and lobate species with which they are at present allied. *P. durietzii*, which belongs to the latter group, has been shown to develop a small, transient, blue-green morphotype from which the fertile, green algal phase is established. Because of the unique sorediate nature of the cephalodia in this species, the behaviour of the mycobiont in *P. durietzii*

★ See footnote on p. 31.

must possibly be considered as anomalous; nevertheless, affinities between the placodioid-lobate species of the *Psoroma sphinctrinum* group and *Pannaria* species such as *P. lurida* should be worthy of further investigation.

These studies show that certain genera in the Peltigerineae, and possibly also in the Pannariaceae, are probably alone amongst the lichens in having mycobionts which possess the ability to combine successfully with two phycobionts. Although the origin, pattern and ultimate form of the interrelationship between different pairs of morphotypes may vary from genus to genus, there is little doubt that the ability to initiate more or less dimorphic thalli, often possessing different ecological factors, has been an important factor in the biological success of the genera concerned.

4. Sexuality

A notable feature of all those lichens so far investigated in which paired morphotypes occur is that sexuality, as indicated by the formation of ascocarps, is restricted to the green algal morphotype. None of the Dendriscocaula, free-living or as part of paired morphotypes, has been observed with either pycnidia or ascocarps, despite a few literature records to the contrary. Nor are these structures present in the blue-green algal morphotypes of *Sticta canariensis*. The failure of the blue-green algal phase to produce ascocarps might, in these cases, be interpreted as implying that this morphotype is of a derivative nature and thus a less successful relationship which has failed to attain sexuality, unlike that of the same mycobiont with a green algal partner. However, in all the morphotypes of *Sticta* examined (caulescent and non-caulescent), the blue-green algal morphotype is the precursor of the green one. Furthermore, the coralloid branchlets of the *Dendriscocaulon*-like forms, at least in those species from New Zealand, are capable of dispersal by fragmentation, new propagules derived in this way forming the initials of characteristic and extensive swards of many individuals (P. W. James, unpublished). Similarly, the blue-green morphotype of *S. canariensis* (i.e. *S. dufourii*) has numerous, flattened, laminal and marginal folioles which are easily severed by abrasion. Thus, as in the Dendriscocaula, the propensity for ascocarp development in this morphotype is largely negated by the development of an efficient system of asexual propagation.

The relationship between sexuality and the obligatory presence of a green phycobiont is a particularly interesting aspect of the work on morphotypes. Although sexuality is confined to the green algal morphotype in *Sticta*, in the genera where pairs of morphotypes are suspected (and at present treated a different and unrelated species) different patterns regarding the algal partner and ensuing sexuality seem to have been established. In *Peltigera*, for instance. there is possible evidence (p. 48) that sexuality freely occurs in both morphotypes and that, in consequence, asexual reproduction is a rather rare feature throughout this genus. A different and very interesting pattern, perhaps more akin to that occurring in *Sticta*, is characteristic of *Solorina*. In this genus the species exhibit a range from wholly internal cephalodia to species in which the cephalodia are fertile in the absence of a green phycobiont. In detail, in *S. bispora*, *S. octospora*, *S. saccata*, and several other species, the influence of the green phycobiont appears to predominate and the blue-green algae present are strictly limited to well defined internal cephalodia within the medulla of these species. In *S. crocea*, the blue-green algae occupy a more or less complete layer beneath the green algal zone. Conversely, in *S. spongiosa* (Plate VF) the mature cephalodia are external; although primarily developed internally from the capture of blue-green algal cells by the cortical layer of the green algal morphotype, they soon break through the lower cortex to form large, irregular, often subcoralloid structures which are frequently more or less buried in the soil substrate (Galløe, 1939). Although these liberated cephalodia are capable of an independent existence and are self-propagating (P.W.James, unpublished) but ascocarp formation still depends on the capture of at least a few green algal cells. The green algal cells may be present throughout the whole life of the thallus or may have been subsequently picked up by a developing cephalodium which had previously lost its green algal complement. The resulting dorsiventral thallus of this morphotype is very different in structure to the cephalodia on which it develops; in extreme cases the developing ascocarp is surrounded by a very narrow collar of tissue containing green algal cells which sits on the cephalodium (in the blue-green algal morphotype). In *S. spongiosa*, sexuality still depends on a green algal partner whereas in an undescribed species (Colombia, Andes, alt. 3680 m, *Grubb* 20471, BM) the ascocarps appear to derive directly from a continuous cerebriform blue-green algal-containing crust. Sections of both the immature and mature ascocarps of this taxon failed to reveal the presence of even a few green algal cells at any stage of their development.

There is clearly considerable difference in the necessity and rôle of a green phycobiont in ascocarp formation in different genera of the Peltigerineae. Clearly in some genera, such as *Pseudocyphellaria*, it is possible to have a complete interchange of algal content without in any way affecting the ascocarp potential of the species concerned (see p. 53 *et seq.*). In other genera such a *Lobaria*, *Sticta*, and *Solorina*, it is difficult to elucidate whether the absence of ascocarp development in the blue-green morphotype is due to some inadequacy of the particular union between the blue-green alga and mycobiont, or due to the development of alternative asexual methods of dispersal.

5. Nomenclature

The data obtained from these lichen morphotypes raise problems for the nomenclature of the group. The provisions governing this in the International Code of Botanical Nomenclature are based on the assumption that each lichen includes a different species of fungus, none of which, unlike their corresponding algal partners, is free living in nature. The discovery of the same fungus in two diverse lichenized states, as well as the occurrence of interrelationships involving the same fungus between species containing green and blue-green algae in the same or different genera, suggests a need for modification of the existing rules to accommodate the new information. At the present time, the occurrence of only a relatively small number of paired morphotypes, coupled with the fact that in all the cases studied in detail the blue-green algal morphotype is consistently sterile, suggests that the logical adjustment to the rules is one which treats the blue-green algal morphotype as a *phase* of the green algal partner which can be fertile. Thus, *Sticta dufourii* is a blue-green algal phase of *S. canariensis*, the latter name assuming priority. Similarly, those species of *Dendriscocaulon* associated with *Sticta* or *Lobaria* are automatically phases of the corresponding green algal morphotypes in the two genera. However, it is clear that the status of the genus *Dendriscocaulon*, although with proven affinities with *Lobaria* and *Sticta*, will probably remain enigmatic as it seems unlikely that appropriate green algal partners will ever be found for all the species in the blue-green algal genus (see pp. 31, 42 and 48).

However, the recent, as yet unexamined, paired morphotypes of *Peltigera*, both possibly known as normally free-living, fertile species, as well as the supposed affinities between *Pannaria pezizoides* and *Psoroma hypnorum*, suggest that the nomenclatural adjustment outlined above may only be an interim solution. The relationships between pairs of fertile species with different algal partners within the same genus is now a distinct possibility (e.g. *Nephroma*, *Peltigera*, *Sticta*, *Pseudocyphellaria*), as is that between species in different genera (perhaps *Pannaria* and *Psoroma*). It should now be possible to validate such observations by means of careful field studies and also by experimental work in growth chambers or a phytotron.

MATERIAL PHOTOGRAPHED

Specimens from which the plates were prepared are listed here. Herbarium abbreviations follow "Index Herbariorum" except that material in the personal herbarium of A. Henssen is cited as "Hsn".

Dendriscocaulon sp., New Zealand, South Island, King County, Mangaataki Station, on moss on shrubs, 1972, *Galloway* 3291 (BM), Plates VIIG and H, VIIIF.—*Dendriscocaulon* sp.

(cf. *Sticta calithamnia* Tayl.), Chile, Prov. Malleco, Parque Nacional Contulmo, on *Nothofagus*, 1973, *Henssen, Vobis & Redón* 24276a (MB), Plate VIIIE.—*Dendriscocaulon umhausense* (Auersw.) Degel., Austria, Tirol, Ötztal, Stuibenfall, on rocks above the falls, 1869, *Auerswald*, Rabenh. Lich. eur. exs. no. 862 (MB), Plate VIIA; Norway, Sör-Tröndelag, Dovre Fjell, Knutshö, on rocks, 1960, *Henssen* 12132a (Hsn), Plate VIIIB.—*Lobaria amplissima* (Scop.) Forss., France, Bretagne, Finistère, Le Gouffre de Huelgoet, 1966, *Henssen, Jahns & Massé*, 18760c (MB), Plate VIIIC and D.—*Psoroma durietzii* P. James & Henss., New Zealand, South Island, South Hokitika Experimental Station, on trunks of *Weinmannia racemosa*, 1927, *Du Rietz & Du Rietz* 1570c (OTA—holotype; BM—isotype), Plates IG, IXC; New Zealand, South Island, Lake Manapouri, on *Nothofagus*, 1927, *Du Rietz & Du Rietz* 2089 (BM, OTA), Plate IXA and D; Australia, Tasmania, Mount Field National Park, on *Nothofagus*, 1963, *James* 7022 (BM), Plates ID–F, IXB, E and F.—*Solorina spongiosa* (Sm.) Anzi, British Isles, Scotland, Perthshire, Killin, Finlarig, 1873 (BM), Plate VF.—*Sticta canariensis* Bory ex Del. (incl. *S. dufourii* Del.), British Isles, Scotland, Kintyre, Campbeltown, Balnabraid Glen, 1973, *James* (BM), Plate VIA–I; Scotland, Island of Mull, near Carsaig, on damp sheltered coastal rocks, 1970, *James* (BM), Plate VC; Ireland, Co. Kerry, Killarney, Galway's Bridge, on *Ilex aquifolium*, 1964, *Henssen & Mitchell* 17511h (MB), Plate VA; Ireland, Co. Kerry, Killarney, Gallavally, 1964, *Henssen & Mitchell* 17523a (MB), Plate VB; Ireland, Co. Kerry, Killarney Lakes, Windy Gap, on mossy boulders by stream in dense woods, 1966, *James & Sheard* (BM), Plate VD and E.—*Sticta dichotomoides* Nyl., Society Islands, Tahiti, Lépine, *Vesco* (syn. *S. damaecornis* var. *linearis* Nyl., H—holotype; BM—isotype), Plates VIIB and E, VIIIA.—*S. filicinella* Nyl., Venezuela, Prov. de Merida, Miraflor, 1846, *Spruce* (BM), Plate VIID.—*S. filix* Laur., New Zealand, South Island, Fiordland, Lake Te Anau, head of Middle Fiord, above Lake Hankinson, near western shore of Lake Thomson, near rest hut, on mossy boulder near and in spray of a waterfall, 1962, *James* 381 (BM), Plates IIA, IIIA–D; New Zealand, North Island, *Lyall* (BM), Plate IA.—*S. hypochroa* Vain., Argentina, Neuquen, Lago Lacar, Pucara, on stones along a stream, 1973, *Henssen & Vobis* 24577a (MB), Plate IC.—*S. latifrons* Rich., New Zealand, Middle Island, *Lyall* (BM), Plate IB; Stewart Island, Port Pegasus Settlement, North Arm, 1969, *Galloway* (BM), Plates IIB–E, IVA–J; New Zealand, Middle Island, *Lyall* (BM—as *S. menziesii*), Plate IB.—*S. marginifera* var. *coralloidea* (Müll. Arg.) Zahlbr., New Caledonia, Monte Mu, on trunks of trees, 1866, *Vieillard* (G—holotype), Plates IIIE–G, VIIC and F.—*S. stipitata* C.Kn. ex Wils., Australia, Queensland, 1891, *Shirley* (G—holotype of *S. shirleyana* Müll. Arg.), Plate VII.

ACKNOWLEDGEMENTS

We are very grateful to Dr D. J. Galloway, Prof. R. Santesson and Mr O. Vitikainen for allowing us to study their most interesting collections; to Dr J. W. Sheard for drawing the attention of one of the authors (P.W.J.) to the *Sticta canariensis* morphotype; to Mrs H. Klappstein and Mrs G. Traute for technical assistance in the preparation of the plates; to the Deutsche Forschungsgemeinschaft, who supported the visit of one of the authors to South America (A.H.); to the curators of herbaria cited in the text for the loan of material in their care (G,H); to Dr D. L. Hawksworth for his assistance in the preparation of the manuscript; and to Dr J.-C. Massé, Dr M. Mitchell and Professor J. Redón for acting as leaders of field trips in Brittany, Ireland and Chile respectively.

REFERENCES

AHMADJIAN, V. (1966). Artificial reestablishment of the lichen *Cladonia cristatella*. *Science, N.Y.* **151**, 199–201.

AHMADJIAN, V. (1967). "The Lichen Symbiosis." Blaisdell, Waltham, Mass.

AHMADJIAN, V. and HEIKKILÄ, H. (1970). The culture and synthesis of *Endocarpon pusillum* and *Staurothele clopima*. *Lichenologist* **4**, 259–267.

BEYERINCK, M. W. (1877). Ueber Pflanzengallen. *Bot. Ztg* **35**, 34–38.

BOWLER, P. A. and RUNDEL, P. W. (1976). Reproductive strategies in lichens. *Bot. J. Linn. Soc.* **70**, in press.

BRODO, I. M. (1973). The lichen genus *Coccotrema* in North America. *Bryologist* **76**, 260–270.

CULBERSON, C. F. (1967). The structure of scrobiculin, a new lichen depside in *Lobaria scrobiculata* and *Lobaria amplissima*. *Phytochemistry* **6**, 719–725.

CULBERSON, C. F. (1969). "Chemical and Botanical Guide to Lichen Products." University of North Carolina Press, Chapel Hill, N.C.

DEGELIUS, G. (1935). Das ozeanische Element der Strauch- und Laubflechtenflora von Skandinavien. *Acta phytogeogr. suec.* **7**, i–xii, 1–411.

DEGELIUS, G. (1941). Contributions to the lichen flora of North America II. The lichen flora of the Great Smoky Mountains. *Ark. Bot.* **30 A** (3), 1–80.

DUGHI, R. (1936). Étude comparée du *Dendriscocaulon bolacinum* Nyl. et de la céphalodie fruticuleuse du *Ricasolia amplissima* (Scop.) Leight. *Bull. Soc. bot. Fr.* **83**, 671–693.

DUGHI, R. (1937). Une céphalodie libre lichénogène: le *Dendriscocaulon bolacinum* Nyl. *Bull. Soc. bot. Fr.* **84**, 430–437.

DUGHI, R. (1944). Sur les relations, la position systématique et l'extension du genre *Dendriscocaulon*. *Annls Fac. Sci. Marseille* **16**, 147–157.

DUGHI, R. (1945). Une nouvelle céphalodie fruticuleuse de *Ricasolia*. Le *Dendriscocaulon Lesdainei*. *Annls Fac. Sci. Marseille* **16**, 239–242.

DUGHI, R. (1954). Sur la taxonomie des lichens. *Bull. Soc Hist. nat. Toulouse* **89**, 97–120.

FORSSELL, K. B. J. (1883). Studier öfver cephalodierne. *Bih. K. svenska VetenskAkad. Handl.* **8**, 1–12.

FORSSELL, K. B. J. (1884a). Lichenologische Untersuchungen I. Ueber die Cephalodien. *Flora, Jena* **67**, 1–8.

FORSSELL, K. B. J. (1884b). Lichenologische Untersuchungen II. Kommen Cephalodien nur bei Archilichenen vor? *Flora, Jena* **67**, 33–46.

FORSSELL, K. B. J. (1884c). 2. Ueber den Bau und die Entwicklung des Thallus bei *Lecanora (Psoroma) hypnorum* (Hoffm.) Ach. *Flora, Jena* **67**, 187–193.

GALLØE, O. (1939). "Natural History of the Danish Lichens", Vol. 6. Munksgaard, Copenhagen.

HALE, M. E. (1957). Conidial stage of the lichen fungus *Buellia stillingiana* and its relation to *Sporidesmium folliculatum*. *Mycologia* **49**, 417–419.

HENSSEN, A. (1963). Eine Revision der Flechtenfamilien Lichinaceae und Ephebaceae. *Symb. bot. upsal.* **18** (1), 1–123.

HENSSEN, A. and JAHNS, H. M. (1973) ["1974"]. "Lichenes. Eine Einführung in die Flechtenkunde." Thieme, Stuttgart.

JAHNS, H. M. (1970). Remarks on the taxonomy of the European and North American species of *Pilophorus*. *Lichenologist* **4**, 199–213.

JAMES, P. W. and HENSSEN, A. (1975). A new species of *Psoroma* with sorediate cephalodia *Lichenologist* 7, in press.

JORDAN, W. P. (1972). Erumpent cephalodia, an apparent case of phycobiont influence on lichen morphology. *J. Phycol.* 8, 112–117.

KAULE, A. (1931). Die Cephalodien der Flechten. *Flora, Jena* 126, 1–44.

KAULE, A. (1934). Über die Cephalodien der Flechten. (2. Beitrag.). *Flora, Jena* 127, 345–361.

KERSHAW, K. A. and MILLBANK, J. W. (1970). Nitrogen metabolism in lichens, II. The partition of cephalodial-fixed nitrogen between the mycobiont and phycobionts of *Peltigera aphthosa*. *New Phytol.* 69, 75–79.

KOMIYA, T. and SHIBATA, S. (1969). Formation of lichen substances by mycobionts of lichens. Isolation of (+)usnic acid and salazinic acid from mycobionts of *Ramalina* spp. *Chem. Pharm. Bull., Tokyo* 17, 1305–1306.

LAMB, I. M. (1947). A monograph of the lichen genus *Placopsis* Nyl. *Lilloa* 13, 151–288.

LAMB, I. M. (1956). *Compsocladium*, a new genus of lichenized ascomycetes. *Lloydia* 19, 157–162.

LAMB, I. M. (1974). The lichen genus *Argopsis* Th. Fr. *J. Hattori Bot. Lab.* 38, 447–462.

MANI, M. S. (1964). "The Ecology of Plant Galls." Junk, The Hague.

MILLBANK, J. W. and KERSHAW, K. A. (1974) ["1973"]. Nitrogen metabolism. *In* "The Lichens" (V. Ahmadjian and M. E. Hale, eds), pp. 289–307. Academic Press, New York and London.

MOORE, B. J. (1969). *Lobaria lobulifera*, a new species from the southeastern United States. *Bryologist* 72, 404–406.

MOREAU, F. (1928). "Les Lichens. Morphologie, Biologie, Systématique." Lechevalier, Paris.

MOREAU, F. (1956). Sur la théorie biomorphogénique des lichens. *Revue bryol. lichén.* 25, 183–186.

NYLANDER, W. (1888). "Lichenes Novae Zelandiae." Schmidt, Paris.

POELT, J. (1970). Das Konzept der Artenpaare bei den Flechten. *Vortr. Bot. Ges.* [*Dtsch. bot. Ges.*], n.f. 4, 77–81.

QUISPEL, A. (1945). The mutual relations between algae and fungi in lichens. *Rec. Trav. bot. néerland.* 40, 413–451.

RICHARDSON, D. H. S. (1974) ["1973"]. Photosynthesis and carbohydrate movement. *In* "The Lichens" (V. Ahmadjian and M. E. Hale, eds), pp. 249–288. Academic Press, New York and London.

RICHARDSON, D. H. S., HILL, D. J. and SMITH, D. C. (1968). Lichen physiology XI. The role of the alga in determining the pattern of carbohydrate movement between lichen symbionts. *New Phytol.* 67, 469–486.

ROSE, F. and JAMES, P. W. (1974). Regional studies on the British lichen flora I. The corticolous and lignicolous species of the New Forest, Hampshire. *Lichenologist* 6, 1–72.

SANTESSON, R. (1952). Foliicolous lichens I. A revision of the taxonomy of the obligately foliicolous lichenized fungi. *Symb. bot. upsal.* 12 (1), 1–590.

SHIBATA, S. (1974). Some aspects of lichentaxonomy. *In* "Chemistry in Botanical Classification" (G. Bendz and J. Santesson, eds), pp. 241–249. Academic Press, New York and London.

THOMSON, J. W. (1950). The species of *Peltigera* in North America north of Mexico. *Am. Midl. Nat.* **44**, 1–68.

TOBLER, F. (1925). "Biologie der Flechten." Borntraeger, Berlin.

TOBLER, F. (1934). "Die Flechten. Eine Einführung in ihre allgemeine Kenntnis." Fischer, Jena.

WERNER, R.-G. (1927). "Recherches Biologiques et Expérimentales sur les Ascomycètes de Lichens." Thèse, Braun, Paris.

WERNER, R.-G. (1931). Histoire de la synthèse lichénique. *Mém. Soc. Sci. nat. Phys. Maroc* **27**, 7–44.

WILSON, F. R. M. (1891). Notes on a remarkable lichen growth in connection with a new species of *Sticta*; with descriptions of both. *Proc. R. Soc. Qd* **7**, 8–11.

WINTER, G. (1877). Lichenologische Notizen. I. Cephalodien von *Sticta* und *Solorina*. *Flora, Jena* **60**, 177–184.

YOSHIMURA, I. (1971). The genus *Lobaria* of eastern Asia. *J. Hattori bot. Lab.* **34**, 231–364.

4 | Algal Taxonomy and the Taxonomy of Lichens: the Phycobiont of *Verrucaria adriatica*

E. TSCHERMAK-WOESS

Botanical Institute, University of Vienna, Austria

Abstract: Identifications of the algal components in lichens are still relatively few. The taxonomic position of the phycobionts of many crustose species is unknown and even that of some foliose genera requires re-examination. Studies on the phycobiont of *Verrucaria adriatica* are reported. In the thallus the phycobiont occurs mostly as isolated rounded cells containing one parietal, chlorophyll-green chloroplast with a single pyrenoid surrounded by a thin sheath of starch. No haustoria have been detected. The phycobiont was isolated from the lichen and found to be *Dilabifilum arthopyreniae*, a species also occurring free-living in the area near the lichen and in other sites. New and characteristic behaviour of the zoospores produced in culture is reported—when coming to rest, the four flagella are reflexed and the proximal parts sink into the protoplast whilst the ends remain free, becoming drawn in later.

The importance of algal taxonomy in lichen taxonomy is discussed. The spectrum of algal genera lichenized with one fungal genus is either very narrow and restricted to one or two algal genera or rather wide and comprising many more algal genera.

INTRODUCTION

Our knowledge of the taxonomic position of the algal components (phycobionts) of lichens is still very incomplete. Synopses of generally well documented reports have been provided in recent years by several authors (Ahmadjian, 1967; Letrouit-Galinou, 1969; Henssen and Jahns, 1973). In the case of many crustose genera in particular surprisingly little information is available (see e.g. Duncan, 1970). Even in some foliose genera a thorough study of the phycobionts is still required. The green phycobiont of *Sticta*, for example, is said to belong to *Palmella*—an ill-defined genus applied by lichenologists earlier

Systematics Association Special Volume No. 8, "Lichenology: Progress and Problems", edited by D. H. Brown, D. L. Hawksworth and R. H. Bailey, 1976, pp. 79–88. Academic Press, London and New York.

this century to many phycobionts now treated in quite different chlorophycean genera.

The appearance of the algal cells may become modified in the process of lichenization and this is particularly striking in the case of algae which are filamentous in the free-living state. Within many lichens their cells are isolated and spherical in contrast to the markedly elongated form they assume when free living. The single plate-like chloroplast is often folded and may appear almost cup-like or even isodiametric within the lichen thallus.

Consequently, in these cases the only method of establishing the taxonomic position of a phycobiont with any certainty is to isolate it from its fungal partner and prepare unialgal cultures of it. In many cases this is not very difficult as phycobionts, in contrast to some other algae, tend not to have extremely specialized nutritional requirements. This, one might speculate, could be a consequence of selection for the ability to live on poor nutrient sources which, of course, favour the establishment of the lichen symbiosis.

PHYCOBIONT OF *VERRUCARIA ADRIATICA*

Verrucaria adriatica★ material utilized in this investigation was collected near Maslinica, Island of Šolta (an island in the Adriatic Sea near Split, Yugoslavia). Unialgal cultures of the phycobiont were obtained by placing small parts of the lichen on artificial seawater agar (Ott, 1965)†. As it made no difference to the growth of the alga the small amount of soil water which Ott recommended be added to the medium was omitted. The alga also grew equally well in liquid seawater media. Cultures were kept in a north-facing window as well as in a phytochamber under a regime of 16 h light (*c.* 800 lux) and 8 h dark, with temperatures of 17 and 14°C, respectively. Grains of iron oxide were identified by treatment with potassium ferrocyanide and hydrochloric acid.

This maritime species is widely distributed and abundant along the rocky shores of the eastern Adriatic Sea. In its locality on Šolta it is associated with *Lichina confinis* and occurs on limestone within the splash-zone. The chorophyll-green cells of the phycobiont within the lichen are orientated in rows vertical

---

★ *Verrucaria adriatica* (Zahlbr.) Zahlbr. (syn. *Dermatocarpon adriaticum* Zahlbr.) is, according to Dr R. Santesson, probably conspecific with *V. amphibia* R. Clem., a species known from many sites on the Atlantic coast of Europe, to judge from material distributed by Zahlbruckner (*Lich. rar. exs.* no. 61). As the holotype of Zahlbruckner's name has not yet been examined by Dr Santesson the name *V. adriatica* is retained and used provisionally in this paper.

† The cultures obtained were unialgal from the start, indicating that the lichen thallus was not overgrown by other algae. The same phenomenon was noted during the establishment of cultures of the phycobiont of *Verrucaria mucosa*.

with respect to the substrate, as they are in some other species of *Verrucaria*. Individual cells are spherical, ovoid, or irregularly ellipsoid and occur either as isolated cells or in adjacent twos or threes (Fig. 1). When parts of the thallus are scraped away and the limestone crystals between them dissolved, filaments of three to five, and exceptionally even ten, cells can be found. In the region around the ascocarps in particular, branching can also be occasionally observed. From

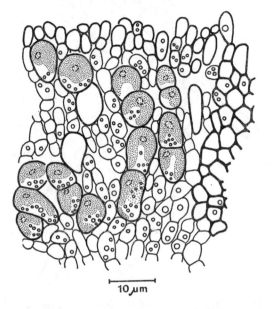

10 μm

FIG. 1. *Verrucaria adriatica*. Section through the thallus of a living specimen. Note that the thick-walled fungal cells at the right are part of the excipulum.

these observations it is clear that the phycobiont is a species of the filamentous Chlorophyceae. Each cell has a single chloroplast with a rather irregular form which appears parietal, folded and with frequent indentations. One pyrenoid surrounded by a thin starch sheath is present in each chloroplast: no starch is present in the stroma, at least in the material studied which was collected in July. No fungal haustoria penetrating into the lumen of the cell could be found and there were no indications of haustoria entering the wall alone (i.e. none of the local thickenings of the algal cell wall associated with this phenomenon in other lichens were seen; see Tschermak, 1941; Plessl, 1963).

When thallus fragments are kept on the agar medium branched filaments consisting of relatively long and narrow cells grow out of the spherical ones (Fig. 2). These filaments have plate-like chloroplasts containing starch throughout the stroma in addition to that surrounding the pyrenoid. One month old subcultures of the phycobiont have rounded colonies growing mainly at the margins with swollen cells in the centre. Zoospores are formed in these swollen

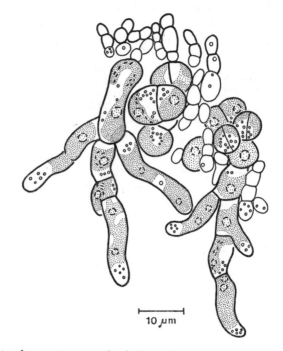

10 μm

FIG. 2. *Verrucaria adriatica.* Portion of a thallus cultured on agar for 9 days showing the cells of the phycobiont growing out to form filaments.

cells (Plate IA). Zoosporangia typically bulge out on one side of the filament only and have a slimy thickening at the apex of the bulge through which the zoospores are later discharged (Plate IIA).

In old cultures, apparently when the nutrients are almost exhausted, no zoosporangia are to be found. Such cultures have densely packed cells in their centres which are short, rounded, and divide in different directions (Plate IB). These changes in cell form are similar to those observed inside the intact lichen thallus.

The zoospores themselves are drop-shaped, with four flagella, each of which has a rather short whiplash region and no eyespot (Fig. 3a–c). They are mobile for only about 10 min and come to rest in a very characteristic manner: the flagella become orientated backwards, whereupon their proximal part sinks into the protoplast while the free ends bend in such a way that from above they appear like the arms of a windmill (Fig. 3d–e). Subsequently the ends are also drawn in and the cells assume a more rounded shape (Fig. 3f). The spore often later becomes covered by a film of tightly adhering grains or plates of iron

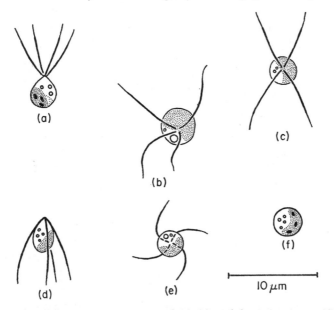

FIG. 3. *Dilabifilium arthopyreniae* zoospores. (a)–(c) While swimming; (d)–(f) while coming to rest. Based partly on living specimens and partly on ones fixed in osmium tetraoxide vapour.

oxide prior to its starting to germinate. Iron oxide is also often deposited on old filaments in a characteristic manner (Fig. 4). No signs of sexual reproduction (i.e. biflagellate motile cells) were encountered in this investigation.

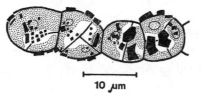

FIG. 4. *Dilabifilium arthopyreniae*. Short-celled living filament from a liquid medium culture with deposits of iron oxide.

Finally, it should be noted that the alga living as the phycobiont in the material of *Verrucaria adriatica* on Šolta also occurs in the free-living state in small depressions and fissures in the limestone in the vicinity of the lichen. In the free-living state it appeared somewhat atypical, probably because it had grown rather slowly, but when grown in culture exhibited the same characters as the isolated phycobiont. A herbarium specimen of *V. adriatica* from an island in the northern Adriatic Sea (Scoglio Galiola, Quarnero, 1910, *Brunnthaler*, det. A. Zahlbruckner, WU) has, at the side of the lichen in a depression of the stone, a green film which, as far as can be ascertained from dried material, consists of this same alga.* In the case of this lichen it would therefore appear that germinating ascospores can relatively easily encounter suitable algal cells and initiate a new lichen thallus.

The characters of the phycobiont of *V. adriatica* agree exactly with those of an alga isolated by Vischer (1953) from material of *Arthopyrenia kelpii*† collected by Klement and Doppelbauer (1952) and subsequently studied in more detail by Tschermak-Woess (1970). The first-named authors considered the alga to be the phycobiont of that species of *Arthopyrenia* and Vischer named the putative phycobiont *Pseudopleurococcus arthopyreniae*. The somewhat ambiguous genus *Pseudopleurococcus*, described by Snow (1899), could not be used for this species after the zoospores had been found and Tschermak-Woess (1970) consequently described the new genus *Dilabifilum* to accommodate *D. arthopyreniae* and two additional species. *Dilabifilum* belongs to the tribe Leptosireae in the family Chaetophoraceae.

Prior to Vischer's investigations, other authors (see Feldmann, 1937, and literature cited by him) had stressed that the phycobiont of *Arthopyrenia halodytes* was a blue-green and not a green alga. The algal species involved being determined as *Gloeocapsa crepidinum*, *Hyella caespitosa* and *Mastigocoleus testarum* by various workers. On the basis of squash preparations of taxa treated by him under *A. halodytes*, comprising both living specimens and type herbarium material, Swinscow (1965) concluded that the phycobiont was the same in all cases. Swinscow sent two specimens to Ahmadjian, who cultured the phycobiont and determined it as *Hyella caespitosa* (see also Ahmadjian, 1958, 1967). As little information on this phycobiont was provided in Swinscow (1965), the present author examined some specimens from coastal limestone and siliceous rocks collected in Anglesey, North Wales, and supplied by Dr A.

---

* In contrast to the free-living alga the cell contents of the phycobiont in this and other herbarium specimens of *V. adriatica* collected early this century are much more faded. For this reason no positive comments as to its systematic position can be made although there is nothing to suggest that the phycobiont is not identical with that in the material from Šolta used in this investigation.

† *Arthopyrenia kelpii* Körb., together with some other species (including *V. consequens* Nyl.) were treated by Swinscow (1965) as synonyms of *A. halodytes* (Nyl.) Arnold.

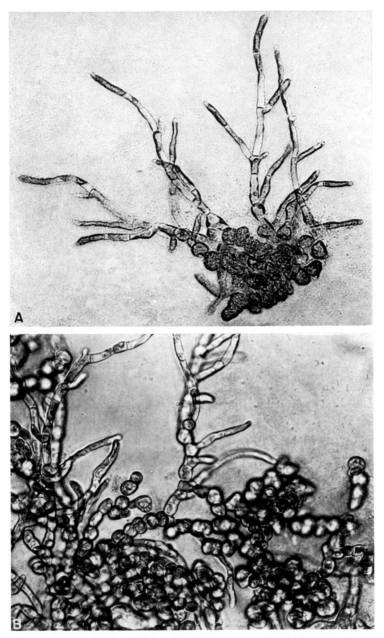

PLATE I

Dilabifilium arthopyreniae. **A**, Part of a young thallus from a living culture with elongated cells from the margin and swollen zoosporangia from the central part of the culture. **B**, Part of an older thallus from a 2-month-old living culture lacking zoosporangia and showing elongated marginal and short, rounded central cells. **A**, × 320; **B**, × 550.

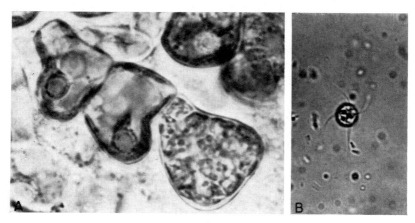

Dilabifilium arthopyreniae. **A,** Two young and one fully developed living zoosporangium. **B,** A zoospore fixed in osmium tetroxide showing the windmill stage. **A,** × 1550; **B,** × 1200.

Fletcher. All contained *H. caespitosa*, a species with many characteristic features including endospores, sporangia occasionally with a stalk-cell, and a tendency to form chroococcoid groups. Many cells of the phycobiont were attacked by one to four small intracellular haustoria. *Dilabifilium arthopyreniae* was also present in one specimen (with the cells encrusted with iron oxide) but was not enveloped by fungal hyphae and not attacked by haustoria. These observations consequently confirm the results obtained by Ahmadjian and Swinscow.

From the preceding studies and independent observations by Koch, who found *D. arthopyreniae* to be abundant in the estuary of the River Weser in Germany, it is evident that this algal species is widely distributed on seashores and that it can endure desiccation as well as marked variations in temperature (such as those which occur on rocky shores of the Adriatic Sea). As might be expected, therefore, *D. arthopyreniae* is also the phycobiont of some other maritime species. It has been isolated from *Verrucaria mucosa* by Fletcher (unpublished) and my subsequent investigations have confirmed this. Koch (unpublished) has found that *Roccella phycopsis* from Mallorca has a phycobiont belonging to the genus *Dilabifilium*. Finally, two further species of *Verrucaria* which occur in fresh water (*V.* cf. *rheithrophila* and *V. aquatica*), have another species of *Dilabifilium*, *D. incrustans*, as their phycobiont (Binz and Vischer, 1965; Tschermak-Woess, 1970). An additional species of *Dilabifilium*, *D. printzii*, is currently only known in the non-lichenized state but has only been discovered once (Vischer, 1933).

That two *Verrucaria* species from maritime habitats and two from freshwater habitats are lichenized with closely related species of phycobionts might perhaps be indicative of parallel phylogenetic development of both partners the ancestors of which were already united in a lichenized association.

SIGNIFICANCE OF ALGAE IN LICHEN TAXONOMY

The classification of the lichen algae themselves does not normally pose any extreme taxonomic difficulties. Nearly all those which have so far been encountered are either identical with or closely related to free-living algae. The question raised by Ahmadjian (1967) as to whether *Trebouxia*, the most widespread lichen phycobiont, occurs free in nature requires further investigation. That associations of blue-green and green phycobionts with a single mycobiont can lead to quite distinct morphological forms has been demonstrated in the recent studies of James and Henssen (Chapter 3) on *Sticta filix* and *Dendriscocaulon*-like forms and other examples. Similar differences also appear to occur in *Solorina spongiosa*. Apart from the problems of classification posed, by such cases, the possibility of algae exerting such a profound influence on

D

thallus form emphasizes the necessity of a thorough knowledge of the systematic position of lichen phycobionts.

What rôle can algal taxonomy play in the taxonomy of lichens? As is well known, classification and nomenclature are difficult in the lichens because of their dual nature. Most lichenologists now agree that, in the first place, classification should be based on the characters of the fungal partner. Whether the features of the vegetative thallus, to which the phycobiont contributes, can be taken to be those of the mycobiont is open to question. An extreme standpoint is taken, for example, by Hale (1961) who stated that ". . . it is doubtful whether the algae have any place in the classification of lichens". In the opinion of the present author they do have some contribution to make, at least in the lower taxonomic ranks and in certain cases also in higher ones.

If one first considers classification at species level it appears that a single species of lichenized fungus is normally associated with a single species of alga. Notable exceptions occur in the case of cephalodia discussed in more detail by James and Henssen (Chapter 3) where blue-green and green algal genera both occur together with a single mycobiont. Several cases have, however, been reported where two or more algal species (of the same algal genus) may in fact occur in a single lichen species: for example in *Coenogonium* (Uyenco, 1965), *Lichina* (Ahmadjian, 1962) and *Lobaria* (Asahina, 1937).* More detailed confirmatory studies are needed in order to clarify the situation here, but in some instances an insufficient knowledge of the natural, non-hereditary (phenotypic) variation of a phycobiont species has led to the erroneous establishment of additional species said to be associated with a single mycobiont. The case of *Synalissa symphorea* (mentioned by Henssen and Jahns, 1974, who perhaps based their conclusions on work by Letrouit-Galinou, 1969) can be explained, at least partially, in this way. Here the *Gloeocapsa* phycobiont usually has blue, red or violet mucilaginous envelopes—a character formerly used to distinguish the algal species but which has later been found to be dependent on changes in environmental conditions within a single *Gloeocapsa* species (Jaag, 1940, 1945; Jaag and Gemsch, 1940; Schiman, 1957). Whether *S. symphorea* can also be lichenized by another *Gloeocapsa* species with yellow envelopes, as stated by Jaag (1945, p. 395), should be investigated again as Schiman (1957), who checked over 200 thalli from different localities and substrates, found none with yellow pigments. In so far as *detailed* taxonomic studies on lichen phycobionts are concerned it appears safe to conclude that the taxonomic position of the phyco-

---

* See also Hawksworth (Chapter 7, p. 154) for further references on the variability of algal partners within single lichen species.

biont provides a valuable diagnostic character of the thallus and mycobiont species.

At ranks above that of species, one might enquire how far the mycobionts of a particular genus or family favour single or multiple phycobiont genera. In many genera the specialization is rather narrow, for example, *Trebouxia* is the only algal genus involved in *Cladonia, Parmelia, Physcia, Usnea* and their respective families, *Gloeocapsa* the only one in *Pyrenopsis* and *Synalissa*, and *Nostoc* the only one in *Collema, Leptogium* and other genera. The association of two algal genera, *Coccomyxa* and *Nostoc*, is characteristic of *Peltigera, Solorina*, and some other genera.

In contrast, the spectrum of phycobiont genera seems to be rather wide in *Verrucaria*, from which about six phycobiont genera are currently known, and in *Lecidea* (incl. *Biatora* and *Psora*), from which about seven phycobiont genera are reported. In these and other genera showing comparable variations in their phycobionts algal taxonomy might prove of value in classification and identification. Studies of sterile material in particular might be facilitated.

ACKNOWLEDGEMENTS

I am grateful to Dr R. Santesson for examining my material of *Verrucaria adriatica* and for information on the taxonomy of this lichen, to Dr A. Fletcher for sending me several specimens of *Arthopyrenia halodytes*, and to Dr D. L. Hawksworth for correcting the English and style of this paper. This investigation was supported by the "Fonds zur Förderung der wissenschaftlichen Forschung" in Austria (Projekt 1030).

REFERENCES

AHMADJIAN, V. (1958). A guide for the identification of algae occurring as lichen symbionts. *Bot. Notiser* **111**, 632–644.

AHMADJIAN, V. (1962). Lichens. *In* "Physiology and Biochemistry of Algae" (R. A. Lewin, ed.), pp. 817–822. Academic Press, New York and London.

AHMADJIAN, V. (1967). A guide to the algae occurring as lichen symbionts: isolation, culture, cultural physiology and identification. *Phycologia* **6**, 127–160.

ASAHINA, Y. (1937). Über den taxonomischen Wert der Flechtenstoffe. *Bot. Mag., Tokyo* **51**, 759–764.

BINZ, A. and VISCHER, W. (1965). Zur Flora des Rheinlaufes bei Basel. *Verh. naturf. Ges. Basel* **67**, 195–217.

DUNCAN, U. K. (1970). "Introduction to British Lichens." Buncle, Arbroath.

FELDMANN, J. (1937). Sur les gonidies de quelques *Arthopyrenia* marines. *Revue bryol. lichén.* **10**, 64–73.

HALE, M. E. (1961). "Lichen Handbook." Smithsonian Institution, Washington D.C.

HENSSEN, A. and JAHNS, H. M. (1973) ["1974"]. "Lichenes. Eine Einführung in die Flechtenkunde." Thieme, Stuttgart.

JAAG, O. (1940). Neuordnung innerhalb der Gattung *Gloeocapsa. Verh. schweiz. naturf. Ges.* **120** [*Locarno*], 157–158.

JAAG, O. (1945). Untersuchungen über die Vegetation und Biologie der Algen des nackten Gesteins in den Alpen, im Jura und im schweizerischen Mittelland. *Beitr. Krypt.-fl. Schweiz.* **9** (3), 1–560.

JAAG, O. and GEMSCH, N. (1940). Beiträge zur Kenntnis der Hüllenfarbstoffe in der Gattung *Gloeocapsa*. *Verh. schweiz. naturf. Ges.* **120** [*Locarno*], 158–159.

KLEMENT, O. and DOPPELBAUER, H. (1952). Über die Artberechtigung einiger marinen Arthopyrenien. *Ber. dt. bot. Ges.* **65**, 166–174.

LETROUIT-GALINOU, M.-A. (1969). Les algues des lichens. *Bull. Soc. bot. Fr., Mém.* **1968**, *Coll. Lich.* 35–37.

OTT, F. D. (1965). Synthetic media and techniques for the xenic cultivation of marine algae and flagellata. *Virginia J. Sci., n.s.* **16**, 205–218.

PLESSL, A. (1963). Über die Beziehungen von Haustorientypus und Organisationshöhe bei Flechten. *Öst. bot. Z.* **110**, 194–269.

SCHIMAN, H. (1957). Beiträge zur Lebensgeschichte homoeomerer und heteromerer Cyanophyceen-Flechten. *Öst. bot. Z.* **104**, 409–453.

SNOW, J. (1899). *Pseudo-Pleurococcus*, nov. gen. *Ann. Bot., Lond.* **13**, 189–195.

SWINSCOW, T. D. V. (1965). Pyrenocarpous lichens: 8. The marine species of *Arthopyrenia* in the British Isles. *Lichenologist* **3**, 55–64.

TSCHERMAK, E. (1941). Untersuchungen über die Beziehungen von Pilz und Alge im Flechtenthallus. *Öst. bot. Z.* **90**, 233–307.

TSCHERMAK-WOESS, E. (1970). Über wenig bekannte und neue Flechtengonidien V. Der Phycobiont von *Verrucaria aquatilis* und die Fortpflanzung von *Pseudopleurococcus arthopyreniae*. *Öst. bot. Z.* **118**, 443–455.

UYENCO, F. R. (1965). Studies on some lichenized *Trentepohlia* associated in lichen thalli with *Coenogonium*. *Trans. Am. microsc. Soc.* **84**, 1–14.

VISCHER, W. (1933). Über einige kritische Gattungen und die Systematik der Chaetophorales. *Beih. bot. Zbl.*, I, **51**, 1–100.

VISCHER, W. (1953). Über primitivste Landpflanzen. *Ber. schweiz. bot. Ges.* **63**. 169–193.

5 | *Vezdaea*, a peculiar Lichen Genus, and its Phycobiont

E. TSCHERMAK-WOESS

Botanical Institute, University of Vienna, Austria

and

J. POELT

Institute of Systematic Botany, University of Graz, Austria

Abstract: The new lichen genus *Vezdaea*, allied to *Micarea*, is described. *Vezdaea* is characterized by (*a*) the structure of the thallus which is subcuticular at first and later develops goniocysts or soredia with conical hyphal outgrowths; (*b*) a primitive ascocarp lacking an excipulum and containing only asci, anastomosing paraphyses enveloping individual asci and a more or less wanting subhymenium; (*c*) thick-walled asci lined with an amyloid layer extending into their bases; (*d*) characteristic conidial ontogeny; (*e*) the phycobiont; and (*f*) rapidly developing and short-lived ascocarps. The genus includes a single species, *V. aestivalis* (syn. *Lecidea aestivalis*, *Pachyascus byssaceus*, etc.), occurring in the north temperate zone epi- and hypophyllously mainly on bryophytes. The development and life-history of this species are described. The phycobiont was isolated in unialgal culture and found to be conspecific with *Leptosira obovata*, a species hitherto unknown in the lichenized state.

INTRODUCTION

In 1970 Vězda discussed a very remarkable lichen which he referred to as *Pachyascus byssaceus*. It was previously described by Vězda (1963) under the name *Catillaria byssacea* and was placed in *Catillaria* as there was no appropriate genus known and the species had biatorine (lecideine) ascocarps with 1-septate ascospores. Later it was transferred (Vězda, 1970) to the more appropriate genus *Pachyascus* which had been described in the intervening years by Poelt and Hertel (1968).

Systematics Association Special Volume No. 8, "Lichenology: Progress and Problems", edited by D. H. Brown, D. L. Hawksworth and R. H. Bailey, 1976, pp. 89–105. Academic Press, London and New York.

In recent years we have been able to collect this lichen on many occasions in Austria and Bavaria, thus facilitating a thorough examination of the species. The results of this investigation are presented here.

MATERIAL AND METHODS

Living material used in this investigation was obtained from (a) Munich, (b) Pleschkogel, alt. c. 1000 m, near Graz, Austria and (c) the type locality near Mokra Horá, Czechoslovakia. In (a) the species was associated with *Lophocolea heterophylla* and *Mnium punctatum*, in (b) with *Camptothecium lutescens* interspersed with lesser amounts of species of *Fissidens* and *Lophocolea* and dead leaves of *Picea*, and in (c) with *C. lutescens* interspersed with *Bryum elegans*. The material from these three sites agreed in all essential characters. Thalli from Munich were, however, smaller and thinner than those from Mokra Horá and had a smaller number of asci (6–10) in each ascocarp.

Unialgal cultures of the phycobiont were obtained by scraping young thalli, which were not overgrown by other algae, off leaves of the host collected in Munich in early October. These were placed on a 1% agar medium containing Bristol's solution modified according to Deason and Bold (1960) and supplemented with thiamine-HCl (0·2 mg/l), biotin (1 μg/l) and vitamin B_{12} (1 μg/l).* Most cultures made in this way were unialgal from the start, whilst material derived from spring and early summer collections was found to be less suitable, other algae often being present. Subcultures were prepared both from suspensions of zoospores and portions of filaments of algal cells. The phycobiont was found to grow equally well on agar and liquid media. Cultures were kept in a north-facing window and in a phytochamber under a regime of 16 h light (c. 800 lux) and 8 h dark, with temperatures of 17 and 14°C, respectively.

TAXONOMY OF THE LICHEN

That "*Catillaria*" *byssacea* has nothing in common with *Catillaria* was emphatically indicated by Vězda (1970), but in our view neither can it be satisfactorily placed in *Pachyascus*. It differs from the type species of *Pachyascus*, *P. lapponicus*, in the form of the ascocarps which are often irregular in lacking an excipulum and the asci which are thick-walled, becoming thinner at the apex (those of *P. lapponicus* are extremely thick walled even at the apex). The asci of *P. lapponicus* also have a clearly visible and relatively highly refractive inner layer. Furthermore, the spores of *C. byssacea* are *not* amyloid (contrary to the observations of Vězda, 1970) and have a finely verrucose ornamentation, whereas those of *P.*

---

* Whether the vitamins in this routinely used medium are really necessary for the satisfactory development of the cultures has not been verified.

lapponicus are strongly amyloid and quite smooth. Further differences may well also exist in the structure of the thallus itself but the scanty type material of *P. lapponicus* precludes a thorough examination of this. *C. byssacea* cannot be placed in *Micarea* (as interpreted by Hedlund, 1892) as that genus has clearly delimited ascocarps, moderately thick-walled asci, and smooth-walled spores, to judge from the species of it we have so far investigated. The separation of these genera is discussed further below (p. 92). In view of these considerations it seems necessary to establish a new genus for *C. byssacea*, for which we propose the name *Vezdaea*★ (dedicated to Dr A. Vĕzda, who first described the remarkable features of *C. byssacea*).

It has been found in the meantime that this species was in fact first described, albeit incompletely, over a century ago under the name *Lecidea aestivalis* by Ohlert (1870) on the basis of material from eastern Prussia (Pelonken). In his original description Ohlert drew attention to the strikingly thick-walled asci and examination of an isotype of Ohlert's name in H-Nyl. established that *L. aestivalis* and *C. byssacea* are conspecific. *L. metamorpha* (Nylander, 1856) may perhaps be an even earlier name for the same taxon but no authentic material has been examined and from the original description it seems likely to represent another species. It should perhaps also be pointed out that James (1965) treated *L. aestivalis* as a synonym of *L. helvola*, but, to judge from material we have examined, this latter taxon seems to represent a corticolous species of *Micarea* quite distinct from *L. aestivalis*. The new combination *Vezdaea aestivalis*† is consequently made to accommodate the type and only species of this genus.

A brief synopsis of the characters of *Vezdaea* is provided below, for a more detailed description and illustrations of the type species see Vĕzda (1970, sub *Pachyascus byssaceus*).

Thallus crustose, at first subcuticular, mainly on bryophytes. After rupturing and disintegrating the host cuticle rounded particles resembling goniocysts or soredia develop. These have many short hyphal processes at the periphery and apparently serve as

---

★ **Vezdaea** Tsch.-Woess & Poelt **gen.nov.**—Thallus crustaceus, in initio subcuticulariter crescens, demum goniocystes vel soredia aculeolata formans, *Leptosiram* continens. Asco-carpia rotundata ad indefinita, ± convexa, albida ad carneipallida ad sordidescentia, in initio leviter lanata, excipulo nullo vel hyphis paucis paraphysibus similibus indicato. Asci aequaliter percrassi apice excepto, amyloidei, tholo solum in stato juvenili provisi. Para-physes valde ramoso-connexae, reticula circum ascos formantes. Sporae octonae, plerumque bicellulares, non amyloideae, hyalinae, demum levissime flavescentes et distincte asperulatae. —Holotypus (monotypus) generis: *Lecidea aestivalis* Ohl.

† **Vezdaea aestivalis** (Ohl.) Tsch.-Woess & Poelt **comb. nov.** (Basionym: *Lecidea aestivalis* Ohl., *Schr. phys.-ökon. Ges. Königsb.* **10**, 16, 1870; synonyms: *Catillaria byssacea* Vĕzda, *Lich. sel. exs.* no. 184, 1963; *Pachyascus byssaceus* (Vĕzda) Vĕzda, *Folia geobot. phytotax.* **5**, 318, 1970).

vegetative propagules. The ascocarps are rounded or very irregular in form, moderately convex, without a margin, whitish at first but later becoming flesh-coloured or (when decaying) greyish. Young ascocarps are covered by a very fine white tomentum and lack an excipulum although occasionally some peripheral hyphae similar to the paraphyses suggesting an excipulum-like structure may occur. A hypothecium (subhymenium) is ± wanting. Young asci, prior to the differentiation of the ascospores, have an outer layer c. 3 μm thick coloured light-blue in a weak and becoming red in a strong solution of iodine, and an inner layer c. 2 μm thick coloured blue in iodine solutions of different concentrations. At the apex of the ascus the outer layer is only c. 1 μm thick and the inner layer is widened to form a more amyloid tholus which is not sharply delimited from its lower parts. In mature asci the inner layer decreases in thickness and is represented only by a thin film; the tholus also disappears. The paraphyses are branched and richly anastomosing, composed of irregular and rather flattened cells, and form a network around individual asci. There are usually eight spores per ascus, the spores normally being 1-septate (but simple and 3-septate spores are also encountered rarely) with a transverse septum. The spores are not amyloid, are hyaline or with a pale yellowish tint, and when fully developed have a fine verrucose ornamentation.

The relationships of *Vezdaea* will be considered in detail in a future publication but its separation from some other genera should also be mentioned briefly here. The genus differs from both *Catillaria* and *Lecidea* particularly in the lack of any clearly differentiated excipulum, the thick-walled asci, the structure of the paraphyses, the verrucose spores, the mode of development, and the subcuticular habit. Only *Micarea*, as defined by Hedlund (1892), appears to be closely allied to *Vezdaea*. Species of *Micarea* vary in the septation of their ascospores so that they are dispersed amongst the genera *Bacidia*, *Catillaria* and *Lecidea* in the system of Zahlbruckner (1926). In *Micarea* the thallus, when fully developed, consists of small gelatinous granules but never starts subcuticularly. The ascocarps are well defined, and, as in *Vezdaea*, an excipulum is missing. The ascus wall in *Micarea* is moderately (1–2 μm) thick in contrast to the 3–5 μm thick wall in *Vezdaea*, and the paraphyses do not form a network around the asci. In all species of *Micarea* examined by us the spores are smooth-walled and permanently hyaline. *Pachyascus* differs from *Vezdaea* primarily in having asci which have walls composed of several distinct layers, and amyloid spores which originate in numbers above eight within each ascus.

Lastly, it should be noted here that while Ohlert (1870) in his description of *Lecidea aestivalis* mentioned some characters clearly agreeing with those of the species under discussion, he did state that the spores were only one-celled. The isotype material examined (p. 91) does, however, have 1-septate spores agreeing in all respects with material referred to "*C. byssacea*". Ohlert either fortuitously saw only one-celled spores (which we have noted rarely), or simply overlooked the septum.

To judge from our many observations, mainly in Upper Bavaria (Germany) and Styria (Austria), *Vezdaea aestivalis* almost exclusively occurs on mosses and liverworts. From its bryophyte hosts it occasionally spreads over on to other lichens and the litter of vascular plants. The bryophyte hosts belong to various families and most frequently are species of, for example, *Camptothecium*, *Fissidens, Lophocolea, Mnium* and *Tortella*. We have not observed any preference for certain species or genera of bryophytes. The habitats in which this species is found are commonly somewhat shaded slopes with colonizing vegetation undergoing rapid succession, particularly in sites subject to rapid erosion such as the sides of newly erected paths and roads. No comments with respect to any preference for particular types of substrate (e.g. rock and earth) can be made here although it can be stated that it does not appear to occur on very poor substrates.

This very inconspicuous lichen may be detected by searching suitable habitats in wet weather and examining bryophytes which have become discoloured to a greenish-grey. The thalli can then be seen with the aid of a good hand lens which will also show the ascocarps (particularly the tomentose young ones). Thicker grey thalli often occurring together with *Vezdaea* usually belong to species of *Bacidia* and *Lecidea*.

From previously published records and our own investigations it is apparent that *V. aestivalis* is widely distributed in the central European belt of deciduous woodlands. Although it has been collected only rarely it has probably been frequently overlooked. Details of the specimens examined by us are as follows:

Austria: Styria, Grazer Bergland, Pailgraben, between Graz-Andritz and Gratkorn, woods in wet gully, 375–400 m, June 1972, *Poelt* (hb. Poelt no. 11034); valley of the River Mixnitz west of Bärenschützklamm near Mixnitz, 600–750 m, May 1972, *Poelt* (hb. Poelt no. 11249); gulley near Kesselfall, south of Semriach, 580–650 m, October 1972, *Poelt* (GZU); mixed forest in valley south of Großstübing, October 1972, *Poelt* (hb. Poelt no. 12777); sides of woodland path, Pleschkogel, above Stift Rein, June 1974, *Poelt* (hb. Poelt no. 12741). Oststeirisches Grabenland, south-facing forest at the side of the River Au, east of Authal, March 1972, *Poelt & Döbbeler* (GZU). **Czechoslovakia:** Moravia, above Mokrá Hora, near Brno, on earth and mosses at the side of woodland paths, 1961, *Vězda* (*Lich. sel. exs.* no. 184, hb. Poelt); *loc. cit.*, *Vězda* (*Krypt. exs. Vindob.* no. 4438, GZU); *loc. cit.*, April 1973, *Vězda & Poelt* (hb. Poelt); Tisnov, ad ruinam arcis Vickov, *c.* 400 m, 1966, *Vězda* (hb. Vězda). Slovakia, Velka Fatra, in valle "Harmanecka dolina," *c.* 800 m, 1962, *Vězda* (hb. Vězda). **France:** Brittany, Finistère, Camoret, Pointe de Peuhir, 50 ft, 1972, *Coppins* (hb. Vězda). **Germany:** Upper Bavaria, forest at slope on left side of the River Inn, beyond Wasserburg, *c.* 430 m, April 1972, *Poelt* (hb. Poelt); München-Großhessellohe, forest at steep slope on right side of the River Isar near

Menterschwaige, July 1972, *Poelt* (hb. Poelt, hb. Döbbeler no. 362); Ammergauer Voralpen, Hörnle above Kohlgrub, 1450–1500 m, north-facing side of woodland paths, May 1972, *Poelt* (hb. Poelt no. 11100). **Ireland:** Co. Galway, Killary Bay, 1876, *Larbalestier* (H). **Poland:** Eastern Prussia, Pelonken, at the base of *Carpinus*, *Ohlert* (H).

LIFE-HISTORY AND BIOLOGY

With respect to the biology of *Vezdaea aestivalis*, the following unexpected phenomenon was noted. In specimens from the Munich locality collected early in October which were apparently young and healthy (i.e. not overgrown or

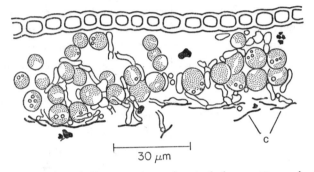

FIG. 1. *Vezdaea aestivalis*. Thallus growing subcuticularly on *Camptothecium lutescens* showing the cuticle (c) lifted off and torn. (From a living specimen; parts of the lichen not discernible are omitted.)

interspersed with other algae or fungi), patches of thalli taken from leaves of *Lophocolea* and *Mnium* on which they grow epi- and hypophyllously (Plate IA) came away with the host cuticle and were firmly attached to it. This was also seen during examination of material from Mokrá Hora and Graz but was less obvious in these, perhaps because they had been collected in early spring and summer and had older and relatively thick thalli on hosts not then growing vigorously (see p. 95). The cuticle is apparently partly torn and dissolved by attack of the mycobiont. Portions of the cuticle, together with the lichen, became spontaneously detached from the leaves when these were picked off the stems of the host. Sections vertical to the lamina of the leaves showed that this lichen commonly grows subcuticularly and adopts this habit on living and partly dead leaves and stems of bryophytes and also on dead leaves of *Picea* lying amongst the moss (Fig. 1). Some of the leaves investigated had rather thin cuticles and the hyphae of the mycobiont were mostly tightly joined to them so that the cuticle could only be distinguished in section with difficulty (although readily apparent in surface view, particularly at the edges). In spite of this the cuticle was also definitely observed in these instances, where, for example, sections had become turned over or the lining of hyphae was discontinuous. While staining techniques only revealed it clearly in surface view, when it was removed from the leaves together with specimens of the lichen, it emerged clearly in sections when these were treated with concentrated sulphuric acid and the lichen was gradually dissolving.

When growing on other lichens, *Vezdaea* is also not freely exposed to the air. It has been noted★ growing from mosses on to the upper surface of *Peltigera canina* where it penetrated the more or less homogeneous outer layer of wall material of the cortex lifting the outermost 1·5–3 μm of it (leaving 5–7 μm behind). At the margins of the lobes of the *Peltigera* it turned to the lower side of the host thallus which has no cortex, growing there beneath and between the hyphae of the medulla.

The growth of *V. aestivalis* is mainly related to the cool and wet periods of the year. There also appears to be some correlation between the periodicity of growth of the lichen and that of the bryophytes on which it occurs. The seasonal pattern of growth in this species may be reconstructed from collections made at different times of the year and the study of fresh material kept in the laboratory. In the autumn the substrate and lichen thalli are both found in a healthy condition and their growth appears to be co-ordinated. Later the thalli become thicker and the substrate ceases to grow and starts to show signs of damage. In the course of the spring and summer protonema and new stems arise from the older parts of the mosses and overgrow the lichen forming a new layer above it. On the latter, meanwhile, propagules of the lichen (ascospores, conidia and soredia) become established having been brought by various vectors to the young emergent substrate. Many germinating ascospores, conidia and extensive fungal hyphae (lacking the phycobiont) can often be found in this stage. On germination the ascospores never give rise directly to a mycelium, but at first form short hyphae from which conidia are produced in the manner illustrated by Vězda (1970). The conidia are rounded at the distal end, truncated-conical at the proximal end where the wall is slightly thickened. They are mostly single-celled but rarely are found to be partitioned into two cells by a very delicate septum (Fig. 2). Only conidia are able to grow out into long thin hyphae.

Within the young, often extensive, mycelium small clusters of newly captured algal cells can occasionally be found. How they arrive could not be observed directly but they most probably travel in the form of zoospores. In fresh raw cultures of the lichen kept in the laboratory which were periodically wetted and dried, the phycobiont in some places eliminated the mycobiont and

---

★ In one of the herbarium specimens from Pleschkogel (see p. 93).

formed zoosporangia. This might well occur under field conditions during wet periods giving rise to the lichen consortium on young host tissues, while under dry conditions transport of soredia may well be most important. Mycelium still free of the phycobiont, and mycelium in initial stages of lichenization, frequently occur on top of the cuticle on both sides of the host leaves. On

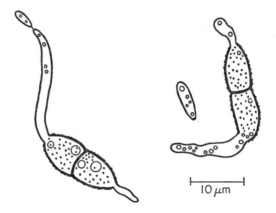

FIG. 2. *Vezdaea aestivalis.* Germinating ascospores showing their finely verrucose walls, conidiophores, and one ripe conidium. (From a living specimen.)

Lophocolea in particular, large parts of the mycelium also occur beneath the cuticle, running along the anticlinal walls of the leaves, attacking some of the living cells, and finally killing them. In the leaves of *Camptothecium lutescens*, on which the *V. aestivalis* from Mokrá Hora is most abundant signs of the cuticle being penetrated and of hyphae below the cuticular layer were encountered less often. At any rate it is probable that the subcuticular growth of the lichen is achieved through an extension of the epicuticular parts of the initial stages of the lichen consortium along the established subcuticular parts of the mycelium.★

The ascocarps are always quite free from the cuticle. They are rather irregular in form and may even occur all round the tip of a leaf. In the living or fresh state the ascocarps are whitish but become brownish when dry or decaying. They are formed within a period of only about 14 days and may also decay very quickly.

Conidiophores arising from free mycelium or lichen thalli mostly occur in clusters at the tips of the leaflets of the host, or along their edges where the cuticle has been torn so that they have access to the air. These conidiophores

---

★ Further investigations into this aspect and also the relationship between the mycobiont and the substrate are currently being planned.

consist of relatively thick hyphae similar to those arising from ascospores. The cell wall of ripe ascospores has a finely verrucose outer layer which is most easily seen in surface view after germination when the amount of fatty substances within them has been reduced (Fig. 2).

In certain places the thalli break up to form soredia or goniocysts. This probably occurs after the cuticle of the host has been chipped off, torn, or

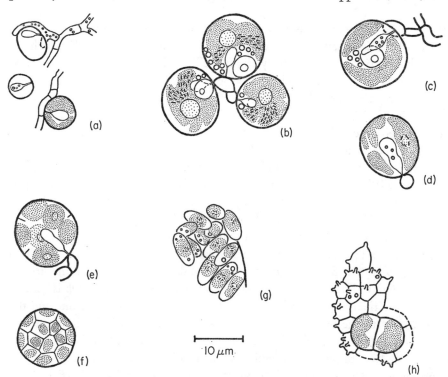

Fig. 3. *Vezdaea aestivalis.* (a) Cells of the phycobiont with adjoining hyphae taken directly from the thallus and showing the characteristic haustoria; (b)–(d) as (a) but kept on the agar medium for 7 days (in (b) and (d) only some starch grains are indicated); (e)–(g) formation of autospores (note that in (e) their delimitations are not completely drawn); (h) a soredium in which three cells (hatched) of the mycobiont are brownish and dead. (From living specimens).

dissolved in many places, but this point requires further examination. Exposure to the open air in the absence of an intervening cuticle appears to induce the formation of soredia. This view is supported by the observation that in the soredia the superficial cells of the mycobiont which tightly enclose those of the phycobiont commonly have one or more conical processes (Fig. 3h; Plate Ic).

These processes occur in all parts of the soredia which do not face the substrate. If the soredia arose whilst the cuticle still covered the lichen these processes would apparently occur only on the sides and not also on the apex of the soredium. When first formed these processes are hollow, but later they become filled with highly refractive material. Their length, diameter at the base, and also the number per cell, vary considerably.

CHARACTERS AND TAXONOMY OF THE PHYCOBIONT

In the lichenized state the phycobiont has chlorophyll-green cells which are mainly spherical or occasionally slightly ellipsoid (4–12·5 μm diam. or 7·5–14·5 × 6–12·5 μm) (Fig. 3a; Plate Ib). One parietal chloroplast occurs in each cell and this contains a single pyrenoid, lacking a compact sheath, but with starch grains in its vicinity. The shape of the chloroplast varies considerably: in most cells it is either cup-like or appears as a band folded back centrally, but in others it has processes and indentations. Every cell, apart from those which are very young, is penetrated by one (rarely two) characteristic haustoria. These haustoria have a stalk-like region one-third to a half of their length and a swollen part which often contains oil droplets (Fig. 3a). By virtue of this swollen region the haustoria of *Vezdaea aestivalis* can be readily seen in living algal cells—a situation which contrasts markedly to that in many other lichens.

When small portions of the lichen are placed on the agar medium the cells of the phycobiont at first enlarge. Starch can be detected throughout the stroma, but, as it fails to form a sheath around the pyrenoid, the pyrenoid itself is less easy to see here than in many other cases. The haustoria also grow at first and this often leads to the production of a new swollen region within the algal cells (Fig. 3b, d). Successive divisions of the protoplast then lead to the production of 4, 8 or 16 autospores from which the haustoria are excluded (Fig. 3e); these autospores are either distinctly elongate-ellipsoid or isodiametric (Fig. 3f–g). Following subsequent enlargement these again form autospores or enlarge more markedly, particularly if they are towards the periphery of the inoculum, to form either zoosporangia or filaments (Plates IIa, IIIa). The mycobiont, as in most other lichens, does not show any further development.

The phycobiont produces branching filaments at the periphery growing partly into the substrate and partly into the air. The cells of these filaments are cylindrical, readily falling apart, and measure 11·5–42(–60) μm long and 6–10 μm wide. The single chloroplast in these cells is elongate and irregularly plate-like in form and has lobes and indentations. One, or more rarely two, lenticular pyrenoids are present in each cell but in some living cultures they could only be seen with some difficulty although they were commonly seen in the majority.

Staining with iodine in potassium iodide solution and other fixing agents enhances the visibility of the pyrenoids. As first noted in living cells, and later demonstrated more clearly by treatment with 2 BX (Darlington and La Cour, 1960) or osmium tetroxide vapour, when seen in profile the pyrenoids are found to consist of several flat subunits (Plate IIIA). A single nucleus occurs in each cell but it is often not located centrally and does not appear to be displaced by other cell organelles (Plate IIIA). The cytoplasm appears rather foam-like but this does not appear to be an indication of ill-health (Plate IIA). It sometimes contains distinct droplets of oil. In general, the branching of the filaments starts by the swelling of a cell just below the cross-wall separating it from the adjoining more apical cell (Plate IIB).

In older parts of the thallus the algal cells become rounded and transformed into zoosporangia with a spherical, ellipsoid, or irregular shape (Plate IIB). The content of these cells becomes markedly yellowish in colour and, in a site not related to the former or still existing filament, a rather flat thickening develops in the wall (Plate IIIA). In older cultures, all the cells in the central parts of the inoculum may be isolated and multiplied by either auto- or zoospores. The zoospores, in common with the autospores, are produced by strictly successive divisions. In vigorously growing cultures 64 or 128 zoospores are formed in each zoosporangium, with fewer in older cultures. The zoosporangia open to liberate the zoospores at the thickened region of the cell wall. The form of the zoospores themselves varies both within and between cultures.* In some instances they appear as truncated drops and are markedly flattened, whilst in others they are more elongate, even pointed at the base, and only slightly flattened (Plate IIIB; Fig. 4). The two flagella are apparently not inserted side by side but arise along a common line perpendicular to the long axis of the cell. This usually results in a characteristic appearance when fixed (Fig. 4; Plate IIIB). The flagella are either as long or slightly longer than the cell body and the whiplash region is rather short (Fig. 4; Plate IIIB). The chloroplast within the zoospores lies slightly obliquely at the side of the cell, the nucleus (seen as a hyaline body free from oil droplets when alive or more clearly when stained) is located just below the point of insertion of the flagella, the stigma lies in the middle or lower region of the cell, and a single pulsating vacuole is also present in the middle of the cell (Fig. 4; Plate IIIB and c).

---

* The factors determining the form of the zoospores was not ascertained. Neither temperature nor the conditions of gaseous exchange (hanging drop as opposed to normal preparations) influence it. Also, the assumption that truncated zoospores might arise only from rather old zoosporangia proved to be incorrect.

The zoospores come to rest after swimming for 2–4 h. An indication, not always seen, that they are about to settle is the appearance of a fine cytoplasmic thread at the base of the cell. This originates during the motile phase and the cells may become attached to particles of detritus for some time, although in general

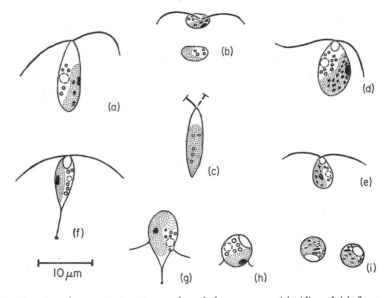

FIG. 4. *Leptosira obovata*. Swimming and settled zoospores: (a), (d) and (e) front view; (b) anterior view; (c) side view; (f) attached to a particle of detritus; (g) with the organelles moved and the major part of the flagella taken into the protoplast; (h) rounded but with flagella as in (g); (i) completely rounded with the flagella entirely taken in. (Living specimens and specimens after treatment with osmium tetroxide vapour.)

this thread does not serve as an organ of attachment (Fig. 4f and g). After further swimming and amoeboid movements, the cell becomes rounded, whereupon the flagella become reflexed and suddenly sink into the protoplast—either in one step or with their ends remaining outwards for a short time and becoming taken in later (Fig. 4h). In some instances the rounding off of the zoospores occurs only after the disappearance of the flagella (Fig. 4g). The size of the zoospores is rather variable as is shown in Fig. 4 and Plate III.

In old agar cultures where no traces of free water are present, both the formation of flagella and motility are suppressed and, instead of zoospores, aplanospores develop in large numbers. After rupturing of the sporangial wall, the preformed thickened parts of them protrude and grow out to form filaments (Plate IIID). Prior to the transformation of the potential zoospores into aplano-

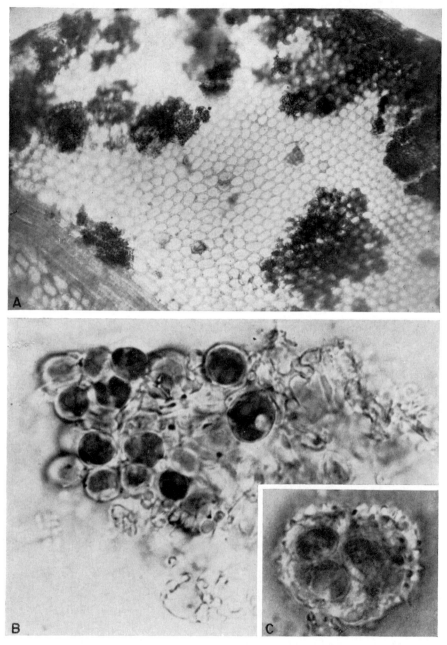

PLATE I

Vezdaea aestivalis. **A**, Young thalli growing epi- and hypophyllously on *Mnium punctatum*.
B, Part of a thallus with rounded phycobiont cells, that to the right showing the pyrenoid
and haustorium in optical section. **C**, Soredium showing the conical processes on the
exposed surface. (Living specimens.) **A**, *c*. × 100; **B** and **C**, × 1600.

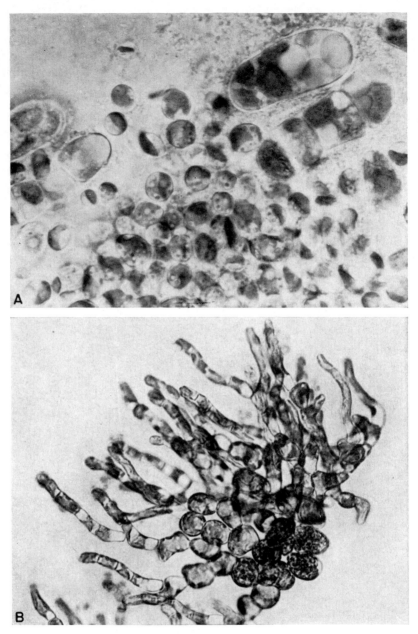

PLATE II

Leptosira obovata. **A,** Phycobiont growing out from pieces of thalli of *Vezdaea aestivalis* kept for about 2 weeks on the agar medium (some of the small cells still contained haustoria). **B,** Portion of the thallus from a subculture (note that the older parts to the right consist of zoosporangia). (Living specimens.) **A,** × 1000 **B,** × 375.

Leptosira obovata. **A,** Isolated cells after treatment with 2 BX showing on the left an immature zoosporangium with part of the thickened wall, and to the right lamellate pyrenoids (l) and the nucleus (n). **B,** Zoospores in front and side views to show the insertion of the flagella (n), position of the pulsating vacuole, and (in that on the extreme right) amoeboid movements (phase contrast after treatment with osmium tetroxide vapour). **C,** Zoospore after slight drying on to the slide and staining in acetocarmine showing the nucleus at the anterior end. **D,** Aplanospores in a living specimen emerging and growing out from sporangia. **E,** A young thallus of a living specimen on agar with the light-coloured appearing parts growing into the substrate and the dark-coloured appearing parts growing into the air. **A,** × 1600; **B** and **C,** × 1100; **D,** × 365; **E,** × 145.

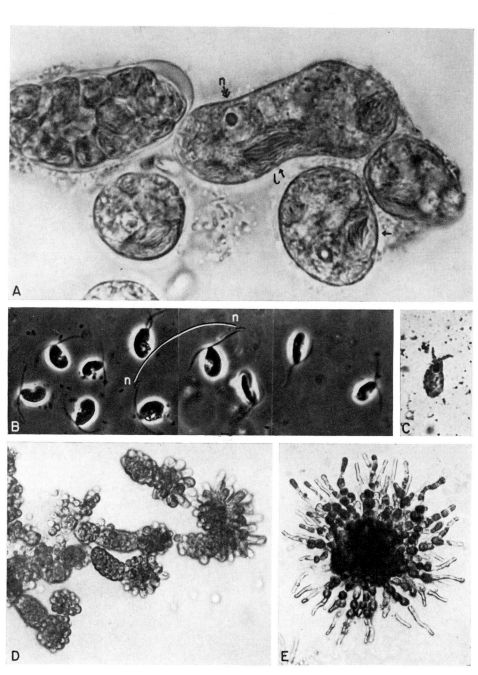

PLATE III

spores the colour of the sporangia changes from its originally yellowish tinge to an intense green.

DISCUSSION

As currently conceived, the family Lecideaceae includes a considerable number of taxa meriting separation as distinct families and genera. These have in the past been assigned to a few very large genera on the basis of a few characters, and only parts of these are connected by true affinities. The genus *Vezdaea* is only very distantly related to *Lecidea* s.s. but interrelationships probably exist between *Vezdaea* and *Micarea*, a genus well characterized by Hedlund (1892) but later misunderstood.

The most conspicuous characters of *Vezdaea* are the thick-walled asci, the amyloid inner layer of the ascus wall which almost reaches their base, the primitive structure of the ascocarps (which perhaps scarcely merit that name), and the subcuticular development of the thallus. To what extent other features, for example the spiny surface of the goniocysts or soredia, the species of phycobiont, the envelopment of individual asci by paraphyses, and the development of conidia, may be taken as diagnostic of the genus must await further studies of other taxa in the group. The asci of *Vezdaea* correspond in their structure to the type illustrated by Chadefaud (1973, p. 144, Fig. 12b) and so appear to represent a rather primitive type. However, an example of a species in which this type occurs has not previously been provided. According to Letrouit-Galinou (1973, p. 37, Fig. 11) only the type of ascus seen in *Tremotylium angolense* is more primitive, having a continuous "intern endoascus" but lacking a tholus and opening in the "jack-in-the-box" manner. In contrast the asci in *Vezdaea* open by a broad rupturing of the apex, whilst the endo- and exoascus remain attached together. Furthermore, the ascocarps in *Vezdaea* develop in a remarkably short time for a lichenized fungus. The question consequently arises as to whether some of the characters of the ascocarps in this genus arose secondarily, in which case they could not be interpreted as primitive.

A future publication dealing with additional species of the group is being prepared and in this it is hoped to provide a more detailed discussion of the classification of this genus.

The characters of the phycobiont of *V. aestivalis* correspond to those of *Leptosira obovata*, a species isolated by Vischer (1933) from the water of bogs near Basel in Switzerland. In this site it in all probability occurred in the free-living and not a lichenized state, attracting attention only when isolated. In order to confirm the identity of the phycobiont, cultures of *L. obovata*, sent by Vischer to the culture collection at Göttingen and which had been maintained there since, were studied and found to be identical in all respects. Contrary to his earlier statements, Vischer later recognized that the chloroplast of L. *obovata* contained a pyrenoid (see Tschermak-Woess, 1953). At the ultrastructural level, Wujek (1971) demonstrated that it is composed of several flat subunits which are separated by lamellae of two or more closely appressed thylakoids. That the pyrenoid had escaped the attention of earlier investigators may be accounted for in that it never has a complete and compact starch sheath and is, at best, surrounded by sparse small starch granules. Similar pyrenoids are known from *Trebouxia*, and Peveling (1970) has demonstrated them at the ultrastructural level in the phycobiont of *Ramalina*. Although they occur in the free-living alga *Chlorokybus* (Geitler, 1942; Rieth, 1972) and some other genera, this type of pyrenoid does not appear to be widespread amongst the Chlorophyceae.

Leptosira obovata has not previously been recorded as occurring in the lichenized state. Interestingly, the closely related *L. thrombii* is known to be the algal partner of *Thrombium epigaeum* (Tschermak-Woess, 1953). In general, algae reproducing by zoospores have this characteristic suppressed when they are in the lichenized state. Thalli of *Thrombium* maintained in the laboratory for some weeks and regularly wetted did, however, form zoosporangia (Tschermak-Woess, 1953). Schiman (1961), by specially searching for them, was also able to detect these in freshly collected thalli. On the basis of the investigations reported here it can be suggested that *L. obovata* is also able to form zoosporangia and zoospores under field conditions.* Under certain conditions isolated cells of *L. thrombii* are surrounded by a mucilaginous envelope but this was not seen in *L. obovata*. Further studies are required to ascertain whether *L. obovata* can be induced to form such a mucilage under special conditions.

One character previously used to separate *L. mediciana* (Borzi, 1883) and other species, the pointed end of the zoospores, does not appear to be reliable in view of the variability of this character in *L. obovata*. The same type of variability has previously been recorded by Reisigl (1964) in *L. polychloris*. The form of the cells in the filaments also proves to be a highly variable character; these may be rounded, cylindrical, or, at the ends of filaments, slightly pointed or expanded. Similar variability probably also occurs in free-living material in water or on soil. The taxonomic concepts within *Leptosira*, which hitherto have been based on these features (cf. Printz, 1964), will require a critical examination in the future.

An unexpected result of our investigations on *V. aestivalis* was the discovery of its habit of growing below the cuticles of bryophytes and even of leaf litter of

---

* Another phycobiont able to produce zoosporangia in the lichenized state is, according to Santesson (1952), *Cephaleuros* (Trentepohliaceae).

Picea. When growing on *Peltigera* it is able to occur below the outermost layer of wall material of the cortex. Previously, only two genera of lichens which are closely related and mainly tropical have been found to live subcuticularly, namely *Raciborskiella* and *Strigula* (Santesson, 1952, p. 35; Brodo, 1974, p. 420). These species occur on vascular plants, the majority of which are angiosperms, although the widespread *S. elegans* also occurs on *Blechnum* as well as angiosperm hosts*. *S. elegans* and *S. nitidula* are the only subcuticular species previously reported from Europe. No subcuticular lichens at all appear to have previously been discovered on bryophytes. A more detailed comparison between the subcuticular growth, the possibility of more intensive attack on the host tissue, as well as the induction of host reactions, with *Raciborskiella* and *Strigula* on the one hand, and *Vezdaea* on the other, would clearly be of interest. Some of these aspects have been touched on by Montagne (1842), who first described the subcuticular habit of these two tropical genera, by Cunningham (1877), whose report is evidently erroneous in some critical as well as other respects, by Ward (1883) in his careful studies, and also by Fitting (1910). According to Santesson (1952: 35–36) ". . . the often assumed parasitical nature of *Strigula* must be further studied"; other statements of the former authors must also be treated with some scepticism. In the case of *Vezdaea* also, the relationships of the lichen to its hosts require more detailed investigation. It is, however, clear that the phycobionts of *Raciborskiella* and *Strigula*, which belong to the genus *Cephaleuros* (Trentepohliaceae), are also able to grow subcuticularly in the non-lichenized state. The subcuticular growth in these two lichen genera consequently is, in the first instance, due to the habit of their symbiotic algae (Ward, 1883; Santesson, 1952, p. 147). In contrast, in the case of *V. aestivalis*, it is the fungal partner which starts to grow subcuticularly.

A further unusual feature of *V. aestivalis* is the occurrence of conical processes on the cells of the mycobiont in the soredia. These occur on all sides of the soredia except those facing the substrate (p. 98). No references to this type of soredia have been located in the literature, but they have, however, been discovered by Geitler (unpublished) in an undetermined sterile lichen with a *Trebouxia* phycobiont, and by one of the present authors (E. T.-W.) on similar isolated soredia in the British Isles (Somerset, near Bristol, Goblin Coombe, on *Betula*). These observations indicate that this type of soredia is probably not extremely rare but very much overlooked.

---

* In addition to many angiosperm hosts of *S. nemathora*, Santesson (1952, p. 154) also cites "*Deplazium*". This is probably a misprint for the genus *Diplazium* of the Filicinae and implies that this species of *Strigula* is also able to occur on a non-angiosperm host.

ACKNOWLEDGEMENTS

We are grateful to Dr A. Vězda for the supply of living specimens and assistance in many other ways, to Mr O. Vitikainen for the loan of herbarium specimens, and to Dr D. L. Hawksworth for correcting the English and style of this paper. This investigation was supported by the "Fonds zur Förderung der wissenschaftlichen Forschung" in Austria (Projekt 1030).

REFERENCES

BORZI, A. (1883). "Studi Algologici", Volume 1. Gaetano Capra, Messina.

BRODO, I. M. (1974) ["1973"]. Substrate ecology. *In* "The Lichens" (V. Ahmadjian and M. E. Hale, eds), pp. 401–441. Academic Press, New York and London.

CHADEFAUD, M. (1973). Les asques et la systematique des Ascomycetes. *Bull. Soc. mycol. Fr.* **89**, 127–170.

CUNNINGHAM, D. D. (1877). On *Mycoidea parasitica*, a new genus of parasitic algae, and the part which it plays in the formation of certain lichens. *Trans. Linn. Soc. Lond., Bot.*, ser. II, **1**, 301–316.

DARLINGTON, C. D. and LA COUR, L. F. (1960). "The Handling of Chromosomes." Allen and Unwin, London.

DEASON, T. R. and BOLD, H. C. (1960). "Phycological studies I. Exploratory Studies of Texas Soil Algae" [University of Texas Publ. no. 6022]. University of Texas, Austin, Texas.

FITTING, H. (1910). Über die Beziehungen zwischen den epiphyllen Flechten und den von ihnen bewohnten Blättern. *Annls Jard. bot. Buitenz.*, *sér.* II, *Suppl.* **3**, 505–518.

GEITLER, L. (1942). Morphologie, Entwicklungsgeschichte und Systematik neuer bemerkenswerter atmophytischer Algen aus Wien. *Flora, Jena, n.f.* **36**, 1–29.

HEDLUND, T. (1892). Kritische Bemerkungen über einige Arten der Flechtengattungen *Lecanora* (Ach.), *Lecidea* (Ach.) und *Micarea* (Fr.). *Bih. K. Svenska Vet. Akad. Handl.* **18** (3), 1–102.

JAMES, P. W. (1965). A new check-list of British lichens. *Lichenologist* **3**, 95–153.

LETROUIT-GALINOU, M.-A. (1973). Les asques des lichens et le type archaeascé. *Bryologist* **76**, 30–47.

MONTAGNE, C. (1842) ["1838–42"]. Botanique—Plantes cellulaires. *In* "Histoire physique, politique et naturelle de l'Ile de Cuba" (R. de La Sagra, ed.), pp. i–x, 1–549. Bertrand, Paris.

NYLANDER, W. (1856) ["1857"]. Prodromus lichenographiae Galliae et Algeriae. *Actes Soc. linn. Bordeaux* **21**, 249–467.

OHLERT, A. (1870). Zusammenstellung der Lichenen der Provinz Preussen. *Schr. phys.-ökon. Ges. Königsb.* **10**, 1–51.

PEVELING, E. (1970). Das Vorkommen von Stärke in Chlorophyceen-Phycobionten. *Planta* **93**, 82–85.

POELT, J. and HERTEL, H. (1968). *Pachyascus lapponicus* nov. gen. et spec., eine bemerkenswerte Flechtengattung unklaren Anschlusses. *Ber. dt. bot. Ges.* **81**, 210–216.

PRINTZ, H. (1964). "Die Chaetophoralen der Binnengewässer." Junk, The Hague.

REISIGL, H. (1964). Zur Systematik und Ökologie alpiner Bodenalgen. *Öst. bot. Z.* **111**, 402–499.

RIETH, A. (1972). Über *Chlorokybus atmophyticus* Geitler 1942. *Arch. Protistenk.* **114**, 330–342.

SANTESSON, R. (1952). Foliicolous lichens I. A revision of the taxonomy of the obligately foliicolous, lichenized fungi. *Symb. bot. upsal.* **12** (1), 1–590.

SCHIMAN, H. (1961). Über die Entwicklungsmöglichkeiten von *Leptosira thrombii* Tschermak-Woess als Algenkomponente in der Flechte *Thrombium epigaeum. Öst. bot. Z.* **108**, 1–4.

TSCHERMAK-WOESS, E. (1953). Über wenig bekannte und neue Flechtengonidien III. Die Entwicklungsgeschichte von *Leptosira thrombii* nov. spcc., der Gonidie von *Thrombium epigaeum. Öst. bot. Z.* **100**, 203–216.

VĚZDA, A. (1963). "Lichenes selecti Exsiccati", Fascicle 8. Instituto Botanico Universitatis Agriculturae et Silviculturae, Brno.

VĚZDA, A. (1970). Neue oder wenig bekannte Flechten in der Tschechoslowakei. I. *Folia geobot. phytotax.* **5**, 307–337.

VISCHER, W. (1933). Über einige kritische Gattungen und die Systematik der Chaetophorales. *Beih. bot. Zbl.*, I, **51**, 1–100.

WARD, H. M. (1883). On the structure, development, and life-history of a tropical epiphyllous lichen (*Strigula complanata*, Fée, fide Rev. J. M. Crombie). *Trans. Linn. Soc. Lond., Bot.*, ser. II, **2**, 87–119.

WUJEK, D. E. (1971). Light and electron microscope observations on the pyrenoid of the green alga *Leptosira. Mich. Academician* **3**, 59–62.

ZAHLBRUCKNER, A. (1926). Lichenes (Flechten) B. Spezieller Teil. *In* "Die natürlichen Pflanzenfamilien", Second edition (A. Engler, ed.), **8**, 61–270. Engelmann, Leipzig.

6 | Studies in the Developmental Morphology of Lichenized Ascomycetes

A. HENSSEN

Department of Botany, University of Marburg, West Germany

Abstract: A special developmental type of ascocarp in the lichenized ascohymenial fungi is compared with the common developmental types of apothecia in *Collema* and *Parmeliella* and perithecia in *Pyrenula*. The special developmental type starts with a globose primordium in which a cavity arises, usually by the rupture of hyphae. Subsequently either perithecia or apothecia are formed containing, at least in young stages, periphyses or periphysoids, respectively. This ontogenetical type is characteristic of the families Verrucariaceae, Porinaceae, Gyalectaceae, Ostropaceae, Thelotremataceae and the genera *Coccotrema* and *Lepolichen*.

Coccotrema and *Lepolichen* are included, *ad interim*, in the new family Coccotremataceae placed in the suborder Pertusariineae of the order Lecanorales. The fruit bodies of *Coccotrema* and *Lepolichen* are considered to be highly modified apothecia. The currently unsatisfactory delimitation of these two genera is mentioned.

Normandina pulchella and its alleged fruit bodies were also studied. The fruit bodies belong to a lichenicolous fungus, *Sphaerulina chlorococca* (Leight.) R. Sant. **comb. nov.** The name *Normandina pulchella* (Borr.) Nyl. is lectotypified and *Pyrenidium actinellum* Nyl. rejected as based on discordant elements. The systematic position of *Normandina pulchella* and *Coriscium viride* is discussed on the basis of anatomical and electron microscopical studies. In both lichens, the hyphae have simple pores and not dolipores. The doliporous septa in *Omphalina hudsoniana* and *Cora pavonia* are documentated by photographs.

INTRODUCTION

A special type of development giving rise to the formation of either perithecia or apothecia seems to be widespread amongst lichenized ascomycetes of the ascohymenial group. This developmental type is known to occur in the gymnocarpous Gyalectales and Ostropales as well as in the pyrenocarpous Verrucariales (Henssen and Jahns, 1973). In recent studies we have found that ascocarps in the

Systematics Association Special Volume No. 8, "Lichenology: Progress and Problems", edited by D. H. Brown, D. L. Hawksworth and R. H. Bailey, 1976, pp. 107–138. Academic Press, London and New York.

family Porinaceae and the genera *Coccotrema* and *Lepolichen* have a corresponding ontogeny. The Porinaceae are included in the pyrenocarpous Sphaeriales.
The systematic position of the genera *Coccotrema* and *Lepolichen* was unclear.
Even the nature of the fruit bodies had still been a matter for discussion (Brodo,
1973). Due to their unusual shape the ascocarps have been interpreted as either
perithecia or apothecia. In the present chapter the systematic position of the
genera *Coccotrema* and *Lepolichen* is investigated on the basis of anatomical and
ontogenetical characters.

The monotypic genus *Normandina* is usually placed in the pyrenocarpous
family Dermatocarpaceae (Zahlbruckner, 1926), a family united by modern
authors with the Verrucariaceae (Hale and Culberson, 1970), the only family
ad interim of the Verrucariales. As a result of morphological studies of the lichen
thallus we came to the conclusion that the fruit bodies described might be not the
true ascocarps of *Normandina* but rather a parasitic fungus (Henssen and Jahns,
1973), a view now confirmed by our investigations of the ascocarp ontogeny of
the fungus. In his studies of lichen parasites, Dr R. Santesson came to the same
conclusion and kindly provided us with a name for the fungus (*Sphaerulina
chlorococca*; see p. 129).

SPECIAL TYPE OF DEVELOPMENT IN COMPARISON WITH THE COMMON
DEVELOPMENTAL TYPE OF APOTHECIA AND PERITHECIA

The peculiarities of the special developmental type are better understood if
compared with the common developmental morphology in lichenized fungi
of the ascohymenial group.

1. Development of the Apothecia in Collema *and* Parmeliella

The genera *Collema* and *Parmeliella* are selected to demonstrate the development
of zeorine and biatorine apothecia. In *Collema* the development starts with the
differentiation of coiled ascogonia from ordinary thallus hyphae as described
almost a century ago by Stahl (1877). The ascogonia are produced below the
main algal zone and the tips of the trichogynes protrude above the thallus
surface (Plate IA). Where ascogonia are lying close together, several may be
included in the developing primordium. The monokaryotic mycelium of the
primordium which surrounds the ascogenous hyphae arises from the stalk cells
of the ascogonium as in the classical ontogeny of an apothecium outlined by
Corner (1929). In this respect, the Collemataceae differ from other lichen families in which the surrounding sheath is a product of the thallus hyphae adjacent
to the ascogonium or might even precede the formation of the ascogonia—the
first stage of the sprouting ascogonial and stalk cells is seen in Plate IA. A dense

web of hyphae is formed in which remnants of the ascogonium and the stalk may still be recognized (Plate Ic). Gradually, the paraphyses and the excipulum are formed (Plate ID and E). The paraphyses are, at first, branched paraphysoids developing directly from the enveloping hyphae; secondarily, true paraphyses grow up from the subhymenial layer*. By the pressure of the growing primordium the thallus bursts and the hymenium becomes exposed (Plate IE). Together with the primordium, the surrounding thallus enlarges to form a margo thallinus. The apothecium is zeorine since the hymenium is protected by a cupular excipulum proprium and also by a thalline margin. In *Collema subnigrescens* the excipulum is relatively thin. Numerous anchoring hyphae are sent from the outer cells into the thallus (Plate ID and F)—a characteristic feature in many members of the family.

In *Parmeliella plumbea*, a member of the Pannariaceae, the apothecium is biatorine. A group of ascogonia with branched trichogynes is differentiated within or above the algal layer of the thallus (Fig. 1A): The occurrence of ascogonia in dense groups is a characteristic feature in the ascohymenial lichen families apart from the Collemataceae. The ascogonia can be very numerous, and in *Pilophorus* several hundred have been counted (Jahns, 1973). The ascogonia are surrounded by generative tissue produced by the adjacent thallus hyphae. The first paraphyses are paraphysoids as in *Collema*. The primordium extends to the thallus surface, splitting the cortical layer (Fig. 1B); vertical growth of the ascocarp continues, true paraphyses and an annular excipulum proprium are formed (Fig. 1C) and hairs are produced by the outer cells of the excipulum (Fig. 1D).

The ontogeny in *Collema* and *Parmeliella* is typical of the development of gymnocarpous apothecia in lichenized fungi. Occasionally, a hemiangiocarpic development is simulated in the Collemataceae if the primordium arises deeply immersed and the hymenium is still covered by thallus tissue (Plate IE). In a true hemiangiocarpic ontogeny, the layer covering the hymenium is a part of either the generative tissue or the thallus with a particular structure. In both the Collemataceae and Pannariaceae the production of paraphysoids precedes the development of paraphyses, a feature widespread amongst lichenized ascomycetes of the ascohymenial group.

2. Development of the Perithecium in Pyrenula

The common developmental type of a perithecium may be seen in *Pyrenula nitidella*. Mature fruit bodies are partly immersed in the thallus, the perithecial wall (peridium) is thick and black and the ostiole a lighter coloured spot (Plate

---

* See pp. 134–135 for definitions of terms.

A. Henssen

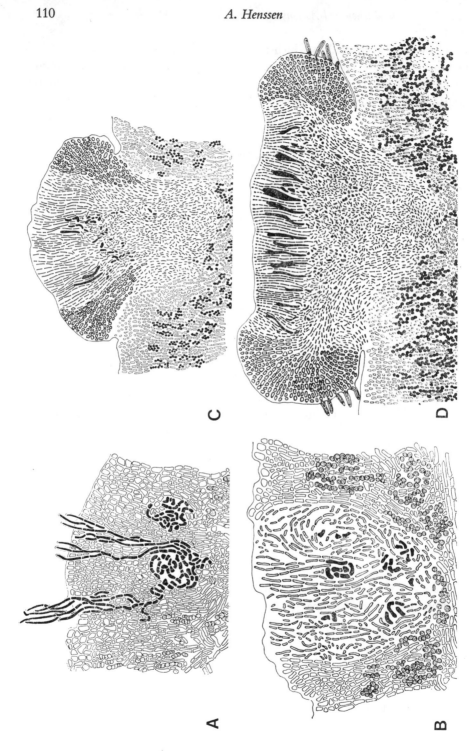

FIG. 1. (Caption on facing page)

IIA). The generative tissue consists of short-celled hyphae. Ascogonia arise in the centre and the surrounding hyphae are stretched, becoming vertically orientated (Plate IIB). At this early stage the formation of the dark peridium starts and oxalate crystals are deposited. Parts of both the adjacent thallus and the substrate are incorporated into the peridium. Ascogenous hyphae accumulate in the basal part of the centrum and true paraphyses grow up between them. Hyphae at the top of the primordium stretch to form the ostiole (Plate IIc) and later periphyses are differentiated in this region (Plate IID and E). The perithecial wall gradually enlarges to a very thick, black and hard layer (Plate IIF). The perithecial cavity is filled with mucilage and includes the thin, indistinctly septate paraphyses and bitunicate asci which arise from a concave hypothecium.

The development of the perithecium in *Pyrenula* corresponds to the ontogeny of the gymnocarpous apothecia in *Collema* and *Parmeliella*: the first paraphyses produced are paraphysoids and the protective tissue around the hymenium is built up gradually.

3. Special Developmental Type giving Rise to Perithecia or Apothecia

In this special developmental type, the generative tissue is a compact structure which differentiates into a central cavity and a cortical layer. Hyphae may grow in from all around the cavity; in the basal part they develop into paraphyses whilst in the upper they differentiate into periphyses in perithecia or periphysoids (i.e. lateral paraphyses) in apothecia. The periphyses and periphysoids are homologous structures but they cannot be named in the same way since the term periphyses is a typographical one confined to hair-like structures within or near the ostioles of perithecia or other pear-shaped organs (Ainsworth, 1971). In the further course of development, a pore arises in the apex of the primordium which may be either a well defined ostiole or a simple breaking up of the cortical layer. The mature ascocarp is either a typical apothecium or perithecium, or a strongly modified structure (e.g. apothecia with a punctiform disc or perithecia lacking a well defined ostiole). The frequent occurrence of modified fruit bodies is a characteristic feature in this developmental type. The apothecia are biatorine, lecideine, or zeorine. Their development is hemiangiocarpous;

FIG. 1. Ascocarp development in *Parmeliella plumbea*. **A,** Group of ascogonia with branched trichogynes in the upper part of the algal layer of the thallus; **B,** primordium with ascogenous hyphae embedded in the generative tissue which differentiates into paraphysoids; **C,** young apothecium with long stipe developed by considerable vertical elongation of the ascocarp; **D,** mature apothecium with hairs at the base of the excipulum. (**A–D** from Henssen and Jahns, 1973).

the layer covering the hymenium in young stages is part of the generative tissue. The ontogeny, therefore, corresponds to the classical hemiangiocarpous development in unlichenized discomycetes (Corner, 1929). In regard to the details, the special type of development differs slightly within the different systematic groups of lichens in which it occurs.

(*a*) *Verrucariaceae*. In the genera *Dermatocarpon* and *Endocarpon*, the generative tissue arises in the upper part of the algal layer (see also Doppelbaur, 1960). The apex of the primordium elongates to the thallus surface, and a group of contorted ascogonia are seen in the lower part (Plate IIIA). The outer hyphae of the generative tissue differentiate into a cortical layer of periclinally arranged hyphae. A cavity arises in the centre and hyphae grow downwards from the limiting cells in the upper part (Plate IIIB) and are the first periphyses formed. The development of the asci starts at an early stage, the asci pushing up between the periphyses (Plate IIIC; Fig. 2D). The cells in the elongated apical part of the primordium stretch, the hyphae become separated, and an ostiole is gradually differentiated. Periphyses are inserted continuously in acropetal direction lining the inner wall of the perithecium and finally filling the ostiole (Plate IIID–F). The ostiole opens schizogenously and may become elongated so that the perithecium has a correspondingly conspicuous neck (Plate IIIE). Paraphyses are not developed (Plate IIIE and F) although they are reported in the literature to be rapidly dissolved. Obviously, the periphysoids or emptied asci have been misinterpreted as paraphyses. The asci are bitunicate, the thin exoascus bursting with the ejaculation of the ascospores (Vězda, 1968; Henssen and Jahns, 1973). The mature perithecia have a multilayered wall which is compressed by the adjacent medullary hyphae. In *Endocarpon* cells of the phycobiont are enclosed within the wall of the young perithecia from which they later pass into the perithecial cavity, divide repeatedly and give rise to the small hymenial algae (Plate IIIF).

The most striking difference of the perithecial development in *Dermatocarpon*

Fig. 2. Development of perithecia in *Pyrenula* and *Dermatocarpon*. **A** and **B**, *Pyrenula nitidella*: **A**, primordium with asci, paraphyses and an apical structure giving rise to the formation of the ostiole; **B**, young perithecium with ostiole containing periphyses and a dark peridium including oxalate crystals. **C** and **D**, *Dermatocarpon rivulorum*: **C**, primordium with cavity in which the periphyses grow in from the upper part of the cortical layer; **D**, primordium with young ascus and periphyses. **E**, *Dermatocarpon miniatum*: young perithecium with ostiole differentiated. (**A–E** from Henssen and Jahns, 1973.)

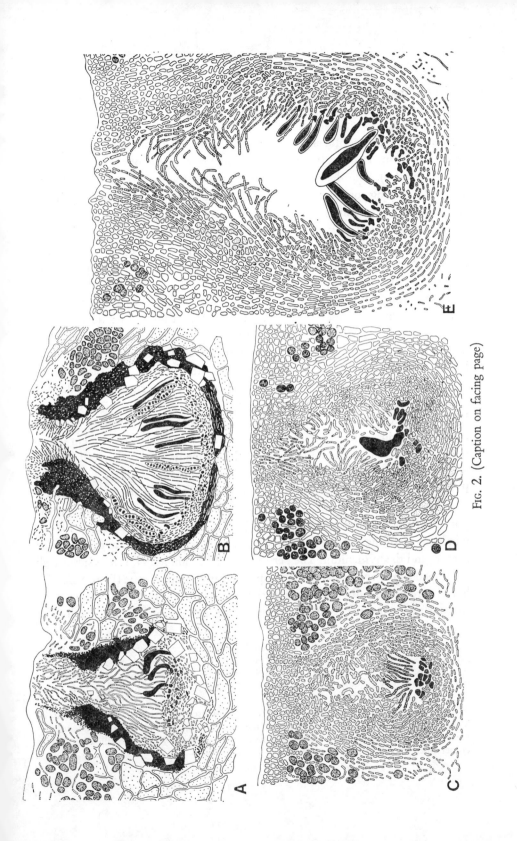

Fig. 2. (Caption on facing page)

and *Endocarpon* to that in *Pyrenula* (Fig. 2) is the formation of a cavity with periphyses preceding the differentiation of the ostiole.

(*b*) *Porinaceae*. The perithecia of *Porina nucula*, *Belonia russula* and *Clathroporina calcarea* are immersed singly (or rarely in pairs) in verrucae of the thallus (Plates IVA, VA and B). The thallus warts correspond to the margo thallinus of apothecia developing together with the enlarging ascocarp. The ascocarp primordium is already surrounded by an elevation of the thallus containing oxalate crystals, a characteristic component of the thallus margin around perithecia in the Porinaceae.* The primordium itself is composed of numerous straight ascogonia ending in unbranched trichogynes and imbedded in a generative tissue of similar vertically arranged hyphae (Plate IVB). The cell rows of the ascogonia and generative tissue are very much alike, the ascogonia being recognizable by the somewhat larger basal cells. During further development the number of vertical hyphae in the centre of the primordium increases while a cortical layer is formed in the outer part (Plate IVE). Some of the vertical hyphae already seem to be true paraphyses with free tips, the others being connected with the cortical layer. The connected hyphae rupture in a distinct zone and a cavity arises separating the paraphyses from the primary covering layer. This process is best followed in *Belonia russula* (Plate VC–F). In the relatively small primordium of *Porina nucula*, the cavity may be recognized in somewhat inclined sections (Plate IVc). Here the remnants of the ruptured hyphae hang down into the cavity (Plate VF) and their elongation is very limited in the Porinaceae. In *Porina nucula* and *Belonia russula* they disappear almost completely; later on, in *Clathroporina calcarea*, they line the upper part of the wall below the pore in the adult fruit body (Plate IVF). The paraphyses elongate to fill the whole cavity of the perithecium and new paraphyses are formed in the marginal part around the hymenium. The perithecial wall is thin in *Porina nucula* but well developed in the two other species, especially in the upper part (Plate VB). The hyphae are periclinally orientated in the basal part and irregularly interwoven towards the summit. Around the pore an anticlinal arrangement of the hyphal

---

* At present there appears to be no nomenclaturally correct name for this family, for, as noted by Santesson (1952), Porinaceae Reichenb. (1828) was based on *Porina* Ach. (a synonym of *Pertusaria* DC. nom. cons.) and not *Porina* Müll. Arg. nom. cons. Santesson (1952) adopted the name Strigulaceae Fr. sensu Zahlbr. for a family including both *Porina* Müll. Arg. and *Strigula* Fr. but these two genera are not now considered to be confamilial (Henssen and Jahns, 1973; Poelt, 1974). Poelt (1974) used the name "Clathroporinaceae Vězda ad int." for this family but the name has not been validly published. In the author's view the name Porinaceae should be conserved and typified by *Porina* Müll. Arg. but its conservation has not yet been proposed.

tips is visible (Plate Vɢ) and, seen from above, these hyphae form a distinct zone around the pore (Plate Vᴀ). The pore itself is produced by a simple breaking up of the cortical layer, underneath which the radiating tips of the paraphyses may be recognized (Fig. 3). Sections of mature perithecia may open widely, simulating apothecia with lateral paraphyses (Plate IVꜰ). The asci in the

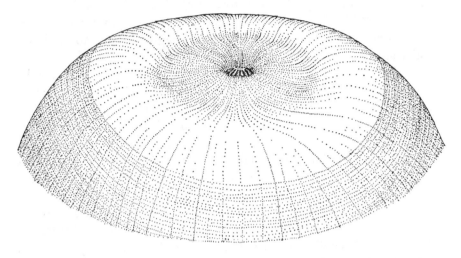

Fɪɢ. 3. *Belonia russula*; summit of a ripe perithecium shown schematically. The radiating tips of the paraphyses are seen through the apical pore (drawn by G. Vobis).

Porinaceae are unitunicate, non-amyloid and provided with a uniformly thin wall.

(*c*) *Gyalectaceae*. The apothecia of the Gyalectaceae, the only family included in the Gyalectales (Henssen and Jahns, 1973), usually have a concave or deeply urn-shaped disc with, frequently, an uneven or irregularly cut margin (Plate VI) and are biatorine or zeorine. In the latter case, the margo thallinus may be rudimentary (Plate IXᴇ and ꜰ) or lack a clear boundary towards the excipulum proprium (Plate IXᴅ).

The generative tissue arises as an outgrowth of the thallus (Plate VIIᴀ) or may be immersed in the thallus or substrate (Plate VIIIᴀ); a group of ascogonia occurs in the centre. The subsequent stages of development resemble those described in the Porinaceae: a cortical layer is formed in the outer part of the primordium, and a cavity arises by the rupture of vertical hyphae (Plates VIIʙ and ɢ, IXᴀ). Further development varies in different species. In *Dimerella* hyphae grow rapidly into the cavity from all around the limiting cortical layer

(Plate VIIc), the basal hyphae differentiate into paraphyses, the upper hyphae are the later periphysoids and additional paraphyses and periphysoids are produced when the primordium enlarges. The paraphyses elongate to form a cone-like group protruding to the cortical layer, the periphyses remaining short as long as the ascocarp is still closed (Plate VIID and E). In the mature apothecia they elongate gradually to form a continuous layer with the paraphyses (Plate VIIF). The cortical layer opens lysogenously by a median pore which expands under the pressure of the enlarging hymenium to produce a biatorine apothecium with an open disc (Plate VIB).

In *Gyalecta ulmi* the apothecium is zeorine and the reddish disc is surrounded by a white irregular margin (Plate VIA) composed of a thick excipulum proprium and a margo thallinus (Plate VIIID). In young stages the paraphyses are arranged in a cone, but, as in *Dimerella*, the periphysoids, however, remain indistinct (Plate VIIIB). When the paraphyses elongate to fill the cavity, the periphysoids form a rim of agglutinated hyphae lining the inner wall (Plate VIIIc, E and F) and in mature apothecia they are restricted to the outermost margin (Plate VIIID). The thallus enlarges simultaneously with the developing ascocarp. At the stage when the first asci are produced, first the cortical layer and then the covering thallus tissue breaks up (Plate VIIIE). The excipulum is thick and composed of irregularly interwoven hyphae (Plate VIIID).

In *Gyalecta jenensis* the excipulum and the thalline margin are closely connected without any sharp boundary (Plate IXc and D). The periphysoids remain short and imbedded in a mucilage forming a rim along the inner boundary of the excipulum against the hymenium. In *G. geoica* the hymenium and excipulum are also separated but the inner wall of the excipulum bears a rim of isodiametric cells (Plate IXE), a structure reported by Vězda (1966) to occur in *Gloeolecta*, another genus of the Gyalectaceae. In *Gyalecta foveolaris* only the upper part of the excipulum with its radiating hyphae is separated from the hymenium (Plate IXB). *Gyalecta peziza* is remarkable in having anticlinally arranged hyphae in a very thick excipulum (Plate IXF). The asci of the Gyalectaceae are unitunicate, amyloid and provided with a thin wall and the paraphyses are distinctly septate and have enlarged apical cells.

Some stages of the Porinaceae and Gyalectaceae have a striking similarity (Fig. 4). The closed primordia with elongating paraphyses look very much alike (Fig. 4A, B and D). The opening of the cortical layer takes place in the same way by the formation of a lysogenous pore (Fig. 4c and E). Only the mature fruit bodies differ, being apothecia with a freely exposed hymenium in the Gyalectaceae and perithecia with a limited opening in the Porinaceae. The enlarged upper part of the peridium in *Belonia* and *Clathroporina* resembles the apical thickening of the excipulum in *Dimerella* (Fig. 4c and f). The lack of a well defined ostiole and periphyses in the mature perithecia of the Porinaceae together with the wide opening of the sections are further similarities to the apothecia of the Gyalectaceae.

(d) *Ostropales.* The Ostropales include the two sub-orders Ostropineae and Graphidineae which differ mainly in the shape of the apothecia and the course of development (Henssen and Jahns, 1973). In the Graphidineae the ascocarps are usually hysterothecia (Plate Xi and j) but rarely roundish apothecia (Plate Xk). The Ostropineae are characterized by a great variety of ascocarp shape: here, for example, flat structures with an irregular margin (Plate Xa and e), urn-shaped apothecia with intersections of different origin (Plate Xb–d and f), apothecia simulating perithecia (Plate Xh) or entirely immersed fruit bodies with only the punctiform disc protruding above the thallus surface (Plate Xg), occur.

The ascocarp development in the Ostropaceae has been described in detail by Gilenstam (1969) for *Conotrema*. The generative tissue is a compact structure in which, according to Gilenstam, one or several ascogonia are differentiated. A cavity arises in the centre of the primordium and a cone of true paraphyses is formed pushing against the primary covering layer which is lined by downward-growing hyphae (Plate XIb). The adjacent thallus and even parts of the sub-strate contribute to the formation of the multilayered margins of the apothecia. The margin "carbonizes" and contains deposited oxalate crystals. In mature apothecia, the periphysoids line the inner wall of the excipulum which is sepa-rated by a split from the hymenium (Plate XIf).

In the Thelotremataceae, the second family of the Ostropineae, the develop-ment often proceeds in a similar way. In *Thelotrema lepadinum* (see also Letrouit-Galinou, 1966) the generative tissue containing a group of ascogonia is deeply immersed in the substrate (Plate XIa). The cortical layer surrounding the cavity of the primordium is strongly developed in the upper part where it forms a thick "roof" ("toit") in the young stages (Plate XId). Periphysoids remain short (Plate XIe) and the apothecial margin is composed of several layers: a rim of periphysoids inserted at right angles to the excipulum, the excipulum itself, and the surrounding thalline margin containing parts of the substrate and deposited oxalate crystals. The split between the excipulum and hymenium is not the only one in the apothecium of this species. A second split arises in adult fruit bodies separating a part of the thalline margin with incorporated substrate from the rest of it (Plate XIc). In some species of *Leptotrema* the split between hymenium

E

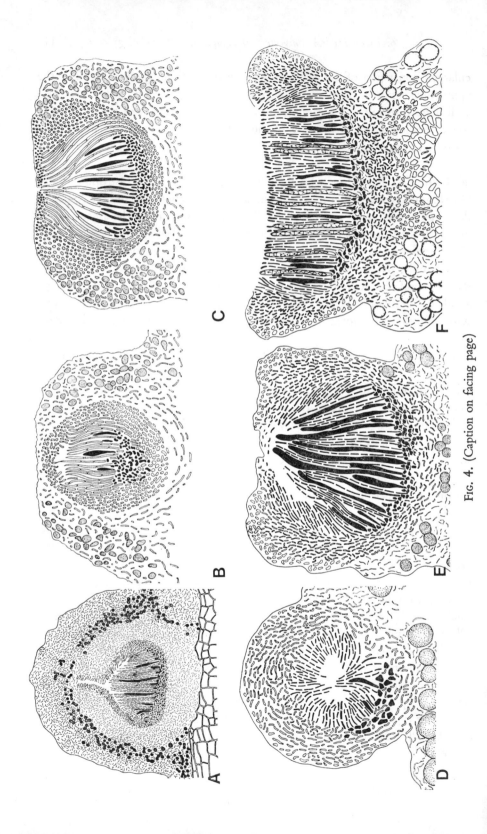

Fig. 4. (Caption on facing page)

and excipulum is still distinct but the periphysoids are strongly reduced (Henssen and Jahns, 1973, Fig. 13.40). In *Thelotrema interpositum*, the periphysoids elongate gradually to form a continuous layer with the paraphyses as seen in *Dimerella*, and a split is not developed. This variant represents a transition to the ontogeny of the second sub-order, the Graphidineae.

In the Graphidaceae, the only family of the sub-order, the ontogeny differs in many respects. The ascocarps are mostly hysterothecia which have two opposite growing points. A distinct cortical layer around the young primordium and a split between hymenium and excipulum—as far as the latter is developed—are lacking.

The course of development is best followed in species without an excipulum such as *Graphina mendax*. In this species the hysterothecia have a broad pruinose disc and a thick thalline rim (Plate XI). The generative tissue, a web of loosely interwoven hyphae, is produced within the substrate (Plate XIIA). After the formation of ascogonia the generative tissue differentiates into a basal part of vertical hyphae and an upper layer with a net-like structure (Fig. 5D). Some of the vertical hyphae seem to be true paraphyses with free tips. The connecting hyphae rupture at a certain level to form a cavity separating the basal layer of paraphyses from the upper hyphal net (Plate XIIB and C; Fig. 5E). The paraphyses elongate gradually and push against the covering layer (Plate XIID). The hysterothecium elongates more strongly at one of the two growing points (Plate XIIC, D and E in the right corner). A net of branched hyphae like the generative tissue is produced here continuously and differentiates in the same way. The ascogenous hyphae simultaneously grow horizontally along the base of the hymenium with the developing hysterothecium. The covering layer disintegrates (Plate XIIE) and the remnants cause the pruina on the surface of the disc of the hysterothecium.

In some other genera of the Graphidaceae such as *Gyrostomum*, a genus with roundish apothecia (Plate XK), no cavity is formed in the primordium. The

---

FIG. 4. Development of the ascocarp in *Gyalecta*, *Belonia* and *Dimerella*. **A**, *Gyalecta ulmi*; primordium with asci, paraphyses, periphysoids and a cortical layer which breaks up by an apical pore—the upper part of the primordium is surrounded by thallus tissue. **B** and **C**, *Belonia russula*: **B**, primordium with cavity, asci and paraphysoids embedded in a thalline verruca; **C**, mature perithecium, peridium enlarged in the upper part. **D–F**, *Dimerella diluta*: **D**, primordium with cavity, paraphyses and periphysoids; **E**, young apothecium with opening cortical layer; **F**, mature apothecium, periphysoids forming a continuous layer with the paraphyses. (**B** and **C** drawn by G. Vobis; **A**, and **D–F** from Henssen and Jahns, 1973.)

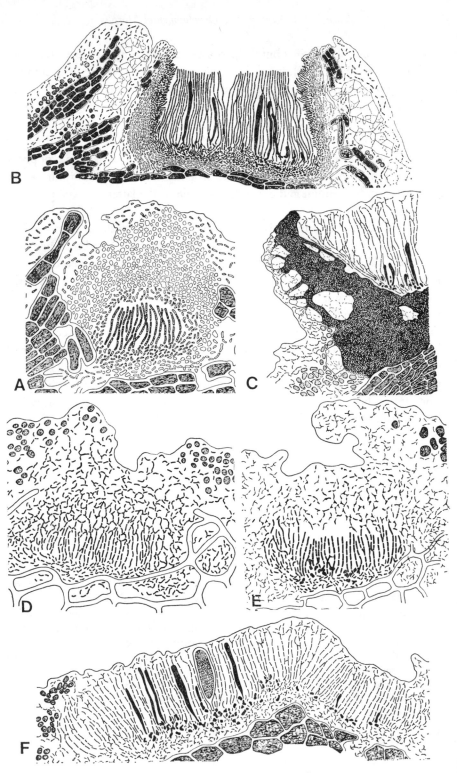

Fig. 5. (Caption on facing page)

vertically arranged hyphae remain in constant contact with the poorly developed covering layer (Plate XIIF) and in this ontogenetical variant the paraphysoids are the only type of paraphyses. The mature apothecium is surrounded by a thick margin including oxalate crystals (Plate XIIG).

According to our present knowledge, there seems to exist a considerable difference in regard to the formation of the cavity in the two sub-orders of the Ostropales: a simple separation of the hyphae comparable to the ontogeny in the Verrucariaceae in the Ostropineae, and a rupture of the hyphae as in the Porinaceae and Gyalectaceae in the Graphidineae. We have to admit, however, that the stage of hyphal rupture is not always easily seen and further studies including other members of the Ostropineae are needed to confirm these results.

In summary, the Ostropales reveal a great variation in the ontogeny of their ascocarps (Fig. 5). The Ostropineae follow the special type of development by producing a primordium with a central cavity and a surrounding cortical layer (Fig. 5A). The mature apothecium contains, typically, a split between the hymenium and the excipulum which is provided with a rim of periphysoids (Fig. 5B) and the paraphyses are true paraphyses. In the Graphidineae, the primordium is not surrounded by a cortical layer, and the cavity, in so far as one is produced, arises by the rupture of hyphae (Fig. 5D and E). The hymenium is covered by a cortical layer originating from the generative tissue which finally disintegrates (Fig. 5C and F); the paraphyses are in part or exclusively paraphysoids.

In regard to other characters, the order is remarkably uniform. The unitunicate non-amyloid asci contain spores with lens-shaped cell lumina, and reveal, frequently, a specific blue–violet stain with iodine. The occurrence of a considerable variation in ontogenetical characters combined with a great consistency in the structure of asci and spores found in the Ostropales is an important fact in regard to our discussion of the systematic position of the genera *Coccotrema* and *Lepolichen*.

FIG. 5. Ascocarp development in the Ostropales. **A** and **B**, *Thelotrema lepadinum*: **A**, primordium with cavity and a cortical layer thickened at the apex; **B**, mature apothecium with paraphyses and periphyses, the latter lining the inner wall of the excipulum. **C**, *Gyrostomum scyphuliferum*; part of a mature apothecium with paraphysoids and a dark margin containing oxalate crystals. **D–F**, *Graphina mendax*: **D**, primordium divided in a basal part of vertical hyphae and an upper part of net-like structure; **E**, primordium with cavity; **F**, mature hysterothecium with disintegrating covering layer. (**A–F** from Henssen and Jahns, 1973.)

COCCOTREMA AND *LEPOLICHEN*:

ASCOCARP ONTOGENY AND SYSTEMATIC POSITION*

Due to the unusual shape of the ascocarps there has been much misinterpretation with regard to consideration of the systematic position of the genera *Coccotrema* and *Lepolichen*. In the system and catalogue of Zahlbruckner (1926, 1921–40), for example, species belonging to the two genera have been included in three different families: *Lepolichen* in the monotypic Phyllopyreniaceae, *Coccotrema* in the Pyrenulaceae, and the genus *Perforaria* (a synonym of *Coccotrema*) in the Pertusariaceae. Two of the families, the Phyllopyreniaceae and Pyrenulaceae, were included in the Pyrenocarpeae whilst the Pertusariaceae belong to the Gymnocarpeae. This indicates that the fruit bodies have been interpreted as either perithecia or apothecia.

The ascocarps have, in fact, a very peculiar morphology. They are globose structures perforated by a pore, and in some species bear protuberances of the thalline margin (Plate XIIIA, B and D). The fruit bodies resemble perithecia like those of *Belonia russula* (Plate VA) as well as the modified apothecia of some *Pertusaria* species (Plate XIIIE) or members of the Thelotremataceae (Plate XH).

The genus *Coccotrema* is placed by modern authors in the Pertusariaceae (e.g. Oshio, 1969), a view challenged recently by Brodo (1973) in his excellent study of the North American species. Brodo was the first to detect hair-like structures lining the inner wall of the excipulum below the pore. He termed these filaments periphyses and the pore an ostiole on account of the great similarity to perithecial anatomy. In a comprehensive survey, Brodo compares the diverse characters of *Coccotrema* with the corresponding features in pyrenocarpous and

---

* *Lepolichen coccophorus* (Mont.) Trev. has been treated as a synonym of *Lepolichen granulatus* (Hook. f. & Tayl.) Müll. Arg. since the studies of Müller Argoviensis in 1889. Dr R. Santesson kindly informed the author, however, that *L. coccophorus* is the correct name of this lichen. According to his studies, the type specimen of *Porina granulata* Hook. f. & Tayl. is identical with *Coccotrema cucurbitula* (Mont.) Müll. Arg. The name would have priority over *C. cucurbitula*. However, *P. granulata* Hook. f. & Tayl. is not the same as *P. granulata* Ach. but a later homonym. Dr Santesson has in preparation a complete schedule of the synonyms of *Coccotrema* and *Lepolichen* species. The synonymy given below has been prepared with the help of his list (see also Brodo, 1973).

Lepolichen Trev., *Spighe e Paglie* 5, (1853); type species: *L. coccophorus* (Mont.) Trev., *op. cit.*: 5 (1853) [synonyms: *Parmelia coccophora* Mont., *Ann. Sci. nat., Bot., sér.* 3, **18**, 309 (1852); *Pertusaria thamnoplaca* Tuck., *Proc. Am. Acad. Arts Sci.* **12**, 175 (1877)].

Coccotrema Müll. Arg., *Mission Scient. Cap Horn, Bot.* **5**, 171 (1889); type species: *C. antarcticum* Müll. Arg., *op. cit.*, 171 (1889) — *C. cucurbitula* (Mont.) Müll. Arg., *Nuovo Giorn. bot. Ital.* **21**, 51 (1889) [synonyms: *Pertusaria cucurbitula* Mont., *in* Gay, *Hist Fisic. Politic. Chile, Bot.* **8**, 200 (1852) et *Ann. Sci. nat., Bot., sér.* 3, **18**, 312 (1852); *Porina granulata* Hook. f. & Tayl., *Fl. Antarct.* **1**, 200 (1844), non Ach., *Syn. Lich.*, 112 (1814)].

discocarpous lichen families but stresses the necessity of ontogenetical studies before a final decision can be made in regard to the systematic position of the genus. According to Brodo, *Coccotrema* has the following features in common with the Pertusariaceae: sunken ascocarps, bivalve dehiscence of the asci, structure of the spores, and the occurrence of β-depsidones. Deviating characters are: the periphyses, distinct excipulum, ascocarps depressed apically and perforated by an ostiole, and the lack of isolichenin in the hymenium.

The monotypic genus *Lepolichen* has not been studied recently. *Lepolichen* is said to differ from *Coccotrema* in its leafy heteromerous thallus. On account of this character, the family Phyllopyreniaceae was erected for it by Zahlbruckner (1926).

Seen from above, the thallus of *Lepolichen* resembles the crustaceous thallus of the Pertusariaceae, e.g. *Pertusaria oculata* (Plate XIIIc and D) and is hard to distinguish from certain *Coccotrema* species like *C. porinopsis*. If turned upside down, horizontally extended narrow lobes are visible which bear "isidia" on the upper surface and whitish rhizines on the lower. In the present investigation we found that lobes contained neither algae nor stages of apothecia or pycnidia. They seem to be merely a kind of prothallus whilst the isidia-like nodules are the true thallus including the principal phycobiont, the cephalodial alga, the fruit bodies and pycnidia. Furthermore, we came to the conclusion that the thallus of *Coccotrema* species is also heteromerous and that the cortex may be even more highly differentiated than in *Lepolichen*. Thus, the limitation of the genera *Coccotrema* and *Lepolichen* is not as clear as has been postulated previously but requires a new consideration.

We have studied the ascocarp development in the Pertusariaceae, *Lepolichen* and *Coccotrema*. The ontogeny in *Lepolichen* and *Coccotrema* follows, on the whole, the special type of development described above. There are, however, some features corresponding to the ontogeny in the Pertusariaceae. We consider the fruit bodies of *Coccotrema* and *Lepolichen* species to be apothecia and therefore use the term periphysoids for the hair-like structures below the pore. Here we restrict ourselves to a short description of the ontogeny in *Coccotrema cucurbitula* and *Lepolichen coccophorus* as a more comprehensive report on the morphology and ontogeny of these genera including other species will appear in a later publication.

We include the Pertusariaceae together with the Trapeliaceae in the suborder Pertusariineae of the Lecanorales (Henssen and Jahns, 1973). In the Pertusariaceae, the genera *Melanaria*, *Ochrolechia*, *Placopsis* and *Varicellaria* are included in addition to the type genus of the family. *Placopsis* is characterized by the presence of cephalodia.

1. Coccotrema cucurbitula

The thallus of *Coccotrema cucurbitula* consists of an upper cortex and a lower part in which the algal cells are uniformly distributed. The ascogonia are differentiated directly by the medullary hyphae and surrounded secondarily by the adjacent thallus hyphae which form the generative tissue as in *Parmeliella plumbea*. The ascogonia are produced in groups between or below the algal cells (Plate XIVA) and the generative tissue is a globose structure with periclinally arranged hyphae in the marginal part, while in the centre, the hyphae form a strand around the ascogenous hyphae bearing the ascus initials (Plate XIVc). The vertical hyphae rupture at a certain level and a central cavity is formed (Plate XIVD), the upper ends of the ruptured hyphae elongating into the cavity. The asci are densely aggregated and often arranged in a concave layer (Plate XIVE and F). By the enlargement of the ascocarp, the algal cells are pushed away but the primordium still remains covered by the cortex of the thallus (Fig. 6B). The hyphae in the upper part of the cortical layer of the primordium grow vigorously to form a short papilla and, finally, the covering thallus cortex breaks away and a pore arises within the papilla, partly by the disintegration and partly by disaggregation of the hyphae. Branched periphysoids are inserted below the pore, reaching downwards almost to the concave layer of the hypothecium (Plate XIVB; Fig. 6c). The paraphyses elongate and fill the cavity in mature apothecia. Some of the paraphyses are branched or connected by anastomoses. The apothecium is zeorine since a thick thallus margin surrounds the excipulum.

2. Lepolichen coccophorus

The first stages of development in *Lepolichen* resemble those of *Coccotrema cucurbitula*: the ascogonia are produced in groups, enveloped by the hyphae of the generative tissue, and a globose primordium is formed with a central cavity lined by radiating hyphae. The further course of development deviates in some features. The papilla is very conspicuous and the periphysoids disappear in the maturing apothecium. The excipulum is less sharply delimited towards the

FIG. 6. Ascocarp development in *Coccotrema* and *Pertusaria*. A–C, *Coccotrema cucurbitula*: A, primordium with ascus initials; B, primordium containing a cavity, paraphysoids and periphysoids; C, young mature apothecium with apical pore and periphysoids. D–F, *Pertusaria oculata*: D, generative tissue with groups of ascogenous hyphae; E, primordium with paraphysoids covered by the thallus cortex; F, part of a mature apothecium with primitive excipulum and epithecium. (A–C drawn by H. Born; D–F from Henssen and Jahns, 1973.)

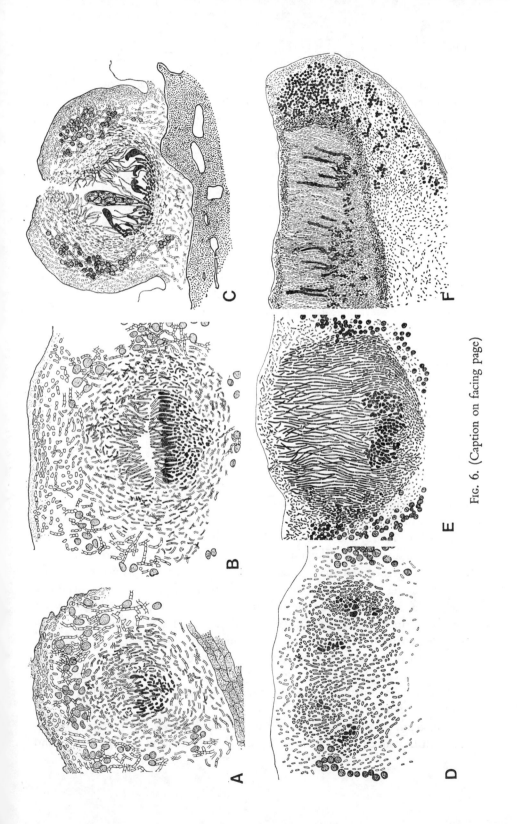

Fig. 6. (Caption on facing page)

thallus tissue, especially in the upper part, and a broad zone of interwoven medullary and excipular hyphae is seen (Plate XVD). The upper part of the cortical layer of the primordium is covered by the thallus cortex as in *C. cucurbitula* and a cone-like structure composed of a hyphal net precedes the formation of the large papilla. At this stage the branched periphysoids are still visible. The cone-like structure at the apex of the primordium develops into the conspicuous papilla which resembles the neck of a perithecium. In the centre a pore arises by the splitting and disintegration of the net-like tissue in a basipetal direction (Plate XVE). The outer delimitation of the papilla is developed from the thallus cortex. Paraphyses elongate and fill the cavity whilst the periphysoids dissolve. The thalline margin of the apothecium grows out and in places produces horn-like protuberances. The constant contact of the upper part of the excipulum and the covering thallus cortex found in *Coccotrema* and *Lepolichen* is a characteristic feature known from the ontogeny of the ascocarp in the Pertusariaceae.

3. Pertusariaceae

The apothecial ontogeny in *Pertusaria pertusa* has been studied by several authors (e.g. Letrouit-Galinou, 1966). *P. pertusa* belongs to the subgenus *Pertusaria* with more or less globose fruit bodies and punctiform discs (Plate XIIIE). The globose hymenium containing the paraphyses imbedded in the hymenial mucilage resembles that of *Lepolichen coccophorus*. It is also covered by the thallus cortex but no papilla perforated by a pore is present. The details of development are best seen in species of the subgenus *Lecanorastrum*, such as *P. oculata*, in which the apothecia have a broad disc (Plate XIIIc). The generative tissue usually contains several groups of ascogonia ending in protruding trichogynes (Plate XVIA) and is produced within the algal layer, remaining covered by the thallus cortex. Like the ascogonia, the ascogenous hyphae are also grouped together and may be scattered throughout the primordium (Plate XVIB). In the outer part of the primordium the hyphae form a primitive excipulum and in the centre they differentiate into paraphysoids (Plate XVIc). Asci are formed at an early stage (Plate XVA). In the further course of development, new paraphysoids arise at the marginal part of the hymenium and these are branched and anastomosing with tips remaining attached to the covering thallus cortex (Plate XVID). In this way the tips of the paraphyses intermingle with the covering thallus cortex to form an epithecium which is easily visible in adult apothecia (Plate XVIF).

In *Placopsis subgelida* the primordium resembles even more the corresponding stages in *Lepolichen* and *Coccotrema* by the development of a cortical layer and vertically arranged central hyphae (Plate XVIE). A rupture of the vertical hyphae followed by the formation of a cavity does not, however, take place in this species.

4. Discussion

Two points require discussion: (*a*) the nature of the fruit bodies in *Coccotrema* and *Lepolichen* and (*b*) the systematic position of these genera.

(*a*) *Nature of the fruit bodies.* The papilla of *Lepolichen* resembles the neck of perithecia (e.g. in *Endocarpon*) in regard to their shape but the internal structure deviates considerably, however. In the Verrucariaceae, the ostiole is a well defined narrow channel lined with periphyses whilst in *Lepolichen* the papillar pore is an irregular breaking-up of the gelatinous inner tissue containing no filaments. Apart from the much larger size, the pore in *Lepolichen* and *Coccotrema* has more similarity to the pore of the Porinaceae.

The ostiole of a perithecium is defined, commonly, as a schizogenous cavity lined by periphyses and ending in a pore (Ainsworth, 1971). According to this definition, neither the opening of *Coccotrema* and *Lepolichen* nor the pore of the Porinaceae would be an ostiole in the strict sense. On the other hand, modified apothecia with a punctiform disc or with a pore are known from several discocarpous lichen families like the Pertusariaceae and Thelotremataceae, respectively. The typification of such ascocarps has caused no difficulties since related species with normal apothecia are known.

In view of these data, the fruit bodies of *Coccotrema* and *Lepolichen* might be called apothecia with more certainty if closely related genera having apothecia were known. A close relationship exists, in our opinion, to the discocarpous Pertusariaceae.

(*b*) *Systematic position.* At first sight, the ontogeny in *Coccotrema* and *Lepolichen* seems to be quite different from the development in the Pertusariaceae (Fig. 6). The formation of a central cavity in the primordium, the production of periphyses, and the development of a pore are the most deviant features. Corresponding features are, on the other hand, the retained contact of the thallus cortex and the excipulum as well as the occurrence of paraphysoids being, obviously, the only type of paraphyses.

Apart from the differing ontogeny, a considerable similarity exists in regard to morphological characters. The structures of the asci and spores are strikingly alike. The mature asci have a thick wall with a dome-like apex (Plate XVB and C) and split into two valvae in the ejaculation of the spores (Plate XVF and G). A small amount of cytoplasm is ejected together with the spores and the thick walls of the asci are seen distinctly after dehiscence. The spores are one-celled, thick-walled and large in all species of *Coccotrema* and *Lepolichen* as they are also in most genera of the Pertusariaceae. *Coccotrema* and *Lepolichen* resemble the Pertusariaceae in regard to the external morphology of the thallus. The correspondence is so pronounced that one of the species, *C. pocillaria*, has even been described as a species of *Ochrolechia* (i.e. *O. pacifica*, see Brodo, 1973). *Coccotrema* and *Lepolichen* have the presence of cephalodia in common with *Placopsis*.

The correspondence in the content of lichen substances has been mentioned above (p. 123). Brodo (1973) stated that isolichenin was absent. An amyloid reaction of the ascus tip may, however, be demonstrated after treatment with KOH-solution.

The considerable morphological similarity of the genera *Coccotrema* and *Lepolichen* to the Pertusariaceae indicates a close relationship to this family. The decisive criteria, in our opinion, are the structure of the asci and spores. This has already been done in the delimitation of the order Ostropales (p. 177), a group with a convergent combination of features—a corresponding morphology and a diverging ontogeny. In the Ostropales transitional forms link the variations in ascocarp ontogeny but although such variants are not yet known to occur in the Pertusariaceae or in the *Coccotrema–Lepolichen* group they might be expected. At the present time the inclusion of these two genera in the suborder Pertusariineae appears to be fully justified. They do, however, differ sufficiently from the Pertusariaceae and Trapeliaceae to be kept in a separate family. We propose, *ad interim*, to include them in a new family, the Coccotremataceae.*

NORMANDINA AND ITS ALLEGED FRUIT BODIES: THE LICHENICOLOUS FUNGUS *SPHAERULINA CHLOROCOCCA*†

Normandina pulchella is widely distributed in the northern and southern hemispheres. Its small squamules are usually found without any traces of fruit bodies or pycnidia. Rarely, the thallus supports densely aggregated black perithecia which have been taken for the fruit bodies of the lichen. The name *Verrucaria*

---

* It should be noted that although the family name Phyllopyreniaceae Zahlbr. (*Nat. Pflanzenfam.* **1** (1*), 68, 1903) was based on *Lepolichen* Trev., this name cannot be taken up for *Coccotrema* and *Lepolichen* as it was not validly published (Arts. 18, 32). The name Phyllopyreniaceae might eventually have to be validated and taken up for the genus *Phyllopyrenia* Dodge, first described in 1948, but that genus is of very uncertain systematic position and we have not yet been able to examine material of it.

† The original descriptions of *Normandina pulchella* and its later synonyms have been based on the lichen thallus together with the lichenicolous fungus. In addition to earlier synonyms (Zahlbruckner, 1921–40) Swinscow (1963) has reported two additional synonyms, *Verrucaria chlorococca* Leight. (placed in *Arthopyrenia* by Smith, 1911) and *Polyblastia armericola*

W. Wats. in the type specimens of which the lichen thallus is deformed and abundantly sorediate. These names can be considered as based on discordant elements and rejected under Art. 70 unless one element can be considered to be a satisfactory type. Dr R. Santesson (*in litt.*) considers the fungus to be a species of *Sphaerulina* (a somewhat hetereogeneous assembly requiring a monographic revision), and suggests the application of the new combination *S. chlorococca* to it, typified by the perithecia embedded in Leighton's type specimen. Similarly, the name *Normandina pulchella* should be typified by material without the lichenicolous fungus. In the original description of *Verrucaria pulchella* Borrer refers to collections made "on mossy trees in Sussex" and "on a mountain near Bantry, growing on *Lichen plumbeus* on stems of heath" collected by Miss Hutchins. Dr D. L. Hawksworth (*in litt.*) reports that both these collections are preserved in Borrer's herbarium in K and an annotation slip indicates that these were used by Sowerby in the preparation of the drawings which accompanied the original description of this species; comparison of the specimens with the published illustrations shows that both were in fact illustrated. The Sussex material devoid of the *Sphaerulina* and accompanied by *Frullania* is illustrated without the *Sphaerulina*. Borrer's name is consequently lectotypified by the Sussex specimen here to preserve current usage of his epithet in accordance with Rec. 7B of the Code.

Normandina pulchella (Borr.) Nyl., *Ann. Sci. nat., Bot., sér.* 4, **15**, 382 (1861) (basionym: *Verrucaria pulchella* Borr., *in* Smith & Sowerby, *Engl. Bot., tab.* 2602, Fig. 1 (1829); type: England, Sussex, on beech, *Borrer*, K–Borrer (left-hand specimen)—lectotype).

Thallus squamulose, glaucous to pale grey or greenish-grey. Squamules shell-like with raised margins, 1–2 mm diam, scattered or aggregated to form colonies. Soredia often present on surface and margins. Thallus heteromerous, upper cortex pseudoparenchymatous or evanescent, distinct lower cortex absent. Medullary hyphae mostly with globose or ellipsoid cells arranged in a net-like structure and surrounding groups of algal cells. Phycobiont *Trebouxia*. Fruit bodies and pycnidia unknown. Epiphytic on other lichens or mosses in localities with somewhat oceanic conditions, widely distributed.

Sphaerulina chlorococca (Leight.) R. Sant. **comb. nov.** (basionym: *Verrucaria chlorococca* Leight., *Lich. Fl. Br.*, ed. 3, 484 (1879); type: Buckinghamshire, Chiltern Hills, Stokenchurch, (on thallus of *Normandina pulchella*), *Larbalestier*, BM—lectotype).

Perithecia immersed in the host thallus, *c.* 0·5 mm broad, with a black wall. Asci distinctly bitunicate, thickened apically, the walls easily bursting, 8-spored. Ascospores usually 7-septate, finally brown (Plate XVIIIE), 25–35(–45) × 6–9 μm, clinging together after ejection. Periphyses well developed, paraphyses lacking. On the thallus of *Normandina pulchella*, widely distributed like the host but only rarely collected.

The ascus type is difficult to recognize on account of the very easily bursting walls. Mature asci with undestroyed walls were obtained only in gently squashed preparations and seem to have a similar structure to the bitunicate ascus in *Dermatocarpon* (Vězda, 1968; Henssen and Jahns, 1973). The groups of eight spores clinging together after ejection are a characteristic feature of this fungus (Swinscow, 1963).

Infested thalli of *Normandina pulchella* resemble fertile thalli of *Dermatocarpon* or *Endocarpon* species. The protruding ostioles of the fungus are visible as dark dots on the thallus surface and are often surrounded by thalline verrucae (Plate XVIID and E). Seen from below the globose perithecia appear to have a clearly parasitic origin (Plate XVIIA). In later stages, the thallus is more or less deformed or might be almost replaced by the abundant production of soredia. The type specimen of *Polyblastia armericola* is on such a highly dissolved thallus (Swinscow, 1963). The deformation and the dissolution of the host thallus suggests a parasitic rather than a parasymbiotic relationship of *S. chlorococca* to *N. pulchella*.

pulchella was first introduced by Borrer, in 1831, and later descriptions of this "lichen" are based on such "fertile" stages. Apart from the structure of the spores, the mature perithecia resemble the ascocarps of *Verrucaria* species. The genus *Normandina* has consequently usually been placed near the genera *Dermatocarpon* and *Endocarpon*.

In regard to the external and internal morphology of the thallus, *Normandina pulchella* corresponds most closely to *Coriscium viride*, a lichen currently considered to be the imperfect stage of the basidiomycete *Omphalina hudsoniana* (Oberwinkler, 1970). The striking similarity between *Normandina* and *Coriscium* led us to suggest the hypothesis that the described fruit bodies of *Normandina* might be the ascocarps of a lichenicolous fungus and that the lichen itself might also be the imperfect stage of a basidiomycete (Henssen and Jahns, 1973).

The first part of the hypothesis, the parasitic nature of the fruit bodies, was demonstrated in the recent investigations presented here and these confirm unpublished results obtained by Dr R. Santesson. The second part of the hypothesis, however, proved to be erroneous; we studied the hyphal septa with the aid of electron microscopy and found them to contain only simple pores so that *Normandina* is unlikely to be a basidiomycete lichen. Simultaneously performed studies on *Coriscium* surprisingly revealed that also in this lichen only simple and no dolipore septa occur.

1. *Developmental Morphology in* Sphaerulina chlorococca

The infection of the *Normandina* thallus takes place, obviously, through the lower side of the thallus. Frequently the slightly brownish hyphae of the *Sphaerulina* can be seen growing along the lower surface of the thallus penetrating and orientating upwards to form the ascocarp initials below the algal layer (Plate XIXB). The initial stage is a dense aggregation of isodiametric cells

---

FIG. 7. Ascocarp development in ascolocular fungi. **A–E**, stages of development in the disintegration type: **A**, perithecium with pseudoparaphyses and apical meristematic plug (*Pleospora herbarum*); **B**, initial stage for the development in the three lines indicated by the arrows: a primordium with disintegrating centre containing ascogonial cells; **C**, stage with radiating central hyphae; **D**, mature perithecium with a schizogenous ostiole and periphyses (species of the "*Pleospora collematum*" group); **E**, primordium with ascus fascicle and a cavity filled by single cells (*Mycosphaerella tassiana*). **F–H**, *Sphaerulina chlorococca*: **F**, initial stage of aggregated cells; **G**, primordium with ascogonial cells, ascogenous hyphae and a tissue composed of vertical hyphae; **H**, mature perithecium with a well-defined ostiole lined by periphyses. (**A–E** from Henssen and Jahns, 1973; **F–H** drawn by G. Vobis.)

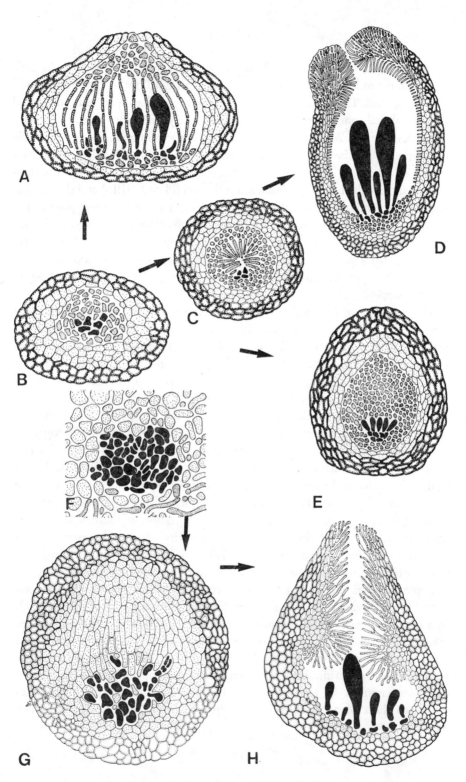

FIG. 7. (Caption on facing page)

forming an almost globose pseudoparenchyma (Plate XVIIIA; Fig. 7F). A dark peridium is produced in the outer part of the primordium by "carbonization" of the cell walls. The centre is divided into two parts, a basal part with imbedded ascogonial cells and an upper tissue with closely connected vertical hyphae (Plate XVIIIB; Fig. 7G). In the elongated apex of the primordium a meristematic plug arises when the first ascal initials are produced (Plate XVIIIc). By subsequent enlargement of the ascocarp the tissue of vertical hyphae becomes separated from the basal layer containing the asci (Plate XVIIID) and the vertical hyphae differentiate to form the first periphyses. The meristematic plug in the summit of the ascocarp gives rise to the ostiole which opens schizogenously and is filled by periphyses inserted gradually in an acropetal direction (Plate XVIIIE and F; Fig. 7H). Apart from the black peridium and the deviating structure of the spores, such stages very much resemble the mature perithecia of *Dermatocarpon* and *Endocarpon*.

The ontogeny of *Sphaerulina chlorococca* corresponds more to the ascolocular than to the ascohymenial group although no species is known to develop precisely in the same way. Henssen and Jahns (1973) referred to two main developmental types amongst the ascolocular fungi—the disintegration type ("*Zerfalltyp*") and the aggregation type. The first stages of development of *Sphaerulina chlorococca* resemble the aggregation type, but later on the correspondence is more to the disintegration type. In the latter type, a well defined ostiole lined with paraphyses may occur. For comparison, some photographs and drawings of the distintegration type are included (Plate XIX; Fig. 7). The development starts with a globose primordium of isodiametric cells. In the centre, the walls of the cells disintegrate (Plate XIXD) and ascogonial cells are differentiated (Fig. 7B). The further development follows different lines. In *Mycosphaerella tassiana*, for example, the disintegration of the cell walls continues and the mass of single cells fills the cavity around the ascus fascicle attached to the base of the cavity (Fig. 7E). This stage somewhat resembles a young primordium of *Sphaerulina chlorococca* (Fig. 7G). In a second line, for example *Pleospora herbarum*, pseudoparaphyses are developed which are connected to the base as well as to the meristematic plug of the ascocarp (Fig. 7A). In a third line, represented in members of the "*Pleospora collematum*" group, the central cells in the upper part of the young primordium stretch and form a radiating structure (Fig. 7c; Plate XIXF). These cells are the first periphyses of the perithecium. The mature ascocarp opens either lysigenously or schizogenously. In the latter case the ostiole is lined with periphyses similar to those seen in the ostiole of *Sphaerulina chlorococca* (Fig. 7D; Plate XIXE).

PLATE I

Apothecium development in *Collema*. **A** and **B**, *Collema subnigrescens*: **A**, group of ascogonia, the trichogynes protruding from the thallus surface (× 400); **B**, budding stage (arrows): hyphae developing from the ascogonia and their stalk cells (× 680). **C**, *C. multipartitum* primordium of densely interwoven hyphae with nests of ascogonia and the stalk (arrow) (× 520). **D–F**, *C. subnigrescens*: **D**, primordium with paraphysoids, young asci and anchoring hyphae growing out from the excipulum (× 400); **E**, later stage with a better developed excipulum (× 360); **F**, part of an adult apothecium with numerous anchoring hyphae (× 220). (**A** reproduced from Henssen and Jahns, 1973.)

PLATE II

Development of the perithecium in *Pyrenula nitidella*. **A**, Habit of thallus and perithecia (× 30). **B**, Primordium with paraphysoids, ascogonia and a differentiating peridium including oxalate crystals (× 400). **C**, Later stage with paraphyses and differentiating ostiole (× 325). **D**, Young mature perithecium, the ostiole lined with periphyses (× 325). **E**, Upper part of the same perithecium (× 520). **F**, Adult perithecium with thick black wall (× 260).

PLATE III

Perithecium development in *Dermatocarpon* and *Endocarpon*. **A**, *Dermatocarpon miniatum*, generative tissue with ascogonia (× 450). **B** and **C**, *Dermatocarpon rivulorum*: **B**, primordium with ascogenous hyphae and periphyses elongating into the central cavity (× 450); **C**, primordium with young ascus (× 450). **D**, *Dermatocarpon miniatum*, ostiole of a mature young perithecium (× 400). **E** and **F**, *Endocarpon pulvinatum*: **E**, adult perithecium with elongated ostiole (× 130); **F**, basal part of a perithecium with asci, periphyses and hymenial algae (× 200). (**F** after Henssen and Jahns, 1973).

PLATE IV

Perithecial development in the Porinaceae. **A–D**, *Porina nucula*: **A**, habit of thallus and perithecia (× 9); **B**, generative tissue containing numerous ascogonia which bear unbranched trichogynes (arrow) (× 730); **C**, primordium with cavity (inclined section, × 390); **D**, mature perithecium surrounded by a thalline margin in which oxalate crystals are deposited (× 260). **E** and **F**, *Clathroporina calcarea*: **E**, primordium with vertical hyphae and a thick cortical layer (× 325); **F**, part of a wide-open perithecium with periphyses (arrow) simulating an apothecium (× 260).

PLATE V

Perithecium development in *Belonia russula*. **A**, Perithecia seen from above (× 40). **B**, thalline verruca with two mature perithecia (× 80). **C**, Primordium with developing cavity, ascogenous hyphae (arrow) and paraphyses in a palisade-like arrangement (× 500). **D**, Upper part of the same section in higher magnification to demonstrate the rupture of the hyphae (× 1300). **E**, Primordium in a thallus verruca with cavity and periphyses (× 160). **F**, the upper part in higher magnification (× 500). **G**, Mature perithecium with paraphyses and asci (× 260).

PLATE VI

Apothecia in the Gyalectaceae. **A**, *Gyalecta ulmi* (× 30). **B**, *Dimerella diluta* (× 30). **C**, *Petractis clausa* (× 18). **D**, *G. jenensis* (× 8). (**C** reproduced from Henssen and Jahns, 1973.)

PLATE VII

Apothecium development in the Gyalectaceae. **A**, *Gyalecta foveolaris*, generative tissue forming a thallus outgrowth (× 520). **B–F**, *Dimerella diluta*: **B**, primordium with cavity (× 450); **C**, primordium with cavity, periphysoids and paraphyses (× 780); **D**, rupture of the cortical layer (× 520); **E**, young apothecium with basal and lateral paraphyses (× 520); **F**, mature apothecium, the periphysoids forming a continuous layer with the paraphyses (× 260). **G**, *Dimerella lutea*, primordium with ascogenous and vertically arranged hyphae surrounded by the cortical layer (× 520).

PLATE VIII

Development of the apothecium in *Gyalecta ulmi*. **A**, Generative tissue imbedded in the substrate (× 520). **B**, primordium with cavity and cone of paraphyses (× 520). **C**, Elongating paraphyses (× 260). **D**, Mature young zeorine apothecium (× 130). **E**, Primordium with rupturing cortical layer (× 200). **F**, closed primordium, some of the paraphyses still connected to the cortical layer (× 260).

PLATE IX

Apothecium development in the Gyalectaceae. **A**, *Gyalecta peziza*, primordium showing the rupturing of the central vertical hyphae (× 260). **B**, *G. foveolaris*, part of a mature apothecium (× 260). **C** and **D**, *G. jenensis*: **C**, breaking up of the cortical layer of the primordium (× 200); **D**, part of a mature apothecium, the excipulum lined with a rim of periphysoids (× 200). **E**, *G. geoica*, part of a mature apothecium, the excipulum with a rim of isodiametric cells (× 520). **F**, *G. peziza*, median part of an adult apothecium demonstrating the broad excipulum (× 200).

PLATE X

Ascocarps in the Ostropales. **A**, *Ocellularia dilatata* (× 8). **B** and **C**, *Thelotrema lepadinum* (× 20, × 10). **D**, *Diploschistes ochrophanes* (× 13). **E**, *Leptotrema santense* (× 7). **F**, *L. bahianum* (× 20). **G**, *L. wightii* (× 3). **H**, *Ocellularia americana*, apothecia simulating perithecia (× 10). **I**, *Graphina mendax* (× 15). **J**, *Graphis afzelii* (× 2). **K**, *Gyrostomum scyphuliferum* (× 17). (**A** and **I** reproduced from Henssen and Jahns, 1973.)

PLATE XI

Development of the apothecia in the Ostropineae. **A**, *Thelotrema lepadinum*, generative tissue with ascogonia developed in the substrate (× 325). **B**, *Conotrema urceolatum*, primordium with paraphyses and periphysoids (× 325). **C–E**, *Thelotrema lepadinum*: **C**, mature apothecium with two splits and periphysoids lining the inner wall of the excipulum (× 200); **D**, primordium with cavity, paraphyses and a thick cortical layer (× 325); **E**, primordium with paraphyses and periphysoids (inclined section, × 260). **F**, *Conotrema urceolatum*, mature apothecium, the excipulum with a rim of periphysoids × 200).

PLATE XII

Ascocarp development in the Graphidineae. **A–E**, *Graphina mendax* (longitudinal sections): **A**, generative tissue developed in the substrate (× 260); **B**, primordium with a differentiating cavity (× 325); **C**, later stage demonstrating the rupture of hyphae (× 520); **D**, elongating paraphyses (× 325); **E**, mature hysterothecium with an older (left) and younger part, the remnants of the covering layer seen upon the hymenium (× 160). **F** and **G**, *Gyrostomum scyphuliferum*: **F**, primordium with paraphysoids and dark covering layer ($\times$ 520); **G**, part of a mature apothecium with paraphysoids and black margin containing oxalate crystals ($\times$ 260).

Ascocarps in the Pertusariineae. **A** and **B**, *Coccotrema cucurbitula*, globose fruit bodies perforated by a pore ($\times$ 40 and $\times$ 8). **C**, *Pertusaria oculata*, lecanorine apothecium with broad disc ($\times$ 9). **D**, *Lepolichen coccophorus*, globose fruit bodies with thalline protuberances and perforated by a pore (arrows), emptied apothecia with wide openings ($\times$ 16). **E**, *P. pertusa*, apothecia with punctiform disc immersed in thalline verrucae ($\times$ 6).

PLATE XIV

Apothecium development in *Coccotrema cucurbitula*. **A**, Group of ascogonia below thallus cortex ($\times$ 650). **B**, Young zeorine apothecium with paraphyses, periphysoids and ostiole ($\times$ 260). **C**, Primordium with vertically arranged hyphae in the centre surrounding the ascogenous hyphae ($\times$ 520). **D**, Formation of the cavity ($\times$ 520). **E**, Primordium with paraphysoids, periphysoids and young asci in a concave layer, the cortical layer still completely closed ($\times$ 260). **F**, A similar stage ($\times$ 520).

PLATE XV

Pertusariineae. **A**, *Pertusaria oculata*, developing ascus (part of the section in Plate XVIB, $\times$ 57). **B**, *Coccotrema cucurbitula*, ascus tip ($\times$ 360). **C**, *P. pertusa*, young and mature ascus ($\times$ 240); **D** and **E**, *Lepolichen coccophorus*: **D**, young apothecium with differentiating papilla and periphysoids (arrow) ($\times$ 120); **E**, mature apothecium with papilla and thalline protuberances (the arrow indicates the thallus cortex covering the papilla) ($\times$ 70). **F**, *P. pertusa*, ascus tip after ejaculation of the spores ($\times$ 600). **G**, *Coccotrema cucurbitula*, ascus tip after the ejection of the spores ($\times$ 700). (**A**, **C** and **F** reproduced from Henssen and Jahns, 1973.)

PLATE XVI

Apothecium development in the Pertusariaceae. **A–D**, *Pertusaria oculata*: **A**, generative tissue with numerous ascogonia and two protruding trichogynes ($\times$ 400); **B**, groups of ascogenous hyphae scattered throughout the primordium ($\times$ 100); **C**, primordium with paraphysoids ($\times$ 200); **D**, part of a young apothecium with excipulum, paraphysoids and epithecium ($\times$ 200). **E**, *Placopsis subgelida*, primordium with cortical layer and vertically arranged central hyphae ($\times$ 200). **F**, *Pertusaria oculata*, margin of an adult apothecium ($\times$ 130).

PLATE XVII

A, *Sphaerulina chlorococca* on *Normandina* seen from below ($\times$ 40). **B**, *Cora pavonia*, upper surface of the thallus ($\times$ 3·5). **C**, *Coriscium viride*, thallus squamules ($\times$ 10). **D** and **E**, Thallus of *Normandina pulchella* infested with *Sphaerulina chlorococca* ($\times$ 8 and $\times$ 10). (**B** reproduced from Henssen and Jahns, 1973.)

PLATE XVIII

Development of the ascocarp in *Sphaerulina chlorococca*. **A**, Primordium with isodiametric cells ($\times$ 2000). **B**, Primordium with dark wall, the centrum divided into an upper tissue of vertical hyphae and a basal part containing the ascogenous hyphae ($\times$ 500). **C**, Primordium with meristematic plug ($\times$ 400). **D**, Primordium with cavity and meristematic plug ($\times$ 260). **E**, Mature perithecium ($\times$ 200). **F**, Ostiole with periphyses ($\times$ 400).

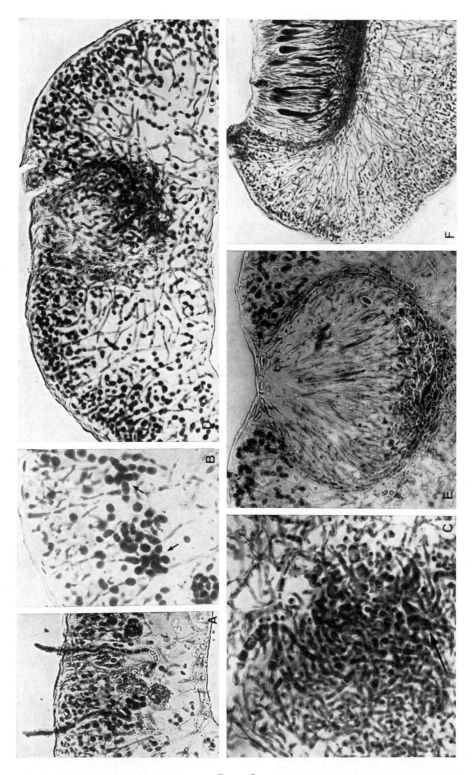

PLATE I

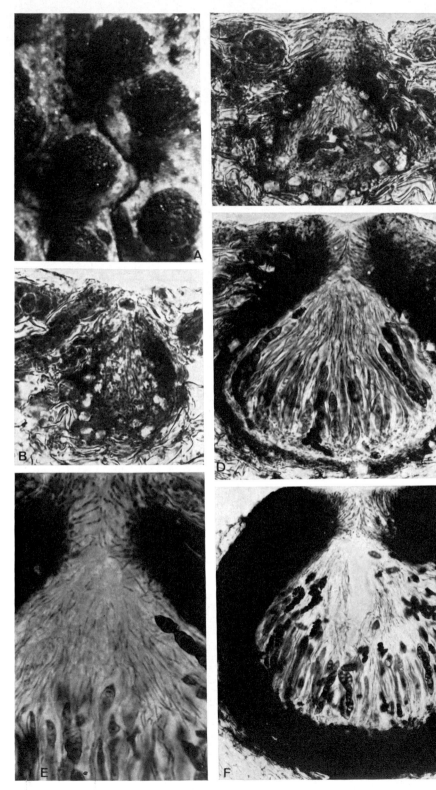

PLATE II

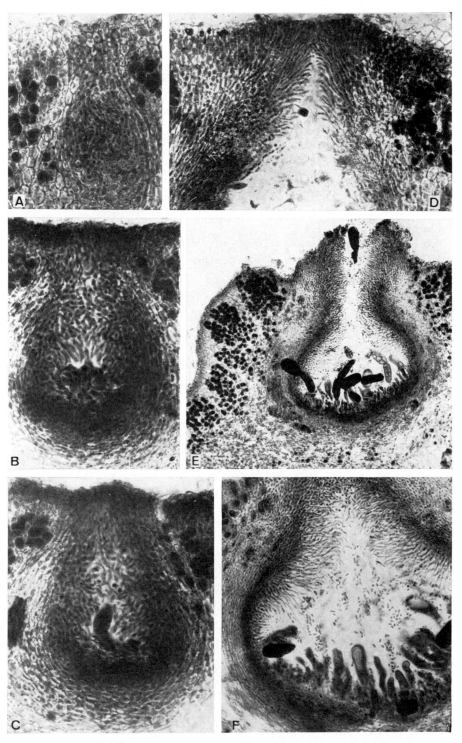

PLATE III

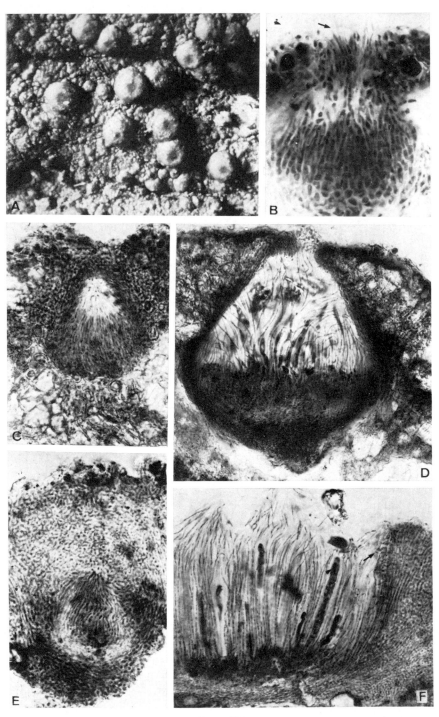

PLATE IV

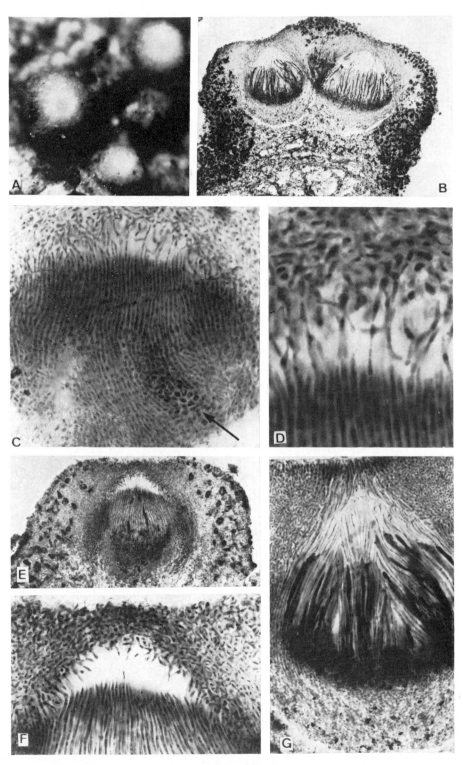

PLATE V

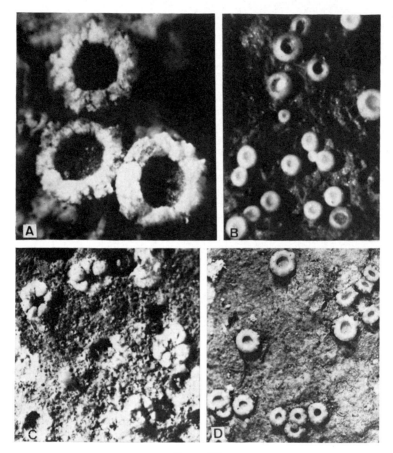

PLATE VI

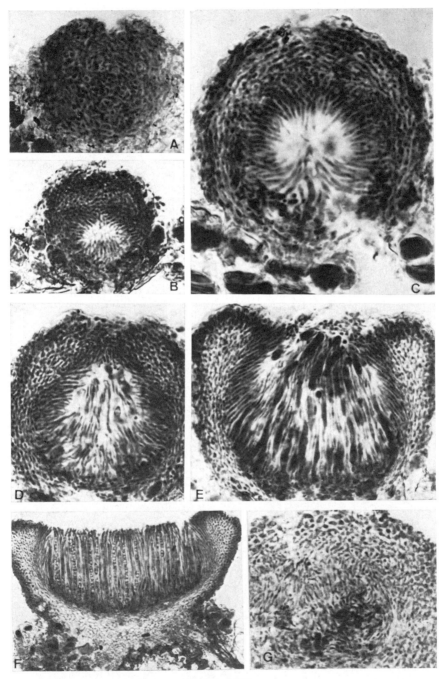

Plate VII

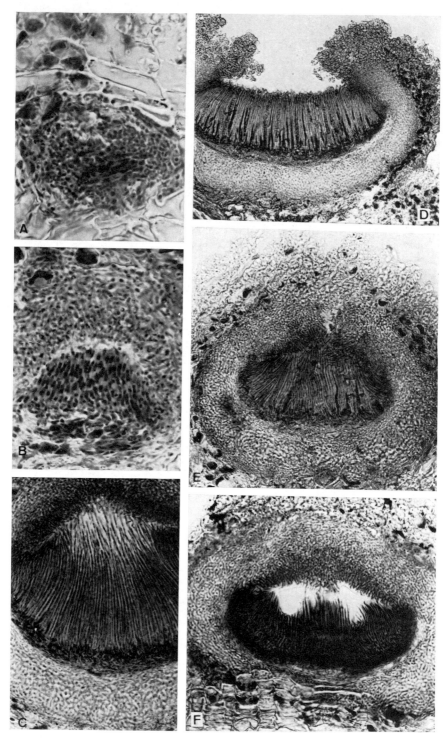

PLATE VIII

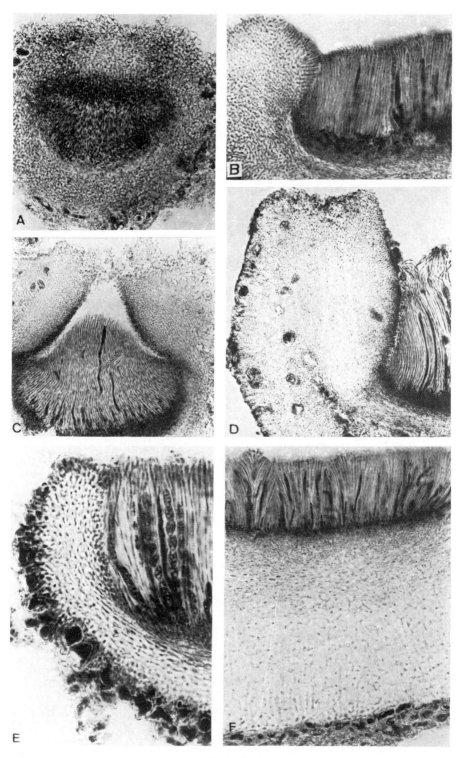

Plate IX

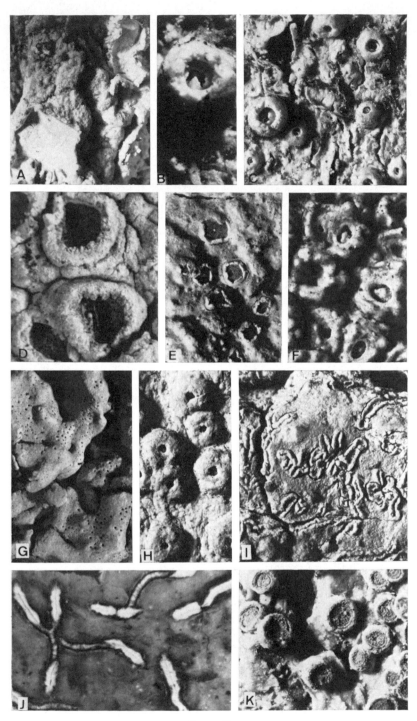

PLATE X

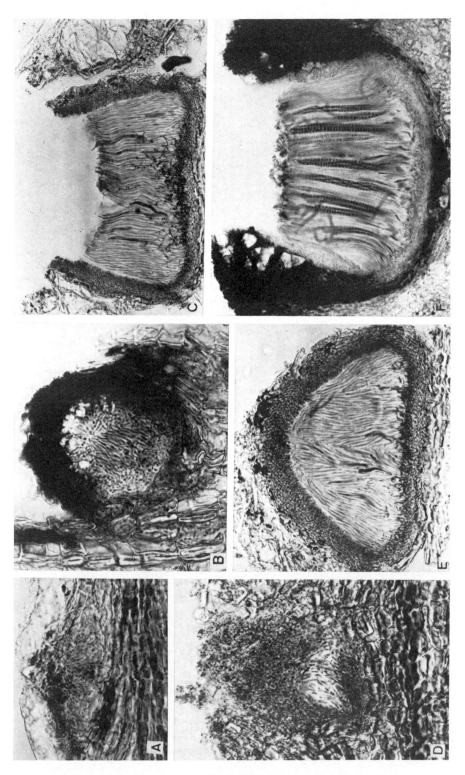

PLATE XI

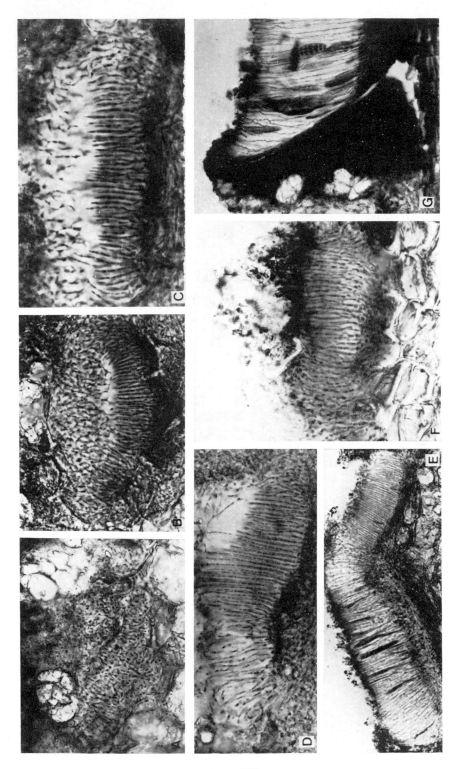

PLATE XII

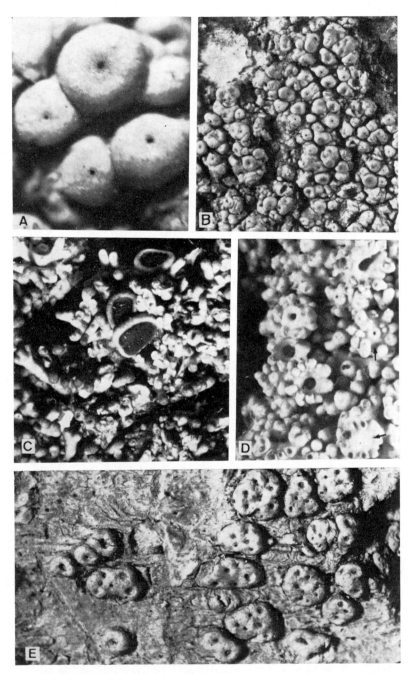

PLATE XIII

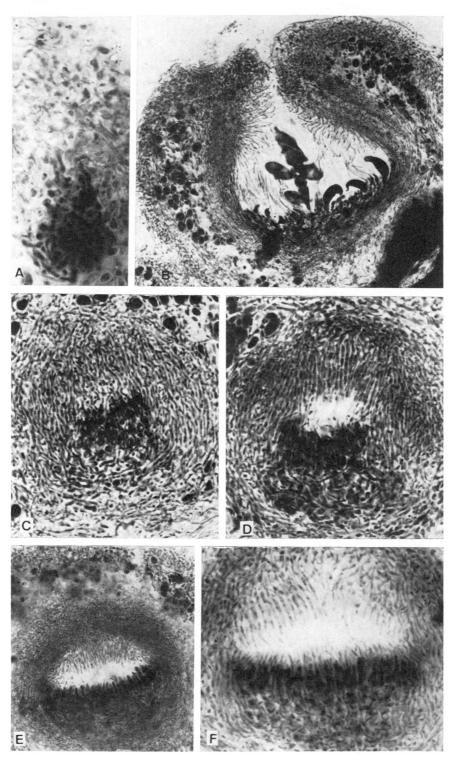

PLATE XIV

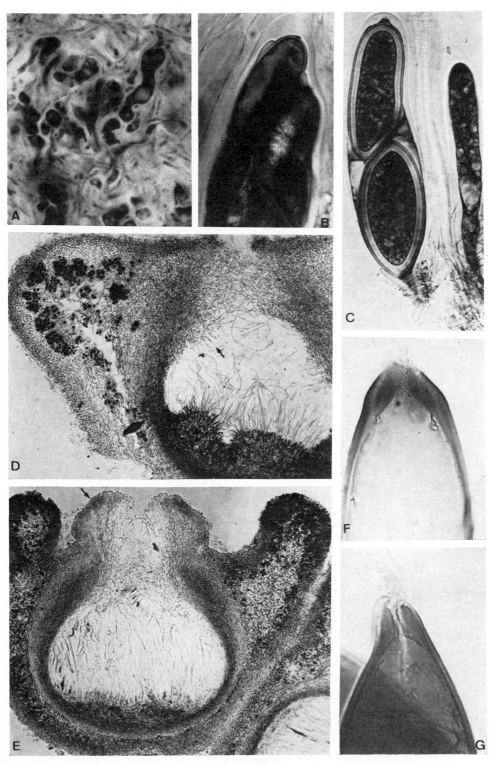

PLATE XV

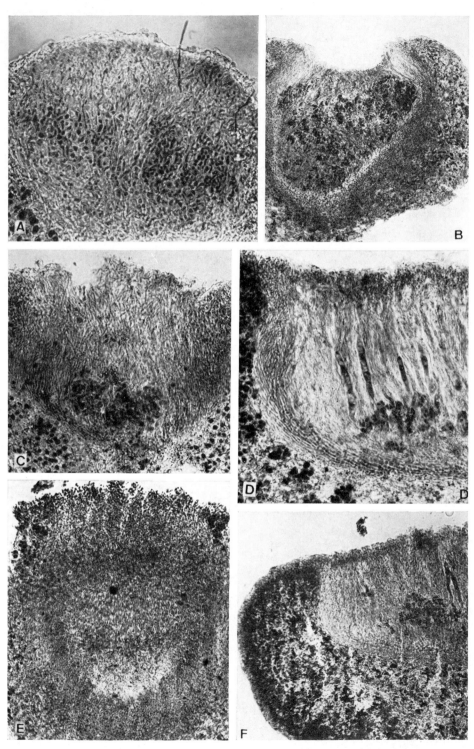

Plate XVI

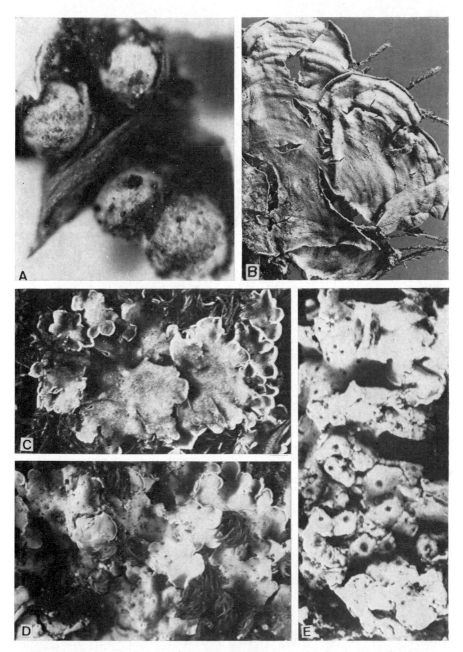

PLATE XVII

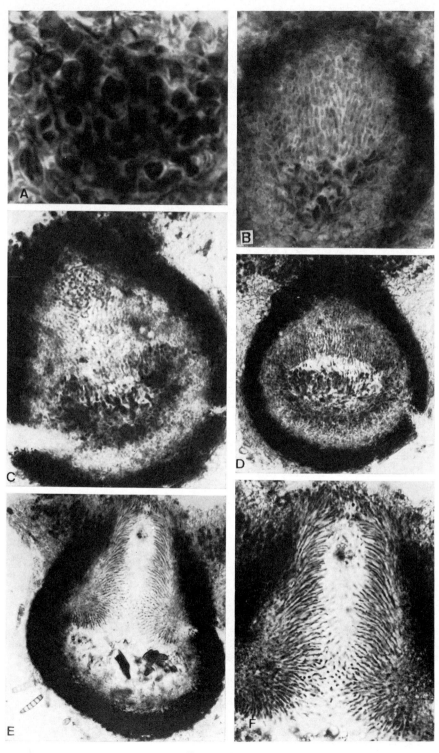

PLATE XVIII

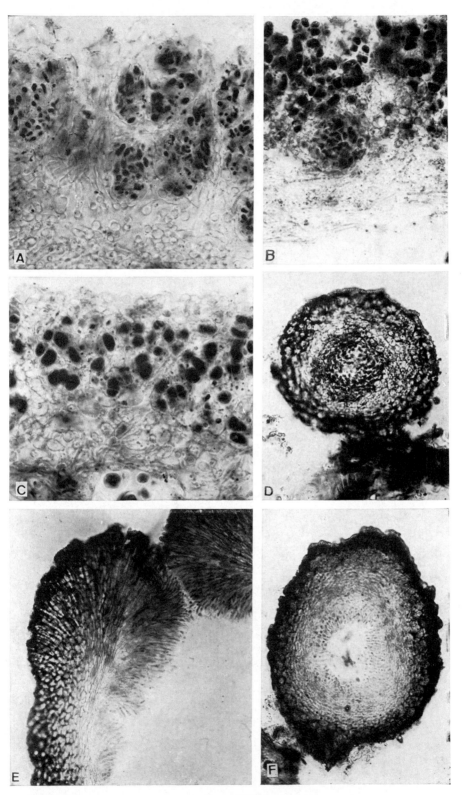

PLATE XIX

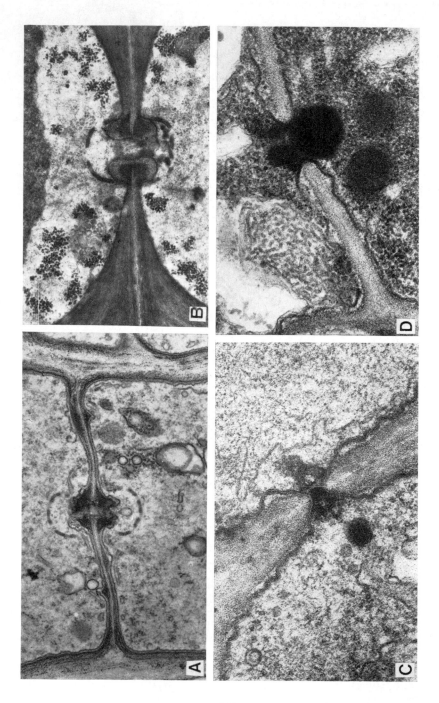

PLATE XX

Ascolocular features in the ontogeny of *Sphaerulina chlorococca* are the differentiation of a meristematic plug in the summit of the primordium and the relatively late production of ascogonial cells when the primordium has already developed a dark peridium. The bitunicate (or rather nonfissitunicate) ascus had been considered to be restricted to the ascolocular fungi but also seems to occur fairly frequently in the ascohymenial group, especially in the lichenized fungi (Henssen and Jahns, 1973).

2. *Reflections on the Systematic Position of* Normandina

The discovery that the alleged fruit bodies of *Normandina pulchella* are in reality a parasitic fungus leads to a reconsideration of the systematic position of the lichen. The genus had been placed in the Verrucariales on account of the perithecial structure. The squamules of the thallus resemble certain species of *Dermatocarpon* and *Endocarpon* but a closer similarity exists to *Coriscium viride*. In both lichens the thallus is shell-like and has a raised margin (Plate XVIIc & D) and they look like minute thalli of *Cora pavonia* (Plate XVIIB), a well known basidiomycete lichen. Even the anatomy corresponds to *Cora* in regard to the thin-walled colourless hyphae surrounding the groups of algal cells. The symbiotic algae differ in the three species. The phycobiont is *Trebouxia* in *Normandina*, *Coccomyxa* in *Coriscium* and a member of the Rivulariaceae in *Cora* (demonstrated by our own culture experiments).

The great correspondence in the morphology of *Normandina pulchella* and *Coriscium viride* has been known a long time, and several authors considered the two lichens to be congenerous (cf. the synonymy in Zahlbruckner, 1921–40). The genus *Coriscium* was separated from *Normandina* by Vainio (1890) on account of a deviant anatomy, *Coriscium* being considered to differ in the presence of an upper cortex. In reality, however, an upper cortex may be present or lacking in both of the species, depending on the age and state of preservation of the thallus (Plate XIXA–c).

Coriscium viride is currently regarded as the imperfect stage of *Omphalina hudsoniana* (Jenn.) Bigelow (syn. *O. luteolilacina* (Favre) Henderson; see Bigelow, 1970, Oberwinkler, 1970). Gams (1962) was the first to report on the new basidiomycete lichen and he found the fruit bodies of the *Omphalina* constantly associated with *Coriscium viride*, a fact confirmed by later investigators (e.g. Poelt and Oberwinkler, 1964; Heikkilä and Kallio, 1966, 1969).

The considerable similarity in regard to the morphology and anatomy of the thallus led us to suppose that *Normandina* might be a basidiomycete lichen like *Coriscium*. The best way to prove the basidiomycete nature of a fungus is by examination of the septal pore with the aid of an electron microscope. While in ascomycetes the septa are typically perforated by simple pores, the hyphae of higher basidiomycetes (Agaricales and Aphyllophorales) are characterized by doliporous septa (see Donk, 1973). Dolipores are found in the monokaryotic and dikaryotic hyphae of the basidiomycetes, heterokaryotic hyphae may also contain simple pores arising by a gradual dissolution of the barrel-shaped protuberances of the dolipores (e.g. Giesy and Day, 1965; Jersild et al., 1967; Casselton et al., 1971).

Electron microscopical investigations of Normandina pulchella did not confirm the suggestion that the lichenized fungus might be a basidiomycete. Material of this lichen has been studied from several localities and simple pores in the hyphal septa invariably occurred (Plate XXc). For comparison, the thallus of Coriscium viride and the fruiting bodies of the associated Omphalina hudsoniana were studied. Somewhat surprisingly the doliporous septa were restricted to Omphalina (Plate XXA). The hyphae in the Coriscium thallus had only septae with simple pores (Plate XXD). The question arose whether lichenization might influence the structure of the fungus hyphae. This is, however, not very likely since doliporous septa are known from the thallus of Cora pavonia (Roskin, 1970, confirmed by our own results). The photograph in Plate XXB is taken from a hypha surrounding a group of algal cells.

Because of these results, it is doubtful whether the mycobiont of the lichen Coriscium really produces the fruit bodies of Omphalina hudsoniana.* An association of Omphalina and Coriscium has been reported repeatedly. There exist, however, no published photographs or drawings demonstrating the alleged development of the Omphalina basidiocarp from hyphae of the Coriscium thallus. Of interest in this connection are the observations of Heikkilä and Kallio (1966) who found Omphalina hudsoniana growing on bare ground or on mosses in the vicinity of Coriscium squamules. The systematic position of Normandina and Coriscium consequently remains unclear. In the system of Zahlbruckner (1926), Coriscium was placed in the family Pyrenidiaceae, a family of doubtful delimitation.†

---

* In recent investigations Henssen and Kowallik (unpublished) obtained conclusive results that the fruit bodies of Omphalina hudsoniana are not produced by the mycobiont of Coriscium viride. The occasionally occurring close connections of the thalli of the ascomycete lichen Coriscium with the stalk of the basidiocarp of Omphalina are only of epiphytic nature.

† The monotypic genus Pyrenidium Nyl. was based on Pyrenidium actinellum Nyl., a name to be rejected under Art. 70 since the description was based on a lichenicolous fungus growing on Leptogium cretaceum (type: Boseley Hill, Kent, Jones, BM—holotype) and neither element constitutes a satisfactory lectotype.

Terms are used here in the sense of Henssen and Jahns (1973).

Generative tissue: the web of monokaryotic hyphae which envelops the ascogonia and differentiates into the first paraphyses and cells of cortical layers in the primordium. The hyphae of this tissue have thin walls, are rich in cytoplasm and therefore stain more strongly in lactic cotton-blue than the hyphae of the surrounding thallus tissues. Apart from the Collemataceae in which it arises from the stalk cells of the ascogonia, the generative tissue develops from the medullary hyphae adjacent to the ascogonia. Frequently it is produced first and the ascogonia are differentiated secondarily from some of the hyphae. Generative tissue is a characteristic feature of the ontogeny of ascohymenial lichen families.

Apothecium, perithecium, hysterothecium and *paraphyses*: these terms are used in the widest sense including the corresponding structures of ascolocular fungi. All kinds of sterile hyphae in the hymenium are termed paraphyses.

True paraphyses: paraphyses of ascohymenial lichens developing from the subhymenial layer and having tips free from the beginning.

Paraphysoids: hyphae arising from the generative tissue in the primordium and later on also in the marginal parts of the ascocarp. Their tips may become secondarily free or may remain in constant contact with a layer covering the hymenium. In some groups of lichens they are the only type of paraphyses, but in most they are substituted by true paraphyses.

Pseudoparaphyses: paraphyses in the ascocarps of ascolocular fungi. These are composed of short cells and fastened at both ends.

Periphyses: hair-like structures below or within the ostiole of a well defined perithecium.

Periphysoids (Vězda, 1966) or *lateral paraphyses* (Gilenstam, 1969): hair-like structures at the inner wall of the excipulum in well defined or modified apothecia.

Unitunicate and *bitunicate*: in regard to ascus structure these terms are employed in the old sense corresponding to the more correct modern designations *nonfissitunicate* and *fissitunicate* (Hertel, 1969).

1. *Material*
The following list of material studied is restricted to the specimens photographed. In the case of photographs taken from more than one specimen of a species the corresponding numbers and letters of the plates are indicated in parentheses. Herbarium abbreviations follow "Index Herbariorum" except that material in the author's personal herbarium is indicated by "Hsn".

Belonia russula Körb., British Isles, Scotland, Forfar, Glen Cova, 1965, *James* (BM).—
Clathroporina calcarea W. Wats., British Isles, Main Argyll, Lismore, 1971, *James* (BM).—

Coccotrema cucurbitula (Mont.) Müll. Arg., Chile, Valdivia, Petrohue, *Henssen* 24303*a* (MB).—*Collema multipartitum* Sm., Sweden, Gotland, Östergarn, *Henssen* 12031c (Hsn). —*C. subnigrescens* Degel., Sweden, Uppland, Vänge, *Henssen* 4856 (Hsn).—*Conotrema urceolatum* Tuck., Canada, Haliburton Co., Dorset, *Henssen* 14653f (Hsn).—*Cora pavonia* (Web.) Fr., Chile, Valdivia, Arigue, *Lechler* 638 (MB) (Plate XVIIв); Uruguay, Sierra de los Animas, *Henssen* 24037 (MB) (Plate XXв).—*Coriscium viride* (Ach.) Vain., British Isles, Scotland, Angus, Kirriemuir, 1973, *Duncan* (Hsn).—*Dermatocarpon miniatum* (L.) Mann, Sweden, Ärlinghundra s:n, 1960, *Nordin* (UPS).—*D. rivulorum* (Arnold) DT. & Sarnth., Norway, Sör-Tröndelag, Knutshö, *Henssen* 5015 (Hsn).—*Dimerella diluta* (Pers.) Trev. Finland, Satakunta, Yläne, *Henssen* 5434 (Hsn).—*D. lutea* (Dicks.) Trev., U.S.A., California, Humboldt Co., Samoa-Peninsula, *Henssen* 13649a (Hsn).—*Diploschistes ochrophanes* Lett., Austria, Kärnten, Gmeineck, *Henssen* 1735 (Hsn).—*Endocarpon pusillum* Hedw., U.S.A., Wyoming, Teton Co., Moran, *Henssen* 13309c (Hsn).— *Graphina mendax* (Nyl.) Müll. Arg., U.S.A., Louisiana, St Charles par., Sarpy, *Henssen* 15066d (Hsn).—*Graphis afzelii* Ach., U.S.A., California, Seminole Co., Altamonte Springs, Dr Schaffert's dome garden, 1956, *Schaffert* 97 (Hsn).—*Gyalecta foveolaris* (Ach.) Schaer., Finland, Lapponia enontekiensis, Kilpisjärvi, *Henssen* Le 502 (Hsn).—*G. geoica* (Wahlenb. ex Ach.) Ach., Sweden, Gotland, Thorsburgen, *Henssen* 6176 (Hsn).—*G. jenensis* (Batsch) Zahlbr., Austria, Steiermark, Grundlsee, *Henssen* 1738 (Hsn) (Plate VIд), 1739 (Hsn) (Plate IXc); Switzerland, Solothurn, Balsthal, *Henssen* 19248b (MB) (Plate IXd).—*G. peziza* (Mont.) Anzi, Finland, Lapponia enontekiensis, Porojärvet, *Henssen* Le 499 (Hsn).—*G. ulmi* (Sw.) Zahlbr., Germany, Allgäu, Ostrachtal, *Müller-Doblies* 66262a (Hsn).—*Gyrostomum scyphuliferum* (Ach.) Nyl., U.S.A., Florida, Dade Co., Everglades, *Henssen* 17258c (Hsn).—*Lepolichen coccophorus* (Mont.) Trev., Chile, Valdivia, Petrohue, *Henssen* 24311a (MB).—*Leptotrema bahianum* (Ach.) Müll. Arg., U.S.A., Florida, Monroe Co., Everglades, *Henssen* 17196r (MB).—*L. santense* (Tuck.) Zahlbr., U.S.A., Louisiana, Evangeline par., Chicot Park, *Henssen* 15086u (Hsn).—*L. wightii* Müll. Arg., U.S.A., Louisiana, Evangeline par., Chicot Park, *Henssen* 15090b (Hsn).—*Normandina pulchella* (Borr.) Nyl., Germany, Baden, Staufen, 1973, *Henssen* (MB) (Plate XXc); Argentina, Rio Negro, Lago Lacar, Pucara, with *Sphaerulina chlorococca*, *Henssen* 24547a (MB) (Plates XVIIa, d and e, XIXв and c).—*Ocellularia americana* Hale ined., U.S.A., Mississippi, Hancock Co., Logtown, *Henssen* 15011e (Hsn).—*O. dilatata* Müll. Arg., U.S.A., Mississippi, Hancock Co., Logtown, *Henssen* 15021d (Hsn).—*Omphalina hudsoniana* (Jenn) Bigelow, Austria, Salzburg, Krimml, 1973, *Kalb* (MB).—*Parmeliella plumbea* (Lightf.) Müll. Arg., British Isles, Scotland, Inverness, *Henssen* 17490a (MB).— *Pertusaria oculata* (Dicks.) Th. Fr., Norway, Sör-Tröndelag, Knutshö, *Henssen* 12129a (Hsn) (Plates XIIIc, XVa, XVIa and в); Norway, Finnmark, Ilfjordfjell, *Henssen* 21065a (MB) (Plate XVIc and d); Finland, Lapponia enontekiensis, Porojärvet, *Henssen* Le 643a (Hsn) (Plate XVIf).—*P. pertusa* (L.) Tuck., France, Bretagne, Morbihan, *Henssen* 21003m (MB).—*Petractis clausa* (Hoffm.) Kremp., Yugoslavia, Istria, *Henssen*, 18132o (Hsn).—*Placopsis subgelida* Nyl., New Zealand, Campbell Island, Mt. Dumas, 1963, *Wise* (BM).—*Porina nucula* Ach., U.S.A., Louisiana, St Martin par., Bayon Benoit Road, *Henssen* 15104c (Hsn) (Plate IVв); U.S.A., Florida, Dade Co., Mathewson Hammock, *Henssen* 17260d (MB) (Plate IVA, c and d).—*Pyrenula nitidella* (Flörke) Müll. Arg., Sweden, Skåne, Dalby, *Henssen* 3967 (Hsn).—*Thelotrema lepadinum* (Ach.) Ach., U.S.A., Washington, Jefferson Co., Olympic National Park, *Henssen* 13543a (Hsn).

2. Methods

Freezing microtome sections were embedded in lactic cotton blue and the coverslips sealed with Glyceel.

For the electron microscopical studies the material was fixed in 4% glutaraldehyde in cacodylate buffer (0·05 M, pH 7·0) followed by post-osmication (overnight), dehydration and embedding in Epon according to standard procedures. Series of thin sections prepared on a LKB ultramicrotome were examined in a Philips EM 300 electron microscope.

ACKNOWLEDGEMENTS

These studies were supported in part by a grant from the Deutsche Forschungsgemeinschaft.

The electron microscopical studies were made by Prof Dr K. Kowallik (Universität Düsseldorf) and form part of a joint paper which will be published elsewhere. The developmental morphology in *Coccotrema* and *Lepolichen* was investigated by Mr H. Born and the photographs and drawings of the anatomical structures presented are taken from his thesis. I am greatly indebted to Prof. R. Santesson (Naturhistoriska Riksmuseet, Stockholm) and to Mr P. W. James (British Museum (Natural History), London) for their advice and helpful discussions. I also thank the curators of BM, H and UPS for the loan of material, Dr M. E. Hale (Smithsonian Institution, Washington, D.C.) for the revision of my Thelotremataccae specimens, Mr G. Vobis for drawing some of the figures, and Mrs R. Schmidt and Mrs H. Klappstein for technical help in preparing the sections and photographs. Dr U. K. Duncan and Mr K. Kalb kindly provided the author with living material of *Coriscium* and *Omphalina*. These studies were supported partly by a grant provided by the Deutsche Forschungsgemeinschaft.

REFERENCES

AINSWORTH, G. C. (1971). "Ainsworth and Bisby's Dictionary of the Fungi" (6th edition). Commonwealth Mycological Institute, Kew Surrey.

BIGELOW, H. E. (1970). *Omphalina* in North America. *Mycologia* **62**, 1–32.

BRODO, I. M. (1973). The lichen genus *Coccotrema* in North America. *Bryologist* **76**, 260–270.

CASSELTON, L. A., LEWIS, D. and MARCHANT, R. (1971). Septal structure and mating behaviour of common A diploid strains of *Coprinus lagopus*. *J. gen. Microbiol.* **66**, 273–278.

CORNER, E. J. H. (1929). Studies in the morphology of Discomycetes. II. The structure and development of the ascocarp. *Trans. Br. mycol. Soc.* **14**, 275–291.

DONK, M. A. (1973). The Heterobasidiomycetes: a reconnaissance—III. *Proc. K. ned. Akad. Wet.*, C, **76**, 1–22.

DOPPELBAUR, H. (1960). Ein Beitrag zur Anatomie und Entwicklungsgeschichte von *Dermatocarpon miniatum* (L.) Mann. *Nova Hedwigia* **2**, 279–286.

GAMS, H. (1962). Die Halbflechten *Botrydina* und *Coriscium* als Basidiolichenen. *Öst. bot. Z.* **103**, 164–167.

GIESY, R. M. and DAY, P. R. (1965). The septal pores of *Coprinus lagopus* in relation to nuclear migration. *Am. J. Bot.* **52**, 287–293.

GILENSTAM, G. (1969). Studies in the lichen genus *Conotrema*. *Ark. Bot., ser.* 2, **7**, 149–179.

HALE, M. E. and CULBERSON, W. L. (1970). A fourth checklist of the lichens of the continental United States and Canada. *Bryologist* **73**, 499–543.

HEIKKILÄ, H. and KALLIO, P. (1966). On the problem of subarctic basidio-lichenes. I. *Rept Kevo Subarct. Res. Stn* **3**, 9–35.

HEIKKILÄ, H. and KALLIO, P. (1969). On the problem of subarctic basidiolichenes. II. *Rept Kevo Subarct. Res. Stn.* **4**, 90–97.

HENSSEN, A. and JAHNS, H. M. (1973) ["1974"]. Lichenes, eine Einführung in die Flechtenkunde." Georg Thieme Verlag, Stuttgart.

HERTEL, H. (1969). Die Flechtengattung *Trapelia* Choisy. *Herzogia* **1**, 111–130.

JAHNS, H. M. (1973). The trichogynes of *Pilophorus strumaticus*. *Bryologist* **76**, 414–418.

JERSILD, R., MISHKIN, S. and NIEDERPRUEM, D. J. (1967). Origin and ultrastructure of complex septa in *Schizophyllum commune* development. *Arch. Mikrobiol.* **57**, 20–32.

LETROUIT-GALINOU, M.-A. (1966). Recherches sur l'ontogénie et l'anatomie comparées des apothécies de quelques discolichens. *Revue bryol. lichén.* **34**, 413–588.

OBERWINKLER, F. (1970). Die Gattungen der Basidiolichenen. *Vortr. bot. Ges.* [*Dtsch. bot. Ges.*], *N.F.* **4**, 139–169.

OSHIO, M. (1969). Taxonomical studies on the family Pertusariaceae of Japan. *J. Sci. Hiroshima Univ., ser. B, div.* 2 (*Bot.*) **12**, 81–163.

POELT, J. (1974) ["1973"]. Classification. *In* "The Lichens" (V. Ahmadjian and M. E. Hale, eds), pp. 599–632. Academic Press, New York and London.

POELT, J. and OBERWINKLER, F. (1964). Zur Kenntnis der flechtenbildenden Blätterpilze der Gattung Omphalina. *Öst. bot. Z.* **111**, 393–401.

ROSKIN, P. A. (1970). Ultrastructure of the host–parasite interaction in the basidiolichen *Cora pavonia* (Web.) E. Fries. *Arch. Mikrobiol.* **70**, 176–182.

SANTESSON, R. (1952). Foliicolous lichens I. A revision of the taxonomy of the obligately foliicolous, lichenized fungi. *Symb. bot. upsal.* **12** (1), 1–590.

SMITH, A. L. (1911). "A Monograph of the British Lichens", vol. 2. British Museum (Natural History), London.

STAHL, E. (1877). "Beiträge zur Entwicklungsgeschichte der Flechten", vol. 1. Felix, Leipzig.

SWINSCOW, T. D. V. (1963). Pyrenocarpous lichens: 5. *Lichenologist* **2**, 167–171.

VAINIO, E. A. (1890). Étude sur la classification naturelle et la morphologie des lichens du Brésil, pars secunda. *Acta Soc. Fauna Flora fenn.* **7** (2), 1–256.

VĚZDA, A. (1966). Flechtensystematische Studien III. Die Gattung *Ramonia* Stiz. und *Gloeolecta* Lett. *Folia geobot. phytotax.* **1**, 154–175.

VĚZDA, A. (1968). Taxonomische Revision der Gattung *Thelopsis* Nyl. (Lichenisierte Fungi). *Folia geobot. phytotax.* **3**, 363–406.

ZAHLBRUCKNER, A. (1921–1940). "Catalogus Lichenum Universalis", 10 vols. Borntraeger, Leipzig.

ZAHLBRUCKNER, A. (1926). Lichenes (Flechten) B. Spezieller Teil. *In* "Die natürlichen Pflanzenfamilien" (A. Engler, ed.), (2nd edition) vol. 8, pp. 61–270. Englemann, Leipzig.

7 | Lichen Chemotaxonomy

D. L. HAWKSWORTH

Commonwealth Mycological Institute, Kew, Surrey, England

Abstract: An historical synopsis of the development of chemotaxonomy in the lichens is presented. After a consideration of biosynthetic and taxonomic principles pertinent to the utilization of chemical data in lichen taxonomy it is shown, by means of examples, that chemical information can make an important contribution to taxonomy at the supraspecific, specific and infraspecific levels. The types of chemical variation which are known in lichens are summarized and the most appropriate taxonomic treatment of these is discussed. It is shown that specific, subspecific or varietal ranks can be used to accommodate different types of chemical variation which are taxonomically significant and that other types are of no taxonomic importance. A range of taxonomic ranks enables more sensitive and acceptable taxonomies to be constructed than treating all major chemical races either as species or as chemical strains. Considerable care is needed in determining the most appropriate rank in each case. Areas in which further work are needed are discussed (particularly the rôle of the phycobiont and the biogenesis of lichen products). A bibliography on this topic is also provided.

INTRODUCTION

Chemical investigations now form an integral part of all serious taxonomic studies in the lichen-forming fungi and, as I have pointed out elsewhere (Hawksworth, 1973a, 1974), any taxonomic revision not considering chemical data is likely to be regarded as incomplete by most contemporary lichenologists. It is now generally accepted that some types of chemical variation, particularly when they are related to morphological differences, merit formal taxonomic recognition whilst others do not. In the majority of cases, however, where chemical variation occurs (or appears to occur) within morphologically similar plants the questions as to "which" and "at what rank" remain a cause of considerable debate.

Systematics Association Special Volume No. 8, "Lichenology: Progress and Problems", edited by D. H. Brown, D. L. Hawksworth and R. H. Bailey, 1976, pp. 139–184. Academic Press, London and New York.

As noted by Poelt (1974), agreement is gradually being reached on the most important criteria for evaluating chemical characters, particularly as a result of the observations of J. Santesson (1970). "Gradually" is the *mot juste* when one considers that the value of chemical characters in lichen taxonomy has been discussed for over a century (see pp. 140–150). However, at a time when chemical data are being studied more widely by practising taxonomists than ever before, firm guidelines are required to ensure that concepts do not vary too markedly between genus and genus and between species and species.

This paper reviews briefly the history of the discussion on the value of chemical criteria in lichen taxonomy, outlines the main types of chemical variation known to occur in lichens and ways in which chemical data contribute to a better understanding of lichen taxa (and their relationships), and proposes some guidelines for the procedures to be adopted in determining the most appropriate taxonomic treatment in particular cases.

HISTORICAL BACKGROUND*

In 1865–66 Nylander carried out preliminary studies showing that chemical reagent tests had considerable potential in lichen taxonomy. First he found iodine to be a useful aid (Nylander, 1865) and later (Nylander, 1866a) discovered that bleaching powder (C) also served as a useful taxonomic tool, for example, in supporting the recognition of *Parmelia perlata* and "*P. olivetorum*" (i.e. *Cetrelia olivetorum*) as distinct species. Shortly afterwards he (Nylander, 1866b) also found that potassium hydroxide (K) was equally valuable. In Britain Leighton was quick to realize the possibilities of chemical reagent tests in lichens and translated Nylander's papers into English (Nylander, 1866c). The value of the K test in particular cases, for example in distinguishing sterile material of *Candelaria concolor* and *Xanthoria candelaria*, appears to have fired Leighton's enthusiasm since he went on to investigate the reactions in *Cladonia* with K (Leighton, 1866a) and discovered that the application of C following K sometimes gave reactions where K and C alone did not (Leighton, 1867). Leighton (1868) stressed that reagent tests were most satisfactorily regarded as aids to identification rather than criteria on which to base species and reported that *Sphaerophorus fragilis* and *S. globosus* could be separated by their reactions to iodine (I).

Some other lichenologists were more sceptical, however, and Fries (1867) and

---

* Only selected papers important to the development of concepts or illustrative of the acceptance and application of these are cited in this section. Some other papers including pertinent information or discussions are referred to elsewhere and listed, together with some not mentioned in the text, on pp. 174–184.

Tuckerman (1868), for example, questioned their value. Meanwhile, Nylander was systematically carrying out reagent test surveys on various genera (e.g. Nylander, 1869a). Perhaps the most critical opponent of chemical tests was Lindsay who, in his first paper on the subject (Lindsay, 1869a), concluded that he could not commend chemical characters but stressed that diagnoses should take account of them. Lindsay, in the thorough manner characteristic of his other work on lichens, carried out a major study in an attempt to arrive at a firm opinion on this subject (Lindsay, 1869b). In this he listed some 130 species in which positive reactions with K, C or ammonium hydroxide were obtained and examined several collections of most of them. As a result of this survey Lindsay concluded that reagent tests afforded ". . . no aid that can be depended on, . . . to the systematist in defining botanical species" and urged other lichenologists to put the assertions of Nylander and Leighton to the test ". . . with a view to a final decision of the question at issue, which is one of great interest to every student of the natural history of lichens". Interestingly some of Lindsay's results appear to have been variable as a result of his inadvertently having discovered some chemical races, four of his five "*Parmelia borreri*" specimens, for example, were C+ red (i.e. *P. borreri* s.s. and (or) *P. subrudecta*) but one was C− (i.e. probably *P. reddenda*). Other variable results may, however, have arisen from his using incorrectly determined material or poor quality reagents, both of which were mentioned by Nylander (1869b) in a response to Lindsay's papers.

Lindsay's views did not deter other lichenologists from employing chemical tests. Chemical characters featured strongly in Leighton's flora (Leighton, 1871) and were taken up by other lichenologists including Stirton and Vainio. Vainio (1890) stressed their value as accessory diagnostic characters which should be treated as other characters but Müller-Argoviensis (1891) was critical of this approach. Crombie, always an ardent follower of Nylander, utilized chemical reactions in his flora (Crombie, 1894) but Schneider (1897), in contrast, considered that no chemical test was absolutely reliable.

The formation of characteristic microcrystals with K was reported by Bachmann (1887); an observation which later was to prove of value for the identification of lichen substances (see p. 142).

The nature of the chemical substances responsible for reagent test colour reactions was now, at the end of the nineteenth century, starting to be established, particularly as a result of the studies of Zopf and Hesse. The first chemotaxonomic studies based on the nature of the chemical components rather than merely on reactions were published by Zopf (1903), who also indicated that some chemical races might be related to ecological preferences (Zopf, 1905).

Zopf went on to produce the first major compendium concerned with the chemical constituents of lichens (Zopf, 1907) and carry out a detailed chemotaxonomic survey of *Cladonia* (Zopf, 1908). Although Zopf's studies provided lichen chemotaxonomy with a much sounder basis, the controversy was not allayed. Lettau (1914), for example, in his comprehensive review of the subject, was not disposed to base new species on chemical characters alone. These attitudes, carried into the first two decades of this century, were summarized by Smith (1921) as follows: "Experienced systematists . . . refuse to accept the tests unless they are supported by true morphological distinctions, as the reactions are not sufficiently constant."

New data continued to be accumulated (see Brieger, 1923) but in the late 1920s and early 1930s chemotaxonomy in the lichens began to fall into disrepute, largely as a result of the indiscriminate use of chemical tests by Gyelnik (and to a lesser extent by Erichsen) who not only used them as the basis for new species descriptions but employed them in the subdivision of genera ignoring other anatomical and morphological characters (e.g. Gyelnik, 1932, 1935). This uncritical approach was not surprisingly resented by almost all Gyelnik's contemporaries (e.g. Magnusson, 1933; Degelius, 1939; see also Hawksworth, 1972a). Des Abbayes (1933) considered that the employment of names in cases where there were apparently intergrading reactions (e.g. *Ramalina farinacea* s.l.) was ". . . contraire à l'esprit scientifique et répugne au simple bons sens".★

The tide began to turn, however, with the studies of Asahina, who introduced a new reagent test, *para*-phenylenediamine (PD) (Asahina, 1934), and developed a series of microcrystal tests (Asahina, 1936–40) which enabled the chemical components of herbarium specimens to be determined more accurately than was previously possible. Asahina became convinced of the taxonomic value of chemical data in lichens and set out his views on this controversial subject quite clearly (Asahina, 1937) employing chemical differences as the sole basis for the recognition of new species. Asahina has continued to adopt this view in a long series of papers continuing to the present day although in some instances he now employs infraspecific ranks, most frequently that of subspecies (e.g. Asahina, 1956, 1964), in preference to the rank of species. Degelius (1937) took a rather different stance, carefully following the principles of Du Rietz (1930) in his taxonomic treatment of chemical races, and recognizing some as constituting varieties and forms.

The microcrystal technique was quickly taken up by both European (e.g. Lamb, 1939; Duvigneaud and Bleret, 1940; des Abbayes, 1949; Dahl, 1950;

★ ". . . contrary to the scientific spirit and repugnant to ordinary good sense" [Trans.].

Krog, 1951) and some North American (e.g. Evans, 1943a, b) lichenologists. Species rank was considered appropriate by Lamb (1947) for morphologically identical plants with entirely different chemical components, but not where a substance was simply present or absent. Dahl (1950) was not prepared to make generalizations but considered that the taxonomic treatment of chemical races in lichens should be discussed independently in each case and any correlations with other systematic, ecological or geographical factors taken into account. Not all lichenologists were ready to accept names based primarily on chemical differences, however, even if there were slight correlations with other factors.

Lamb (1948) suggested the use of the concept of "active" and "inactive" (acid-free) phases for presence–absence situations and later (Lamb, 1951a, b) advocated the use of the term "chemical strain" and the numbering of each chemical race within morphologically defined units, rather than providing the races with formal taxonomic names. Neelakantan *et al.* (1951) independently suggested the use of the term "strain".

In his review of most aspects of lichenology, des Abbayes (1951) was inclined to accept chemical differences as valid criteria in the same way that different algal components were used, but emphasized problems which could arise due to variations in the concentration of particular compounds and the care needed.

The first large-scale evidence for geographical correlations with chemical races was presented by Hale (1952) in his study of *Dimelaena oreina* in North America. Here Hale adopted the "chemical strain" terminology of Lamb rather than treat the races as distinct species, drawing a comparison between the use of the term strain here and that in the non-lichenized fungi. In the same year Dahl (1952) established the value of chemical data at the supraspecific level, showing, for example, that this supported (*a*) the infrageneric classification of *Cladonia*, (*b*) the separation of the *Cetraria glauca* group (i.e. *Platismatia*) from the *C. chlorophylla* group (i.e. *Cetraria* s.s.), and (*c*) the treatment of *Hypogymnia* and *Parmelia* as distinct genera.

In 1954 a new account of the chemical components of lichens was published by Asahina and Shibata (1954), the first attempt at a synthesis of the available information since that of Brieger (1923), and this appears to have stimulated further interest in lichen chemotaxonomy. In the succeeding years, facilitated by the application of paper chromatography to the identification of lichen products (Wachtmeister, 1952, 1956, 1959; Mitsuno, 1953, 1955; Asahina, 1956; Hess, 1958) which enabled the constituents of herbarium specimens to be examined more accurately than by microcrystal tests alone, many studies investigating chemical races in lichens began to appear. Some lichen depsides and depsidones fluoresce in ultraviolet (u.v.) light and this was also now being found of value in the preliminary sorting of some chemical races (e.g. Hale, 1956a; Černohorský, 1950, 1957). The tendency of workers outside Japan was to continue to treat chemical races as unnamed "chemical strains" (e.g. W. Culberson, 1958; W. Culberson and C. Culberson, 1956; Hale, 1955, 1956b, c, 1957) but, as Runemark (1956) pointed out, this approach was really ". . . only a way to avoid the actual problem". Runemark stressed the value of chemical data as supporting characters and considered that their taxonomic value varied in a way comparable to that of morphological criteria, so that different treatments would sometimes be appropriate in different genera and even occasionally within different species of the same genus.

Tétényi (1958) suggested the use of the term "chemovar(iety)" for chemical races in medicinal vascular plants but this view was criticized by Lanjouw (1958) on the grounds that a special terminology implied that chemical characters were intrinsically different to those derived from other spheres and considered that they should be treated in ordinary infraspecific ranks.

Up to this time no attempt had been made to establish that the age of lichen thalli did not influence the chemical components. In order to rectify this situation C. Culberson and W. Culberson (1958) examined 138 specimens of different ages of *Lasallia papulosa* (i.e. *Umbilicaria papulosa*) and found that the chemical constituents were constant so verifying their value as taxonomic characters.

In his monograph of *Nephroma* in North America, Wetmore (1960) accorded equal weight to chemical, morphological and anatomical characters. Hale (1961) emphasized that lichenologists should not ". . . overlook the possibility that strain formation is a forerunner of morphological differentiation and a process of speciation". In his monograph of *Cladonia* subgen. *Cladina*, however, Ahti (1961) adopted a conservative approach utilizing the "chemical strain" notation except for geographical colour races where the rank of variety was employed (e.g. *C. tenuis* var. *leucophaea*) and sporadically occurring colour races where the rank of form was used (e.g. *C. subtenuis* f. *cinerea*). Ahti stressed that ideally coloured-compound races and other chemical races should be treated in a comparable manner. W. Culberson (1961a) discussed the taxonomic treatment in lichens and concluded that where replacement of one chemical component by another was involved species rank was perhaps the most appropriate. He later extended his observations on the *Parmelia borreri* and *P. bolliana* groups (W. Culberson, 1962) accepting taxa involving replacements in species rank and noting parallel relationships in the two groups. The need for caution and the danger of placing too much emphasis on chemical criteria was noted by Riedl (1962). Extra-nomenclatural systems for denoting chemical races in lichens were, however, rejected by Thomson (1963) because of their cumbersome nature and he considered that, even if they were eventually proved experimentally to be of no value in species delimitation, naming them would not be a disadvantage as the appropriate units could be relegated to synonymy.

The first studies involving mass-sampling techniques were made by Hale (1963), who examined the chemical constituents of 16 398 specimens of the *Cetraria ciliaris* group collected at random in 218 woodland stands in the Appalachians. Hale demonstrated that in this case the distribution of the olivetoric acid and the alectoronic acid races were not related to environmental factors, treating these (and a further North American race not occurring in the Appalachians) as "chemical strains". Contrary to Ahti's (1961) views, W. Culberson (1964) treated the usnic acid deficient race of *Haematomma coccineum* (i.e. *H. ochroleucum*) as a distinct species (i.e. *H. porphyrium*; see p. 167).

Targé and Lambinon (1965) employed the rank of "chemovar." (see p. 144) in their study of the *Parmelia borreri* group, but in the same year Hale (1965a), on the basis of correlations of chemical replacements, morphological characters and geographical distribution, adopted the rank of species in this group as had W. Culberson (1961a, 1962) previously. Species rank was also used by Hale (1965b) in his monograph of *Parmelia* subgen. *Amphigymnia* for races involving very different chemicals when these were correlated with major distributional differences, but he emphasized the need for exhaustive field and herbarium studies before any definitive solution to the problem could be attained. Neelakantan (1965), however, pointed out that ". . . small differences in chemical structures should not be over-emphasized and new species created on that basis, but they have to be recognized as providing varieties" taking into account the biogenesis of the compounds involved (this approach is discussed at greater length by Neelakantan and Rao, 1967). Biogenetic considerations were also discussed in the *Cladonia chlorophaea* group by Shibata and Chiang (1965a), these authors suggesting that on this basis the race with grayanic acid (*C. grayi*) was distinct from other taxa in this group with more closely related compounds (e.g. *C. merochlorophaea, C. cryptochlorophaea*).

Many lichenologists were not yet convinced that chemical characters should be used as the basis for formally described taxa in any rank. This view may be illustrated by the paper of Almborn (1965), who emphatically stated that chemical characters should not be used to distinguish any taxonomic units and only the "chemical strain" terminology adopted for the distinction of races within morphologically indistinguishable plants. That is, chemical data should be regarded as taxonomic tools rather than as bases for recognizing new species.

Relatively little attention had been given to the possible evolutionary significance of chemical components at the supraspecific level up to this time but Hale (1966) established that there was a high degree of association between the occurrence of o-methylated depsides and depsidones and "structurally advanced" genera (e.g. *Cladonia*). Subspecific rank was first used in North America for chemical races by Imshaug and Brodo (1966) but, as these authors were careful to point out, they used subspecies rank only where there were correlations with ecological and (often subtle) morphological characters; races without such correlations being simply treated as "strains".

W. Culberson and C. Culberson (1967a) discovered that habitat selection by the chemical races of the *Ramalina siliquosa* group occurred on a promontory in North Wales, demonstrating local ecological sorting of the chemical races. These authors suggested that the races were probably adapted physiologically, the chemical components being the phenotypic expression of this. As a result of a careful analysis of chemical and morphological characters in 1042 specimens from four sites in Europe (in Brittany, France, Scotland and Wales) together with comparable investigations on herbarium specimens, W. Culberson (1967a) recognized six distinct species based primarily on chemical criteria, although the compounds involved appeared to be rather closely related biogenetically (see pp. 165–166, Fig. 4). Comparable approaches in the taxonomic treatment of chemical races were adopted in the *R. farinacea* group (W. Culberson, 1967b) and the *Cetraria ciliaris* group (W. Culberson and C. Culberson, 1967b).

The use of any infraspecific rank for chemical races in lichens was opposed by Hale (1967) on the grounds that it diminished the importance of chemical characters without sufficient prior study. Lamb (1967), however, stressed the need for caution in the naming of chemical races suggesting that so-called accessory substance races be referred to as "chemical phases" (see p. 143 and Lamb, 1948, 1964). In common with Thomson (1963), Hale (1968a) considered that by naming chemical races and perhaps eventually having to reduce them to synonymy "... we will have paid a small price to gain so much new knowledge on biochemical variation in a plant group". Species rank was accepted for the chemical races in *Pseudevernia* by Hale (1968b; see also pp. 165, 171) and in their monograph of *Cetrelia* and *Platismatia* W. Culberson and C. Culberson (1968) stated that "Morphologically similar or identical populations with different but constant medullary constituents are considered distinct species". Krog (1968), in contrast, refused to accept species ("chemospecies") based solely on chemical criteria and used the "chemical strain" terminology, except where there were correlations with slight morphological differences when the rank of "variety" was employed. The term "chemotype" was recommended by R. Santesson

(1968) for "chemically characterized parts of morphologically indistinguishable individuals" of "not indicated or not determined taxonomic rank or having no taxonomic value".

A further advance in microchemical techniques, the development of thin-layer chromatography (t.l.c.), was also now being taken up by lichenologists (Ramaut, 1963, 1967; Bachmann, 1963; J. Santesson, 1965, 1967; W. Culberson and C. Culberson, 1967a) and proving more sensitive than either microcrystal tests or paper chromatography. Solvent systems and plates have subsequently been standardized (C. Culberson and Kristinsson, 1970; C. Culberson, 1972) and t.l.c. is now the technique most widely used by lichen taxonomists. (A useful introduction to this method is given by Menlove, 1974.)

Some preliminary studies by Hawksworth (1968) indicated that habitat selection was occurring in chemical races of *Ramalina subfarinacea* but, as closely allied compounds were involved, these were treated as varieties. Cases where chemical races involving closely related compounds correlated with distributional tendencies in *Pseudevernia furfuracea* s.l. and *R. farinacea* (see p. 165 were also accepted in varietal rather than species rank by Hawksworth (1969).

The studies of C. Culberson and Kristinsson (1969; see also Kristinsson, 1974; Leuckert *et al.*, 1972; Chapter 8) on the *Cladonia chlorophaea* group revealed a complex situation of race formation in which a replacement series of compounds was the most important, some replacement races also having sporadically additional "accessory" compounds in them (e.g. fumarprotocetraric acid). Fourteen races were distinguished primarily on the basis of replaced compounds, six of which were unnamed (one of these described in species rank here). A particularly interesting and somewhat confusing pattern was discovered by Kristinsson (1969) in the *Cetraria islandica* complex in Iceland as a result of a study of some 3000 individuals from 100 localities. Whereas *C. islandica* and *C. ericetorum* (syn. *C. islandica* subsp. *crispa*) are usually readily separated on both morphological and chemical grounds in Europe, in Iceland (and perhaps other arctic and subarctic areas) the morphological and chemical boundaries appear to break down and gene-exchange was suggested to occur between PD− and PD+ races (see p. 164).

W. Culberson (1969a) carried out some preliminary studies of the *Ramalina siliquosa* group on a single pyramidal boulder in Portugal which suggested that microhabitat selection was also taking place in the chemical races there. The situation in this species was discussed further by W. Culberson (1969b) in a review of the subject in which he pointed out the quandary of the taxonomist in such situations, concluding that "It would seem that only the rank of species defines accurately the nature of this [*R. siliquosa*] type of variation and associated behaviour." Lambinon (1969) could not accept such a thesis, however, because mycologists and phanerogamists do not recognize chemical races as constituting taxa, even where geographical correlations are known, and cited *Claviceps purpurea* as an example. Krog (1969) concluded that ". . . there is very little to be gained by the introduction of chemical species" but acknowledged their value at the supraspecific level and recommended the "chemical strain" terminology and a system of registration for such strains. In the case of the *R. siliquosa* complex, Krog considered that there were two species present on morphological grounds, one with four and one with two "chemical strains" (see pp. 165–166).

So much new information was becoming available by this time that C. Culberson (1969) prepared a most valuable catalogue of this and issued a supplement the following year (C. Culberson, 1970). Some reports conflict as a result of chemical and (or) morphological misidentifications; the need for a continuing cataloguing of the components of type specimens in particular is clearly apparent. The listing of published reports of compounds in particular species has proved an invaluable aid to lichen taxonomists. One of the first outcomes of this was a consideration of the supraspecific classification of lichens in the light of the 209 chemical substances reported in 2315 species of lichens (W. Culberson and C. Culberson, 1970). These authors demonstrated clearly the importance of chemical data as "Most genera and families that are well defined morphologically and appear to represent natural taxa show highly uniform chemistries of several to many biogenetically related substances or sets of substances."

In examining the *Parmelia borreri* group in Scandinavia, Krog (1970a) accepted the species concepts of W. Culberson (1962) and Hale (1965a), based on correlations between chemical, morphological and distribution criteria, noting that from ". . . the primary results of chemical analyses, a differentiation into species with distinct morphological characters has been accomplished" (see p. 162). Jørgensen and Ryvarden (1970), contrary to the views of W. Culberson and C. Culberson (1968), did not accept the chemically differentiated *Cetrelia cetrarioides* and *C. olivetorum* as distinct species, treating them as a single species (*C. olivetorum* s.l.) with two chemical strains. Hawksworth (1970) followed W. Culberson (1964) in treating *Haematomma lapponicum* and *H. ventosum* as distinct on chemical and distributional grounds but regarded usnic acid present and absent races of the latter as chemotypes because of the absence of ecological and geographical correlations (see p. 168). A comparable colour variant of *H. ochroleucum* was treated as a variety (var. *porphyrium*) by Laundon (1970) because there were ecological correlations, although *porphyrium* was recognized as a distinct species by W. Culberson on chemical grounds (see p. 167).

In an important review of chemical race/ecological correlations in lichens, W. Culberson (1970a) stressed his view that the belief that visible rather than invisible (chemical) characters were more fundamentally important in taxonomy was not well founded and that consequently ". . . an *a priori* rejection or relegation of chemical variations to a preconceived rank (subspecies, 'chemovarietas', etc.) is philosophically indefensible". In the same year Poelt (1970) published the first of two detailed papers on the "Artenpaare" concept, applying this to some instances where chemical race formation was also important (e.g. *Pseudevernia*). J. Santesson (1970) demonstrated the value of chemical data at the generic and family level in the Teloschistaceae (see p. 159) and also pointed out that chemotypes could arise as a result of at least six different pathways, some of which would clearly be expected to be more important taxonomically than others (see p. 153; Fig. 2).

Chemical characters were used extensively by Yoshimura (1971) in his monograph of *Lobaria* in Eastern Asia according to the premise that they should be evaluated in a comparable way to morphological characters and when ". . . a chemical character alone deserves a taxonomic rank, the rank can be variable in accordance with the nature of the chemical character, for which the biogenesis of the chemical substance may give considerable help because biosynthetic pathways are controlled by genes". Comparable views were also expressed by Hawksworth (1971a) at the First International Mycological Congress.

In the 1970s chemical data have come to occupy an increasingly important position in lichen systematics. Most studies have continued to show their value at the supraspecific (e.g. Hawksworth, 1972a) and specific levels, and below, and to reveal complex situations (e.g. the 13 combinations of phenolic compounds found in the *Usnea florida* group by Fiscus, 1972). Poelt (1972) considered the application of the "Artenpaare" concept to chemical races and, by discussing the possible treatments of the *Cetrelia olivetorum* complex in the light of this, concluded species rank to be the preferred system. The approach of considering each case on its merits as previously recommended by some workers and using specific, infraspecific or no rank as appropriate was advocated by Hawksworth (1972b, 1973a, b).

W. Culberson (1973) and W. Culberson and C. Culberson (1973), as a result of careful investigations in the *Parmelia perforata* group, made a further major contribution to discussions of the taxonomic treatment of chemical races. They showed that by applying the "Artenpaare" concepts and taking into account

F

ecological and geographical data the group could be interpreted as comprising four (one unknown and presumed but possibly extinct) primary species (asco-carp bearing) which had given rise to four secondary species (reproducing asexually) with similar chemistries by parallel evolution (Fig. 6). W. Culberson (1973) considered that treatment in infraspecific categories was not incompatible with the data but ". . . would pay for the illusion of taxonomic conservatism with a complex nomenclature and a devaluation of the real evolutionary dis-tance among morphs and races that is shown by the evidence . . .".

Recently valuable summaries of lichen products and their biosynthesis have been prepared by several workers (see below), and Nourish and Oliver (1974) have provided a critique of techniques available for their identification. A synopsis of microcrystal tests is included in Yoshimura (1974).

Although we have now immeasurably more information on chemical races than was the case in discussions in the 1860s, equally great differences of opinion remain. In contrast to W. Culberson's (1973) work mentioned above, Dahl and Krog (1973), for example, retain the "chemical strain" approach for all cases without morphological correlations but recognize the value of chemical data in taxonomy, while Dodge (1973) continues to ignore all chemical information completely (apart from mentioning K reactions in some *Usneae*).

BIOSYNTHETIC CONSIDERATIONS

Detailed reviews of the available information and hypotheses concerning the biosynthesis of lichen products have been included in several recent publications (W. Culberson and C. Culberson, 1968; C. Culberson, 1969; Huneck, 1968, 1971, 1974; J. Santesson, 1973; Mosbach, 1974) and these should be consulted for comprehensive information on this aspect.

Biosynthetic routes are fundamental to any discussions as to the value of chemical criteria in lichen systematics but it is only since 1964 that developments in techniques involving radioactively labelled precursors have been applied in the lichens. It is not surprising to find, therefore, that, as emphasized by Mosbach (1974), the information available on this aspect is meagre and some analogies have to be taken from investigations in non-lichen-forming fungi if a comprehensive picture is to be suggested. As the greatest affinity of the major orders of lichen-forming fungi probably lies with groups of non-lichen-forming fungi unknown in culture and which have consequently not been the subject of biosynthetic studies it is clear that uncertainties must remain.

The biosynthetic routes leading to the major groups of lichen products are summarized in Fig. 1. Photosynthetic products pass from the phycobiont as mobile carbohydrates (e.g. glucose, ribitol). Most lichen products then appear to arise from an "acetate–malonate" pathway in which acetyl-CoA* and *n*-malonyl-CoA units condense and undergo various modifications to produce either aliphatic acids (e.g. protolichesterinic acid) or aromatic polyketides (e.g. depsides and depsidones). Terpenes probably originate from the conversion of

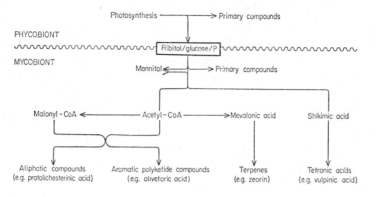

FIG. 1. Biosynthetic routes leading to the major groups of lichen products. (Adapted from J. Santesson, 1973.)

acetyl-CoA to mevalonic acid by a "poly-isoprene" biosynthetic route but no studies of the biosynthesis of these compounds in lichens appear to have been published. The "shikimic acid" pathway has received more attention from lichen biochemists and in this case a condensation of two C_6–C_3 units leads to a terphenyl structure which, as a result of ring fusion, gives rise to the various tetronic acid (pulvic acid) derivatives.

From this it is evident that the biogenesis of aliphatic products, aromatic polyketides, terpenes and tetronic acid derivatives all involve different enzyme systems each of which may reasonably be expected to be controlled by a considerable number of different genes. Thus a group producing tetronic acid derivatives alone might be expected to have marked genetic differences from a similar group producing aromatic polyketides alone.

The major group of compounds employed by lichen chemotaxonomists is the aromatic polyketides. In this case the "acetate–malonate" pathway leads to the production of orsellinic acid, units of which are built up to form the aromatic polyketides. The biosynthetic route for most of such compounds remains almost completely unknown and it would clearly be desirable from the taxonomist's standpoint for a great deal of biochemical work to be undertaken on the biosynthesis of aromatic polyketides. From the studies so far carried out it is

---

* CoA = coenzyme A.

apparent that in situations where one is dealing with major groups of orsellinic acid derivatives (e.g. dibenzonfurans v. anthraquinones v. depsides) considerable genetic difference (and hence taxonomic importance) can probably be inferred but this may not always be the case. It is perhaps also reasonable to assume that orcinol depsides and depsidones are more closely allied to one another than to β-orcinol depsides and depsidones on the basis of their distribution within the lichens. Norstictic acid is probably a precursor of stictic acid, giving rise to the latter by a one-step methylation reaction, and so the occurrence of norstictic acid in stictic acid-containing plants is not likely to be of much taxonomic importance. Depsidones may arise from their corresponding depsides by the action of a "depside dehydrogenase" and this view is supported by the occasional joint occurrence of a depsidone and its depside precursor (C. Culberson, 1964, 1965).

J. Santesson (1970, 1973; Fig. 2) has summarized the ways in which chemical races may arise from the biogenetic standpoint. It is evident from Fig. 2 that, for example, types 1, 3 and 4 are more similar to each other than 1 is to 5a, 5b or 6. In particular cases taxonomic treatments must attempt to consider the probable biogenesis of races and discussions of biogenetic relationships of compounds have consequently come to feature in taxonomic discussions (e.g. Runemark, 1956; Shibata and Chiang, 1965a, b; W. Culberson, 1967a; W. Culberson and C. Culberson, 1967a, b, 1968). In view of the lack of biochemical work at the present time, however, hypotheses rather than experimentally demonstrated biosynthetic routes must, unfortunately, predominate in such discussions. The possibility that the same compound might arise by more than one biosynthetic route in different taxa, while perhaps unlikely, cannot be entirely ruled out in our present state of knowledge.

Some other biosynthetic considerations also have taxonomic implications. W. Culberson (1961b) pointed out that the stipulation of Art. 13 Note 4 of the Stockholm International Code of Botanical Nomenclature (Lanjouw *et al.*, 1952) that the names of lichens be considered as referring to the fungal component (mycobiont) alone had considerable implications for lichen taxonomy, most of which were not readily apparent. Some of the problems arising from this relating to the production of different thallus forms by the same mycobiont with different algal genera are discussed by James and Henssen (Chapter 3) and so are not considered in detail here. Whether or not the phycobiont has any rôle in determining the production of chemical components is also pertinent to the acceptability of chemical data in taxonomy—in the case of some cephalodiate species at least it seems that they do play an important rôle (Chapter 3). If the phycobionts do, in general, have a rôle in determining the chemical components of lichens, as suggested by some authors (e.g. Asahina, 1937, 1967; Hess,

1959), the justification for their use as taxonomic criteria should at least be open to discussion as this stipulation remains in the most recent (Seattle) edition of the Code (Stafleu *et al.*, 1972)*. Fox and Huneck (1969) considered that as lichen acids

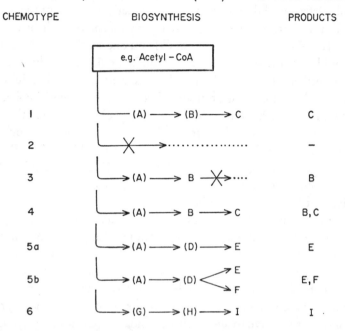

CHEMOTYPE   BIOSYNTHESIS   PRODUCTS

FIG. 2. Biogenesis of chemical races in lichens. (After J. Santesson, 1973.)

used for the delimitation of species seemed to be a result of symbiosis the practice of describing new fungi on that basis ". . . seems irrational unless the understanding exists that the species in question produces a particular substance as a result of its particular environment".

Ahmadjian (1967), rejecting an earlier unconfirmed report, noted that "No isolated mycobiont has produced a depside or a depsidone" but that some isolated mycobionts had produced pigments and crystals some of which could not be detected in the corresponding thalli. This may be partly due to cultural conditions but, following an unsuccessful attempt (Kurokawa *et al.*, 1969), Komiya and Shibata (1969) reported the production of salazinic and usnic acids in an isolated mycobiont of a chemical race of *Ramalina siliquosa* ("*R. crassa*"). It is consequently now clear that isolated mycobionts can form at least depsidones

---

* As this provision is for *nomenclatural* purposes, one can pedantically argue that, as nomenclature cannot prescribe a taxonomy, they can be employed in making taxonomic decisions.

in addition to substances such as anthraquinones, dibenzofurans and usnic acids. If chemical race formation arises from the occurrence of different physiological strains of mycobionts, as postulated by Ahmadjian (1967), and is indicative of genetically determined differences in the enzymatic processes of the mycobiont, as these data suggest, such characters are clearly acceptable in lichen taxonomy.

Peoples (1973) claimed that "enzymes" from one species mixed with those of others could lead to the production of different lichen products but in the absence of further information his conclusions must be treated as dubious.

There is evidence that, at least in some lichen species, material collected from different localities has different algal strains as phycobionts and in other instances a single thallus or different thalli from a single population may include more than one species from the same algal genus (see Ahmadjian, 1967; Piattelli and De Nicola, 1968; Wang-Yang and Ahmadjian, 1972).* The ability of a few isolated mycobionts to form lichen products in pure culture cannot rule out the possibility that in nature, in the intact thallus, the alga exerts some influence. Algae from various chemical races of the *Ramalina siliquosa* complex were isolated and studied by Ahmadjian who did not find any variations that could be associated with the chemical races of the donor lichens (W.Culberson, 1969c). It is conceivable that pertinent differences in the algae might not be revealed except by extensive biochemical studies and it is clear that a great deal more research into this aspect is needed before this possibility can be entirely ruled out. For the moment we have to agree with Ozenda and Clauzade (1970) that this is possible but not probable. Notwithstanding the observations of Jordan (1972), however, it is now clear that in the case of some species with erumpent cephalodia the algal genus present can exert a profound influence on the biosynthesis of lichen products (see Chapter 3, pp. 61–63).

While some doubt must still surround questions as to the rôle(s) of the algal components in lichens in the production of chemical races, environment can now be stated to have no qualitative effect on the chemical constituents produced (although quantitative variations may sometimes be related to environmental variables, see pp. 168–169). This can be stated categorically since (a) many common species with constant chemical components grow in a wide range of habitats in climatically different regions, (b) the same chemicals occur in species restricted to quite different habitats and (c) studies of large numbers of specimens, in *some* cases where chemical races occur, have failed to show any correlation with environmental factors. Where chemical races are environ-

---

* Some reports in the latter category are, however, in need of critical study by algal taxonomists, as pointed out by Tschermak-Woess (Chapter 4, p. 86).

mentally limited, historical factors and (or) physiological adaptations may be expected to be the important limiting ones.

Interestingly, none of the basidiomycete lichens appears to produce any characteristic lichen products (Follmann, 1972). The fungal, and often also the algal, partners are, however, generally quite different from those of the asco-mycete lichens and so differences in their metabolic activities are perhaps to be expected.

TAXONOMIC CONSIDERATIONS

Almborn (1965) stated that "It is essential that lichen taxonomy be founded on firm principles not deviating too much from those generally accepted in other plant groups." As pointed out by W. Culberson and C. Culberson (1968), however, what is taxonomically important in one group of organisms may not be so in another. It is now coming to be generally accepted by systematists that concepts of particular taxa should be based on as many characters as possible, whether or not they are visible to the naked eye (e.g. chromosome numbers, chemical components). The concepts of what constitutes differences meriting separations at various ranks of the taxonomic hierarchy continue to vary even within the major groups of plants. The ranks of species and below, however, are now fairly clearly defined as regards the vascular plants, at least from the theoretical standpoint (Du Rietz, 1930; Davis and Heywood, 1963). Despite the impossibility of carrying out incompatibility studies in lichens and in most non-lichen-forming fungi, concepts of what constitutes a species, subspecies, variety or form derived from vascular plant taxonomy can be applied in both the lichens and non-lichenized fungi (Hawksworth, 1974). There has been a tendency in both mycology and lichenology to base new species on single characters but this approach, at least in some groups of lichens, appears to be contrary to the con-cepts of vascular plant taxonomists. R. Santesson (*in* Shibata, 1974) noted that "For me it is not an acceptable taxonomy to base a species on one character only, which is made by lichenologists describing 'chemical species'."

Poelt (1972), in rejecting a system for the *Cetrelia olivetorum–C. cetrarioides* complex based on the utilization of subspecific and varietal categories, noted that such a system ". . . wäre begründbar, wird aber der Isolierung der abgelei-teten Sippen und den Verbreitungsunterschieden nicht gerecht, und ist zudem sehr schwerfällig; . . ."*. As mentioned above (p. 149), Poelt considered species rank for the chemical races here the most favourable in our present state of knowledge.

---

* ". . . would be justifiable, but will not do justice to the isolation of derived populations and differences in distribution, and moreover is too clumsy" [Trans.].

The primary aim of taxonomy is perhaps to provide a means, by way of latinized names, of communicating suites of morphological and other characters which a particular individual or population has in common with the nomenclatural type of the name. The information content of a name must be suited to the needs of most of those who wish to use it. Ideally the units recognized should also reflect phylogenetic relationships but the ranks at which the taxonomic system does this can be varied. The basal taxonomic rank of species in the case of the lichens must have an information content conveying significant morphological, anatomical and chemical properties, particularly where these are related to ecological and geographical distributional factors.

Lichen chemotaxonomists are not the only botanists who use lichen names and so lichenologists must be wary of devising taxonomic systems which cannot readily be used by others. A situation in which in order to name a specimen to species rank, an ecologist has to examine much of his material both by thin-layer chromatography and study it under the scanning electron microscope is not perhaps in the best long-term interests of lichenology, but may be being approached in some cases (see Hale, 1972). Such a system may reflect a much more thorough knowledge of the characteristics of a particular taxon but in time it could result in entries like "*Cladonia* sp." and "*Parmelia* sp." becoming even more frequent in non-taxonomic lichenological and ecological publications than is the case at present.

A sensitive taxonomy reflecting relationships, including chemical data, can be achieved by the careful application of clear concepts of taxonomic ranks. In some cases the rank of species appears to be the most suitable one for certain chemical races but it is interesting that, in many instances where chemical races meriting this rank occur, subtle (or overlooked) morphological characters often seem to be correlated with the chemical differences. A taxonomy utilizing infraspecific categories may perhaps be somewhat clumsy, as noted by Poelt (1972, see above), but if it is in the best interests of most users of lichen taxonomy it should not be rejected merely on that basis. Most lichenologists in North America in particular seem to have something of an aversion to infraspecific categories and tend to treat both morphological and chemical variation worthy of taxonomic recognition only at species rank, not recognizing infraspecific taxa. Interestingly this is not true of North American phanerogamists. Such a "species or nothing" approach, however, can only lead to a relatively insensitive taxonomic system. This view perhaps originates in no small measure from the use of infraspecific categories in the past for taxonomically meaningless environmental modifications (see Hawksworth, 1973b).

The remainder of this paper will endeavour to demonstrate how chemical

TABLE I. The principle types of specific and infraspecific chemical variation in lichens and guide-lines for proposed taxonomic treatments (see text)

1. *Replacement of one substance by one or more biogenetically distinct substances*
   A. Correlated with morphological and (or) ecological differences—**species** (e.g. *Cladonia grayi*, *Parmeliopsis hyperopta*)
   B. Correlated with major geographical differences—**species** (e.g. *Thamnolia subuliformis*)
   C. Not correlated with morphological, ecological or geographical differences—**none**

2. *Replacement of one substance by one or more biogenetically closely related substances*
   A. Correlated with morphological and distributional differences—**species** (e.g. *Parmelia discordans*, *P. subrudecta*)
   B. Correlated with local geographical differences or tendencies—**variety** (e.g. *Pseudevernia furfuracea* var. *ceratea*)
   C. Correlated with ecological or microhabitat differences—**variety** (e.g. *Ramalina curnowii–R. siliquosa* varieties)
   D. Not correlated with morphological, ecological or geographical differences—**none**

3. *Presence of one or more unreplaced substances*
   A. Correlated with major geographical differences—**species** (e.g. *Haematomma lapponicum*)
   B. Correlated with differences in ecological amplitude—**variety** (e.g. *H. ochroleucum* var. *porphyrium*)
   C. Correlated with local distributional differences or tendencies—**variety** (e.g. *Cladonia tenuis* var. *leucophaea*)
   D. Not correlated with morphological, ecological or geographical differences—**none** (e.g. fumarprotocetraric acid in *Cladonia chlorophaea* group, usnic acid in *Haematomma ventosum*)

4. *Variations in concentration of particular substances*
   A. Correlated with light intensity—**none** (e.g. parietin in *Xanthoria parietina*, usnic acid in *Cladonia subtenuis**)
   B. Correlated with the heavy metal content of the substrate—**none** (e.g. "oxydated" thalli of *Lecidea lapicida*)
   C. Not known to be correlated with any environmental or distributional factors—**none** (e.g. fumarprotocetraric acid in *Usnea fasciata*)

* f. *subtenuis*, not f. *cinerea* which is in category 3D (Ahti, *in litt.*).

data can contribute to the production of workable taxonomies by means of particular examples. These examples have been selected so that between them the principal types of chemical variation known to occur in the lichens (Table I) are considered.

The usage of specific and infraspecific ranks adopted in the following sections follows the principles of Du Rietz (1930), Davis and Heywood (1963) and Hawksworth (1974).

SUPRASPECIFIC CHEMICAL VARIATION

Lichenologists have generally accepted, since the important paper of Dahl (1952, see p. 143), that chemical data can, by correlation with morphological characters, contribute to the assessment of affinities and classification at the supraspecific level. W. Culberson and C. Culberson (1970) provide an excellent account of the value of chemical components at the family and generic levels. These authors considered, for example, that the separation of the Ramalinaceae from the Usneaceae was supported by chemical data as no species of the Usneaceae s.s. produce orcinol *meta*-depsides whilst these are richly represented in the Ramalinaceae (this separation was also supported by Poelt (1974) and Henssen and Jahns (1973) on other grounds). In the case of the Teloschistaceae, R. Santesson and J. Santesson (J. Santesson, 1970) elegantly demonstrated that a remodelling of the classification adopted by Zahlbruckner on both morphological and chemical grounds (the occurrence of parietin) led to a much more satisfactory concept of the family (Fig. 3).

At the generic level chemical data clearly supports the segregation of *Hypogymnia* and *Menegazzia* from *Parmelia* and has been used to support the separation of *Anaptychia* into two genera on the basis of differences in the ascospores (W. Culberson, 1966). In the Parmeliaceae, W. Culberson and C. Culberson (1968) found that two new genera could be recognized on a combination of chemical and morphological characters; all species of one (*Platismatia*) produce caperatic acid in the medulla and have large ascospores and a thin hypothecium, whilst those of the other (*Cetrelia*) do not produce caperatic acid and have relatively small ascospores and a thick hypothecium (subhymenium). Also, in *Cetrelia*, orcinol depsides and depsidones occur; *Platismatia* has one β-orcinol depsidone (fumarprotocetraric acid) unknown in *Cetrelia* (Table II).

Within *Cladonia* the correlation of supraspecific groupings with the chemical components has been demonstrated by Dahl (1952) and they perhaps support the recognition of *Cladina* at generic rank (Nourish and Oliver, 1974). As pointed out by Nourish and Oliver, however, caution is required in such cases where unstudied, poorly known and possibly intermediate taxa occur. In the case of

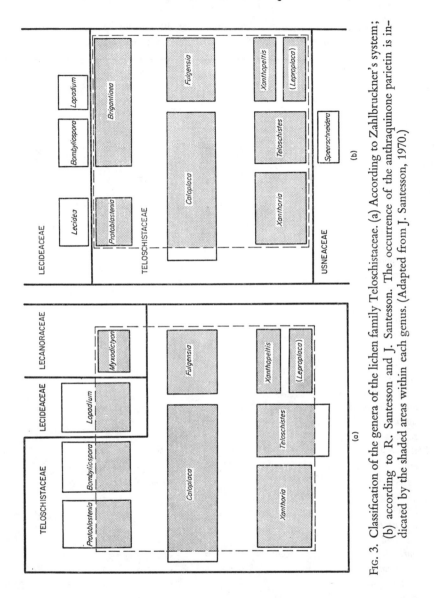

FIG. 3. Classification of the genera of the lichen family Teloschistaceae. (a) According to Zahlbruckner's system; (b) according to R. Santesson and J. Santesson. The occurrence of the anthraquinone parietin is indicated by the shaded areas within each genus. (Adapted from J. Santesson, 1970.)

Alectoria there is also a good correlation between chemical and other characters at the subgeneric level (Hawksworth, 1972a) but the genus has not been split by Hawksworth into four distinct genera as this would separate groups which are probably closely related, and no legitimate name in generic rank is available for the bulk of the species. As emphasized elsewhere (Hawksworth, 1974), changes

TABLE II. The constant chemical components of all known species of *Cetrelia* and *Platismatia* (after W. Culberson and C. Culberson, 1968)

Species	attranorin	caperatic acid	alectoronic acid	anziaic acid	α-collatolic acid	fumarprotocetraric acid	imbricaric acid	microphyllinic acid	olivetoric acid	perlatolic acid
Cetrelia										
alaskana	+	.	.	.	.	.	+	.	.	.
braunsiana	+	.	+	.	+	.	.	.	.	+
cetrarioides s.s	+	.	.	.	.	.	.	.	.	+
cetrarioides s.l.	+	.	.	.	.	.	+	.	.	.
chicitae	+	.	+	.	+	.	.	.	.	.
collata	+	.	.	.	.	.	+	.	.	.
davidiana	+	.	.	.	.	.	.	.	+	.
delavayana	+	.	.	.	.	.	.	.	.	+
isidiata	+	.	.	+	.	.	.	.	.	.
japonica	+	.	.	.	.	.	.	+	.	.
nuda	+	.	+	.	+	.	.	.	.	.
olivetorum	+	.	.	.	.	.	.	.	+	.
pseudolivetorum	+	.	.	.	.	.	.	.	+	.
sanguinea	+	.	.	+	.	.	.	.	.	.
sinensis	+	.	.	.	.	.	+	.	.	.
Platismatia										
erosa	+	+	.	.	.	.	.	.	.	.
formosana	+	+	.	.	.	.	.	.	.	.
glauca	+	+	.	.	.	.	.	.	.	.
herrei	+	+	.	.	.	.	.	.	.	.
interrupta	+	+	.	.	.	.	.	.	.	.
lacunosa	+	+	.	.	.	+	.	.	.	.
norvegica	+	+	.	.	.	.	.	.	.	.
regenerans	+	+	.	.	.	.	.	.	.	.
stenophylla	+	+	.	.	.	.	.	.	.	.
tuckermanii	+	+	.	.	.	.	.	.	.	.

in generic names are not something to be introduced lightly when one is dealing with large numbers of species with familiar generic names. It is naturally gratifying to find that chemical data tends to correlate with supraspecific categories below the rank of genus based primarily on non-chemical characters but whether additional supporting information of this nature should be used to introduce narrower generic concepts is a matter for taxonomic judgement.

A further rôle for chemical data at the supraspecific level is in attempts to establish the generic position of sterile, particularly crustose, lichens whose ascocarps are unknown. The inclusion of the sterile genus *Leproplaca* in the Teloschistaceae (see Fig. 3) is supported by the presence of parietin, for example, and the newly described sterile species *Schismatomma niveum* was placed in *Schismatomma* primarily on a consideration of the phycobiont and chemical components (James, 1971).

SPECIFIC CHEMICAL VARIATION

Species were defined by Du Rietz (1930) as "The smallest natural populations permanently separated from each other by a discontinuity in the series of biotypes . . .". Each species should be separated from all others by marked discontinuities, ideally in several unrelated characters (see p. 155). Within each genus at least species should be separated from one another by characters of a comparable order (although this criterion may be difficult to apply in groups which have not been subject to recent comprehensive monographic studies). If taxonomically significant characters are of a lesser order, then an infraspecific rank might be more appropriate.

In many lichens differences in chemical components and morphological characters are correlated and in such cases species rank is clearly appropriate. There are some instances where an examination of chemical characters has led to previously overlooked morphological characters being found to correlate with them. One of the best examples of this is to be found in the *Parmelia borreri* group (W. Culberson, 1961a, 1962; Hale, 1965a; Krog, 1970a; Hawksworth, 1972b, 1973a). In this case atranorin is consistently produced in the cortex but the medullary components can differ (Table III). *P. borreri* contains the tridepside gyrophoric acid whilst *P. subrudecta* has the corresponding depside lecanoric acid; the former has a black rather than pale tan underside to the lobes and a more southerly distribution in Europe. As there is a correlation here between a replaced chemical component, morphological and distributional characters (Table I category 2A) species rank is clearly appropriate and now generally accepted. *P. reddenda* presents a different problem for, apart from a tendency to form ascocarps more rarely, it appears to be morphologically very similar to *P. borreri*. In this case fatty acids are produced instead of depsides in the medulla which may be taken to indicate fundamental biogenetic differences (see Fig. 1). If the European distribution of *P. reddenda* is considered it appears to have a somewhat western tendency in contradistinction to the more southerly

TABLE III. Correlation of chemical and morphological characters in the *Parmelia borreri*
species complex in Europe

Species	Cortex	Medulla	Colour of underside	Pseudo-isidia	Apothecia	Substrate
P. borreri	atranorin	gyrophoric acid	dark	absent	common	trees
P. subrudecta	atranorin	lecanoric acid	pale	absent	rare	trees and rocks
P. reddenda	atranorin	fatty acids	dark	absent	rare	trees
P. stictica	atranorin	gyrophoric acid	dark	present	rare	rocks

tendencies of *P. borreri*. In view of the different types of chemical components
produced in the medulla and the distributional factors, I consider that *P. reddenda*
should also be regarded as a species despite the apparent absence of clear morphological differences.

Parmelia discordans and *P. omphalodes* are extremely similar morphologically
but they differ in one medullary component, the former having protocetraric
and the latter salazinic acid (W. Culberson, 1970b; Table IV). These compounds
are closely related (see Fig. 4) but *P. discordans* tends to have a more western
(oceanic) distribution in Europe and also a panniform habit, smoother lobes and
duller thallus colour. The morphological characters are more subtle than in the
P. borreri–P. subrudecta species pair but together with the chemistry and distributional factors support their retention at species rank as advocated by
Culberson.

In the *Cladonia chlorophaea* group, the cryptochlorophaeic and merochloro-

TABLE IV. Chemical components of *Parmelia discordans*
and *P. omphalodes* in Europe

Chemical components	*P. discordans*	*P. omphalodes*
Cortex		
atranorin	+	±
Medulla		
lobaric acid	+	+
protocetraric acid	+	−
salazinic acid	−	+

± = sometimes present

phaeic acid races were considered rather remote from the grayanic acid race by Shibata and Chiang (1965a) on biogenetic grounds. Ahti (1966) suggested that subtle morphological correlations might also occur and the subsequent studies of Kristinsson (1971) established that, at least in North Carolina, there was a relationship to the size of the soredial granules. For this reason it is evident that the race with grayanic acid (*C. grayi*) merits retention at specific rank (Table I category 1A) whilst the other two races are perhaps more appropriately accorded lesser status. The psoromic acid race of *C. cariosa* reported by Dahl (1950) was also found to be correlated with morphological differences when more specimens became available and so Kristinsson (1974) is justified in treating it as a distinct species (*C. dahliana*).

Probably in the foregoing examples the situation has been one of replacements of compounds correlated with morphological and (or) distributional characters. There are instances, however, where biogenetically different compounds are replaced and where these correlate with distributional but not any clear morphological differences (Table I category 1B). *Thamnolia vermicularis* has two chemical races, one containing thamnolic acid (*T. vermicularis* s.s.) and one containing both baeomycesic and squamatic acids (*T. subuliformis*). The distribution of these two chemical races has been studied by Sato (e.g. Sato, 1968) who has found that the former seems to predominate in the southern hemisphere while the latter predominates in the northern hemisphere. In view of the very different chemical compounds involved and the major geographical distributional differences it is appropriate to follow W. Culberson (1963) and regard these chemical races as separate species despite the lack of evidence for any clear morphological correlations (but see Poelt, 1969).

There also appear to be a few rare instances where an unreplaced chemical not correlated with any morphological differences but associated with ecological or geographical differences may merit treatment at species rank. An example in this category (Table I category 3A) is provided by *Haematomma lapponicum* and *H. ventosum* which differ primarily in that the former lacks thamnolic acid whilst the latter contains it (Table V). When the world distributions of these taxa are considered, the former is found to be circumboreal while the latter is exclusively European (see map in Hawksworth, 1973b). Furthermore, *H. ventosum* has a much wider altitudinal range in Europe than the more arctic–alpine *H. lapponicum* and from a consideration of their distributions may perhaps have evolved from *H. lapponicum*. In this case, following W. Culberson (1964) and Thomson (1968a), species rank appears to be most appropriate (Hawksworth, 1970, 1973b). This view is also supported by the occasional occurrence of barbatic acid and zeorin in *H. lapponicum* but not apparently in *H. ventosum* and by the supposition that physiological differences also occur (based on the greater altitudinal amplitude of *H. ventosum*).

TABLE V. Chemical components reported in *Haematomma lapponicum* and *H. ventosum*

Chemical components	H. lapponicum	H. ventosum
barbatic acid	±	—
decarboxythamnolic acid[a]	—	±
divaricatic acid	+	+
haemoventosin[b]	±[b]	±[b]
thamnolic acid	—	+
(+)-usnic acid	+	±
zeorin	±	—

± = sometimes present
[a] Possibly an artifact due to the decomposition of thamnolic acid on chromatography
[b] The red pigment of the apothecia; present only in fertile specimens

INFRASPECIFIC CHEMICAL VARIATION

1. Subspecies

A subspecies was defined by Du Rietz (1930) as ". . . a population of several biotypes forming a more or less regional facies of a species. . . . The various subspecies of a species are continuously intergrading into each other, their delimitation thus being much more arbitrary than that of the species." In recent years the concept of subspecies has gained considerable popularity amongst vascular plant taxonomists for geographical facies of species but has not yet been taken up to any significant extent by cryptogamic taxonomists, perhaps largely because its use requires an intimate knowledge of the variation in populations. There is one notable exception in lichenology, the studies of Imshaug and Brodo (1966) on the *Lecanora pallida* complex. As already mentioned above (p. 146), these authors employed this rank for chemical races which were correlated with ecological, regional and often subtle morphological traits. Their usage of this rank within *L. caesiorubella* (where five subspecies were distinguished) seems to accord well with the usage of this category by vascular plant taxonomists (see Davis and Heywood, 1963).

Another instance where subspecies rank may be most appropriate is the *Cetraria islandica–C. ericetorum* (syn. *C. islandica* subsp. *crispa*) complex. In Europe the narrow-lobed plants usually lack fumarprotocetraric acid (*C. ericetorum*) whilst the broader lobed plants contain this acid (*C. islandica*).

However, detailed studies in Iceland by Kristinsson (1969) revealed that there a complex situation occurred in which morphological and chemical traits were poorly correlated and there were indications of possible gene exchange between the chemical races. The accommodation of such situations within the formal taxonomic framework is not easy but, as suggested by Kristinsson, "Perhaps the best solution here is to . . . treat these taxa as subspecies."

As already mentioned above (p. 142), subspecies rank has been applied to some chemical races by Asahina. From the limited information available on many of these it is probable that most do not fall within the concept of a subspecies as outlined above but merit treatment in other infraspecific ranks where they are taxonomically significant. The rank of subspecies was employed by Laundon (1965) for *Haematomma lapponicum* but, as will be evident from the discussion of this example above (p. 163), specific rank seems to be the most appropriate in this case.

2. Variety

A variety was defined by Du Rietz (1930) as ". . . a population of one or several biotypes forming a more or less distinct facies of a species". The use of the rank of variety to accommodate some types of chemical races has been a feature of the work of several European lichen taxonomists in recent years but has not gained general acceptance outside Europe. However, this seems to be the most satisfactory rank for a number of types of chemical races.

In the case of *Pseudevernia furfuracea*, two chemical races occur, one with the depside olivetoric acid in the medulla and one with the corresponding depsidone physodic acid in the medulla. The probability that a depsidone is closely related biogenetically to its corresponding theoretical precursor depside has already been mentioned (p. 152) and in this case there is a report of their joint occurrence (C. Culberson, 1965). No morphological correlations occur in this case but there are clear distributional tendencies in Europe, the olivetoric acid race predominating in northern and the physodic acid race in southern Europe (Hale, 1956c, 1968b; Hawksworth and Chapman, 1971). In view of the biogenetic affinity of the compounds and the nature of the geographical distribution, the races are more or less distinct regional facies of a species (Table I category 2B) and so the rank of "variety" appears to be particularly appropriate; i.e. the olivetoric acid race is called var. *ceratea* (syn. var. *olivetorina*, *P. olivetorina*).

A more complex but in many ways comparable situation exists in the *Ramalina siliquosa* group studied by W. Culberson (1967a, 1969a, c), W. Culberson and C. Culberson (1967a) and Follmann (1973). In common with Krog (1969) and many earlier workers, Hawksworth (1972b) suggested that two species should be recognized on morphological grounds, *R. curnowii* and *R. siliquosa*. Fletcher (1975) has recently reported that the cortical characters also support the recognition of these taxa as distinct and each with various chemical races. The chemical races in these two species show quantitative distributional tendencies and habitat selection and so clearly merit some taxonomic recognition. As noted by W. Culberson (1967a), however, the chemicals involved are very closely related biogenetically (Fig. 4) and so the rank of variety again

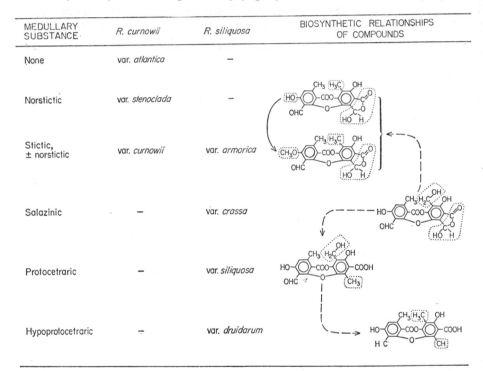

MEDULLARY SUBSTANCE	*R. curnowii*	*R. siliquosa*	BIOSYNTHETIC RELATIONSHIPS OF COMPOUNDS
None	var. *atlantica*	—	
Norstictic	var. *stenoclada*	—	
Stictic, ± norstictic	var. *curnowii*	var. *armorica*	
Salazinic	—	var. *crassa*	
Protocetraric	—	var. *siliquosa*	
Hypoprotocetraric	—	var. *druidarum*	

FIG. 4. Chemotaxonomy of the *Ramalina siliquosa–R. curnowii* complex. [Suggested biosynthetic relationships of the compounds (hatched line = possible; solid line = probable) follows W. Culberson (1967a).]

appears to be the most appropriate one here. It should perhaps be noted that some additional chemical races appear to occur in this complex (Imshaug and Harris, 1971; Follmann, 1974) which are poorly understood and require more detailed study. Habit selection involving closely allied compounds has also been reported in *Cladonia polycarpoides–C. polycarpia* (W. Culberson, 1969b) and *Ramalina subfarinacea* (Hawksworth, 1968) and in these instances also the rank of variety seems to be the most appropriate.

Some chemical races involving unreplaced compounds also appear to fulfil the requirements of varietal rank. *Cladonia tenuis* var. *leucophaea*, differing from var. *tenuis* in the absence of usnic acid, also has a more oceanic distribution in Europe than var. *tenuis* and this led Ahti (1961) to treat it in this rank (see also Nourish and Oliver, 1974). Ahti suggested that the species *Parmeliopsis ambigua* and *P. hyperopta* might be a comparable case but as usnic acid occurs in the former and is apparently replaced by atranorin in the latter (C. Culberson, 1969) and there are also ecological, slight distributional and perhaps subtle morphological differences, this appears to fall within Table I category 1A rather than 2c and so species rank is probably acceptable in this case although the situation may be more complex here than work so far published indicates (Rundel, unpublished).

An example of ecological preferences related to an unreplaced compound is seen in *Haematomma ochroleucum* where in Europe the usnic acid-containing race (var. *ochroleucum*) occurs predominantly on siliceous rocks whilst a race lacking this acid (var. *porphyrium*) occurs commonly on both siliceous rocks and trees. These habitat preferences led Laundon (1970) to follow earlier European workers in using the rank of variety rather than that of species or no rank. W. Culberson (1964) treated these races as distinct species but saw only one North American specimen which was of the usnic acid absent race; if further studies establish that there is a major geographical difference as well, this example might fall in Table I category 3A rather than 3B and merit reconsideration for specific rank.

3. Form

The rank of form was defined by Du Rietz (1930) as ". . . a population of one or several biotypes in a species-population (not forming distinct regional or local facies of it) and differing from the other biotypes of this species-population in one or several distinct characters". In lichenology, this rank has come to be applied to numerous taxonomically meaningless environmental modifications. This is perhaps unfortunate in some ways as it has led to its being held in disrepute although there may be instances where a single distinct, genetically determined, character which it might be desirable to distinguish taxonomically occurs sporadically throughout the range of a species. Some types of chemical variation, for example the rare pigment-deficient races of *Caloplaca verruculifera* (Søchting, 1973) and *Solorina crocea* (Krog, 1970b), may well fall within the concept of "forma" as used by vascular plant taxonomists. In the case of lichenology and mycology, because of the previous misuse of the rank, it should be avoided wherever possible (Hawksworth, 1974). Most examples which might be considered as belonging to this rank scarcely merit any taxonomic recognition and are considered separately below.

TAXONOMICALLY INSIGNIFICANT CHEMICAL VARIATION

The major part of the chemical variations in lichens appear to have little taxonomic significance. In some instances this may be due to inadequate study where attempts to relate the chemical variations to ecological, distributional or morphological characteristics have simply not been carried out. Many such races may be referred to as "chemotypes" as proposed by R. Santesson (1968; see p. 146).

1. Accessory Substances

Accessory substances are ones which occur sporadically in a species, usually in addition to other constituents which may be showing replacement patterns of chemical variation. In some cases their reported presences and absences may perhaps be merely a reflection of the techniques employed but technique sensitivity does not appear to be responsible in all instances. Accessory substances are particularly well documented for *Cladonia* (see C. Culberson, 1969) where, as demonstrated by Kristinsson (1971) in the *C. chlorophaea* group, it is replacement chemical variation rather than the occurrence of accessory fumarprotocetraric acid which seems to be taxonomically important.

Haematomma ventosum has a race which produces usnic acid and here the production of the acid appears to be genetically determined as plants with and without it occur side by side (Hawksworth, 1970). No correlations with other factors seem to occur and so no taxonomic significance has been attached to this type of variation (Table I category 3D). A comparable situation to that in *H. ventosum* seems to occur in the parietin-lacking chemotype of *Caloplaca verruculifera* (Søchting, 1973). Similarly, *Alectoria virens* has a chemical race with the depsidone virensic acid which is scattered through its range and not correlated with morphological or ecological variations; virensic acid is not replaced by any other constituent and so no taxonomic recognition for this race appears to be justified (Hawksworth, 1971b).

Evernia prunastri may well be comparable to the situation in *H. ventosum*, but in this case the position is more complex as the concentration of usnic acid varies more markedly due apparently to illumination effects (see below); the usnic acid "absent" race ("*E. herinii*"; Duvigneaud, 1940) appears to have no taxonomic significance (see C. Culberson, 1963a, b; Hawksworth, 1972b).

2. Variations in Concentration

Some pigmented compounds deposited in lichen cortices in particular show variations in their concentration. Such variations appear to be a response to the light intensity perhaps serving as a method of protecting the algal layer from the incident light (Hawksworth, 1973b; J. Santesson, 1973). The cases of usnic acid in *Cladonia subtenuis* f. *subtenuis* and parietin in *Xanthoria parietina* have been investigated by Rundel (1969) and Hill and Woolhouse (1966), respectively. This phenomenon appears to be rather widespread in some lichen genera but is not discussed further here as this aspect has recently been reviewed by Hawksworth (1973b). This type of variation (Table I category 4A) simply reflects a phenotypic response of the genotype to the degree of illumination and is clearly without taxonomic significance.

Some variations in concentrations of medullary components seem to occur in, for example, some species of *Cladonia* where if there is sufficient fumarprotocetraric acid a K+ brownish reaction sometimes occurs. Lamb (1964) also reported this in some Antarctic *Usnea* species (e.g. *U. fasciata*) where differing concentrations of the same acid gave a range of reactions with K and PD. Lamb referred to these as "active" (K+ brown, PD+ intense red), "semi-active" (K−, PD+ intense red) and "inactive" (K−, PD−) phases and they are clearly without taxonomic significance (Table I category 4B).

Lastly, some comment should perhaps be made of a third type of variation in concentration, that due to the uptake of heavy metals from the substrate giving the thallus an "oxydated" appearance (e.g. in *Lecidea lapicida*). This type of variation (Table I category 4C) is also almost always clearly of no taxonomic significance and is discussed further by Hawksworth (1973b).

3. Localization of Chemical Components

The chemical constituents of lichens used by taxonomists are mainly deposited as crystals on the surfaces of fungal hyphae (see Hale, Chapter 1, Plate IX). In many genera the cortical layers contain one or more components and the medulla has one or more different compounds. The cortical component is usually atranorin, lichexanthone or a pigmented compound (e.g. usnic acid, vulpinic acid, parietin) whilst most colourless depsides and depsidones tend to be confined to the medulla. As will be evident from the preceding sections it is the variation in the medullary components which are of the greatest taxonomic significance.

Further localization of compounds may cause some problems for the lichen chemotaxonomist. Restriction of depsidones to the apical parts of *Cladonia* subgen. *Cladina* podetia has long been recognized (see Ahti, 1961) but this phenomenon is by no means confined to that genus. *Alectoria tenuis*, for example, has frequently been considered to comprise PD− and PD+ chemical races (with or without fumarprotocetraric acid) but an examination of large numbers of specimens of this species by Brodo and Hawksworth (unpublished) established that with the possible exception of one very old herbarium specimen all (including the holotype) contained fumarprotocetraric acid at least in some of the apical branches (many specimens have some apices PD− and others PD+).

The restriction of brightly coloured pigments to the apothecia of some lichens (e.g. rhodocladonic acid in *Cladonia* subgen. *Cocciferae*, haemoventosin in *Haematomma ventosum*) is well known. An even greater degree of localization was found to occur in *Letharia californica* (i.e. *L. columbiana*) by W. Culberson (1969d). In this case norstictic acid was demonstrated in the hymenium of 47% of the fertile specimens examined but was not detected in the vegetative thallus of any he investigated.

In *Cladonia strepsilis* Nuno (1973) has recently reported that, in addition to baeomycesic and squamatic acids and strepsilin which occurred in the basal squamules and podetia of all specimens studied, barbatic acid was also present in the podetia although it could scarcely be detected in the basal squamules even of specimens bearing podetia with large amounts of this acid. Examination of this case by more sensitive chemical methods (see Nourish and Oliver, 1974; Chapter 8) is required. Baeomycesic, barbatic and squamatic acids are probably closely related biogenetically and occur together in several other species.

From the examples cited above, it is evident that lichenologists must endeavour to examine different tissues carefully before using chemical data in their taxonomies, in order to prevent localized components being overlooked. An interesting aspect of localization is why a single mycobiont, if it is the biont primarily responsible for the production of chemical components (see p. 152), should produce one chemical in parts of a thallus but not in others.

DISCUSSION

The preceeding sections show how different types of chemical variation can be accommodated into various taxonomic ranks. The use of a range of ranks enables a more sensitive taxonomic system to be constructed than the "extreme" approach of either utilizing species rank or not granting any chemical race a formal taxonomic name in any rank. By adoption of the principles outlined above as "guidelines" for consideration of the future taxonomic status of particular chemical races (i.e. treating them in a comparable way to morphological variations as advocated by several earlier authors, see pp. 140–150) a satisfactory taxonomy can be achieved which will at the same time have an information content of taxonomically significant chemical variation and meet the requirements of most lichenologists who are not themselves chemotaxonomists (see p. 156).

The sensitivity of the system proposed here and its relationship to Poelt's "Artenpaare" concept may be illustrated by reference to the genus *Pseudevernia*. Poelt (1972) considered that on chemical and morphological grounds (i.e. frequency of apothecia v. frequency of asexual reproductive structures) the primary species *P. intensa* with lecanoric acid had given rise to the secondary species *P. consocians* with the same acid whilst *P. furfuracea* had arisen from an unknown or extinct primary species (Fig. 5). Lecanoric acid is probably not

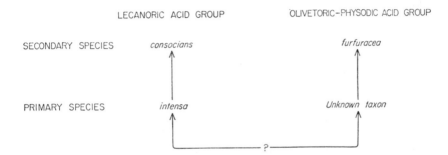

FIG. 5. Application of the "Artenpaare" concept to the main chemical races of the *Pseudevernia furfuraceae–P. intensa–P. consocians* species complex. (After Poelt, 1972.)

closely related to olivetoric acid biogenetically and so *P. consocians*, taking into account the geographical differences, would come in category 1B of Table I meriting treatment at specific rank (the fertile *P. intensa* meriting this rank on morphological grounds). Within *P. furfuracea* local distributional tendencies are correlated with biogenetically closely allied compounds (category 2B) and varietal rank for these within this species appears to be the most appropriate (see p. 165). In the case of the *Cetrelia olivetorum* complex discussed by Poelt (1972) imbricaric and olivetoric acids are probably not closely related biogenetically so here, on morphological and chemical grounds and taking into account distributional differences, the rank of species as proposed by W. Culberson and C. Culberson (1968) appears to be the most appropriate. In the more complex situation in the *Parmelia perforata* group (W. Culberson, 1973; W. Culberson and C. Culberson, 1973), in which parallel evolution appears to have taken place (Fig. 6), a combination of specific and subspecific ranks (suggested as a possibility by W. Culberson, 1973) might lead to the most acceptable taxonomy.

C. Culberson and Hale (1973) found that in the 19 species of *Parmelia* sect. *Hypotrachyna* four major chemical groups could be recognized involving closely allied compounds which had strong geographical correlations while the morphological types did not show such geographical correlations. These authors consider the main groups to have arisen by mutation and hybridization of chemical components to form other compounds and that morphological

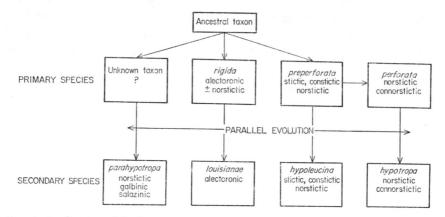

FIG. 6. Application of the "Artenpaare" concept to the *Parmelia perforata* species complex as interpreted by W. Culberson. (After W. Culberson, 1973.)

parallel evolution has taken place within the chemical groups to produce the extant taxa in the different regions.

So far most lichen taxonomists have adopted names only for chemical races where the chemical variation undeniably has some taxonomic and evolutionary significance. This approach is clearly desirable and when chemical variations are discovered it is essential that any possible correlations with ecological, distributional and morphological traits are investigated before new taxa are erected. The "guidelines" proposed in this paper cannot lead to acceptable taxonomies for cases which have not been the subject of adequate studies.

Subjective assessment must remain a part of the taxonomic treatment of chemical races just as it is a part of the taxonomic treatment of morphological variations. In addition to a consideration of the correlations of ecological, distributional and morphological characters, subjective assessment also comes into a consideration of the biogenetic affinity of the chemical compounds involved. As pointed out above (p. 150), a great deal more research by lichen biochemists into the biogenesis of lichen products is needed before much help from this field can be expected.

Other experimental work which should ideally be carried out in connection with lichen chemotaxonomy at the present time in order to resolve unanswered or only partly answered questions is (*a*) a study of the whole-thallus physiology in different chemical races under a range of environmental conditions, (*b*) a re-consideration of the rôle of the phycobiont(s) in different chemical races (see p. 154) and (*c*) long-term transplant experiments with ecologically characterized chemical races to eliminate entirely the possibility that in some case environmental factors may influence medullary components.

Poelt (1974) has pointed out the importance of reagent tests with iodine in some groups of lichens (e.g. *Lecidea* sect. *Silacea*) but the causes of these are almost completely unknown. Lichen chemotaxonomists should perhaps start to investigate the nature of the compounds causing these in more detail, in view of their apparent taxonomic importance. Other areas in which additional chemotaxonomic data might be forthcoming and which have yet not been taken up by lichenologists to any significant extent are electrophoretic studies of enzymes and other proteins (presumably feasible with isolated mycobionts at least), serological investigations using agglutination techniques (e.g. Barrett and Howe, 1968), studies on components of phycobionts (e.g. carotenoids) and critical studies on polysaccharides (see C. Culberson, 1969, 1970, for details of available data). Pigments in ascocarps or portions of ascocarps (cf. p. 170) also perhaps merit further study.

Debates as to the taxonomic treatment of chemical races in lichens will be watched with interest by mycologists who are now starting to conduct chemotaxonomic surveys in non-lichen-forming fungi. Compared with the lichens, remarkably little information is available on the constituents of non-lichen-forming fungi (Shibata *et al.*, 1964; Turner, 1971; Wolf, 1973). Taxonomic studies including chemical surveys are now starting to appear in fungal literature (e.g. Arpin, 1969), however, and it is now known that herbarium specimens of non-lichen-forming fungi can be examined by thin-layer chromatography and yield taxonomically interesting data (e.g. Nespiak *et al.*, 1973). As the nature of the chemicals becomes determined and increasing numbers of specimens are studied, chemotaxonomy can be expected to assume an increasingly important rôle in the taxonomy of other groups of fungi.

ACKNOWLEDGEMENTS

I am very grateful to my colleagues for discussions of various aspects of the principles outlined in this paper, in particular to Professor W. L. Culberson and his students for the care they took in commenting on the manuscript of a paper on this topic I presented to the First International Mycological Congress at Exeter in 1971; to Dr T. Ahti, Dr M. E.

Hale, Mr P. W. James and Dr P. W. Rundel for their comments on drafts of the present paper; and to Professor S. Shibata for kindly sending me a copy of the proofs of one of his papers, unpublished when this chapter went to press.

REFERENCES*

DES ABBAYES, H. (1933). Notes critiques sur quelques lichens Armoricains. *Revue bryol. lichén.* **6**, 68–77.

DES ABBAYES, H. (1949). Some new *Cladoniae* (Lichens) from Panama. *Bryologist* **52**, 92–96.

DES ABBAYES, H. (1951). "Traité de Lichenologie". [*Encycl. Biol.* **41**, i–x, 1–217]. Lechevalier, Paris.

AHMADJIAN, V. (1967). "The Lichen Symbiosis." Blaisdell, Waltham, Mass.

AHTI, T. (1961). Taxonomic studies on reindeer lichens (*Cladonia*, subgenus *Cladina*). *Annls bot. Soc. zool.-bot. fenn.* '*Vanamo*' **32** (1), i–iv, 1–160.

AHTI, T. (1966). Correlation of the chemical and morphological characters in *Cladonia chlorophaea* and allied lichens. *Annls bot. fenn.* **3**, 380–390.

ALMBORN, O. (1965). The species concept in lichen taxonomy. *Bot. Notiser* **118**, 454–457.

ARPIN, N. (1969). Les caroténoïdes des Discomycètes: essai chimiotaxinomique. *Bull. mens. Soc. linn. Lyon* **38**, *Suppl.*, 1–169.

ASAHINA, Y. (1934). Über die Reaktion von Flechten-Thallus. *Acta phytochim.* **8**, 47–64.

ASAHINA, Y. (1936–40). Mikrochemischer Nachweis der Flechtenstoffe I–XI. *J. Jap. Bot.* **12**, 516–525, 859–872; **13**, 529–536, 855–861; **14**, 39–44, 244–250, 318–323, 650–659, 767–773; **15**, 465–472; **16**, 195–193.

ASAHINA, Y. (1937). Über den taxonomischen Wert der Flechtenstoffe. *Bot. Mag., Tokyo* **51**, 759–764.

ASAHINA, Y. (1956). "Lichens of Japan. Vol. III. Genus *Usnea*." Research Institute for Natural Resources, Tokyo.

ASAHINA, Y. (1964). Lichenologische Notizen (§193). *J. Jap. Bot.* **39**, 165–171.

ASAHINA, Y. (1967). Lichenologische Notizen (§203). *J. Jap. Bot.* **42**, 1–9.

ASAHINA, Y. and SHIBATA, S. (1954). "Chemistry of Lichen Substances." Japan Society for the Promotion of Science, Tokyo.

BACHMANN, E. (1887). Mikrochemische Reaktionen auf Flechtenstoffe. *Flora, Jena* **70**, 291–294.

BACHMANN, O. (1963). Dünnschichtchromatographische Trennung von Flechtensäuren der β-Orcin-Gruppe. *Öst. bot. Z.* **110**, 103–107.

BARRETT, J. T. and HOWE, M. L. (1968). Hemagglutination and hemolysis by lichen extracts. *Appl. Microbiol.* **16**, 1137–1139.

---

* In addition to references cited in the text of this Chapter this list includes some further papers containing information pertinent to the use of chemistry in lichen systematics; these are prefixed by an asterisk (*). A comprehensive account of substances reported for different species is provided by C. Culberson (1969, 1970) and further information on microchemical methods can be traced through the works of Hawksworth (1971c), C. Culberson (1972), J. Santesson (1974), Hale (1974), Nourish and Oliver (1974), and Menlove (1974).

BRIEGER, W. (1923). Synthetische Versuche auf dem Gebiete der Flechtenstoffe und ihre Bausteine. *In* "Handbuch der biochemischen Arbeitsmethoden, Abteilung I: Chemische Methoden" (E. Aberhalden, ed.), **10**, 205–438. Urban and Schwarzenberg, Berlin.

*BYSTREK, J. (1972). "Zarys Lichenologii." Pánstwowe Wydawnictwo Naukowe, Warsaw.

ČERNOHORSKÝ, Z. (1950). Fluorescence of lichens in ultra-violet light. Genus *Parmelia* Ach. *Stud. bot. Čechosl.* **11**, 1–3.

ČERNOHORSKÝ, Z. (1957). Fluoreszenz der Flechten im ultravioletten Licht. II. Gattung *Cladonia* (Hill) Vain. *Preslia* **29**, 1–4.

CROMBIE, J. M. (1894). "A Monograph of Lichens found in Britain", **1**. British Museum (Natural History), London.

CULBERSON, C. F. (1963a). Sensitivities of some microchemical tests for usnic acid and atranorin. *Microchem. J.* **7**, 153–159.

CULBERSON, C. F. (1963b). The lichen substances of the genus *Evernia*. *Phytochemistry* **2**, 335–340.

CULBERSON, C. F. (1964). Joint occurrence of a lichen depsidone and its probable depside precursor. *Science, N.Y.* **143**, 255–256.

CULBERSON, C. F. (1965). A note on the chemical strains of *Parmelia furfuracea*. *Bryologist* **68**, 435–439.

CULBERSON, C. F. (1969). "Chemical and Botanical Guide to Lichen Products." University of North Carolina Press, Chapel Hill, N.C.

CULBERSON, C. F. (1970). Supplement to "Chemical and Botanical Guide to Lichen Products". *Bryologist* **73**, 177–377.

CULBERSON, C. F. (1972). Improved conditions and new data for the identification of lichen products by a standardized thin-layer chromatographic method. *J. Chromatogr.* **72**, 113–125.

CULBERSON, C. F. and CULBERSON, W. L. (1958). Age and chemical constituents of individuals of the lichen *Lasallia papulosa*. *Lloydia* **21**, 189–192.

CULBERSON, C. F. and HALE, M. E. (1973). Chemical and morphological evolution in *Parmelia* sect. *Hypotrachyna*: product of ancient hybridization? *Brittonia* **25**, 162–173.

CULBERSON, C. F. and KRISTINSSON, H. (1969). Studies on the *Cladonia chlorophaea* group: a new species, a new *meta*-depside, and the identity of "novochlorophaeic acid". *Bryologist* **72**, 431–443.

CULBERSON, C. F. and KRISTINSSON, H. (1970). A standardized method for the identification of lichen products. *J. Chromatogr.* **46**, 85–93.

CULBERSON, W. L. (1958). The chemical strains of the lichen *Parmelia cetrarioides* Del. in North America. *Phyton* **11**, 85–92.

CULBERSON, W. L. (1961a). *Parmelia pseudoborreri* Asahina, lichen nouveau pour la flore d'Europe, et remarques sur les "espèces chemiques" en lichénologie. *Revue bryol. lichén.* **29**, 321–325.

CULBERSON, W. L. (1961b). Proposed changes in the International Code governing nomenclature of lichens. *Taxon* **10**, 161–165.

CULBERSON, W. L. (1962). Some pseudocyphellate *Parmeliae*. *Nova Hedwigia* **4**, 563–577.

CULBERSON, W. L. (1963). The lichen genus *Thamnolia*. *Brittonia* **15**, 140–144.

CULBERSON, W. L. (1964). A summary of the lichen genus *Haematomma* in North America. *Bryologist* **66**, 224–236.

CULBERSON, W. L. (1966). Chemistry and taxonomy of the lichen genera *Heterodermia* and *Anaptychia* in the Carolinas. *Bryologist* **69**, 472–487.

CULBERSON, W. L. (1967a). Analysis of chemical and morphological variation in the *Ramalina siliquosa* species complex. *Brittonia* **19**, 333–352.

CULBERSON, W. L. (1967b). Chimie et taxonomie des lichens du groupe *Ramalina farinacea* en Europe. *Revue bryol. lichén.* **34**, 841–851.

CULBERSON, W. L. (1969a). The behaviour of the species of the *Ramalina siliquosa* group in Portugal. *Öst. bot. Z.* **116**, 85–94.

CULBERSON, W. L. (1969b). The chemistry and systematics of some species of the *Cladonia cariosa* group in North America. *Bryologist* **72**, 377–386.

CULBERSON, W. L. (1969c). The use of chemistry in the systematics of the lichens. *Taxon* **18**, 152–166.

CULBERSON, W. L. (1969d). Norstictic acid as a hymenial constituent of *Letharia*. *Mycologia* **61**, 731–736.

CULBERSON, W. L. (1970a). Chemosystematics and ecology of lichen-forming fungi. *A. Rev. Ecol. Syst.* **1**, 153–170.

CULBERSON, W. L. (1970b). *Parmelia discordans*, lichen peu connu d'Europe. *Revue bryol. lichén.* **37**, 183–186.

*CULBERSON, W. L. (1972). Disjunctive distributions in the lichen-forming fungi. *Ann. Mo. bot. Gdn* **59**, 165–173.

CULBERSON, W. L. (1973). The *Parmelia perforata* group: niche characteristics of chemical races, speciation by parallel evolution, and a new taxonomy. *Bryologist* **76**, 20–29.

CULBERSON, W. L. and CULBERSON, C. F. (1956). The systematics of the *Parmelia dubia* group in North America. *Am. J. Bot.* **43**, 678–687.

CULBERSON, W. L. and CULBERSON, C. F. (1967a). Habitat selection by chemically differentiated races of lichens. *Science, N.Y.* **158**, 1195–1197.

CULBERSON, W. L. and CULBERSON, C. F. (1967b). A new taxonomy for the *Cetraria ciliaris* group. *Bryologist* **70**, 158–166.

CULBERSON, W. L. and CULBERSON, C. F. (1968). The lichen genera *Cetrelia* and *Platismatia* (Parmeliaceae). *Contr. U.S. natn. Herb.* **34**, 449–558.

CULBERSON, W. L. and CULBERSON, C. F. (1970). A phylogenetic view of chemical evolution in the lichens. *Bryologist* **73**, 1–31.

CULBERSON, W. L. and CULBERSON, C. F. (1973). Parallel evolution in lichen-forming fungi. *Science, N.Y.* **180**, 196–198.

DAHL, E. (1950). Studies in the macrolichen flora of south west Greenland. *Meddr Grønland* **150** (2), 1–176.

DAHL, E. (1952). On the use of lichen chemistry in lichen systematics. *Revue bryol. lichén.* **21**, 119–134.

DAHL, E. and KROG, H. (1973). "Macrolichens of Denmark, Finland, Norway and Sweden." Universitetsforlaget, Oslo, Bergen and Tromsø.

DAVIS, P. H. and HEYWOOD, V. H. (1963). "Principles of Angiosperm Taxonomy." Oliver and Boyd, Edinburgh and London.

DEGELIUS, G. (1937). Lichens from southern Alaska and the Aleutian Islands, collected by Dr E. Hultén. *Acta Horti gothoburg.* **12**, 105–144.

DEGELIUS, G. (1939). Die Flechten von Norra Skaftön. *Uppsala Univ. Åsskr.* **1939** (11), 1–206.

DODGE, C. W. (1973). "Lichen Flora of the Antarctic Continent and Adjacent Islands." Phoenix Publishing, Canaan, N.H.

DU RIETZ, G. E. (1930). The fundamental units of biological taxonomy. *Svensk bot. Tidskr.* **24**, 333–428.

DUVIGNEAUD, P. (1940). L'acide usnique et les espèces dites "chimiques", en lichénologie, cas de *Evernia prunastri* (L.) Ach. et *Evernia herinii* nov. spec. *Bull. Soc. r. bot. Belg.* **72**, 148–154.

★DUVIGNEAUD, P. and BLERET, L. (1940). Notes de microchimie lichénique II. Sur la valeur systématique de *Cladonia pycnoclada* (Pers.) Nyl. em. des Abb. *Bull. Soc. r. bot. Belg.* **72**, 155–159.

★EIGLER, G. and POELT, J. (1965). Flechtenstoffe und Systematik der Lobaten Arten der Flechtengattung *Lecanora* in Holarktis. *Öst. bot. Z.* **112**, 285–294.

EVANS, A. W. (1943a). Asahina's microchemical studies on the *Cladoniae*. *Bull. Torrey bot. Club.* **70**, 417–438.

EVANS, A. W. (1943b). Microchemical studies on the genus *Cladonia*. Subgenus *Cladina*. *Rhodora* **45**, 417–438.

FISCUS, S. A. (1972). A survey of the chemistry of the *Usnea florida* group in North America. *Bryologist* **75**, 299–304.

FLETCHER, A. (1975). Key for the identification of British marine and maritime lichens I. Siliceous rocky shore species. *Lichenologist* **7**, 1–52.

★FOLLMANN, G. (1969). Flechtensymbiose und Flechtenstoffe. *Bull. Soc. bot. Fr., Mém.* **1968**, *Coll. Lich.*, 161–167.

FOLLMANN, G. (1972). Zur vergleichenden Phytochemie der Basidiolichenen. *Willdenowia* **6**, 427–430.

FOLLMANN, G. (1973). Beobachtungen zum Vorkommen spanischer Flechten I. Der Formenkreis um *Ramalina siliquosa* (Huds.) A. L. Smith. *Philippia* **2**, 3–12.

FOLLMANN, G. (1974). Beobachtungen zum Vorkommen spanischer Flechten III. Eine neue Stranchflechte aus dem Verwandtschaftskreis von *Ramalina crassa* (Nyl.) Mot. *Philippia* **2**, 67–72.

★FOLLMANN, G. and HUNECK, S. (1969). Mitteilung über Flechteninhaltstoffe LXI. Zur Chemotaxonomie der Flechtenfamilie Ramalinaceae. *Willdenowia* **5**, 181–216.

FOX, C. H. and HUNECK, S. (1969). The formation of roccellic acid, eugenitol, eugenetin, and rupicolon by the mycobiont *Lecanora rupicola*. *Phytochemistry* **8**, 1301–1304.

FRIES, Th. M. (1867). Lichenes spitsbergenses. *K. svenska VetenskAkad. Handl., n.s.* **7** (2), 1–53.

★GRAHAM, W. L. (1969). The occurrence of the lichen complex, *Cetraria ciliaris*, in the straits region of Michigan. *Mich. Bot.* **8**, 67–71.

GYELNIK, V. (1932). *Alectoria* Studien. *Nyt Mag. Naturv.* **70**, 35–62.

GYELNIK, V. (1935). Conspectus *Bryopogonum*. *Feddes Repert.* **38**, 219–255.

HALE, M. E. (1952). Studies on the lichen *Rinodina oreina* in North America. *Bull. Torrey bot. Club* **79**, 251–259.

HALE, M. E. (1955). *Xanthoparmelia* in North America I. The *Parmelia conspersa–stenophylla* group. *Bull. Torrey bot. Club* **82**, 9–21.

HALE, M. E. (1956a). Fluorescence of lichen depsides and depsidones as a taxonomic criterion. *Castanea* **21**, 30–32.

HALE, M. E. (1956b). Chemical strains of the *Parmelia conspersa-stenophylla* group in south central United States. *Bull. Torrey bot. Club* **82**, 218–220.

HALE, M. E. (1956c). Chemical strains of the lichen *Parmelia furfuracea. Am. J. Bot.* **43**, 456–459.

HALE, M. E. (1957). "Lichen Handbook." Smithsonian Institution, Washington, D.C. [Mimeographed.]

HALE, M. E. (1961). "Lichen Handbook." Smithsonian Institution, Washington, D.C.

HALE, M. E. (1963). Populations of chemical strains in the lichen *Cetraria ciliaris. Brittonia* **15**, 126–133.

HALE, M. E. (1965a). Studies on the *Parmelia borreri* group. *Svensk bot. Tidskr.* **59**, 37–48.

HALE, M. E. (1965b). A monograph of *Parmelia* subgenus *Amphigymnia. Contr. U.S. natn. Herb.* **36**, 193–358.

HALE, M. E. (1966). Chemistry and evolution in lichens. *Israel J. Bot.* **15**, 150–157.

HALE, M. E. (1967). "The Biology of Lichens." Arnold, London.

HALE, M. E. (1968a). Biochemical systematics in lichens: another viewpoint. *Internat. Lichenol. Newsletter* **2** (1), **104**.

HALE, M. E. (1968b). A synopsis of the lichen genus *Pseudevernia. Bryologist* **71**, 1–11.

HALE, M. E. (1972). *Parmelia pustulifera*, a new lichen from southeastern United States. *Brittonia* **24**, 22–27.

HALE, M. E. (1974). "The Biology of Lichens" (2nd edition). Arnold, London.

HAWKSWORTH, D. L. (1968). A note on the chemical strains of the lichen *Ramalina subfarinacea. Bot. Notiser* **121**, 317–320.

HAWKSWORTH, D. L. (1969). The lichen flora of Derbyshire. *Lichenologist* **3**, 105–193.

HAWKSWORTH, D. L. (1970). The chemical constituents of *Haematomma ventosum* (L.) Massal. in the British Isles. *Lichenologist* **4**, 248–255.

HAWKSWORTH, D. L. (1971a). Types of chemical variation in lichens and their taxonomic treatment. *In* "Abstracts, First International Mycological Congress, Exeter, 1971" (G. C. Ainsworth and J. Webster, eds), p. 41. First International Mycological Congress, Exeter.

HAWKSWORTH, D. L. (1971b). Chemical and nomenclatural notes on *Alectoria* (Lichenes) III. The chemistry, morphology and distribution of *Alectoria virens* Tayl. *J. Jap. Bot.* **46**, 335–342.

HAWKSWORTH, D. L. (1971c). A brief guide to microchemical techniques for the identification of lichen products. *Bull. Br. Lichen Soc.* **2** (28), 5–9.

HAWKSWORTH, D. L. (1972a). Regional studies in *Alectoria* (Lichenes) II. The British species. *Lichenologist* **5**, 181–261.

HAWKSWORTH, D. L. (1972b). The natural history of Slapton Ley Nature Reserve IV. Lichens. *Fld Stud.* **3**, 535–578.

HAWKSWORTH, D. L. (1973a). Some advances in the study of lichens since the time of E. M. Holmes. *Bot. J. Linn. Soc.* **67**, 3–31.

HAWKSWORTH, D. L. (1973b). Ecological factors and species delimitation in the lichens. *In* "Taxonomy and Ecology" (V. H. Heywood, ed.), pp. 31–69. Academic Press, London and New York.

HAWKSWORTH, D. L. (1974). "Mycologist's Handbook." Commonwealth Mycological Institute, Kew.

HAWKSWORTH, D. L. and CHAPMAN, D. S. (1971). *Pseudevernia furfuracea* (L.) Zopf and its chemical races in the British Isles. *Lichenologist* **5**, 51–58.

HENSSEN, A. and JAHNS, H. M. (1973) ["1974"]. "Lichenes. Eine Einführung in die Flechtenkunde." Thieme, Stuttgart.

HESS, D. (1958). Über die Papierchromatographie von Flechtensäuren. *Planta* **52**, 65–76.

HESS, D. (1959). Untersuchungen über die Bildung von Phenolkörpen durch isolierte Flechtenpilze. *Z. Naturf.* **14b**, 345–347.

HILL, D. J. and WOOLHOUSE, H. W. (1966). Aspects of the autecology of *Xanthoria parietina* agg. *Lichenologist* **3**, 207–214.

*HILLMANN, J. and GRUMMANN, V. J. (1957). Flechten. *Krypt.-Fl. Mark Brandenb.* **8**, i–x, 1–898.

HUNECK, S. (1968). Lichen substances. *In* "Progress in Phytochemistry" (L. Reinhold and Y. Liwschitz, eds), pp. 223–346. Wiley Interscience, London, New York and Sydney.

HUNECK, S. (1971). Chemie und Biosynthese der Flechtenstoffe. *Fortschr. Chem. org. Naturstoffe* **29**, 209–306.

HUNECK, S. (1974) ["1973"]. Nature of lichen substances. *In* "The Lichens" (V. Ahmadjian and M. E. Hale, eds), pp. 495–522. Academic Press, New York and London.

IMSHAUG, H. A. and BRODO, I. M. (1966). Biosystematic studies on *Lecanora pallida* and some related lichens in the Americas. *Nova Hedwigia* **12**, 1–59.

IMSHAUG, H. A. and HARRIS, R. C. (1971). Notes on the taxonomic relationships and chemistry of *Ramalina subdecipiens* J. Stein. *Bryologist* **74**, 369–371.

JAMES, P. W. (1971). New or interesting British lichens: 1. *Lichenologist* **5**, 114–148.

JORDAN, W. P. (1972). Erumpent cephalodia, an apparent case of phycobial influence on lichen morphology. *J. Phycol.* **8**, 112–117.

*JØRGENSEN, P. M. (1974). *Dirinaria applanata* (Fée) Awasthi on the Açores, and a comment on the taxonomical value of sekikaic acid in the genus. *Port. Acta Biol.*, **B, 12**, 1–5.

JØRGENSEN, P. M. and RYVARDEN, L. (1970). Contribution to the lichen flora of Norway. *Årbok Univ. Bergen, mat.-nat. ser.* **1969** (10), 1–24.

KOMIYA, T. and SHIBATA, S. (1969). Formation of lichen substances by mycobionts of lichens. Isolation of (+) usnic acid and salazinic acid from mycobionts of *Ramalina* spp. *Chem. Pharm. Bull., Tokyo* **17**, 305–306.

KRISTINSSON, H. (1969). Chemical and morphological variation in the *Cetraria islandica* complex in Iceland. *Bryologist* **72**, 344–357.

KRISTINSSON, H. (1971). Morphological and chemical correlations in the *Cladonia chlorophaea* complex. *Bryologist* **74**, 13–17.

KRISTINSSON, H. (1974). Two new *Cladonia* and one *Cetraria* species from Iceland. *Lichenologist* **6**, 141–145.

*KROG, H. (1950). Litt om lavsyrer, noen saermerkte organiske stoffer i lav. *Blyttia* **8**, 91–98.

KROG, H. (1951). Microchemical studies on *Parmelia*. *Nytt Mag. Naturv.* **88**, 57–85.

KROG, H. (1968). The macrolichens of Alaska. *Norsk Polarinst. Skr.* **144**, 1–180.

KROG, H. (1969). Kjemiens rolle i lavsystematikken. *Blyttia* **27**, 65–79.

KROG, H. (1970a). The Scandinavian members of the *Parmelia borreri* group. *Nytt Mag. Bot.* **17**, 11–15.

KROG, H. (1970b). Forekomst av en upigmentert *Solorina crocea* i Finnmark. *Blyttia* **28**, 181–182.

KUROKAWA, S., SHIBATA, S. and KOMIYA, T. (1969). Isolation of algal and fungal components from lichens and their chemical products. *Miscnea bryol. lichen.*, *Nichinan* **5**, 8–9.

LAMB, I. M. (1939). A review of the genus *Neuropogon* (Nees & Flot.) Nyl., with special reference to the Antarctic species. *J. Linn. Soc., Bot.* **52**, 199–237.

LAMB, I. M. (1947). A monograph of the lichen genus *Placopsis*. *Lilloa* **13**, 151–288.

LAMB, I. M. (1948). Further data on the genus *Neuropogon*. *Lilloa* **14**, 139–168.

LAMB, I. M. (1951a). Biochemistry in the taxonomy of lichens. *Nature, Lond.* **168**, 38.

LAMB, I. M. (1951b). On the morphology, phylogeny, and taxonomy of the lichen genus *Stereocaulon*. *Can. J. Bot.* **29**, 522–584.

LAMB, I. M. (1964). Antarctic lichens I. The genera *Usnea, Ramalina, Himantormia, Alectoria, Cornicularia*. *Br. Antarct. Surv. sci. Rept.* **38**, 1–34.

LAMB, I. M. (1967). Chemotaxonomy in the lichens. *Internat. Lichenol. Newsletter* **1** (3), 1–4.

LAMBINON, J. (1969). "Les Lichens. Morphologie, biologie, systématique, écologie." Les Naturalistes Belges, Bruxelles. [Reprinted with minor changes from *Naturalistes belg.* **49**, 205–208, 449–558 (1968).]

LANJOUW, J. (1958). On the nomenclature of chemical strains. *Taxon* **7**, 43–44.

LANJOUW, J. *et al.*, eds (1952). International Code of Botanical Nomenclature adopted by the Seventh International Botanical Congress, Stockholm, July 1950. *Regnum Vegetabile* **3**, 1–228.

LAUNDON, J. R. (1965). Lichens new to the British flora: 3. *Lichenologist* **3**, 65–71.

LAUNDON, J. R. (1970). Lichens new to the British flora: 4. *Lichenologist* **4**, 297–308.

LEIGHTON, W. A. (1866a). Notulae lichenologicae. No. XI. On the examination and rearrangement of the *Cladoniei*, as tested by hydrate of potash. *Ann. Mag. nat. Hist., ser.* 3, **18**, 405–420.

★LEIGHTON, W. A. (1866b). Notulae lichenologicae. No. IX. *Ann. Mag. nat. Hist.*, ser. 3, **18**, 169–171.

LEIGHTON, W. A. (1867). Notulae lichenologicae. No. XII. On the *Cladoniei* in the Hookerian herbarium at Kew. *Ann. Mag. nat. Hist.*, ser. 3, **19**, 99–124.

LEIGHTON, W. A. (1868). Notulae lichenologicae. No. XVIII. *Ann. Mag. nat. Hist.*, ser. 3, **20**, 439–442.

LEIGHTON, W. A. (1871). "The Lichen Flora of Great Britain, Ireland, and the Channel Islands." Privately published, Shrewsbury.

LETTAU, G. (1914). Nachweis und Verhalten einiger Flechtensäuren. *Hedwigia* **55**, 1–78.

★LEUCKERT, C., POELT, J. and SCHULZ, G. (1970). Chemotaxonomische Probleme in der Flechtengattung *Pertusaria*. *Vortr. Bot. Ges.* [*Dtsch. bot. Ges.*], n.f. **4**, 45–80.

LEUCKERT, C., ZIEGLER, H. G. and POELT, J. (1972). Zur Kenntnis der *Cladonia chlorophaea*-Gruppe und ihrer Problematik in Mitteleuropa. *Nova Hedwigia* **22**, 503–534.

LINDSAY, W. L. (1869a). On chemical reaction as a specific character in lichens. *J. Linn. Soc., Bot.* **11**, 36–63.

LINDSAY, W. L. (1869b). Experiments on colour-reaction as a specific character in lichens. *Trans. Proc. bot. Soc. Edinb.* **10**, 82–98.

MAGNUSSON, A. H. (1933). Gedanken über Flechtensystematik und ihre Methoden. *Acta Horti gothoburg.* **8** (1), 49–76.

MENLOVE, J. E. (1974). Thin-layer chromatography for the identification of lichen substances. *Bull. Br. Lichen. Soc.* **2** (34), 3–5.

MITSUNO, M. (1953). Paper chromatography of lichen acids I. *Pharm. Bull., Tokyo* **2**, 170–173.

MITSUNO, M. (1955). Paper chromatography of lichen substances II. *Pharm. Bull., Tokyo* **3**, 60–62.

MOSBACH, K. (1974) ["1973"]. Biosynthesis of lichen substances. *In* "The Lichens" (V. Ahmadjian and M. E. Hale, eds), pp. 523–546. Academic Press, New York and London.

★MOTYKA, J. (1947). Lichenum generis *Usnea* studium monographicum. Pars generalis. *Annls Univ. Mariae Curie-Skłodowska, C*, **1**, 277–476.

MÜLLER-ARGOVIENSIS, J. (1891). Kritik über Dr. Wainio's "Étude". *Flora, Jena* **74**, 383–389.

NEELAKANTAN, S. (1965). Recent developments in the chemistry of lichen substances. *In* "Advancing Frontiers in the Chemistry of Natural Products" [International Monographs on Advanced Chemistry **1.**], pp. 35–84. Hindustan Publishing, Dehli.

NEELAKANTAN, S. and RAO, P. S. (1967). Place of chemistry in lichen taxonomy. *Bull. natn. Inst. sci. India* **34**, 168–178.

NEELAKANTAN, S., SESHADRI, T. R. and SUBRAMANIAN, S. S. (1951). Indian strains of the lichen *Parmelia tinctorum*. *J. scient. ind. Res.* **10B**, 199–200.

NESPIAK, A., NOCULAK, A. and SIEWIŃSKI, A. (1973). Bemerkungen über fluorszierende Stoffe der Schleierlinge und ihre Auswertung für die Systematik. *Acta mycol.* **9**, 205–216.

NOURISH, R. and OLIVER, R. W. A. (1974). Chemotaxonomic studies on British lichens I. *Cladonia* subgenus *Cladina*. *Lichenologist* **6**, 73–94.

NUNO, M. (1973). Miscellaneous notes on *Cladonia* (1). *Miscnea bryol. lichen., Nichinan* **6**, 126–127.

NYLANDER, W. (1865). Ad historiam reactionis iodi apud Lichenes et Fungos notulae. *Flora, Jena* **48**, 465–468.

NYLANDER, W. (1866a). Circa novum in studio lichenum criterium chemicum. *Flora, Jena* **49**, 198–201.

NYLANDER, W. (1866b). Quaedam addenda ad nova criteria chemica in studio lichenum. *Flora, Jena* **49**, 233–234.

NYLANDER, W. (1866c). Hypochlorite of lime and hydrate of potash, two new criteria in the study of lichens [transl. W. A. Leighton]. *J. Linn. Soc. (Bot.)* **9**, 358–365.

★NYLANDER, W. (1866d). Lichenes Lapponiae orientalis. *Not. Fauna Fl. fenn. Förhandl.* **8**, 101–192.

NYLANDER, W. (1869a). De reactionibus in *Alectoriis*. *Flora, Jena* **52**, 444.

NYLANDER, W. (1869b). Remarks to Dr Lindsay's paper "On chemical reaction as a specific character in lichens". *J. Bot., Lond.* **7**, 214–215.

OZENDA, P. and CLAUZADE, G. (1970). "Les Lichens, étude biologique et flore illustrée." Masson et Cie, Paris.

PEOPLES, A. G. (1973). Biochemical evidence for the placement of *Cetraria tuckermanii* (Oakes) and *Cetraria lacunosa* (Ach.) in the genus *Platismatia*. *Mich. Academician* **6**, 243–248.

PIATTELLI, M. and DE NICOLA, M. G. (1968). Anthraquinone pigments from *Xanthoria parietina* (L.). *Phytochemistry* **7**, 1183–1187.

POELT, J. (1969). "Bestimmungschlüssel europäischer Flechten." Cramer, Lehre.

POELT, J. (1970). Das Konzept der Artenpaare bei den Flechten. *Vortr. bot. Ges.* [*Dtsch. bot. Ges.*], *n.f.* **4**, 187–198.

POELT, J. (1972). Die taxonomische Behandlung von Artenpaare bei den Flechten. *Bot. Notiser* **125**, 77–81.

POELT, J. (1974) ["1973"]. Classification. *In* "The Lichens" (V. Ahmadjian and M. E. Hale, eds), pp. 599–632. Academic Press, New York and London.

★RAMAUT, J. L. (1959). Les critères biochemiques et la taxonomie des lichens. *Naturalistes belg.* **40**, 262–274.

★RAMAUT, J. L. (1962). Réactions thallines, microcristallisations et chromatographie de partage sur papier en lichénologie. *Naturalistes belg.* **43**, 359–370.

RAMAUT, J. L. (1963). Chromatographie en couche mince des depsidones du β-orcinol. *Bull. Soc. chim. belg.* **72**, 97–101.

★RAMAUT, J. L. (1965a). Reflexions sur la valeur chimiotaxonomique des substances licheniques à basses concentrations: le cas de l'acide usnique chez *Evernia prunastri* (L.) Ach. *Phytochemistry* **4**, 199–202.

★RAMAUT, J. L. (1965b). Une nouvelle mise en doute de la valeur chimiotaxonomique de l'acide usnique: le cas de *Parmelia ecaperata* Müll. Arg. et de *P. concrescens* Wainio. *Bull. Jard. bot. État Brux.* **35**, 67–72.

RAMAUT, J. L. (1967). Séparation en chromatographie sur couches minces de l'atranorine et de l'acide usnique. *J. Chromatogr.* **31**, 580–582.

★RAMAUT, J. L., LAMBINON, J. and TARGÉ, A. (1962). Le problème de l'acide usnique chez *Evernia prunastri* (L.) Ach. *Lejeunia*, *n.s.* **12**, 1–11.

★RAMAUT, J. L., SCHUMACKER, R., LAMBINON, J. and BAUDUIN, C. (1966). Dosage spectrophométrique et signification chimiotaxonomique de l'acide usnique chez *Cladonia tenuis* (Floerke) Harm., *C. leucophaea* des Abb. et *C. impexa* Harm. *Bull. Jard. bot. État Brux.* **36**, 399–414.

★RÄSÄNEN, V. (1941). Kemia jäkäläsystematiikan palveluksessa. *Luonnon Ystävä* **45**, 168–178.

RIEDL, H. (1962). Flechtenchemie und Flechtentaxonomie. *Taxon* **11**, 65–68.

RUNDEL, P. W. (1969). Clinal variation in the production of usnic acid in *Cladonia subtenuis* along light gradients. *Bryologist* **72**, 40–44.

★RUNDEL, P. W (1972). Notes on the ecology and chemistry of *Ramalina montagnei* and *R. denticulata* in the U.S. Virgin Islands. *Bryologist* **75**, 69–72.

RUNEMARK, H. (1956). Studies in *Rhizocarpon* I. Taxonomy of the yellow species in Europe. *Op. bot. Soc. bot. Lund.* **2** (1), 1–152.

SANTESSON, J. (1965). Studies on the chemistry of lichens 24. Thin layer chromatography of aldehydic aromatic lichen substances. *Acta chem. scand.* **19**, 2254–2255.

SANTESSON, J. (1967). Chemical studies on lichens 4. Thin layer chromatography of lichen substances. *Acta chem. scand.* **21**, 1162–1172.

SANTESSON, J. (1970). Neuere Probleme der Flechtenchemie. *Vortr. bot. Ges.* [*Dtsch. bot. Ges.*], *n.f.* **4**, 5–21.

SANTESSON, J. (1973) ["1974"]. Chemie der Flechten. *In* "Lichenes. Eine Einführung in die Flechtenkunde" (A. Henssen and H. M. Jahns), pp. 152–185. Thieme, Stuttgart.

SANTESSON, J. (1974) ["1973"]. Identification and isolation of lichen substances. *In* "The Lichens" (V. Ahmadjian and M. E. Hale, eds), pp. 633–652. Academic Press, New York and London.

SANTESSON, R. (1968). Lavar. Some aspects on lichen taxonomy. *Svensk Natur.* **1968**, 176–184.

SATO, M. (1968). The mixture ratio of the lichen genus *Thamnolia* in Tasmania and New Guinea. *J. Jap. Bot.* **43**, 328–334.

SCHNEIDER, A. (1897). "A Text-book of General Lichenology." Clute, Binghamton, New York.

*SHIBATA, S. (1965). Biogenetical and chemotaxonomical aspects of lichen substances. *In* "Beitrage zur Biochemie und Physiologie von Naturstoffen" [Festschrift Kurt Mothes zum 65. Geburtstag.], pp. 451–465. Gustav Fischer, Jena.

SHIBATA, S. (1974). Some aspects of lichen chemotaxonomy. *In* "Chemistry in Botanical Classification" (G. Bendz and J. Santesson, eds), pp. 241–249. Academic Press, New York and London.

SHIBATA, S. and CHIANG, H.-C. (1965a). The structure of cryptochlorophaeic acid. *Phytochemistry* **4**, 133–139.

SHIBATA, S. and CHIANG, H.-C. (1965b). Some chemotaxonomical aspects of phenolic compounds in lichens. *Bull. natn. Inst. sci. India* **30**, 151–157.

SHIBATA, S., NATORI, S. and UDAGAWA, S. (1964). "List of Fungal Products." University of Tokyo Press, Tokyo.

SIPMAN, H. J. M. (1973). The *Cladonia pyxidata–fimbriata* complex in the Netherlands, with description of a new variety. *Acta bot. neerl.* **22**, 490–502.

SMITH, A. L. (1921). "Lichens." Cambridge University Press, Cambridge.

SØCHTING, U. (1973). Anatomical and cytological characteristics of unpigmented *Caloplaca verruculifera* from Denmark. *Bot. Tidsskr.* **68**, 152–156.

STAFLEU, F. A. *et al.*, eds (1972). International Code of Botanical Nomenclature adopted by the Eleventh International Botanical Congress, Seattle, August 1969. *Regnum Vegetabile* **82**, 1–426.

*SWINSCOW, T. D. V. (1964). The classification of lichens. *Advmt Sci., Lond.* **21**, 241–244.

TARGÉ, A. and LAMBINON, J. (1965). Etude chimiotaxonomique du groupe de *Parmelia borreri* (Sm.) Turn. en Europe occidentale. *Bull. Soc. r. bot. Belg.* **98**, 295–306.

*TAVARES, C. N. (1954). Química e taxonomia nos líquenes. *Gazeta de Física* **3**, 17–20.

TÉTÉNYI, P. (1958). Proposition à propos de la nomenclature des races chimiques. *Taxon* **7**, 40–41.

THOMSON, J. W. (1963). Modern species concepts: lichens. *Bryologist* **66**, 94–100.

THOMSON, J. W. (1968a). *Haematomma lapponicum* Räs. in North America. *J. Jap. Bot.* **43**, 305–310.

*THOMSON, J. W. (1968b) ["1967"]. "The Lichen Genus *Cladonia* in North America." University of Toronto Press, Toronto.

*TRASS, H. (1968). Voprosy khimicheskoĭ taksonomii v sovremennoĭ likhenologii. *Trans. Tartu St. Univ.* **211**, *Pap. Bot.* **8**, 172–184.

TUCKERMAN, E. (1868). Can lichens be identified by chemical tests? *Am. Nat.* **2**, 104–107.

TURNER, W. B. (1971). "Fungal Metabolites." Academic Press, London and New York.

VAINIO, E. A. (1890). Étude sur la classification naturelle et al morphologie des lichens du Bresil. Pars prima. *Acta Soc. Fauna Flora fenn.* **7** (1), i–xxx, 1–247.

★VERSEGHY, K. (1962). Die Gattung *Ochrolechia*. *Beih. Nova Hedwigia* **1**, 1–146.

WACHTMEISTER, C. A. (1952). Studies on the chemistry of lichens. 1. Separation of depside components by paper chromatography. *Acta chem. scand.* **6**, 818–825.

WACHTMEISTER, C. A. (1956). Identification of lichen acids by paper chromatography. *Bot. Notiser* **109**, 313–324.

WACHTMEISTER, C. A. (1959). Flechtensäuren. *In* "Papierchromatographie in der Botanik" (H. F. Linskens, ed.), pp. 135–141. Springer, Berlin.

WANG-YANG, J.-R. and AHMADJIAN, V. (1972). A morphological study of the algal symbionts of *Cladonia rangiferina* (L.) Web. and *Parmelia caperata* (L.) Ach. *Taiwania* **17**, 170–181.

★WETHERBEE, R. (1969). Population studies in the chemical species of the *Cladonia chlorophaea* group. *Mich. Bot.* **8**, 170–174.

WETMORE, C. M. (1960). The lichen genus *Nephroma* in North and Middle America. *Publs Mich. St. Univ. Mus., ser. biol.* **1** (11), 369–452.

WOLF, F. A. (1973). Synthesis of various products, especially pigments by fungi. *J. Elisha Mitchell sci. Soc.* **89**, 184–205.

★XAVIER FILHO, L. (1973a). A quimiotaxonomia dos liquens. *Publs Univ. Pernambuco, Bot., sér. D*, **2**, 1–40.

★XAVIER FILHO, L. (1973b). "Methologia do estudios dos liquenes." Recife. [Not seen.]

YOSHIMURA, I. (1971). The genus *Lobaria* of Eastern Asia. *J. Hattori bot. Lab.* **34**, 231–364.

YOSHIMURA, I. (1974). "Lichen Flora of Japan in Colour." Hoikusha Publishing, Osaka.

★YOSHIMURA, I. and HAWKSWORTH, D. L. (1970). The typification and chemical substances of *Lobaria pulmonaria* (L.) Hoffm. *J. Jap. Bot.* **45**, 33–41.

ZOPF, F. W. (1903). Vergleichende Untersuchungen über Flechten in Bezug auf ihre Stoffwechselproduckte. Erste Abhandlung. *Beih. bot. Zbl.* **14**, 95–126.

ZOPF, F. W. (1905). Biologische und morphologische Beobachtungen en Flechten. I. *Ber. dt. bot. Ges.* **23**, 497–504.

ZOPF, F. W. (1907). "Die Flechtenstoffe in chemischer, botanischer, pharmakologischer und technischer Beziehung." Gustav Fischer, Jena.

ZOPF, F. W. (1908). Beiträge zur einer chemischen Monographie die Cladoniaceen. *Ber. dt. bot. Ges.* **26**, 51–113.

8 | Chemotaxonomic Studies on the *Cladonia chlorophaea-pyxidata* Complex and Some Allied Species in Britain

R. NOURISH and R. W. A. OLIVER*

*Department of Chemistry and Applied Chemistry,
University of Salford, England*

Abstract: In the present paper studies on the *Cladonia chlorophaea–pyxidata* complex and some allied species are presented and discussed, with special reference to the techniques of lichen mass spectrometry and high-pressure liquid chromatography which have been used together for the first time in routine chemical analyses of lichens.

INTRODUCTION

The idea that chemical characters could be used as valid taxonomic criteria was conceived over a century ago but has gained widespread acceptance only in the last decade (see Chapter 7). Advances in analytical chemistry permitting detailed studies on the nature and distribution of the characteristic secondary lichen products involved, have undoubtedly contributed to this viewpoint. The techniques currently available for the identification of lichen products have been reviewed by Nourish and Oliver (1974).

The present paper provides a detailed account of two of the latest techniques, lichen mass spectrometry and high pressure liquid chromatography, to illustrate practically the progress and problems in lichen chemotaxonomic studies. A number of reports have been concerned with the application of mass spectrometry to lichen material in the direct mode, a technique termed lichen mass spectrometry (l.m.s.) by Santesson (1969). Santesson made a special study of the occurrence of certain xanthones, anthraquinones (Santesson, 1970) and other pigments in a variety of species (Santessson, 1974). Similarly, Bohman (1969)

Systematics Association Special Volume No. 8, "Lichenology: Progress and Problems", edited by D. H. Brown, D. L. Hawksworth and R. H. Bailey, 1976, pp. 185–214. Academic Press, London and New York.

* Any correspondence in connection with this paper should be addressed to Dr Oliver.

TABLE I. Summary of the lichen compounds in the *Cladonia chlorophaea–pyxidata* complex and in allied species

Species	Perlatolic acid	Imbricaric acid	Atranorin	Homosekikaic acid	4-O-methylcrypto-chlorophaeic acid	Sekikaic acid	Merochlorophaeic acid	Cryptochlorophaeic acid	Grayanic acid	Fumarprotocetraric acid	Usnic acid	Rangiformic acid	Protolichesterinic acid	Evans' substance H	Unidentified substance	Comments (see footnote)
Group 1: *C. chlorophaea* sensu lato																
C. chlorophaea "strain 1"	—	—	—	—	—	—	—	—	—	rc	—	—	—	—	—	1
C. chlorophaea "strain 2"	—	—	—	r	—	—	—	—	—	r	—	—	—	—	—	2
C. chlorophaea "strain 3"	—	r	—	—	—	—	—	—	—	—	—	—	—	—	—	3
C. chlorophaea "strain 4"	—	—	—	—	—	—	—	—	—	r	—	r	r	—	—	4
C. chlorophaea "strain 5"	—	—	—	—	—	—	—	—	—	—	—	—	—	—	—	
C. cryptochlorophaea	—	—	—	—	—	—	—	rc	—	rc(±)	—	—	—	—	—	
C. grayi	—	—	—	—	—	—	—	—	rc	rc(±)	—	—	—	—	—	
C. merochlorophaea "strain 1"	—	—	—	rc	rc	rc	rc	—	—	rc(±)	—	—	—	—	—	5
C. merochlorophaea "strain 2"	—	—	—	—	r	—	r	—	—	rc(±)	—	—	—	—	—	
C. perlomera	r	—	—	—	—	—	—	—	—	—	—	—	—	—	—	6

Group 2: Species closely related to C. *chlorophaea* sensu lato

	C1	C2	C3	C4	C5	C6	C7	C8	C9	C10	C11
C. *conista* sensu Evans	—	—	—	—	—	rc	—	—	—	—	—
C. *conistea* (C. *conista* sensu Dahl)	rc	—	—	—	—	rc	—	—	—	—	r
C. *cyathomorpha*	r	—	—	—	—	rc	—	—	—	—	c
C. *fimbriata*	—	—	—	—	—	rc	—	—	—	—	—
C. *fimbriata* f. *fibula*	—	—	—	—	r	—	r	—	—	—	—
C. *fimbriata* var. *major*	r	—	—	—	r	rc	—	—	—	—	—
C. *fimbriata* var. *simplex*	r	—	—	—	—	r	r	—	—	—	—
C. *pocillum*	rc(±)	—	—	—	—	rc	—	—	—	—	—
C. *pyxidata*	rc(±)	—	—	—	—	rc	—	—	—	—	—

Group 3: C. *coniocraea–ochrochlora* complex

	C1	C2	C3	C4	C5	C6	C7	C8	C9	C10	C11
C. *coniocraea*	—	—	—	—	—	rc	—	—	—	—	—
C. *cornuta*	—	—	—	—	—	rc	—	—	—	—	—
C. *ochrochlora*	c(±)	—	—	—	—	rc	—	—	—	—	—
C. *subulata*	—	—	—	—	—	rc	—	—	—	—	—

[1] Three specimens known only from Iceland (Culberson and Kristinsson, 1969).
[2] Two specimens known from Iceland and Wyoming, U.S.A. (Culberson and Kristinsson, 1969)
[3] Very rare (Ahti, 1966; Krog, 1968).
[4] One report (Ahti, 1966) from British Columbia.
[5] var. *novochlorophaea* Sipman (1973).
[6] Ten specimens known from North America (Culberson and Kristinsson, 1969).

Key: r = reported; c = confirmed for the British material examined in the present study; (±) = accessory substance.

investigated the distribution of anthraquinones in *Caloplaca*. The main research interests in this laboratory have been in the application of l.m.s. to studies on *Cladonia* (Newton, 1970; Hall, 1971; Nourish, 1972). Only one report has been published to date on the application of high pressure liquid chromatography (h.p.l.c.) to lichens (Culberson 1972b) although the potentialities of this technique have been indicated in the field of medical biochemistry (Oliver, 1972).

In the present study l.m.s. and h.p.l.c. have been used in conjunction with thin layer chromatography (t.l.c.) to form part of a detailed reappraisal of the genus *Cladonia* in the British Isles with special reference to the *Cladonia chlorophaea–pyxidata* complex and allied species. Following the original studies of Asahina (1941, 1943), there has been a considerable revival of interest in this perplexing group which contains a large number of chemical variants. These are listed in group 1, Table I, and a study of the comment column shows that some taxa are apparently very rare. The distribution of the major named races in this group have recently been reported for North America, Central Europe, Finland and the Netherlands (Table II) but their taxonomic status is still far from clear since many authors are uncertain as to the extent and significance of some observed morphological correlations, such as the size and frequency of podetial soredia (Kristinsson, 1971). There have been no published reports to date on the British representation of group 1, *Cladonia chlorophaea* s.l., and the only taxon recognized in addition to *C. chlorophaea* s.s. has been *C. grayi* on account of its negative reaction with *para*-phenylenediamine. Extensive studies have been carried out on the morphology, including SEM techniques, distribution and ecology of all the British taxa listed in Table I (Nourish and Oliver, unpublished), but the main purpose of the present paper is to report the chemical findings to date and show how the techniques of l.m.s. and h.p.l.c. have been used to obtain unambiguous chemical information for the species studied. The incorrect or incomplete identification of various compounds in the "cup lichens" due to the insensitivity or non-specificity of the microchemical methods employed has undoubtedly hampered a clear understanding of the group in earlier studies. Homosekikaic acid and sekikaic acid, for example, visible as a single compound, "novochlorophaeic acid", in microcrystal tests and sometimes difficult to resolve in naturally occurring concentrations using t.l.c., were finally identified by mass spectrometry (Culberson and Kristinsson, 1969). Specimen sampling procedures are also extremely important in chemotaxonomic studies. A recent paper (Leuckert et al., 1972) which present a large number of analytical results for mixed material underlines the fact that chemical analyses must be performed on single podetia to avoid unnecessary confusion in the study of taxa which are perplexing enough in nature. There is no doubt that,

TABLE II. Reported intra-population relationships of the major chemical "strains" of *C. chlorophaea* s.l. and preliminary results concerning the British population

Location	Chemical "strains" as % of total					No. of specimens tested	Technique(s) used	Reference
	C. chlorophaea	*C. cryptochlorophaea*	*C. grayi*	*C. merochlorophaea* "strain 1"	*C. merochlorophaea* "strain 2"			
U.S.A.								
Connecticut	5	8	87	—	—	—	m.c.t.	Evans (1944)
Wisconsin	51	28	16	5	—	—	m.c.t.	Thomson (1968)
Michigan	35	19	38	8	—	500	m.c.t., some t.l.c.	Wetherbee (1969)
N. Carolina	4	23	73	—	—	86[a]	t.l.c.	Kristinsson (1971)
N. America	41	27	26	6	—	364[a]	m.c.t., t.l.c.	Bowler (1972) including data from Krog (1968) Moore (1968) and Taylor (1968)
Europe								
British Isles (preliminary indication)	42	25	9	19	5	240	l.m.s., t.l.c., h.p.l.c.	Nourish and Oliver (present study)
Central Europe	14	17	24	32	13	260[b]	t.l.c., p.c.	Leuckert *et al.* (1972)
Finland	common	rare	common	common	—	200	m.c.t., p.c.	Ahti (1966)
Netherlands	31	25	1	28	15	433	m.c.t.	Sipman (1973)

[a] Data abstracted from distribution maps, actual sample number probably much larger.
[b] 25% of the collections tested consisted of mixed material.
m.c.t. = microcrystal tests.   p.c. = paper chromatography.

owing to the new experimental situation, lichenology will see a vast increase in the number of reports on lichen products, and it is to be hoped that they follow the excellent guidelines laid down by Culberson (1970).

METHODS AND MATERIALS

1. Solvents

General purpose reagent qualities of 2,2,4-trimethylpentane, glacial acetic acid and chloroform were found to be satisfactory for use in h.p.l.c. Aromatic components in hexane which absorbed strongly in the ultraviolet region were removed by treatment with concentrated sulphuric acid (hexane:H_2SO_4, 3:1 v/v with vigorous shaking for 4 h). The hexane fraction was washed repeatedly with water, dried over anhydrous magnesium sulphate, and twice distilled using a 3 m column. The purified hexane was collected at 68–69°C. Propan-2-ol was dried by passing through anhydrous magnesium sulphate, distilled (3 m column) and collected at 81–82°C. Solvent purity was checked by ultraviolet absorption spectrophotometry.

2. Chromatography

(a) High pressure liquid chromatography. A diagram of the chromatograph constructed for use in the present study is given in Fig. 1. The "pump" consisted of two solvent coils, made from $\frac{3}{8}$ inch diam stainless steel tubing, and a nitrogen cylinder with a pressure regulator and gauge. The coils were filled (at atmospheric pressure) with the required, degassed, solvent system from the refill solvent reservoir by means of the two-way liquid valves. The solvent overflow chamber was made of glass which permitted a visual check that air had been expelled from the system. With the liquid valves in the second position (completing the high pressure circuit) the "pump" was driven by constant nitrogen pressure which acted directly on the solvent in the coils. The system provided silent, pulse-free operation for pressures between 0 and 2500 psi. The two coils used in series had a volume sufficiently great to enable routine analyses for several hours but the 25 ml coil could be used in isolation, for the experimental testing of new solvent systems, thus obviating solvent wastage. The coil, valves and all the other unspecified components were stainless steel type 316. The injection port was a modified Swagelok coupling with a rubber septum. The usual solvent systems quickly attacked standard silicone rubber septa, but non-aqueous septa (Du Pont-820508) proved reliable. Samples were injected with a fixed needle syringe directly on to the column with the pump in operation. However, reproducible results were also obtained by stopping the liquid flow momentarily at the control valve and injecting at reduced pressure. This procedure considerably prolonged the life of the injection septa, reduced plunger and needle bending and permitted the use of syringes with replaceable needles. The columns used in this study were constructed from 500 mm lengths of precision bore stainless steel tube (3 mm internal diam, 6 mm outside diam) dry packed with Corasil II and were used singly, or in series to provide a total length of 1·5 m. The column end coupling was fitted with dished double woven mesh (5 μm) for minimum resolution loss and zero unswept volume, and the eluant conveyed to the flow cell by a minimum length of P.T.F.E. tubing (0·013 inch bore $\frac{1}{16}$ inch outside diam) to avoid peak broadening.

The variable wavelength ultraviolet spectrophotometer (Cecil C. E. 212) with eight sensitivity ranges from 2·0 to 0·01 full scale, enabled analysis to be carried out at optimum

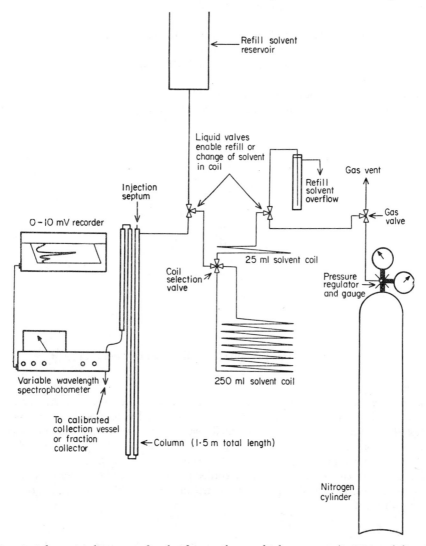

Fig. 1. Schematic diagram of pulse-free, coil type high pressure (0–2500 psi) liquid chromatograph.

wavelength within the range 220–400 nm, with a maximum sensitivity of 0·0005 absorbance unit. The special design features of the flow-through cell used in the present study are given in Fig. 2, namely an optical path length of 10 mm, a capacity of only 8 μl and a solvent flow arrangement which reduces the risk of bubble formation.

A high pressure liquid chromatograph of the type described but with a safety inter-locking system on gas and liquid circuits to prevent dangerous operation is now available commercially from Cecil Instruments, Cambridge.

(b) *Thin layer chromatography*. The preparation of crude acetone lichen extracts for analysis by h.p.l.c. and t.l.c. techniques and the detection of spots was carried out according to the methods described by Culberson and Kristinsson (1970), Culberson (1972a) and Nourish and Oliver (1974).

Solvent A: Toluene:dioxane:acetic acid 180:45:5 (v/v/v).
Solvent B: Hexane:ether:formic acid 130:80:20 (v/v/v).

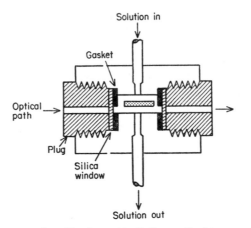

FIG. 2. Diagram of small volume (8 μl) flow cell of 1 cm path length.

3. Spectrometry

The mass spectra were measured at 200°C approximately 30 s after sample insertion on an A.E.I. M.S. 12 spectrometer using the direct inlet system and an ionizing beam energy of 70 eV. The upper portions of finely ground lichen podetia were submitted for analysis in most cases but fragments of the basal squamules of *C. cyathomorpha* were also tested.

The u.v. spectra of individual solvents and solvent systems for h.p.l.c. and of methanolic solutions of reference compounds were recorded using a calibrated (potassium chromate solution; Haupt, 1952) Unicam S.P. 700 spectrophotometer with matched silica cells of 1 cm path length.

4. Specimens Studied

Over 800 British collections were studied including all those in BM. A complete list is included in Nourish (1976). Thin layer chromatography was used for primary chemical screening of all collections. According to the size of the collection between one and three podetia were treated individually and the residues of hot acetone extracts of each podet-ium chromatographed in two solvent systems. Further podetia were analysed in the case of the large collections which appeared morphologically heterogeneous the for species being studied or when it was (subsequently) discovered that a given collection was chemically heterogeneous. In some of the latter cases a minute fragment from every podetium was further analysed by h.p.l.c. in the investigation of apparent morphological correlations with specific chemical types.

To date, h.p.l.c. has been applied to all specimens in *C. chlorophaea* s.l. (Table I, group 1) which were found to contain compounds other than, or in addition to, fumarprotocetraric acid. These analyses have provided qualitative and quantitative data for the specimens concerned.

Lichen mass spectrometry has been applied as a further, sensitive, screening method as follows:

(*a*) to all specimens of *C. cyathomorpha* to check for the presence of atranorin,

(*b*) to 10% of the specimens found to contain fumarprotocetraric acid alone, to check for the presence of other compounds in low concentration,

(*c*) to 20% of the specimens in group 1, Table I found to contain compounds in addition to fumarprotocetraric acid so as to check the validity of the chromatographic identifications.

RESULTS AND DISCUSSION

The results of the chemical analyses on the specimens studied are summarized in Tables I and IV. In this section the results obtained using the newer analytical methods are discussed.

1. Lichen Mass Spectrometry (l.m.s.)

Theoretically, any taxonomic study encompassing this technique requires the prior extraction and purification of the lichen compounds reported to occur in the group concerned, in order that mass spectra may be obtained for these "standard compounds", and a reference collection compiled. Following the detailed interpretation of each spectrum in terms of the molecular structure of the compound which gave rise to it, a fragmentation path may be constructed which should account for all the major ion peaks in the spectrum. However, a large number of mass spectra together with proposed fragmentation patterns have already been published, in connection with structural studies on lichen products (e.g. Huneck *et al.*, 1968) and these can serve as reference spectra which can be used, tentatively, to assign the presence of major ion peaks in lichen mass spectra to particular compounds. A guide to the available collections of mass spectra is included in Oliver and Lomax (1971). Nourish and Oliver (1974) found that the mass spectra of certain compounds (particularly depsides which occur widely in the genus *Cladonia*) vary considerably according to the instrument in which they were measured and the experimental conditions of the measurement. Consequently it has become a practice in this laboratory to measure spectra of all compounds under standardized conditions. The major ion peak indices (m/e > 120) for the compounds involved in the present study are listed in Table III. Many of the compounds in classes II–VI give rise to particularly characteristic mass spectra which each include a molecular ion peak (M^+) in addition to several major fragment ion peaks. In contrast, the lichen depsides (class I) invariably give rise to very small molecular ion peaks or consist of major fragment ions only and in these cases the relative abundances of the various fragment ions play a more important rôle in the characterization of the compound concerned.

Lichens often contain complex mixtures of compounds from several different chemical classes. Consequently their amenability to l.m.s. varies considerably according to the comparative volatility, and the thermal and ionic stability of their constituents under conditions of high vacuum in the mass spectrometer ionization chamber. In a physicochemical study of the subgenus *Cladina* it was shown that the reproducibility of l.m.s. is affected by (*a*) temperature of volatilization of the sample, (*b*) the instrument used and (*c*) the time interval between the insertion of the sample and the recording of its spectrum (Nourish and Oliver, 1974).

The major advantage of the mass spectrometric technique is that it enables the detection of very small quantities (nanograms) of lichen compounds in minute fragments (<1 mg) of lichen thalli. These properties are particularly useful in the study of irreplaceable type material and for the accurate screening of species which may contain only trace amounts of lichen compounds. In these respects it has been possible to establish that usnic acid *is* absent in *C. tenuis* var. *leucophaea* (Nourish and Oliver, 1974) and that squamatic acid *is* absent in *C. deformis* (Nourish and Oliver, unpublished) in the pairs *C. tenuis–leucophaea* and *C. deformis–gonecha*, respectively. Similarly, the characteristic mass spectrum of atranorin has enabled the selection of the lectotype for *Lichen rangiferinus* (i.e. *Cladonia rangiferina*) to be made from less than 0·5 mg of podetium (Nourish, 1972; Nourish and Oliver, 1975) while in the present study it has been established, contrary to an earlier report (Ahti, 1966), that *C. cyathomorpha* does not contain atranorin in addition to fumarprotocetraric acid but does contain a compound which has not yet been identified (see p. 207 for analytical data).

The various chemical types in the *Cladonia chlorophaea* complex proved to be particularly amenable to l.m.s. Examples of the lichen mass spectra of *C. grayi*, *C. merochlorophaea* "strain II" and *C. merochlorophaea* "strain I" are given in Figs 3–5, respectively, which illustrate the emphasis which can be laid on the precision of lichen mass spectral identifications.

The spectrum of *C. grayi* (Fig. 3) presents unambiguous evidence for the presence of grayanic acid in the specimen which gave rise to it since the major ion peaks in the l.m.s. compare almost exactly in terms of position and relative abundance with the spectrum of the pure compound (Table III) and moreover the spectra contain the characteristic molecular ion peak for grayanic acid at m/e 414. Lichens which contain compounds in classes II–VI (Table III) often give rise to lichen mass spectra which may be interpreted in a similarly unambiguous fashion. In the study of critical lichen material, e.g. type specimens, high resolution measurements on single ion peaks accurate to 2 p.p.m., normally permit completely unambiguous determinations but in routine work such measurements are regarded as impracticable and unnecessary. Only a minority of lichens containing depsides (Table III, class I) give rise to l.m.s. which exhibit molecular ions with relative abundances sufficient to merit the accurate identification of specific lichen compounds. An example is presented in Fig. 4, the l.m.s. of *C. merochlorophaea* "strain II". Here the molecular ion peaks at m/e 418 and m/e 446 suggest the presence of sekikaic and homosekikaic acids respectively. Such a premise is supported by the existence of additional major fragment ion peaks which also appear in the ten peak indices of the pure compounds. In contrast, the identification of merochlorophaeic and 4-O-methyl cryptochlorophaeic acids in the l.m.s. of a specimen of *C. merochlorophaea* "strain I" (Fig. 5) is based on the presence of major fragment ion peaks only. Such an assignment must be regarded as tentative in accurate chemotaxonomic studies unless supporting evidence is available from other analytical methods.

Lichen mass spectra may also be used to give an indication of the relative quantities of compounds in the material examined. Theoretically this may be achieved by determining the relative height of the molecular and fragment ion peaks assigned to each constituent. In practice, the occurrence of ion–ion interactions in the mass spectrometer and other effects observed in the mass spectrometry of complex mixtures can make such calculations difficult or impossible. The most reliable quantitative measurements are made by comparing the mass spectra of synthetic mixtures of known proportions with the l.m.s. concerned. In the present study accurate quantitative measurements were more conveniently determined by high pressure liquid chromatography.

2. High Pressure Liquid Chromatography

The analytical technique most widely used in lichenology today is certainly t.l.c. The thorough experimental and bibliographical work of C. F. Culberson (1969, 1970, 1972a) and Culberson and Kristinsson (1970) has enabled lichenologists to check morphological identifications and to carry out original taxonomic research on chemical guidelines. However, for detailed chemotaxonomic studies t.l.c. has a number of limitations which are probably most familiar to workers who have applied the technique to large numbers of lichen specimens.

TABLE III. Mass spectral major peak indices for compounds reported or identified in the *Cladonia* species studied, arranged in order of volatility within each chemical class

Chemical class/name	m.p. °C	Molecular formula	Molecular mass (nominal)	Major ion peak index above m/e 120 m/e (relative abundance %)
IA. *para*-Depsides				
Perlatolic acid	107	$C_{25}H_{32}O_7$	444 No M$^+$	164(100) 150(41·0) 220(35·0) 182(31·0) 238(20·0) 206(20·0) 124(18·0) 163(18·0) 168(14·0) 224(10·0)
Imbricaric acid[b]	126	$C_{23}H_{28}O_7$	416 M$^+$ (0·2)	138(100) 124(48·0) 221(30·0) 152(26·0) 194(25·0) 139(24·0) 123(20·0) 164(20·0) 220(14·0) 151(13·0)
Atranorin	196	$C_{19}H_{18}O_8$	374 M$^+$ (3·0)	164(100) 136(59·0) 196(58·0) 179(52·0) 165(22·0) 197(7·0) 374(3·0)
Squamatic acid[a]	219	$C_{19}H_{18}O_9$	390 M$^+$ (1·5)	191(100) 164(58·0) 136(55·0) 209(48·0) 165(28·0) 182(28·0) 190(12·0) 192(10·0) 390(1·5)
IB. *meta*-Depsides				
Homosekikaic acid	133	$C_{24}H_{30}O_8$	446 M$^+$ (1·0)	193(100) 192(95·0) 236(70·0) 210(41·0) 254(14·0) 194(10·0) 179(8·0) 446(1·0)
4-O-methylcryptochlorophaeic acid	142	$C_{26}H_{34}O_8$	474·5 No M$^+$	235(100) 207(10·0) 222(7·0) 196(6·0) 252(2·0) 178(2·0) 191(2·0)
Sekikaic acid	150	$C_{22}H_{26}O_8$	418 M$^+$ (5·0)	193(100) 194(12·0) 192(9·0) 208(8·0) 135(6·0) 164(6·0) 418(5·0) 179(4·0) 226(3·0) 210(3·0)
Merochlorophaeic acid	164	$C_{24}H_{30}O_8$	446 No M$^+$	207(100) 222(2·0) 196(2·0)
Cryptochlorophaeic acid	182	$C_{25}H_{32}O_7$	460·5 No M$^+$	221(100) 222(19·0) 138(7·0) 182(7·0) 177(6·0) 137(6·0) 164(5·0) 139(5·0) 165(4·0)

II. Depsidones				
Grayanic acid	186	$C_{23}H_{26}O_7$	414 M⁺ (83·0)	165(100) 370(84·0) 414(83·0) 312(65·0) 217(50·0) 257(49·0) 396(45·0) 244(42·0) 229(40·0) 285(35·0)
Fumarprotocetraric acid	250	$C_{22}H_{16}O_{12}$	472 No M⁺	No spectrum obtained at temperatures between 130 and 300°.
III. Dibenzofuran				
Didymic acid[a]	172	$C_{22}H_{26}O_5$	370 M⁺ (25·0)	352(100) 326(98·0) 241(50·0) 370(25·0) 254(25·0) 353(24·0) 239(24·0) 295(24·0) 267(22·0) 269(22·0)
IV. Usnic acids				
Usnic acid	203	$C_{18}H_{16}O_7$	344 M⁺ (55·0)	233(100) 260(60·0) 344(55·0) 232(31·0) 217(29·0) 234(20·0) 215(13·0) 261(11·0) 345(10·0)
V. Aliphatic acids				
Rangiformic acid	104	$C_{21}H_{38}O_6$	386 No M⁺	322(100) 324(38·0) 182(37·0) 200(32·0) 196(32·0) 279(30·0) 214(27·0) 368(26·0)
Protolichesterinic acid[a]	107	$C_{19}H_{32}O_4$	324 M⁺ (51·0)	279(100) 155(51·0) 324(51·0) 280(33·0) 142(26·0) 261(19·0) 123(17·0) 139(14·0) 138(12·0) 121(11·0)
VI. Triterpenes				
Zeorin[a]	250	$C_{30}H_{52}O_2$	444 M⁺ (5·9)	189(100) 121(59·0) 207(54·0) 135(51·0) 149(51·0) 161(48·5) 190(45·0) 426(43·0) 191(42·0) 147(33·5)
Ursolic acid[a]	285	$C_{30}H_{48}O_3$	456 M⁺ (5·6)	248(100) 203(63·0) 133(44·0) 207(42·0) 249(34·0) 189(16·0) 190(15·0) 219(13·0) 204(12·0) 208(11·0)

a Specimens containing any of these compounds are either of uncertain position or have been referred to other species in Cladonia (see Tables I, IV and text).

b Compound isolated from an extract of Cetrelia cetrarioides. The mass spectrum awaits complete molecular interpretation and is therefore provisional.

For example, major problems are connected with the loading of the sample spot at the origin of the thin layer plate and with associated concentration effects. An overloaded spot can give rise to large, diffuse or tailing spots for major constituents thus masking the presence of compounds occurring in lower con-

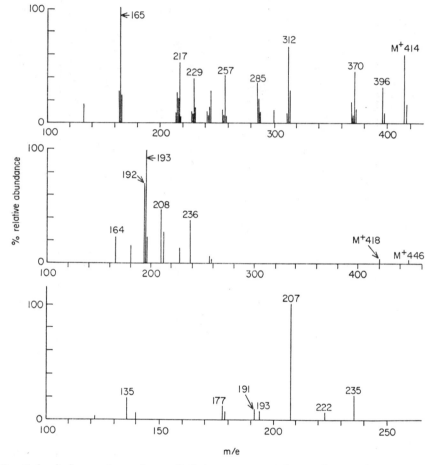

FIG. 3. (top). l.m.s. of a specimen of *Cladonia grayi* at 200°C, confirming the presence of grayanic acid.

FIG. 4. (middle). l.m.s. of a specimen of *Cladonia merochlorophaea* "strain II" at 200°C, indicating the presence of homosekikaic and sekikaic acids in an approximately 1:1 ratio.

FIG. 5. (bottom). l.m.s. of a specimen of *Cladonia merochlorophaea* "strain I" at 200°C, which may be interpreted as providing evidence for the presence of merochlorophaeic and 4-O-methylcryptochlorophaeic acids in an approximately 5:1 ratio.

centrations. In contrast an underloaded spot, whilst producing theoretically ideal spots for major constituents, may provide no evidence for the existence of other compounds. These statements have particular significance for closely related compounds with similar R_F values. In the case of species with a described chemistry, the assignment of minor spots on t.l.c. plates to specific compounds can be an extremely subjective matter, governed more by preconception than scientific evidence. Concentration effects, similarly, have a marked effect on spot coloration following the normal visualization treatment of the spots with sulphuric acid sprays and a period of heating. A detailed study of the chromogenic effects of sulphuric acid at different temperatures and concentrations has recently been made by Jensen (1973) in connection with studies on assay methods for oestriol. A similarly intensive study for lichen compounds has yet to be carried out and is urgently required.

Other features of the t.l.c. technique are the existence of secondary solvent fronts. Extracts containing large concentrations of fumarprotocetraric acid sometimes give rise to a faint spot at the level of the secondary solvent front in addition to the normal spot for this acid. Re-chromatography of these spots shows them to be fumarprotocetraric acid. It is also necessary to bear in mind that all the components in a crude lichen extract are not necessarily the secondary metabolic products of the lichen. Prominent elongated spots in the hR_F ($R_F \times 100$) class 70–80 (u.v. $^{++}$350 nm; H_2SO_4 red turning white after several hours) were often detected especially in the case of fresh material. The red spot proves to be attributable to chlorophyll A and the white spots are probably due to its oxidative breakdown products on the highly active surface of the silica gel. Other components or artefacts may also be detected resulting from the analysis of material previously contaminated with *para*-phenylenediamine or other reagents used in thallus colour tests, and, particularly in the case of the tests on primary thalli, compounds originating from the substratum.

In spite of the recent advances in *in situ* densitometry, the t.l.c. technique does not generally give rise to precise quantitative data. In preparative work, isolated bands of a new compound still require further solvent extraction or elution through a silica column before final purification can be effected.

Many of the shortcomings of t.l.c. are obviated by the technique of h.p.l.c. which was first applied to a number of pure lichen compounds and crude extracts by Culberson (1972b). A large number of reports have discussed the theoretical and practical considerations involved in h.p.l.c. (e.g. Knox, 1972; Snyder, 1972) and an exact empirical relationship between this tech nique and t.l.c. has been established in one case (Geiss and Schlitt, 1973). One oi the basic differences between h.p.l.c. and t.l.c. is that the former utilizes a support medium composed of small (20 μm) uniformly spherical particles. These columns allow a very much better separation of closely related compounds than t.l.c., and since the eluant is passed through the column under pressure the separations can

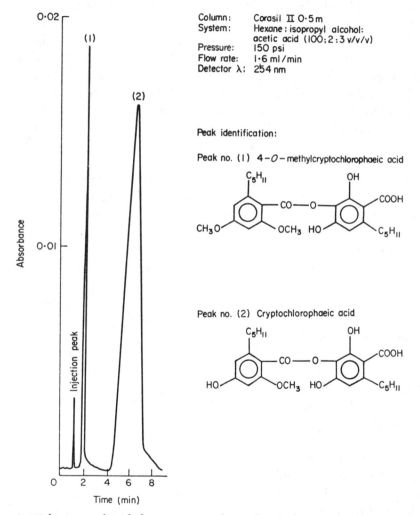

Column: Corasil II 0·5 m
System: Hexane : isopropyl alcohol:
acetic acid (100 : 2 : 3 v/v/v)
Pressure: 150 psi
Flow rate: 1·6 ml/min
Detector λ: 254 nm

Peak identification:

Peak no. (1) 4-O-methylcryptochlorophaeic acid

Peak no. (2) Cryptochlorophaeic acid

FIG. 6. High pressure liquid chromatogram of a synthetic mixture of the *meta*-depsides (1) 4-O-methylcryptochlorophaeic acid and (2) cryptochlorophaeic acid which occur in the *Cladonia chlorophaea* complex.

be achieved rapidly. Accurate quantification of the separated compounds is achieved to the nanogram level by the use of u.v. spectrophotometers. Some of the chemical types in the *C. chlorophaea* complex provide excellent examples of the qualitative and quantitative manner in which h.p.l.c. may be used to advantage in the study of lichen material. Figures 6, 7 and 8 show the separation of

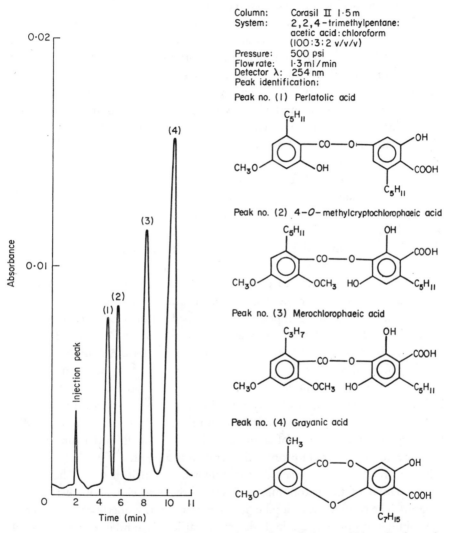

Column: Corasil II 1·5 m
System: 2,2,4-trimethylpentane:
acetic acid:chloroform
(100:3:2 v/v/v)
Pressure: 500 psi
Flow rate: 1·3 ml/min
Detector λ: 254 nm
Peak identification:

Peak no. (1) Perlatolic acid

Peak no. (2) 4-O-methylcryptochlorophaeic acid

Peak no. (3) Merochlorophaeic acid

Peak no. (4) Grayanic acid

FIG. 7. High pressure liquid chromatogram of a synthetic mixture of (1) perlatolic acid, (2) 4-O-methylcryptochlorophaeic acid, (3) merochlorophaeic acid and (4) grayanic acid which occur in the *Cladonia chlorophaea* complex.

synthetic mixtures of eight lichen compounds encountered in the present study. The depside cryptochlorophaeic acid is readily separated from the corresponding methylated compound, 4-O-methylcryptochlorophaeic acid (Fig. 6) using a relatively short column and low pressures. Figure 7 shows the separation of four compounds, three of which, perlatolic, merochlorophaeic and 4-O-methyl-cryptochlorophaeic acids, have been reported to occur together in nature. Since all the chemical variants found amongst the *C. chlorophaea* material studied have been analysed by h.p.l.c., and perlatolic acid has not been detected, it would appear that *C. perlomera* is unknown in the British Isles at the present time. The retention time for the depsidone grayanic acid (Fig. 7), which does occur in Britain in the chemical type *C. grayi*, enables it to be distinguished from most other compounds in the complex. The separation of two compounds which are widely distributed in the genus *Cladonia*, atranorin and usnic acid, is shown in Fig. 8. Synthetic mixtures of homosekikaic and sekikaic acids which differ only by —$(CH_2)_2$— in the *para* position were also readily separated on the h.p.l.c. system as shown in Fig. 8. The separation of these two compounds in a time of 7 min from a crude lichen extract is shown in Fig. 9. It should be noted that this chromatogram was recorded with the detector wavelength adjusted to 263 nm. The analysis was carried out at this wavelength because the complete electronic absorption spectrum of the pure compounds sekikaic and homosekikaic acids showed maximum molar extinction coefficient values at this position. The adjustment of the monitor to positions of optimum wavelength for a given compound(s) permits extremely sensitive, quantitative screening programmes to be carried out for the compound(s) concerned. In the case of critical lichen material the identity of the components separated by h.p.l.c. may be further confirmed by mass spectrometry of the collected fractions. A complete scheme for the chemical analysis of apparently new or interesting lichen material is presented in Fig. 11 and this has been used in the present study. For example, the isolation and identification of didymic acid in specimens previously identified as *C. subulata* (Table IV) was carried out in this way. Original analysis by t.l.c. had suggested the presence of homosekikaic acid in addition to fumarproto-cetraric acid which would prompt the erroneous suggestion that the specimens might be related to *C. nemoxyna*, unknown in the British Isles. The precise determination of the taxonomic position of these specimens is currently in progress. The routine use of h.p.l.c. in the present work underlines a number of those limitations of t.l.c. methods described earlier. A number of specimens of *C. merochlorophaea* "strain 1" and *C. merochlorophaea* "strain 2" appeared to contain only merochlorophaeic acid and homosekikaic acid respectively. H.p.l.c. analyses subsequently revealed the presence of 4-O-methylcrypto-chlorophaeic acid and sekikaic acid, respectively, in low concentrations relative to the major constituents.

A check on the linearity of the u.v. detector at 1×10^{-2} absorbance units full scale for pure sekikaic acid is given in Fig. 9b and shows that for routine work, quantitative determinations may be calculated from uncorrected peak heights without introducing significant errors.

The separation of merochlorophaeic and 4-O-methylcryptochlorophaeic

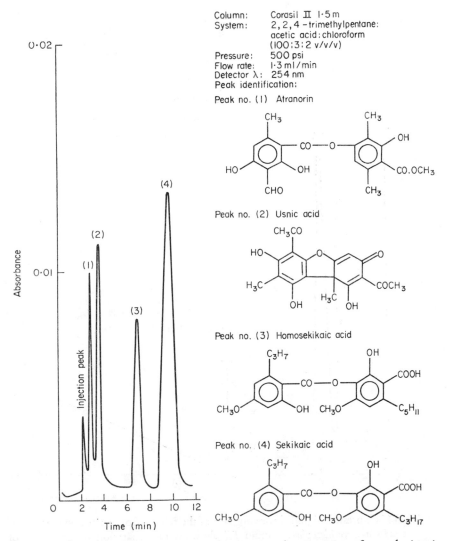

FIG. 8. High pressure liquid chromatogram showing the separation of a synthetic mixture of (1) atranorin, and (2) usnic acid which occur widely in the genus *Cladonia*, and (3) homosekikaic acid, and (4) sekikaic acid which occur together in the *Cladonia chlorophaea* complex.

acids from an extract of *C. merochlorophaea* "strain 1" (Fig. 10) illustrates the sensitivity which can be achieved in h.p.l.c. studies. Peak (1) given by 4-O-methylcryptochlorophaeic acid represents 0·0005 absorbance units which is equivalent to less than eight nanograms of the depside concerned. The sensi-

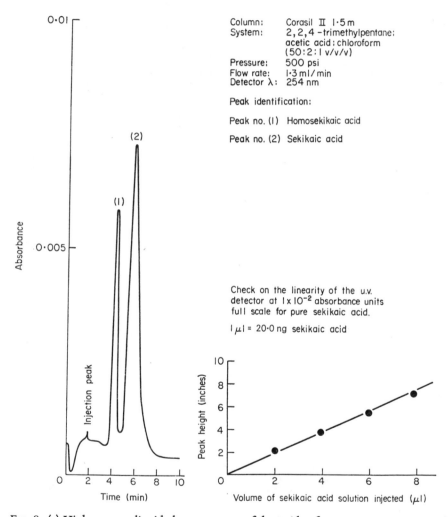

Column: Corasil II 1·5 m
System: 2,2,4 –trimethylpentane: acetic acid: chloroform (50:2:1 v/v/v)
Pressure: 500 psi
Flow rate: 1·3 ml/min
Detector λ: 254 nm

Peak identification:

Peak no. (1) Homosekikaic acid

Peak no. (2) Sekikaic acid

Check on the linearity of the u.v. detector at 1 x 10⁻² absorbance units full scale for pure sekikaic acid.

1 μl = 20·0 ng sekikaic acid

FIG. 9. (a) High pressure liquid chromatogram of the residue from an acetone extract of *Cladonia merochlorophaea* "strain II" showing the separation of an excess of sekikaic acid over homosekikaic acid in the ratio 5:4. Peak (2) represents approximately 100 nanograms of sekikaic acid.

(b) Calibration of the ultraviolet detector for quantitative determinations of sekikaic acid.

tivity of the technique has enabled micro-population studies to be carried out on herbarium collections. In some cases a minute fragment from *every* lichen podetium in collections of chemically mixed lichen has been analysed in order to study possible morphological correlations. Those studied have shown that even when two chemical types are growing side by side on the same substratum, mixed chemical types do not occur. The study of one collection in particular

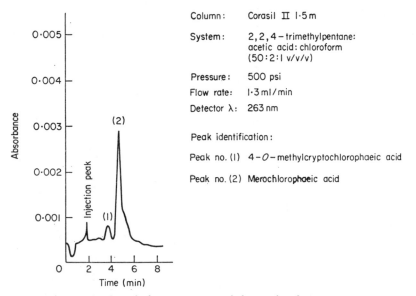

FIG. 10. High pressure liquid chromatogram of the residue from an acetone extract of *Cladonia merochlorophaea* "strain II" showing the separation of an eightfold excess of merochlorophaeic acid over 4-O-methylcryptochlorophaeic acid. A 1 μl aliquot from a solution of the residue in chloroform was applied to the column.

(Angus: Brechin, Rossie Moor, Nov. 1945, *Duncan*, hb. Duncan), which consisted of *C. merochlorophaea* strains 1 and 2 provided the material which led initially to the above conclusion. In this collection the two strains were so closely mingled that in some cases podetia from each strain were united at the points where they touched. However, the chemical identity of each strain was preserved—no "hybrids" were detected. These chemical analyses have contributed much to our understanding of morphological criteria in the complex.

The technique of h.p.l.c. has also permitted the detailed chemical analysis of different parts of individual lichen podetia. At least ten samples from both strains of *C. merochlorophaea* have been studied in this way. In British specimens,

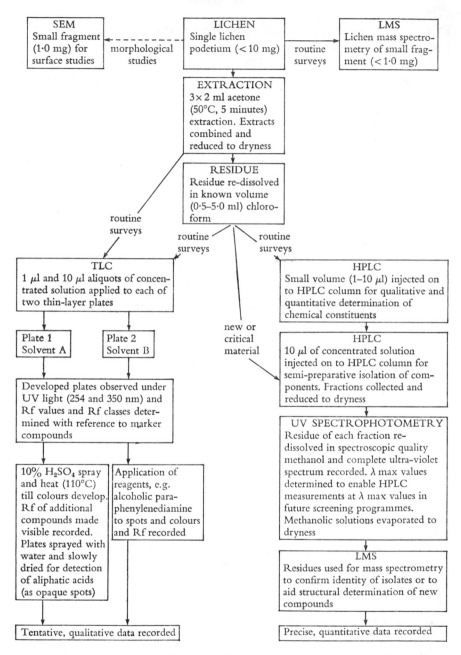

FIG. 11. Scheme for the chemical analysis of lichen material in chemotaxonomic studies.

the most frequently occurring ratios of the lichen compounds are as follows: merochlorophaeic acid:4-O-methylcryptochlorophaeic acid, 4:1 (w/w); sekikaic acid:homosekikaic acid, 4:3 (w/w) although various exceptions have been noted above. These ratios generally remain constant throughout the length of the lichen podetium but quantitative differences are observed. The compounds are found in particularly high concentration at the podetial tip, especially in the secondary podetium of "proliferous" material. Quantitative measurements of compounds in a large number of species in *Cladonia* are included in Nourish (1976).

Following this discussion of some of the practical advantages of h.p.l.c. in lichen studies, it should perhaps be noted that there are a number of disadvantages in comparison with t.l.c. Firstly, a commercial h.p.l.c. is comparatively expensive at the outset and requires a number of carefully purified and costly solvents of spectroscopic quality for routine operation. Secondly, because detection is achieved by u.v. spectrophotometry the solvents used must be optically transparent at the optimal absorption wavelength of the lichen compounds concerned. Many of the solvents found to be so useful in the chromatography of lichen products, e.g. benzene, toluene, dioxane, acetone and acetic acid "cut-off" in the critical u.v. region and either cannot be used or may only be employed in low concentrations. The most successful solvent systems developed in the present study, from experiments involving a large number of solvents in various combinations, are given in Figs 6–10. These solvent systems cut-off in the u.v. region 220–245 nm. However, these problems will doubtless be overcome as specific solvent systems are developed for particular classes of lichen products. The present solvent systems permit chromatography of those lichen compounds which fall into the hR_F classes 30–80 with respect to t.l.c. A suitable (more polar) system for the analysis of specimens containing compounds with low R_F values such as fumarprotocetraric acid has yet to be devised.

It is hoped that lichenologists with access to chemical laboratories will make use of the increased speed, resolution and sensitivity of this quantitative chromatographic method in detailed chemotaxonomic studies.

3. Analytical Data for Unidentified Compounds

Values of hR_F are presented as follows: hR_F of the compound/hR_F of sekikaic acid, hR_F of atranorin.

C. cyathomorpha: t.l.c. solvent A: 19/43, 72. solvent B: 25/53, 78, u.v. quench, brown/purple with 10% H_2SO_4 and heating (110°C) for 15 min. The compound has not yet been satisfactorily chromatographed by h.p.l.c. Major ion peaks (> m/e 120) in the l.m.s.: 149(100) 137(30·0) 256(27·0) 123(25·0) 129(21·0) 136

TABLE IV. Summary of accessory substances, unnamed chemical types and re-identifications arising from the chemical analysis of the British collections studied

Species (original morphological identification)	Total number studied	% confirmed as this species (without accessory substances)	% referred to the species named						% and chemical composition of specimens with accessory substances and/or of uncertain position
			C. conista	C. cryptochlorophaea	C. grayi	C. merochlorophaea "strain 1"	C. merochlorophaea "strain 2"	C. glauca–C. squamosa group	
Group 1									
C. chlorophaea	217	41.0	—	23.5	8.5	18.5	4.0	—	0.5 fumarprotocetraric acid and zeorin. 4.0 fumarprotocetraric acid and atranorin.
Group 2									
C. conista	27	26.0	74.0	—	—	—	—	—	—
C. cyathomorpha (including holotype)	6	100.0	—	—	—	—	—	—	—
C. fimbriata	193	80.0	10.0	2.0	—	0.5	2.0	3.0	0.5 atranorin only. 2.0 fumarprotocetraric acid and unidentified compound[a].

C. pyxidata	153	77·0	—	5·0	1·0	2·0	—	—	12·0 fumarprotocetraric acid and atranorin. 1·0 atranorin only. 2·0 no compounds detected.
C. pyxidata var. pocillum (including isotype)	20	95·0	—	—	—	—	—	—	5·0 fumarprotocetraric acid and atranorin (isotype specimen).
Group 3									
C. coniocraea	62	87·0	1·5	—	—	—	1·5	10·0	—
C. cornuta	19	84·0	—	—	—	—	—	11·0	5·0 fumarprotocetraric acid and ursolic acid.
C. ochrochlora	81	90·0	—	—	—	—	1·0	5·0	4·0 fumarprotocetraric acid and atranorin.
C. subulata	61	82·0	—	—	—	—	—	11·0	7·0 didymic acid and fumarprotocetraric acid.

[a] See experimental section for analytical data.

(20·0) 167(14·0) 185(13·0) 213(13·0) 284(12·0) 236(10·0). M⁺(?) 410(3·0). This spectrum has not yet been definitely assigned to the compound in t.l.c.

C. *fimbriata*: The compound may be related to protocetraric acid. Solvent A: 6/42, 73, Solvent B: 22/53, 78 u.v. quench, spot uniformly dark with 10% H_2SO_4 treatment.

Other faint spots detected in various samples by t.l.c. and which may or may not be attributable to secondary lichen products are not included here since further characterization is in progress.

4. *The British Representatives of the* Cladonia chlorophaea–pyxidata *Complex and their Taxonomic Treatment*

In Table I, group 1, C. *cryptochlorophaea*, C. *merochlorophaea* "strain 1" (var. *merochlorophaea*) and C. *merochlorophaea* "strain 2" (var. *novochlorophaea*★ Sipman) are reported in the British Isles for the first time and the presence of C. *chlorophaea* s.s. and C. *grayi* is confirmed. No evidence has been obtained for the existence in Britain of the five remaining chemical types in this group, which have been reported from very few localities in the world. It can be seen from Table II that the dominant species in the British population is C. *chlorophaea* s.s. and the least common is C. *merochlorophaea* "strain 2". The latter species was originally thought to have a pronounced northern distribution, but recent collections from Bedfordshire (Sandy, Sandy Heath, 4 October 1973, *Nourish & Nourish*, hb. Nourish 775; Flitwick Moor, March 1974, *Nourish*, hb. Nourish 869) has shown this not to be the case. All the specimens examined to date had been collected from peat or acid substrata and this species may be more common in such environments (and therefore in Britain) than these preliminary results would suggest. The composition of the British populations of C. *chlorophaea* s.l. (Table II) compares most closely with that reported for the Netherlands by Sipman (1973) and this may perhaps be explained by the similar climatic conditions in the two countries.

The analytical results for the species in groups 2 and 3 (Table I) show a remarkably uniform chemistry. The exceptions are C. *conistea*, which is shown to be more common in the British Isles than C. *conista* s.s., and C. *cyathomorpha* mentioned previously. In addition, a number of species were found to contain accessory atranorin in low concentrations: C. *pyxidata*, C. *pocillum* and C. *ochrochlora*. Specimens originally identified as C. *fimbriata* and found to contain atranorin could be accurately referred to C. *conistea* and this compound has not been detected in the two collections of C. *fimbriata* var. *major* examined to date.

★ This choice of varietal name is perhaps rather unfortunate since it is based on a misnomer: "novochlorophaeic acid" comprises sekikaic and homosekikaic acids.

Unnamed chemical types of uncertain position in the three groups studied, with the frequency of occurrence of specimens having accessory compounds, are given in Table IV and are expressed as a percentage of the total number of each species studied. The number of re-identifications which have been made as a result of these chemical analyses are also included in this Table. The fairly high proportion of specimens which fall into this category indicate that (a) the morphology of the C. *chlorophaea–pyxidata* complex is not fully understood and (b) some of the species in the group are polymorphic.

Opinions on the taxonomic treatment of a number of species in the groups 1–3 (Table I) vary widely but the chemical types in group 1 have been the subject of particular controversy. Asahina (1940, 1943) recognized the four species C. *chlorophaea* s.s., C. *grayi*, C. *cryptochlorophaea* and C. *merochlorophaea* on an entirely chemical basis. Evans (1944) supported this view as did Shibata and Hsuch-Ching Chiang (1965) following studies on the structure and biogenetical relationships of the three main lichen products involved. The latter authors attached particular significance to C. *grayi* on the grounds that the biogenetical scheme for the synthesis of its major constituent, the depsidone grayanic acid, was entirely different from that proposed for the depsides in the group (see Chapter 7). The most recently formally named, essentially chemical, taxa have been assigned the rank of (a) species: C. *perlomera* (Culberson and Kristinsson, 1969) and (b) variety C. *merochlorophaea* var. *novochlorophaea* (Sipman, 1973) respectively. A number of authors have attempted to assign morphological characters to the various chemical types (Ahti, 1966; Krog, 1968; Kristinsson, 1971; Leuckert *et al.*, 1972; Sipman, 1973) while others have used distributional (Kristinsson, 1971; Bowler, 1972) or ecological data (Wetherbee, 1969) to support recognition of various taxa at species level. In contrast Dahl (1950), Lamb (1951), Luttrell (1954), Imshaug (1957) and Thomson (1968) have suggested that the rank of variety or chemical strain is to be preferred. The present authors share the opinion of Hawksworth (1971; see also Chapter 7) that accurate taxonomic assessments should be made only after a detailed analysis of all the criteria mentioned above. Furthermore, in view of the very large number of morphological characters which have been associated with the cup lichens in general (e.g. degree of cortication, size and frequency of soredia and/or corticate granules, size and shape of podetial scyphi, gross apothecial characters) it is perhaps surprising that numerical taxonomic methods have not been applied in this case as they have in groups of similar complexity in other botanical spheres (e.g. Kendrick and Weresub, 1966). Such studies are therefore being carried out in this connection during the course of the present work. Further, the application of scanning electron microscopy to these lichens has yielded a considerable amount of original information concerning the histological distribution of lichen compounds and the surface structures of apothecia, corticate and sorediate podetia. In view of these studies and further work on the distributional and ecological aspects currently in progress, we do not wish to propose any taxonomic or nomenclatural revisions at the present stage of our work.

In chemistry, the great advances in analytical methodology have been accompanied by the production and interpretation of data in a digital form and it may well be that a more numerical approach to lichenology will help to solve some current problems of taxonomy.

ACKNOWLEDGEMENTS

We are particularly grateful to Mr P. W. James for his invaluable advice and assistance throughout the course of this work and to Mr B. J. Coppins, Dr U. K. Duncan, Miss J. E. Menlove and Mr R. Ross (Keeper of Botany, British Museum (Natural History), London) for the provision of lichen material. One of us (R. N.) gratefully acknowledges grants from the Botanical Research Fund and the Dr J. W. Linnell Trust which enabled this work to be initiated and the subsequent award of a research studentship from the Science Research Council. The interest and advice on chromatographic matters of Dr P. Bristow (I.C.I. Pharmaceuticals Division, Macclesfield) is also acknowledged with thanks.

REFERENCES

AHTI, T. (1966). Correlation of the chemical and morphological characters in *Cladonia chlorophaea* and allied lichens. *Ann. bot. fenn.* **3**, 380–389.

ASAHINA, Y. (1940). Chemismus der Cladonien unter besonderer Berucksichtigung der japanischen Arten. 1. *Cladonia chlorophaea* und verwandte Arten. *J. Jap. Bot.* **16**, 709–727.

ASAHINA, Y. (1941). Chemismus der Cladonien unter besonderer Berucksichtigung der japanischen Arten. Nachtrag zu *Cladonia chlorophaea* und verwandte Arten. *J. Jap. Bot.* **17**, 431–437.

ASAHINA, Y. (1943). Chemismus der Cladonien unter besonderer Berucksichtigung der japanischen Arten (Fortsetzung). *J. Jap. Bot.* **19**, 47–56, 227–244.

BOHMAN, G. (1969). Anthraquinones from the genus *Caloplaca*. *Phytochemistry* **8**, 1829–1830.

BOWLER, P. A. (1972). The distribution of four chemical races of *Cladonia chlorophaea* in North America. *Bryologist* **75**, 350–354.

CULBERSON, C. F. (1969). "Chemical and Botanical Guide to Lichen Products." University of North Carolina Press, Chapel Hill, N.C.

CULBERSON, C. F. (1970). Supplement to "Chemical and Botanical Guide to Lichen Products". *Bryologist* **73**, 177–377.

CULBERSON, C. F. (1972a). Improved conditions and new data for the identification of lichen products by a standardised thin layer chromatographic method. *J. Chromatog.* **72**, 113–125.

CULBERSON, C. F. (1972b). High speed liquid chromatography of lichen extracts. *Bryologist* **75**, 54–62.

CULBERSON, C. F. and KRISTINSSON, H. (1969). Studies on the *Cladonia chlorophaea* group: A new species, a new metadepside and the identity of "novochlorophaeic acid". *Bryologist* **72**, 431–443.

CULBERSON, C. F. and KRISTINSSON, H. (1970). A standardised method for the identification of lichen products. *J. Chromatog.* **46**, 85–93.

DAHL, E. (1950). Studies in the macrolichen flora of South West Greenland. *Meddr Grønland* **150** (2), 1–176.

EVANS, A. W. (1944). Supplementary report on the *Cladoniae* of Connecticut. *Trans. Conn. Acad. Arts Sci.* **35**, 519–626.

GEISS, F. and SCHLITT, H. (1973). Thin-layer chromatography as a pilot technique for rapid column chromatography (Summary). *J. Chromatogr.* **82**, 5–6.

HALL, B. (1971). "Some Studies on the Application of Mass Spectrometry to the Chemical Taxonomy of British Lichens of the Genus *Cladonia.*" B.Sc. dissertation, University of Salford.

HAUPT, G. W. (1952). An alkaline solution of potassium chromate as a transmittancy standard in the ultraviolet. *J. Res. Nat. Bur. Stds* **48**, 418.

HAWKSWORTH, D. L. (1971). Types of chemical variation in lichens and their taxonomic treatment. *In* "Abstracts, First International Mycological Congress, Exeter, 1971" (G. C. Ainsworth and J. Webster, eds), p. 41. First International Mycological Congress, Exeter.

HUNECK, S., DJERASSI, C., BECHER, D., BARBER, M., ARDENNE, M. VON, STEINFELDER, K. and TUMMLER, R. (1968). Flechteninhaltstoffe—XXXI. Massenspektrometrie und ihre Anwendung auf structurelle und stereochemische Probleme—CXXIII. Massenspektrometrie von Depsiden, Depsidonen, Depsonen, Dibenzofuranen und Diphenylbutadien mit positiven und negativen Ionen. *Tetrahedron* **24**, 2707–2755.

IMSHAUG, H. A. (1957). Alpine lichens of Western United States and adjacent Canada 1. The macro lichens. *Bryologist* **60**, 177–272.

JENSEN, S. B. (1973). "Some Studies on the Reaction between Oestriol and Aqueous Sulphuric Acid Mixtures." Ph.D. thesis, University of Salford.

KENDRICK, W. B. and WERESUB, L. K. (1966). Attempting neo-Adansonian computer taxonomy at the ordinal level in the Basidiomycetes. *Syst. Zool.* **15**, 307–329.

KNOX, J. H. (1972). An outline of the theory of chromatography as applicable to high speed liquid chromatography. *Perkin-Elmer Analytical News* **7**, 8–14.

KRISTINSSON, H. (1971). Morphological and chemical correlations in the *Cladonia chlorophaea* complex. *Bryologist* **74**, 13–17.

KROG, H. (1968). The macrolichens of Alaska. *Norsk Polarinst. Skr.* **144**, 1–180.

LAMB, I. M. (1951). On the morphology, phylogeny and taxonomy of the lichen genus *Stereocaulon*. *Can. J. Bot.* **29**, 522–584.

LEUCKERT, C., ZIEGLER, H. G. and POELT, J. (1972). Zur Kenntis der *Cladonia chlorophaea* Gruppe und ihrer Problematik in Mitteleuropa. *Nova Hedwigia* **22**, 503–534.

LUTTRELL, E. S. (1954). The Cladoniaceae of Virginia. *Lloydia* **17**, 275–306.

MOORE, B. J. (1968). The macrolichen flora of Florida. *Bryologist* **71**, 161–265.

NEWTON, R. D. (1970). "Chemical and Taxonomic Studies on some *Cladonia* species of Lichens." B.Sc. dissertation, University of Salford.

H

NOURISH, R. (1972). "A Chemotaxonomic Study of British Lichens of the Subgenus *Cladina* by Physico-Chemical Techniques." M.Sc. thesis, University of Salford.

NOURISH, R. (1976). "Chemotaxonomic Studies on British Lichens of the Genus *Cladonia*." Ph.D thesis, University of Salford. [In preparation.]

NOURISH, R. and OLIVER, R. W. A. (1974). Chemotaxonomic studies on British lichens I. *Cladonia* subgenus *Cladina*. *Lichenologist* 6, 73–94.

NOURISH, R. and OLIVER, R. W. A. (1975) ["1974"]. Chemical studies on some lichens in the Linnean herbarium and lectotypification of *Lichen rangiferinus* L. (em. Ach.). *Biol. J. Linn. Soc.* 6, 259–268.

OLIVER, R. W. A. (1972). High pressure, high resolution, liquid chromatography of complex physiological fluids. *Perkin-Elmer Analytical New*, 7, 15–19.

OLIVER, R. W. A. and LOMAX, M. E. (1971). "A Guide to the Published Collections and Bibliographies of Molecular Electronic Spectra." Perkin-Elmer, Beaconsfield.

SANTESSON, J. (1969). Chemical studies on lichens, 10. Mass spectrometry of lichens. *Arkiv. Kemi* 30, 363–377.

SANTESSON, J. (1970). Chemical studies on lichens, 20. Anthraquinones in *Caloplaca*. *Phytochemistry* 9, 2149–2166.

SANTESSON, J. (1974) ["1973"]. Identification and isolation of Lichen substances. *In* "The Lichens" (V. Ahmadjian and M. E. Hale, eds), pp. 633–652. Academic Press, New York and London.

SHIBATA, S. and CHIANG, H.-C. (1965). The structures of cryptochlorophaeic, acid and merochlorophaeic acid. *Phytochemistry* 4, 133–139.

SIPMAN, H. J. M. (1973). The *Cladonia pyxidata-fimbriata* complex in the Netherlands with description of a new variety. *Acta bot. neerl.* 22, 490–502.

SNYDER, L. R. (1972). A rapid approach to selecting the best experimental conditions for high-speed liquid column chromatography. Part II—Estimating column length, operating pressure and separation time for some required sample resolution. *J. Chromat. Sci.* 10, 369–379.

TAYLOR, C. J. (1968). "Lichens of Ohio, Part 2. Fruticose and Cladoniform Lichens" [Ohio Biological Survey Notes no. 4.]. Ohio State University, Columbus, Ohio.

THOMSON, J. W. (1968) ["1967"]. "The Lichen Genus *Cladonia* in North America." University of Toronto Press, Toronto.

WETHERBEE, R. (1969). Population studies in the chemical species of the *Cladonia chlorophaea* group. *Mich. Bot.* 8, 170–174.

9 | Ecological Aspects of Dispersal and Establishment in Lichens

R. H. BAILEY

Department of Extra-mural Studies, University of London, England

Abstract: The discharge and dispersal of propagules from non-lichenized fungi has been extensively studied. This situation does not prevail amongst the lichenized fungi which have, none the less, a great variety of propagules dispersed in many ways. The relatively small amount of work as yet undertaken upon the dispersal of lichen propagules has concentrated on confirming the assumptions made by earlier lichenologists concerning modes of dispersal. Our knowledge of dispersal by means of wind, water, animals and, in the case of ascospores, by active discharge, is reviewed with the intention of emphasizing gaps in our present knowledge.

Studies on the removal of propagules from their dispersal agents and deposition on surfaces suitable for their germination has received little attention, despite the undoubted ecological importance of these processes. Aspects of deposition and germination are considered and areas in need of further research highlighted.

INTRODUCTION

Methods of dispersal in non-lichenized fungi have received extensive study and a number of reviews of the field have been published (e.g. Ingold, 1953, 1965, 1971). Dispersal studies on the non-lichenized genera have centred around the active discharge of ascospores although other aspects have received attention. Despite the greater variety of propagules amongst the lichenized fungi relatively few experimental or observational data have been published upon their modes of dispersal, germination and establishment. Smith (1921), Tobler (1925) and Ridley (1930) provided summaries of the sparse data and contemporary ideas concerning the dispersal and establishment of lichens.

More recently general works on lichens have included reviews bringing these

Systematics Association Special Volume No. 8, "Lichenology: Progress and Problems" edited by D. H. Brown, D. L. Hawksworth and R. H. Bailey, 1976, pp. 215–247. Academic Press, London and New York.

earlier statements up to date: for example, des Abbayes (1951), Barkman (1958), Smith (1962), Haynes (1964) and Hale (1967). Ahmadjian (1967) reviewed germination of lichen ascopores *in vitro* and provided a brief summary of existing literature on culture techniques (Ahmadjian, 1974). Recently Pyatt (1974) considered lichen propagules specifically and reviewed their dispersal and germination. All these works contain extensive bibliographies.

This paper attempts not only to review progress and to highlight areas requiring research, but also to explain why studies on lichen dispersal and establishment have fallen behind similar studies on other groups, most particularly the non-lichenized fungi. Despite the phenomenal increase in our knowledge of lichens (see Ahmadjian and Hale, 1974; Ferry *et al.*, 1973), techniques for maintaining long-term cultures on sterile media and for transplanting lichens, as part of ecological or taxonomic studies, are still imperfectly worked out. That lichens can be grown on agar media and sterile soil has been demonstrated recently by many workers, for example Bertsch and Butin (1967) and Ahmadjian and Heikkilä (1970), using resynthesized symbionts of *Endocarpon pusillum* and Dibben (1971) working under phytotron conditions with macerated tissues. Kershaw and Millbank (1969) have shown that whole lichens can be maintained when transplanted to growth chambers in urban areas and several workers have developed techniques for transplanting lichens from one natural site to another, principally as part of studies on air pollution (see Brodo, 1961; Richardson, 1967; Hoffman, 1971; Hawksworth, 1971). Despite these varied achievements, simple, successful culture techniques for a wide range of whole lichen thalli have not yet been developed. Such techniques are an essential prerequisite for advance in many aspects of lichen biology, not least the study of dispersal and establishment. The lack of such techniques, allied to the generally slow growth rate of lichen thalli, must largely explain the dearth of experimental data on ascospore discharge from lichenized fungi and our scanty knowledge of the various ecological factors of importance to the establishment of lichens in nature, nutrient and water requirements, the affects of aspect, exposure and light, of texture and chemistry of the substratum and of the separate requirements of the two symbionts. Answers to these problems are beginning to emerge, pieced together from studies on isolated symbionts, investigation of the physiology and biochemistry of thallus samples and from careful field observations of lichen communities as well as from laboratory investigations upon the germination of propagules and of lichen physiology. Much of lichen ecology, including dispersal and establishment, is still very much at a nineteenth century stage despite the many papers that have been published using modern ecological techniques. This paper concentrates, therefore, on aspects of dispersal and establishment in nature rather than on the laboratory studies of ascospore discharge such as have attracted the attention of students of the non-lichenized fungi. The many lacunae in our knowledge of, and techniques for studying, lichens explains also why much of this paper is concerned with simple observational data gathered from varied sources.

TYPES OF PROPAGULE

Lichen propagules can be divided into three types: 1, vegetative diaspores reproducing the whole symbiosis; 2, sexual or asexual spores reproducing the fungus alone; 3, bodies reproducing the alga alone.

Table I lists the various structures that have at one time or another been considered as propagules of all or part of the lichen symbiosis. The anatomy,

TABLE I. Types of lichen propagule

Whole symbiosis	Fungal partner alone		Algal partner alone
	Sexual	Asexual	
Whole thalli	Ascospores	Conidia	Aplanospores
Thallus fragments—differentiated			
—undifferentiated	Basidiospores		Zoospores
Schizidia			Unlichenized
			hormocysts
Lobules			(Soredia)
Isidia			(Isidia)
Soredia			(Thallus fragments)
Lichenized hormocysts			
Ascospores discharged with			
hymenial algae			

morphology and variation of these structures has been extensively reviewed and illustrated (e.g. Jahns, 1974; Poelt, 1974a) and will not be given detailed consideration in this paper. Further information on specific propagules can be found as follows: (*a*) *whole thalli* (erratic lichens) (Smith, 1921); (*b*) *thallus fragments* (Poelt, 1974a); (*c*) *schizidia* (Poelt, 1965, 1974a); (*d*) *lobules*—the isidia squamiformia of Du Rietz (1924) and the regeneration lobules of Thomson (1948, 1950) and Lindahl (1960) are examples of lobules as understood in this paper; (*e*) *isidia* and *soredia* (Du Rietz, 1924; Maas Geesteranus, 1948; Jahns, 1974; Poelt, 1974a); (*f*) *hormocysts* (Henssen, 1969; Jahns, 1974); (*g*) *ascospores* (Hale, 1961, 1967; Pyatt, 1974)—ascospores are varied in size and shape: the polarilocular type (e.g. Fig. 1), and its modifications, characteristic of the

Teloschistaceae (e.g. *Caloplaca*, see Wade, 1965), and found in some other genera, e.g. species of *Rinodina* (Sheard, 1967), is confined to the lichenized fungi; Letrouit-Galinou (1974) summarizes our knowledge of ascocarp ontogeny and anatomy; (*h*) *basidiospores*—only a small minority of lichenized fungi are basidiomycetes (Poelt, 1974b); no general study of their spores has been published; (*i*) *conidia* (Lindsay, 1859; Smith, 1921)—the term is used in this paper to cover a variety of "spores" thought to be asexually produced in varied structures called pycnidia; (*j*) *algal propagules*—those produced in culture are described by Ahmadjian (1967), see also Chapters 4 and 5.

DISPERSAL PROCESSES

It is commonly assumed that lichen propagules are dispersed passively by wind, water and animals as well as by active means. However, it must constantly be borne in mind that only the vegetative diaspores disperse the whole lichen symbiosis. Fungal and algal propagules may or may not serve to establish a whole lichen elsewhere. When so great a variety of propagules is present within a group of plants (and when some reproduce only part of the symbiosis) there is, inevitably, uncertainty as to which are most important in nature. Further work must be undertaken to elucidate this point.

Any effective dispersal mechanism for propagules must include the processes of liberation from the parent thallus, transport to and deposition on a suitable substratum and germination on, or attachment to, that substratum. Liberation and transport are here considered to be dispersal processes in a strict sense while deposition and germination or attachment are considered under the heading of establishment.

Liberation and transport are frequently, but not invariably, accomplished by the same agent although many propagules are undoubtedly affected by more than one agent in the course of transport to a suitable site for establishment. Amongst the vegetative diaspores only soredia and hormocysts are self-liberating in that their mode of growth detaches them from the parent plant. All spores are self-liberating and ascospores (excluding those of the Caliciales) are self-transporting in as much as they are actively discharged to a small distance from the ascocarp. In the Caliciales the ascus wall disintegrates prior to the maturation of the ascospores so the ascospores, fragments of ascus walls and paraphyses lie in a powdery mass within the ascocarp, the mazaedium.

1. Wind as a Dispersal Agent
(*a*) *Field studies on airborne propagules.* Du Rietz (1931) recorded soredia and

thallus fragments in snow falls and noted that the abundance of airborne propagules seemed to be more or less proportional to the abundance, in the region, of the lichen species concerned, Pettersson (1940), Bailey (1966a) and Rudolph (1970) noted the occurrence of soredia in traps sampling the air spora, and Pettersson noted that thallus fragments were, in fact, very much more abundant than soredia. Sernander (1918) recorded fragments of *Ramalina fraxinea* in calcareous tuff; Bond (1969) showed that glacial ice in Colorado contained fragments of four arctic–alpine lichen species that had, presumably, been carried to the glacier by wind.

Recent studies by the author show that a variety of spore types referable to lichens are to be found in the air spora of Sutton, Surrey (Fig. 1) and Ingold

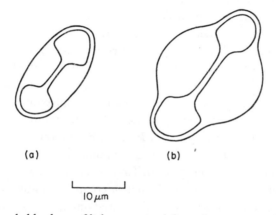

(a)   (b)

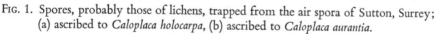

10 μm

FIG. 1. Spores, probably those of lichens, trapped from the air spora of Sutton, Surrey; (a) ascribed to *Caloplaca holocarpa*, (b) ascribed to *Caloplaca aurantia*.

(1965) noted the occurrence of spores attributed to *Pertusaria* species in the air spora. The occurrence of spores of *Pyrenula nitida* on moss some 2 km from the nearest known site for the species was recorded by Bailey (1967), and it is most likely that these spores were carried by wind. Spores (attributed to *Buellia* sp.) as well as soredia were trapped in the air over Antarctica by Rudolph (1970). Miyoshi (1901) described the case of *Sagedia macrospora* (*Arthopyrenia* sp.) where whole perithecia which became dislodged from the thallus were carried away by wind, ejecting spores whenever and wherever they became wet. Barkman (1958) suggested that spores might be transported in thermal micro-currents developed along bark surfaces heated by the sun.

The transport of erratic lichens presents a special case of wind-blown lichen propagules. Both Smith (1921) and Ridley (1930) presented accounts of this phenomenon in the genera *Anaptychia, Lecanora, Parmelia* and *Platismatia*. Erratics appear to arise as fragments of thallus that become detached as a result of external mechanical action, e.g. by frost, insolation or animals, and continue independent growth. In all erratic morphotypes of macrolichens ascocarp formation is very much reduced or eliminated. The detachment of portions of thallus in species such as *Lecanora atra* and *L. calcarea* is readily observable as a lifting and peeling of the central portion of the thallus in older specimens. (In Britain such crustose species have not been recorded as erratics although the lifting and peeling process is common and could provide many fragments for dispersal by various agencies.) An erratic lichen could itself act as a propagule if it were to lodge and grow in a suitable place but more important is its potential for scattering other propagules whilst mobile. Few recent records of erratic lichens have been published. Erratic specimens of the fruticose lichen *Evernia prunastri* were observed by the author at Dungeness, Kent. These developed as ball-like thalli attached to low-growing *Prunus spinosa* amongst shingle. On reaching a certain size in this exposed habitat they seemed liable to become detached by wind and to be carried away. *E. prunastri* is a sorediate species and when travelling at speed (speeds of approximately 20 m/s were observed) could well scatter both soredia and thallus fragments along its path. Erratic specimens of *Chondropsis* and *Parmelia* have recently been observed in New Zealand and Australia (P. W. James, unpublished).

At present studies on the air spora are largely conducted by mycologists or by medical workers concerned with the pollen count and insufficient attention is given to spores that might be those of lichenized fungi, or to soredia. The certain reference of any lichen propagule to a particular species can only be made after its germination and growth into a recognizable plant—which is not, as yet, technically possible amongst lichenized fungi. These two points account for our slender knowledge of lichen micropropagules in the air spora. Macrofragments of lichen thalli in the atmosphere are probably very local but studies in forested areas could well reveal their presence. It has been noted that the fruitcose species *Ramalina menziesii* (Peirce, 1898) and *Usnea "barbata"* (von Schrenk, 1898) are broken by wind and rain (although, under experimental conditions, wind speeds of 50 m.p.h. are needed) and that fragments can entangle themselves around twigs and conifer needles.

(*b*) *Laboratory studies on wind liberation*. Brodie and Gregory (1953) demonstrated that soredia could be removed from dry podetia of *Cladonia pyxidata* by wind speeds of about 2 m/s. They suggested that such cupulate podetia are more important with respect to wind than to water liberation of soredia.

Bailey (1966a) investigated the liberation of soredia by wind from *Lecanora conizaeoides*, *Hypogymnia physodes* and *Pertusaria amara* and the removal of small soredium-like thallus fragments from *Cladonia impexa*. Investigations were carried out to determine the effects of humidity prior to the experiment, and of windspeed, on the liberation of soredia. Using *L. conizaeoides* a regular pattern of liberation was demonstrated, related to the initial water content of the thallus. Wind blowing over thalli that had been stored dry liberated soredia immediately but liberation soon ceased as blowing continued. Wet thalli, by contrast, shed few soredia immediately but reached a peak of liberation after a short period of drying. Wet thalli liberated more soredia overall than did thalli that had been stored dry. Soredia were liberated and trapped approximately 1·5 m away from the point of removal within 1 min at windspeeds of 2·9 m/s. Soredia were also liberated from *H. physodes* and *P. amara* and soredium-like thallus fragments from *Cladonia impexa*. It was found that water droplets, acting simultaneously with wind, provided a more effective situation for liberating soredia from *L. conizaeoides* and *P. amara* than wind alone. This, however, was not the case with *H. physodes* under the conditions of the experiments.

Work in progress indicates that spores from the mazaedia of *Cyphelium inquinans* can be liberated at least by windspeeds exceeding 5 m/s and that spores actively discharged from ascocarps of *Lecanora conizaeoides* into air currents will be carried along therein (Bailey, unpublished). The relatively high windspeed needed to liberate spores from the mazaedia of *C. inquinans* could well be related to the surface sculpturing of the spores.

Wind would seem to be the most likely agent for large-scale transport of lichen propagules over long distances. Degelius (1954) considered that wind was probably the most important dispersal agent for thallus fragments of *Collema* species. Spores, which are small and light, are perhaps best suited to this form of transport although farinose soredia and small thallus fragments could also be transported by wind. Thomson (1972) noted that arctic winds of 70 to 80 m.p.h. would be sufficiently strong to lift and carry lichen diaspores. Too little is known at the present time about the frequency of lichen propagules in the atmosphere for us to judge which type of propagule is most likely to be of importance in long-distance dispersal by wind and further research is much needed.

2. Water as a Dispersal Agent

Water can act as a liberating and transport agent for lichen propagules either as reflected splash droplets or when flowing as trickles over surfaces or in streams and rivers.

(a) *Splash dispersal.* A wide range of plant propagules are dispersed by way of reflected splash droplets. Many fungi, bryophytes and flowering plants have adaptations for this means of liberation and transport (Brodie, 1957) and lichens are no exception. Brodie (1951) reported that soredia could be splashed up to 3 ft from cupulate podetia of *Cladonia* but gave no experimental details. Bailey (1966a) demonstrated that soredia could be carried at least 0·75 m by water droplets falling from a height of 1 m into podetia of *Cladonia chlorophaea.* Under the same experimental conditions soredia were removed from flat thalli of *Lecanora conizaeoides* to the same distance. However, while fewer soredia were removed from *Cladonia* cups in these experiments a greater proportion were carried further than 30·5 cm compared with those removed from *Lecanora* thalli under similar conditions, indicating, perhaps, that the cupulate structure of the podetium in *C. chlorophaea* promotes reflection of splash droplets to a greater distance than from the flat surface of *L. conizaeoides.* The same paper reports the liberation of soredia by reflected splash droplets from *Lecanora conizaeoides* growing on branches. Removal was found to be more effective from dry than wet thalli and from thalli growing on branches of small diameter rather than on flat surfaces. Water droplets of approximately 2·2 mm diameter falling 3·66 m transported substantial numbers of soredia to a distance of 0·75 m. (A medium-sized raindrop has a diameter of 2 mm and a steady terminal velocity of 650 cm/s (Mason, 1962). On falling 3·66 m a 2·2 mm diameter drop achieves some 90% of its terminal velocity and, therefore, the conditions of the experiment approximate to natural conditions.)

No studies seem to have been undertaken on splash dispersal of lichen spores. For most spores which are actively discharged from ascocarps splash dispersal must necessarily be secondary. Pyatt (1968a) noted that rotifers inhabiting the apothecial surfaces of *Xanthoria parietina* ingest and subsequently defaecate viable ascospores of that species, and suggested that the animals could be splashed from the apothecial surfaces and so transport spores. The present author has noted that ascospores of *Cladonia coccifera* are frequently discharged into water films covering the ascocarp surface from whence they are carried into water contained within the cupulate podetia. These suspended spores could, presumably, be dispersed in splash droplets. The possibility that spores of species in the Caliciales, such as *Cyphelium inquinans* and *C. notarisii*, that habitually grow in open situations, are splash dispersed would seem worthy of investigation. The spores of these species might well be dry dispersed by raindrops, as well as being dispersed within the reflected droplets. Dry dispersal of non-lichenized fungus spores by raindrops was studied by Hirst and Stedman (1963).

(*b*) *Dispersal by water flow.* Ridley (1930) laid extensive emphasis on dispersal of phanerogamic plants by rain wash, by rivers and by floods. An examination of flood debris along roadsides or by stream and river edges will frequently reveal specimens of many lichen species. Thomson (1972) noted the occurrence of lichen fragments in the flow lines of arctic melt water. Unlike the propagules of flowering plants, however, many lichens in such situations (epiphytic species for example) will normally need to be secondarily dispersed by another agent before reaching a site suitable for growth. None the less, dispersal of whole lichens and large fragments by this means could provide centres from which further dispersal could be effected by other agencies. Individual propagules of all sorts could undoubtedly be transported in these categories of flowing water. However, this method of transport is unlikely to be important in carrying propagules to a suitable habitat for establishment except amongst aquatic lichens, whose propagule liberation and transport mechanisms remain unstudied. Westman (1973) reported that lichen fragments are sometimes found in arctic drift ice and suggested that these might be carried to the shore by birds.

Barkman (1958) suggested that isidia may be transported by rainwater draining down tree trunks. Bailey (1968a) reported that soredia can be liberated from thalli of *Lecanora conizaeoides* and *Lepraria incana* by water trickles under laboratory conditions approximating to those in the field and that soredia, probably those of *L. incana*, were found in rainwater draining down trunks of *Fraxinus excelsior*. Soredia in such trickles could easily lodge upon irregularities of the bark. Sorediate lichens on vertical surfaces, for example on churchyard memorials, can be seen in downward streaks suggesting rainwash. Observations on these streaks do indicate that extension downwards is more rapid than extension upwards or sideways which in turn suggests that rainwash is aiding simple growth as the means of extending the lichen covered area (Fig. 2). R. A. Armstrong (unpublished) reports evidence from growth rate studies suggesting that specimens of *Parmelia glabratula* on vertical rock faces get younger as one descends the rock face and considers that this indicates initial transport to the top of the rock surface by birds and subsequent downward transport of propagules in rainwash. There seems no reason why many lichen propagules should not be transported downwards in draining rainwater nor why this should not be an important agent of liberation and transport over a wide variety of substrata.

3. Active Liberation

Lichen ascospores from all genera, except those of the Caliciales, are actively discharged from their ascocarps to distances that enable them to escape from the boundary layer of still air over the ascocarp surface.

A certain amount of experimental work has been undertaken on active discharge of ascospores from lichen apothecia which is summarized below. However, work so far published is fragmentary, inconclusive and greatly in need of critical re-evaluation based on further experimental work. No

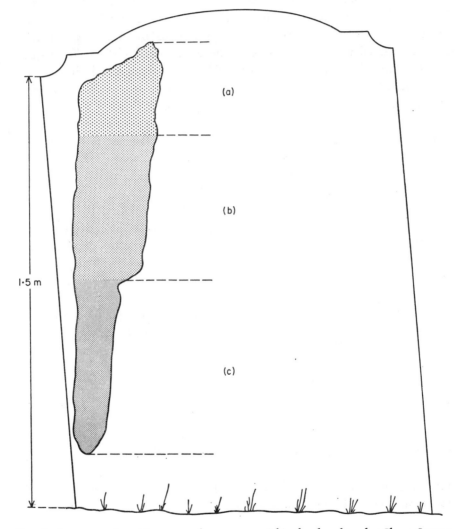

FIG. 2. *Lecanora conizaeoides* on a sandstone memorial in the churchyard at Shere, Surrey, showing downward extension presumed to be accelerated by draining rainwater. (a) zone of thick (>1·5 mm) continuous thallus with occasional ascocarps; (b) zone of thin (<1·5 mm) continuous thallus lacking ascocarps; (c) zone of thin discontinuous thallus. (Diagrammatic representation.)

comparison of discharge from apothecia and perithecia has yet been undertaken in the lichenized fungi and the liberation of basidiospores and conidia from lichenized fungi has not been studied.

The processes causing discharge from mature asci have been the subject of some discussion. It is normally accepted that moistening the ascocarps will initiate discharge (Scott, 1959). However, opinions vary as to whether discharge occurs as a result of drying and consequent rupturing of the asci (Ahmadjian, 1961) or continually while the asci are moist (Bailey and Garrett, 1968). Evidence presented by Pyatt (1974) suggests that discharge occurs whenever the asci are placed under stress following changes in the water balance, whether by loss of water when drying or absorption of water while moist.

Various factors have been studied with reference to ascospore discharge: seasonality of spore production, effect of light, temperature, rain and humidity and pollution by sulphur dioxide and the distance of discharge (Bailey and Garrett, 1968; Pyatt, 1974).

(a) *Seasonality of spore production*. This phenomenon has been extensively studied in the laboratory where spore production is equated with the ability of ascocarps to discharge spores under experimental conditions. It seems unlikely, however, that this ability is unrelated to an ability to discharge spores under natural conditions although spores may be produced, but not discharged, both *in vitro* and *in vivo*. Des Abbayes (1951), reviewing information then available, concluded that spore production in Western Europe is most favoured during the spring. Verseghy (1965) concluded that in dry years spore production was most intensive in spring but that in wet years there was also an intensive period of spore production in autumn. Pyatt (1969) presented data indicating that for a wide variety of lichens spore discharge continues throughout the year but that it tends to be more abundant from September until February. Other workers including Werner (1927), Scott (1959), Bailey and Garrett (1968) and Garrett (1971) have presented conflicting data on seasonality of discharge.

(b) *Effect of light*. Pyatt (1974) reported that light tends to increase spore discharge in most species. The same author (1968b, 1974) reported the existence of a definite endogenous rhythm of spore discharge in *Lecidea macrocarpa* kept under 12/12 h light/dark regimes. Bailey and Garrett (1968) reported that there appears to be no effect of light upon spore discharge in *Lecanora conizaeoides*.

(c) *Effect of temperature*. Pyatt (1974) reported that extremes of temperature depressed discharge. Bailey and Garrett (1968) recorded a slight increase in the number of ascocarps from *Lecanora conizaeoides* discharging spores at 6°C compared with 20°C and an increase in the total number of spores discharged at the lower temperature. This observation could, however, be due to faulty experimental technique and needs further study.

(*d*) *Effect of rain and humidity*. It is generally agreed that moisture is necessary in the form either of water on the surface of the ascocarps, or of moisture absorbed by the ascocarp from beneath in order to initiate spore discharge. There is, however, no agreement on the length of time that must elapse after moistening before discharge occurs. Scott (1959) found a variable time lag of up to 6 h in *Peltigera praetextata* and attributes this to variable moisture content of the ascocarps at the start of the experiment. He relates this to the state in nature by speculating that perhaps brief periods of precipitation are only of slight value in initiating spore discharge. Bailey and Garrett (1968) and Garrett (1971) found that the majority of species they studied discharged ascospores within 1 h of moistening. Garrett (1971) reported that a number of species would discharge spores from ascocarps held on vaselined slides in an atmosphere of 100% r.h. These observations suggest that brief periods of rain could possibly be of importance in effecting discharge of ascospores.

(*e*) *Effect of sulphur dioxide pollution*. Pyatt (1974) reported that an atmosphere polluted by sulphur dioxide tends to depress spore discharge in *Graphis elegans*, *Lecidella elaeochroma*, *Pertusaria pertusa* and *Xanthoria parietina*.

(*f*) *Distance of discharge*. This appears to be generally dependent upon size of projectile. Lichen ascospores vary in size from about 1 μm in *Acarospora* to 510 μm in *Bacidia marginalis* (Hale, 1967). The size of projectile depends, however, not only upon the size of spore but upon the number of spores per projectile. The maximum distance of horizontal discharge for ascospores alone so far observed under experimental conditions appears to be 4·5 cm achieved by *Rhizocarpon umbilicatum* (Bailey and Garrett, 1968). Bertsch and Butin (1967) recorded a distance of 5·0 cm for the combined ascospore/algal propagule of *Endocarpon pusillum*. Amongst the non-lichenized fungi there is sometimes a clear relationship within any one species between number of spores per projectile and distance travelled (e.g. Ingold and Hadland, 1959; Walkey and Harvey, 1966). Such a relationship is not immediately clear in the lichenized ascomycetes so far studied.

The active discharge of lichen ascospores is currently a field filled with varying reports. Further work attempting to relate aspects of discharge to ecological factors and to the ontogeny and anatomy of ascocarps is urgently needed.

Under the category of active dispersal it seems reasonable to include "transport by growth". It is well known that some foliose and squamulose lichens growing outwards radially will die at the centre leaving only an outer circle of outward-growing lobes. Jahns (1974) recorded circles of this type up to 1·0 m diam from the saxicolous species *Parmelia centrifuga*. Such rings can fragment and produce a number of separate thalli, which occasionally regain their radial form, some distance from the original site of the parent thallus. Such rings are less easily detected in grassland but seem not infrequent in *Peltigera canina* growing in grassland habitats. A comparable form of dispersal can be noted in some *Alectoria*, *Ramalina* and *Usnea* species where long fronds growing pendulously from one branch or piece of bark attach themselves to another, eventually becoming established as an independent plant.

4. Animals as Dispersal Agents

The whole field of the relationship between animals and lichens is poorly understood. The reviews of lichen dispersal mentioned above (pp. 215–216) all contain reference to dispersal by animals. Only one extensive review of the relationship between animals and lichens has been published, that of Gerson (1973) in which he considered lichen-arthropod associations. Peake and James (1967) provided a short review of the relationship between molluscs and lichens.

(*a*) *Dispersal by arthropods.* Darbishire (1897) quoted examples of springtails (Collembola) moving over the surface of *Pertusaria amara* and becoming coated with soredia. Barkman (1958) noted that "soredia may also be carried all over the trunk by small arthropods" and went on to mention "small hairy spiders, mites and beetles" covered with soredia. Gressitt *et al.* (1965) noted the presence on the back of large flightless weevils (*Gymnophilus* spp.) of lichens which are themselves inhabited by orabatid mites. They suggest that the mites could disperse these plants from weevil to weevil. The weevils themselves could, presumably, transport thallus fragments and deposit these at suitable sites about the forest. (Subsequent papers by Gressitt and co-workers concerning 24 lichens growing on beetles were listed by Gerson, 1973.) The occurrence of soredia on black lawn ants (*Acanthomyops niger*) travelling between the outermost branches of *Crataegus* and the ground was noted by Bailey (1970). Larvae of the lacewing (*Nodita pavida*) construct and carry about "packets". Skorepa and Sharp (1971) noted that these packets are often constructed exclusively of lichens. A lobe of *Physcia orbicularis* was positively identified in one "packet" and others had the general "appearance of *Lepraria*". Weber (1974) reported larvae of a case moth (*Cebyza leucoteles*) and of an unidentified larva camouflaged with lichen fragments and soredia.

Peake and James (1967) noted that mites are wasteful feeders and cover themselves with fragments of their food. Lists of mites and springtails that feed on lichens are to be found in Richardson (1975). A wide variety of insects and mites were listed, with full bibliographic details, by Gerson (1973) as feeding upon lichens including beetles, moths, psocids, termites, stoneflies, earwigs and mites of both the Cryptostigmata and Prostigmata. Hale (1972) reported a case of the collembolid *Hypogastrura packardii* feeding on *Parmelia baltimorensis*. It seems reasonable to assume that any animal feeding upon lichens will liberate thallus fragments which can be wind transported even if the animal does not itself act as a transport agent.

No studies on the dispersal of lichen propagules by flying insects, comparable to that of Stewart and Schlicting (1966) on the dispersal of algae and protozoa by aquatic insects (mostly dragonflies and damselflies), seem to have been published.

(b) *Dispersal by molluscs*. Apart from the beetles considered above, molluscs and shell-bearing reptiles would seem to be the only animals with sufficiently stable an exoskeleton, and long enough life-span, to act as lichen substrata. Peake and James (1967) recorded 34 species of lichen from the shells of dead molluscs. However, only *Arthropyrenia halododytes*, *Verrucaria maura*, *V. microspora*, *V. mucosa* and *V. striatula* were recorded from live animals—specimens of *Littorina* and *Patella*, which could presumably provide somewhat limited transport; requiring only the liberation of ascospores or thallus fragments to achieve dispersal.

Molluscs feeding upon lichens could liberate fragments capable of germination and development. These fragments could be carried on body slime or passed out as faeces. Peake and James (1967) noted that surprisingly large particles of vegetable material are frequently recognizable in the gut and that digestion rarely effects a complete breakdown. Coker (1967) reviewed the evidence for and against the view that lichen substances discourage or prevent gastropod grazing upon lichens. He reported that the slug *Lehmania marginata* was found grazing upon *Lobaria pulmonaria* in Rassall Ashwood (W. Ross-shire, Scotland) and on *Hypogymnia physodes* and *Pertusaria pertusa* (*P. amara* was not affected) in Mens Wood (W. Sussex, England). Grazing of *Caloplaca decipiens*, but not *C. heppiana*, by molluscs at Cleeve Hill (Gloucestershire, England) was noted by Bailey (1969). A. Fletcher (unpublished) reports grazing of the marine species *Verrucaria amphibia* and *V. striatula* by *Littorina* and *Patella*, thallus fragments so liberated could then be transported by wave action.

(c) *Dispersal by birds*. It is possible that birds contribute to the dispersal of lichens in three ways: by adhesion of propagules to their feet, feathers, etc.; by breaking of lichen thalli, thus liberating thallus fragments, soredia or isidia which could then be transported by wind; and by carrying lichens or lichen-covered material to their nest sites. Little information is available on any of these aspects. Bailey (1976) reported that soredia were found adhering to the feet of birds trapped for ringing in north London and (quoting data supplied by P. W. James) that thallus fragments, soredia and spores were found on the Royal Albatross in the Auckland Islands, New Zealand.

Von Schrenk (1898) recorded that he had seen birds building their nests with *Usnea* "*barbata*" in good growing condition. *Lecanora conizaeoides* and *Hypogymnia physodes* have been noted by the author on twigs of blackbirds' nests in Sutton (Surrey, England). Richardson (1975) listed several species incorporating lichens in nests including the ground-nesting golden plover which uses *Thamnolia vermicularis* s.l. as camouflage around its nest, i.e. in a site where the lichen could establish itself immediately (see also Sauer, 1962).

(d) *Dispersal by mammals*. While it would seem likely that mammals disperse lichens there appear to be no published records of dispersal by mammals other than man.

Laundon (1967) recorded the occurrence of lichens accidentally or deliberately introduced to London by man. While many lichens arrive in London on stone, ornamental trees, firewood and peat blocks for use in gardens there is no case known of these introductions establishing themselves in the polluted air of that city. However, many lichens are thought to have been transported by man and to have established themselves elsewhere. Degelius (1957) considered that several species including *Hypogymnia tubulosa*, *Parmelia exasperatula* and *Usnea hirta* (all found by Lynge, 1940, in one hedge at Husavik, Iceland) might have been transported to Iceland by man on unbarked timber or nursery stock and makes the same suggestion for *Xanthoria lobulata* found on one tree only at an agricultural research station. Bailey (1968b) postulated that *Lecanora conizaeoides* was introduced to Iceland by man, probably on unbarked timber. It is suggested by Ahti (1965) that *Lecanora conizaeoides* might have been transported by man to Newfoundland and New Zealand while Thomson (1963) considered that *Physcia clementei* and *P. millegrana* could have been transported to California by man from Europe and the Eastern United States respectively. When the S.S

Great Britain was brought from the Falkland Islands to Bristol (England) in 1970 a large number of lichens were found on the vessel (Brown, 1971). However, this unwitting attempt to enrich the British lichen flora with what was probably the largest number of living lichen species transported to this country in one journey failed as many died even before the ship was cleaned and painted. The occurrence of specimens of *Usnea* (Rose, 1973) and *Lecanora dispersa* (Bailey, unpublished) on usable motor cars also provides ample opportunities for propagule transport! Lichen-covered branches are used in the construction of decorative bird tables in Devonshire (England) and sold to tourists, who then take them to other parts of the country (D. L. Hawksworth, unpublished).

Many lichens arrive in Britain each year on cork bark (bark of *Quercus suber*) imported for decorative purposes. Some species observed on this substratum, for example, *Parmelia caperata* and *P. perlata*, appear to arrive alive and could no doubt establish themselves, although there is no evidence as to whether they have or have not done so.

Man is undoubtedly responsible for the liberation of thallus fragments from a wide variety of lichens making them available for transport by other means. It has been noted (Anon., 1965; Champion, 1965; Mitchell, 1974) that allergies caused by *Lecanora conizaeoides* and other lichens occur amongst the families of forest workers, indicating that lichen fragments can be carried on clothing for some distance. Rhodes (1931) considered that man was responsible for fragmenting and transporting *Cladonia arbuscula* and *Cornicularia aculeata* at Hartlebury Common (Worcestershire, England). This must be a widespread phenomenon affecting terricolous species of all growth forms. Several authors have commented upon the effect of human pressure in removing lichens from rock faces used for climbing: Hawksworth (1969) noted the absence of *Umbilicaria* species from boulders used by rock climbers on the Derbyshire "Edges" while Bailey and Stott (1973) noted the lack of any lichen cover on many boulders used as seats and climbs at Thurstaston Common (Cheshire, England). The fragments liberated by such activities could presumably be transported further afield by other agents. In New Zealand the mountain lichen *Siphula decumbens* is found to extend from above the tree line into forests only along the edges of paths frequented by red deer and by climbers. There seems little doubt that it has established itself from thallus fragments adhering to the hair and clothes of the two species (D. J. Galloway, unpublished). Thomson (1972) postulated that short-range dispersal of *Cetraria delisei* could have been effected by Eskimos gathering the plant for food and fuel, and also that lichens eaten by caribou could be dispersed by way of fragments remaining undigested in the remains of animals killed by wolves. Grazing mammals, whether feeding directly upon lichens or upon other species in a lichen-containing community, could undoubtedly liberate thallus fragments, subsequently transported by wind or water.

(e) *Dispersal by other animals.* Imagination can provide possible, even probable, examples of dispersal by animal groups not yet considered, just as the existing evidence and data concerning dispersal by many agencies can be imaginatively augmented. However, few observational or experimental data are available.

Rotifers and nematodes were recorded by Gressitt (1966) as part of an epizooic symbiosis considered above. Both could act as transport agents and possibly in the liberation of propagules. Pyatt (1968a) noted that rotifers ingested and defaecated viable spores of *Xanthoria parietina* and could, therefore, act as transport agents if they moved, in water films, off the parent plant.

Hendrickson and Weber (1964) recorded the occurrence of *Dirinaria picta* on giant tortoises in the Galapagos Islands; the tortoise could act as transport while other agencies liberated propagules. It is, perhaps, significant that *D. picta* was noted by these authors as "one of the most ubiquitous of Galapagos lichens".

ESTABLISHMENT OF LICHEN PROPAGULES

Establishment, as here understood, can be divided into deposition of propagules upon a surface suitable for growth and their germination on, or attachment to, that surface.

1. Deposition

Lichen propagules are unlikely to be liberated from their substratum by an agent that itself provides a suitable substratum for germination and development. Most propagules will be transported after liberation by animals, water or wind and must be deposited onto a suitable substratum.

Three modes of deposition from wind are available: sedimentation under gravity, whenever the windspeed provides insufficient energy to counteract gravitational forces; "wash-out" by rain; and impaction upon surfaces. Theoretical and natural aspects of deposition processes were reviewed by Gregory (1973).

"Gravity slides" (microscope slides left exposed so that any airborne particles will settle thereon under the influence of gravity) normally accumulate lichen soredia and thallus fragments if exposed at suitable sites. However, gravity alone is effective in depositing particles from air only in still air. When wind causes even slight turbulence small particles, e.g. *Lycopodium* spores, will be deposited on both the upper and lower surfaces of exposed objects (Gregory, 1973). Visual inspection of threads of spiders' webs and of the slime tracks of molluscs in areas with a reasonable lichen flora will frequently reveal soredia and lichen thallus fragments (Bailey, 1970). Gregory (1951) noted that, under experimental conditions, *Lycopodium* spores were more efficiently impacted upon cylinders of small diameter than upon those of larger diameter. This observation may or may not be relevant under natural conditions but certainly suggests that spiders' webs and small twigs could act as efficient trapping surfaces. In western Britain twigs as little as 2 years old frequently bear lichen thalli, determinable to species level, even in situations where rainwash could not have brought them to the site. Degelius (1964) noted several species occurring on 2-year-old twigs of *Fraxinus excelsior*. In both cases sedimentation, impaction from wind or deposition from animals must have placed the propagules in position.

Garrett (1972) investigated the electrostatic charges on discharged lichen spores and suggested that spores of *Bacidia rubella*, *Sarcogyne regularis* and *Xanthoria parietina* are positively, and those of *Opegrapha atra* negatively, charged on liberation. While this could affect deposition of spores on naturally charged surfaces, Garrett was unable to demonstrate any help or hindrance to deposition on spiders' webs.

Von Schrenk (1898) recorded that filaments of *Usnea "barbata"* are likely to entwine themselves around needles of *Picea*. The author has observed detached filaments of *Alectoria fuscescens* wrapped around bark projections of *Quercus* in the Forest of Dean (Gloucestershire, England). Peirce (1898) noted that fragments of *Ramalina menziesii* (often fragments of considerable size) become more or less securely fastened to branches where they continue to thrive without forming any holdfast.

2. Germination or Attachment

Germination is used to describe the initiation of growth from small thallus fragments, isidia, soredia and spores. Attachment is used to describe the initial appearance of a "holdfast" from large thallus fragments that might or might not have become detached from the parent plant.

(*a*) *Thallus fragments and whole lichens*. Filaments of *Alectoria* and *Usnea* species on Mt Ruwenzori (Kenya) are reported to establish themselves on branches where they come to rest; the filament darkens at the point of attachment and grows on as a new plant (A. Pentecost, unpublished). A similar mode of establishment has been noted in Britain for *Ramalina fraxinea* by the author and for *Usnea ceratina* (F. Rose, unpublished). In *Ramalina fraxinea* the process of attachment may occur before the branches are liberated from the parent thallus. Several species of *Alectoria* are known to attach themselves to the substratum by means of haptera. This could lead to the development of separate plants if breaks occurred between the various points of attachment.

Under phytotron conditions Dibben (1971) established thalli of *Baeomyces, Cladonia* and *Pycnothelia* on soil from fragments produced by maceration of mature thalli, this is perhaps the only experimental evidence that thallus fragments can reproduce the whole lichen symbiosis. (See, however, Chapter 3, pp. 60, 71.)

(*b*) *Lobules.* Kershaw and Millbank (1970) showed that it is possible to obtain growth, under laboratory conditions, of the lobules occurring on the under-surface of *Peltigera aphthosa* var. *variolosa*. Lobules increased some 6-fold in area over a period of 7 months.

(*c*) *Soredia.* Many attempts have been made to germinate soredia and establish viable lichens therefrom. Tobler (1911) germinated soredia of *Cladonia* on sterile soil and observed development for some months but the results do not appear to have been repeated. McWhorter (1921) noted that small lichen thalli would develop on "almost any moss colony" subjected to alternate periods of moisture and drought; soredia would seem the most likely source of these thalli. (See also Chapter 3, p. 47.)

Margot (1973) reported that under laboratory conditions 70% of the soredia of *Hypogymnia physodes* showed normal multiplication of both symbionts when cultured on an agar medium containing Bold's mineral medium (Ahmadjian, 1967), trace elements and soil extract, for one month. A further 27% showed growth of the fungus alone and 3% showed no growth. Treatment with sulphur dioxide reduced the percentage germination. Nienburg (1919) illustrated the development of *Hypogymnia physodes* and *Physcia tenella* from soredia.

(*d*) *Ascospores.* There are many reports of successful germination of lichen asco-spores under a variety of laboratory conditions. These were reviewed in 1951 by des Abbayes and a wide variety of reports have been published since that time (e.g. Ahmadjian, 1961, 1967; Bailey, 1966b; Garrett, 1968). Much of this interest stems from the recognition of the dual nature of the lichen thallus and consequent attempts to resynthesize lichen thalli from their algal and fungal components. These attempts were reviewed by Ahmadjian (1969).

Despite the considerable success that has been achieved in germinating lichen ascospores under varied conditions their importance as propagules of the whole lichen symbiosis in nature is not fully understood. Until success in re-establishing the whole lichen symbiosis from the two symbionts has been achieved to the stage that viable ascospores are produced by the resynthesized plant the importance of ascospores as propagules in nature must necessarily remain in doubt. So far, while considerable progress has been made (Ahmadjian, 1966; Bertsch and Butin, 1967; Ahmadjian and Heikkilä, 1970) resynthesis of the complete cycle has only been achieved with one species (*Endocarpon pusillum*), a species from which hymenial algae are discharged together with the ascospores. Successful spore-to-spore resynthesis of other species, most particularly those that do not discharge hymenial algae, is needed before the value of lichen ascospores as propagules can be regarded as certain.

There are many lichens, especially amongst the crustose species, that produce no soredia, isidia or other vegetative diaspores. While any lichen can be fragmented this is hardly likely to be of frequent occurrence amongst endolithic species, yet these are widespread and almost invariably bear ascocarps which may be presumed to produce viable ascospores. Resynthesis of lichen thalli from germinating ascospores and free-living algae must, in all probability, occur with considerable frequency in nature. It is, however, so far unobserved and few reports presuming its occurrence have been published. One such report is that of Degelius (1964) who claimed the existence of a series of *Pleurococcus* (as "*Protococcus*") cells from the completely free living through various stages of lichenization, to a true lichen, *Bacidia chlorococca*, on twigs of *Fraxinus excelsior*. Ahti and Vitikainen (1974) noted that it was often difficult to make the distinction between free-living cells of *Pleurococcus viridis* and lichenized cells attributed to *B. chlorococca*. This loosely lichenized association may, therefore, be of especial interest in studies on resynthesis under natural conditions.* Ahmadjian (1967) and Pyatt (1974) have reviewed a number of factors affecting ascospore germination in the laboratory. Nutrient levels appear to be of importance, low nutrient media tending to favour germination. pH was also found to have an effect for some species and sulphur dioxide pollution was found to reduce germination. A further study on the effects of pollution on germination was conducted by Kofler *et al.* (1969) and concerned the effects of dust upon spore germination. It was found that dust from calcium carbide and iron alloy factories strongly inhibited germination of the spores of *Physconia pulverulenta* but that those of *Xanthoria parietina* and, above all, of *Lecanora dispersa* were much more resistant. The basic nature of the dust was considered to be an important factor. Scott (1959, 1964) found that an aqueous extract of the phycobiont was necessary before unequivocal proof of germination could be obtained with spores of *Peltigera horizontalis* and *P. praetextata*. Bark extract, and com-

---

* See also Chapter 5, pp. 95–96.

pounds like erythritol, glycerol and pectin accelerated germination of spores of *Xanthoria parietina* (Am Ende, 1950).

(*e*) *Conidia*. Moller (1887, 1888) claimed to achieve germination of conidia from a total of ten lichen species. It is now considered possible that his cultures derived from contaminants or from hyphal fragments (Ahmadjian, 1969). Hedlund (1895) described germination from two other species and claimed that germination was sometimes initiated before the conidia were released. However, the reproductive rôle of conidia is still unproven. (See also Chapter 5, p. 95.)

(*f*) *Other propagules*. No experimental studies on the establishment of isidia, schizidia, basidiospores or hormocysts are known to the author. However, Nienburg (1919) describes a sequence of stages in the development of *Pseudevernia furfuracea* from isidia to 10-year-old thalli.

3. Competition as a Factor in Establishment

Lichens are in competition with other organisms for space, light and nutrients. Barkman (1958) considered that competition among cryptogamic epiphytes in general was of four types: (1) mechanical destruction, (2) suffocation, (3) competition for light, and (4) chemical action. He considered the third to be of greatest importance and pointed out that parasitism could also be a factor.

The endophloedic fungus *Polymorphum rugosum* (syn. *Dichaena faginea*) is capable of "undermining" endophloedic and epiphloedic lichens by extending its mycelium under their thalli, subsequently the formation of fungal fruiting bodies breaks the upper layers of bark and fragments the lichen thallus (Barkman, 1958). This could be one way in which thallus fragments of endophloedic crustose lichens are produced. Such examples of mechanical destruction do not seem to be shown by lichens themselves.

Many examples of lichens crowding out bryophytes and vice versa were noted by Barkman (1958) with reference to both suffocation (prevention of gaseous exchange) and to competition for light. Most field workers can quote examples from their own experience. *Pertusaria albescens* var. *corallina*, for example, is frequently to be seen overgrowing bryophytes with the hyphae of the hypothallus densely enclosing the shoots of mosses. McWhorter (1921) noted examples of actual parasitization of bryophytes by lichens, chiefly *Cladonia* and *Lepraria* spp., causing severe damage to the bryophyte. Penetration of hyphae from the lichen *Icmadophila ericetorum* between cells of the moss *Mnium hornum* was reported by Coker (1966). In the latter instance an area of moss was moribund for an incomplete ring around the lichen. Such putatively parasitic penetration of competitors by mature thalli might or might not indicate a method by which propagules are able to establish themselves in the face of competitors. However, the work of Bonnier (1888, 1899) suggests that germinating ascospores are capable of destroying moss protonemata.

Chemical antibiosis was suggested as a possible factor inhibiting growth around mature thalli by Heilman and Sharp (1963). *Lecidea albocaerulescens* was noted as inhibiting the growth of *Anomodon attenuatis*, *Hedwigia ciliata*, *Porella*

TABLE II. Some angiosperm seeds whose germination and growth are inhibited by lichen extracts

Seed	Lichen	Effects reported	Source
Triticum cv "Capitole Vilmorin"	*Roccella phycopsis*	Reduced germination, reduced growth of coleoptile, retarded germination	Rondon (1966)
Cucumis	*Umbilicaria papulosa*	Reduced root growth, death with high concentrations	Miller *et al.* (1965)
Festuca ovina, *F. rubra*	*Peltigera canina*	Reduced root growth, delayed germination	Pyatt (1967)
Lolium perenne	*Peltigera canina*	Reduced root growth	Pyatt (1967)
Phaseolus vulgaris	*Sticta weigelii*	Delayed germination, retarded growth rate	Follmann and Nakagava (1963)

platyphylla and *Sematophyllum* sp. *Thelotrema subtile* was said to exert a similar effect upon *Frullania eboracensis* although this species can overgrow some lichens. In support of the hypothesis that lichens compete with other plants by means of chemical antibiosis one may quote the inhibition or retardation of growth in certain angiosperm seeds, under experimental conditions, by extracts of some lichens (Table II). Miller *et al.* (1965) presented a report on the antibacterial and antifungal action of extracts from *Umbilicaria papulosa* (together with a review of previous work). These results all concern lichen extracts from mature thalli under experimental conditions, although Pyatt (1967) suggested the existence of an apparent inhibition of unnamed grasses under natural conditions. The relevance of these antibiotic activities to the establishment of lichen propagules in nature does, however, need further investigation.

DISCUSSION

Efficient methods of dispersal are of value to any organism. Dispersal is necessary to extend the range of a species and to ensure that propagules are available to colonize the various ecological niches in which the species can flourish throughout its range. Most animals can rely upon their own movements for extension of their range, while most plants must rely for dispersal upon detachable propagules transported by external means. Dispersal ensures that genetic variation arising in one place becomes available elsewhere, thereby maintaining maximum genetic plasticity within the species, although perhaps at the expense of sophisticated adaptation to one particular habitat. Efficient dispersal and consequent wide range will also serve to check the ravages of pests and natural calamities.

Lichens are undoubtedly widely dispersed organisms which are able to spread throughout the world and colonize newly exposed surfaces with speed and efficiency. In central London newly laid concrete will have a 20% cover of *Lecanora dispersa* within 4 years. Kristinsson (1972) noted three species of lichen on Surtsey, Iceland, in 1970, 3 years after the last effusive phase of eruption in 1967. Degelius (1964) observed that twigs of *Fraxinus excelsior* were colonized within 2 years. Yet despite their efficiency in dispersal, as evidenced by speed of colonization, lichens have evolved no morphological adaptations that are obvious aids to dispersal. The only possible adaptation is the remarkable sculpturing on the spores of some species in the Caliciales. Tibell (1971) suggested that this sculpturing could be an adaptation to dispersal as the rough surfaces of the spores would facilitate their removal from the mazaedium by small arthropods.

Undoubtedly the most important evolutionary trend with respect to dispersal in lichens is the development of soredia. Hale (1967) published a world distribution map of the sorediate *Parmelia cristifera* and the esorediate *P. latissima* (two species that appear to differ only in the presence or absence of soredia). This indicated a pantropical distribution for *P. cristifera* while *P. latissima* is limited to the tropical New World and one collection from eastern India. Further studies on such pairs of species were made by Poelt (1970), who also discussed the taxonomic implications (Poelt, 1972). Culberson (1973) discussed possible evolutionary development of sorediate from esorediate species in the *Parmelia perforata* group, which contains several pairs of esorediate species with apothecia and sorediate species found only rarely, if ever, with apothecia (see pp. 171–172).

In most organisms the extensive development of vegetative means of reproduction would inevitably lead to a loss of genetic plasticity. Very possibly this happens in lichens also. However, the lichen plant is not necessarily a genetically uniform organism from one part of the thallus to another, as would be the case with a bryophyte or vascular plant. A single lichen thallus can be developed from hyphae derived from several spores and algae descended from several individuals. Vegetative propagules from different parts of the same thallus could thus be genetically varied in a manner impossible with higher plants. This problem is discussed further by Jahns (1974) and Henssen and Jahns (1973). It can be observed that lichen ascospores germinating on glass slides produce hyphae that anastomose. The union of hyphae from separate spores derived from the same ascocarp can also, if rarely, be observed (Bailey, unpublished). While we know too little about the cytology and cytogenetics of lichen fungi for any definitive statement to be made it seems possible that nuclear exchange between hyphae derived from different spores could occur. If this were indeed the case, the various vegetative diaspores could well be propagules with a greater degree of genetic variation than is normally the case with vegetative reproductive bodies of other plants. Together with this possibility of a "pseudo-sexual" means of producing genetic variation in vegetative diaspores one must also consider the development of the ascospore as a propagule for the whole symbiosis. In three genera, *Endocarpon*, *Staurothele* and *Thelenidia*, hymenial algae of the same species as those of the thallus are discharged with the asco-spores. While too little is known about the reproduction of lichen algae the ascospores at least would normally be regarded as sexually produced; giving a partially sexual whole lichen propagule in these three genera. *Endocarpon* and *Staurothele* are members of the Verrucariaceae and while the exact taxonomic position of *Thelenidia* seems uncertain (Poelt, 1974b), it appears that this type of propagule has evolved in at least two families of lichenized fungi. Pyatt (1973) reported that some 4% of spores actively discharged from *Pertusaria pertusa* had attached to them a few algal cells and assumed that these were derived from the ascocarp (a similar phenomenon was also stated by him to occur in *Lecidella elaeochroma*). This report merits further investigation.

In practical terms there are many matters of detail remaining to be investigated amongst the lichenized fungi. However, two major problems immediately attract attention. Which amongst the variety of propagules is of greatest importance in dispersal, and by what means is long-distance dispersal normally accomplished?

The study of "species pairs", i.e. pairs of sorediate and related esorediate species, suggests that soredia are far more effective as dispersal agents than ascospores. Hale (1967) also noted that sorediate and isidiate species usually have a greater numerical abundance in lichen communities than their simple frequency in a genus would suggest, and pointed out that of the 18 statistically most frequent lichen species on deciduous trees in southern Wisconsin (out of a total of 55) 13 bore soredia or isidia and only five lacked these propagules. Sixteen of the 18 rarest species lacked any obvious vegetative diaspore (Hale, 1955). Such evidence, and the abundance of soredia and isidia in the world lichen flora, strongly suggests that vegetative diaspores are of great importance in lichen dispersal. It is also noticeable that development of soredia and isidia in a species is frequently accompanied by suppression of ascocarp formation.

General studies on lichen distribution have not yet yielded a great deal of information on the effectiveness of propagules or methods of dispersal. *Lepraria incana* with no known propagules except soredia is recorded from almost every 10 km grid square for which records have been returned to the British Lichen Society's Distribution Maps Scheme. So also, however, is *Lecanora dispersa* which is dependent solely upon spores, or perhaps thallus fragments, for its dispersal (M. R. D. Seaward, *in litt.*). *Thamnolia vermicularis* s.l. with no known propagule other than thallus fragments, broken at random, is known throughout the world (Sato, 1963), usually from high mountains.

It will always be difficult to assess the relative efficiency of the various propagules until simple and effective germination and culture techniques for lichen propagules and thalli become available. Only then will we be able to assess not only the relative efficiency of establishment and its relation to substrata and environmental factors but also the species composition of the air spora which could be our greatest step forward in understanding this problem.

More studies on the colonization of fresh surfaces are also needed. Degelius (1964) calculated that 84% (47 species) of lichens colonizing twigs of *Fraxinus excelsior* possessed vegetative diaspores (70%, 39 species possessed ascospores). Young twigs present an abundant and readily available series of uncolonized surfaces with an easily determined age pattern. More studies like that of Degelius could well accumulate evidence as to which methods of dispersal are more important in nature, although such studies can never be definitive for so many factors other than dispersal are involved in colonization.

Studies on the colonization of other surfaces by lichens are scarce, perhaps because large areas of newly exposed natural surfaces are rare and new artificial surfaces are not of sufficient size to distinguish between the different means by which propagules arrive. Treub (1888) noted that the first plants to colonize Krakatau were blue-green algae, 3 years after the eruption of August 1883 which removed the whole plant cover from the Krakatau islands. Lichens appeared only at a later stage. Boedijn (1940) listed the 13 lichen species from Krakatau that had been noted by van Leeuwen in 1922. All of these were epiphytic on Krakatau, although some, at least in Britain, can be terricolous. The lichen flora of 13 species compares unfavourably with, for example, the 26 mycetozoa, 121 basidiomycetes, 207 angiosperms or 59 pteridophytes recorded by Boedijn. Boedijn suggested that this is due to the "lack of dispersal opportunities". The fact that no terricolous lichens were reported suggests that dispersal to the islands may have been by driftwood rather than by airborne propagules. Surtsey, off the south-west coast of Iceland, is an island that appeared in 1963 since which time a certain amount of new lava has continued to be deposited. The last effusive phase of eruption was in 1967. Studies on lichen colonization of the lava have been undertaken by Kristinsson (1970, 1972, and unpublished). Lichens were first detected in the summer of 1970 when three species were found on the island: *Placopsis gelida*, *Trapelia coarctata* and *Stereocaulon vesuvianum* (Kristinsson, 1972). By August 1973, 11 species had reached the island. Amongst these were four, *Lepraria incana*, *Placopsis gelida*, *Stereocaulon capitellatum* and *S. vesuvianum*, which were widespread and evenly distributed over the island. These plants are presumed to have arrived as windborne propagules for it is unlikely that transport by birds would have produced so widespread and uniform a distribution. *Xanthoria candelaria*, however, also occurred and was limited to the area around an artificial pond; propagules are thought to have been brought to the island by birds, washed off into the pond and splashed out (H. Kristinsson, unpublished). Kristinsson (1972, and unpublished) also noted evidence of *Trapelia coarctata* being transported in rainwash. Lichens established themselves first in and then around holes in the lava, where, undoubtedly, surface texture is important in providing a "foothold" for lichen propagules. In Britain an especially noticeable example of rough surfaces providing a "foothold" can often be seen on the smooth bark of *Fagus sylvatica* where any cuts or abrasions, initials carved by passers-by for example, are rapidly colonized by *Lepraria incana* which only seldom establishes itself on the bark above the splayed bole. The cuts are usually insufficiently deep to protect the plant from sunlight although they may provide a more favourable microhabitat in other ways. It is noticeable, also, that the leaf and bud scars of young twigs are the first areas to be colonized by epiphytic lichens.

Considerable work has been undertaken on the interrelationship of lichens and air pollution but little of this has been centred on the effects of air pollution upon establishment. Amongst non-lichenized fungi Saunders (1966) showed that *Diplocarpon rosae* is most sensitive to sulphur dioxide pollution during ascospore germination. Amongst lichens Margot (1973) showed that low levels of sulphur dioxide damage the phycobiont of *Hypogymnia physodes* in germinating soredia. These results, together with Laundon's observation (1967) that lichen communities are able to maintain themselves at levels of pollution in which they cannot establish themselves afresh and the observations of Kofler *et al.* (1969) on the effects of dust upon ascospore germination, suggest that the moment of establishment is perhaps one at which pollution is an especially important factor.

The list of things we do not know about lichen dispersal is long. There are no experimental or observational data on the liberation of spores from lichenized basidiomycetes. The liberation and transport of isidia has hardly been studied although Barkman (1958) suggested that isidia, being hydrophilic, are likely to be transported in water trickles. We know little of how thallus fragments are produced in nature. Poelt (1965) described schizidia as a form of thallus fragment that is, at least partly, self liberating. Von Schrenk (1898) noted that fragments of *Usnea* "*barbata*" were produced more easily under the influence of both rain and wind. Degelius (1964) suggested that thalli loosen and expand in wet weather and so fragment in time. Scott (1974) also considered that this process, which he called "hygroscopic flexing", was a means of fragment production, and mentioned abrasion by sand as a further agent of fragmentation. We know that fragments can be produced by feeding or trampling animals but there our knowledge stops. Animals are probably extremely important as short-range dispersal agents, but we still need careful observations on the species of animal as well as the species of lichen, in all cases where animal dispersal is suspected. At present such observations are largely lacking. The dispersal of lichen algae, even the relation of lichen algae to free-living forms, is a little-known field. Conidia, their discharge, their germination and even whether they are genuinely propagules or have some other function, also require attention.

Perhaps one last, major, problem still remains. All the evidence suggests that lichens have extremely efficient methods of dispersal. Yet we know nothing of the distance to which dispersal is effective. It is difficult, if not impossible, to trace the passage of individual lichen propagules in wind or water. Studies on the air spora away from land might be helpful. Ecological studies on colonization from isolated sources would also yield useful information. One such study (R. C. Tapper, unpublished) suggests that dispersal of soredia from *Evernia prunastri* and *Ramalina farinacea* is effective only up to 30 m and 20 m respectively from the source, although these distances are far less than those that might be expected.

There is still room for more work: distribution and colonization studies, investigations of the air spora, culture techniques, detailed studies on the liberation and transport of specific propagules and on their germination. All these need research and could add much to our knowledge of lichen ecology.

ACKNOWLEDGEMENTS

I would like to thank the various workers noted for allowing me to make reference to their unpublished data, Mr P. W. James for reading and commenting upon the manuscript, Dr P. W. Rundel and Mr P. A. Stott for discussions on this topic, Miss V. Clarkson who typed the final version of the manuscript and, above all, my wife without whose help and encouragement I would do no scientific work.

REFERENCES

DES ABBAYES, H. (1951). "Traité de Lichenologie." [*Encycl. Biol.* **41**, i–x, 1–217.] Leche-valier, Paris.

AHMADJIAN, V. (1961). Studies on lichenized fungi. *Bryologist* **64**, 168–179.

AHMADJIAN, V. (1966). Artificial reestablishment of the lichen *Cladonia cristatella*. *Science, N.Y.* **151**, 199–201.

AHMADJIAN, V. (1967). "The Lichen Symbiosis." Blaisdell, Waltham, Mass.

AHMADJIAN, V. (1969). Lichen synthesis. *Öst. bot. Z.* **116**, 306–311.

AHMADJIAN, V. (1974) ["1973"]. Methods of isolating and culturing lichen symbionts and thalli. *In* "The Lichens" (V. Ahmadjian and M. E. Hale, eds), pp. 653–659. Academic Press, New York and London.

AHMADJIAN, V. and HALE, M. E. (Eds.) (1974) ["1973"]. "The Lichens." Academic Press, New York and London.

AHMADJIAN, V. and HEIKKILÄ, H. (1970). The culture and synthesis of *Endocarpon pusillum* and *Staurothele clopima*. *Lichenologist* **4**, 259–267.

AHTI, T. (1965). Notes on the distribution of *Lecanora conizaeoides*. *Lichenologist* **3**, 91–92.

AHTI, T. and VITIKAINEN, Ö. (1974). *Bacidia chlorococca*, a common toxitolerant lichen in Finland. *Mem. Soc. fauna fl. fenn.* **49**, 95–100.

AM ENDE, I. (1950). Zur ernährungsphysiologie des pilzes der *Xanthoria parietina*. *Arch. Mikrobiol.* **15**, 185–202.

ANON. (1965). Dermatitis from lichens. *Br. med. J.* **1965** (2), 1447.

BAILEY, R. H. (1966a). Studies on the dispersal of lichen soredia. *J. Linn. Soc. Bot.* **59**, 479–490.

BAILEY, R. H. (1966b). Notes upon the germination of lichen ascospores. *Revue bryol. lichén.* **34**, 852–853.

BAILEY, R. H. (1967). Notes on Gloucestershire lichens—1. *J. N. Gloucestershire nat. Soc.* **18**, 154–156.

BAILEY, R. H. (1968a). Dispersal of lichen soredia in water trickles. *Revue bryol. lichén.* **36**, 314–315.

BAILEY, R. H. (1968b). *Lecanora conizaeoides* in Iceland. *Lichenologist* **4**, 73.

BAILEY, R. H. (1969). Notes on Gloucestershire lichens—4. *J. N. Gloucestershire nat. Soc.* **20**, 19–20.

BAILEY, R. H. (1970). Animals and the dispersal of soredia from *Lecanora conizaeoides* Nyl. ex. Cromb. *Lichenologist* **4**, 256.

BAILEY, R. H. (1976). Lichen reproduction, dispersal and establishment. *In* "Lichen Biogeography and Ecology" (M. R. D. Seaward, ed.), Academic Press, London and New York (in press).

BAILEY, R. H. and GARRETT, R. M. (1968). Studies on the discharge of ascospores from lichen apothecia. *Lichenologist* **4**, 57–65.

BAILEY, R. H. and STOTT, P. A. (1973). A contribution to the lichen flora of the Wirral Peninsula, Cheshire. *Naturalist, Hull* **1973**, 101–105.

BARKMAN, J. J. (1958). "Phytosociology and Ecology of Cryptogamic Epiphytes." Van Gorcum, Assen.

BERTSCH, A. and BUTIN, H. (1967). Die Kultur der erdflechte *Endocarpon pusillum* im Labor. *Planta* **72**, 29–42.

BOEDIJN, K. B. (1940). The mycetozoa, fungi and lichens of the Krakatau group. *Bull. Jard. bot. Buitenz.*, ser. *3*, **16**, 358–429.

BOND, E. K. (1969). Plant disseminules in wind blown debris from a glacier in Colorado. *Arctic Alpine Res.* **1**, 135–139.

BONNIER, G. (1888). Germination des spores des lichens sur les protonémas des mousses et sur des algues différents des gonidies du lichen. *C. r. Séanc. Soc. Biol.*, sér. *8*, **5**, 541–543.

BONNIER, G. (1899). Germination des lichens sur les protonémas des mousses. *Revue gén. Bot.* **1**, 165–169.

BRODIE, H. J. (1951). The splash cup dispersal mechanism in plants. *Can. J. Bot.* **29**, 224–234.

BRODIE, H. J. (1957). Raindrops as plant dispersal agents. *Proc. Indiana Acad. Sci.* **66**, 65–73.

BRODIE, H. J. and GREGORY, P. H. (1953). The action of wind in the dispersal of spores from cup-shaped plant structures. *Can. J. Bot.* **31**, 402–410.

BRODO, I. M. (1961). Transplant experiments with corticolous lichens using a new technique. *Ecology* **42**, 838–841.

BROWN, D. H. (1971). Southern hemisphere lichens transplanted in Bristol. *Bull. Br. Lichen Soc.* **1** (29), 5.

CHAMPION, R. H. (1965). Wood-cutters' disease: contact sensitivity to lichen. *Br. J. Dermatology* **77**, 285.

COKER, P. D. (1966). The destruction of bryophytes by lichens, fungi, myxomycetes and algae. *Trans. Br. bryol. Soc.* **5**, 142–143.

COKER, P. D. (1967). Damage to lichens by gastropods. *Lichenologist* **3**, 428–429.

CULBERSON, W. L. (1973). The *Parmelia perforata* group: niche characteristics of chemical races, speciation by parallel evolution and a new taxonomy. *Bryologist* **76**, 20–29.

DARBISHIRE, O. V. (1897). Die deutschen Pertusariaceen mit besonderer Berücksichtigung ihrer Soredienbildung. *Bot. Jb.* **22**, 593–671.

DEGELIUS, G. (1954). The lichen genus *Collema* in Europe. *Symb. bot. upsal.* **13** (2), 1–499.

DEGELIUS, G. (1957). The epiphytic lichen flora of the birch stands in Iceland. *Acta Horti gothoburg.* **22**, 1–51.

DEGELIUS, G. (1964). Biological studies of the epiphytic vegetation of twigs of *Fraxinus excelsior*. *Acta Horti gothoburg.* **27**, 11–55.

DIBBEN, M. J. (1971). Whole lichen culture in a phytotron. *Lichenologist* **5**, 1–10.

DU RIETZ, G. E. (1924). Die soredien und isidien der flechten. *Svensk bot. Tidskr.* **18**, 371–396.

DU RIETZ, G. E. (1931). Studier över vinddriften på snöfält i de skandinaviska fjallen. *Bot. Notiser* **1931**, 31–46.

FERRY, B. W., BADDELEY, M. S. and HAWKSWORTH, D. L. (Eds) (1973). "Air Pollution and Lichens." University of London, Athlone Press, London.

FOLLMANN, G. and NAKAGAVA, M. (1963). Keinhemmung von angiospermensamen durch Flechtenstoffe. *Naturwissenschaften* **50**, 696–697.

GARRETT, R. M. (1968). Observations on the germination of lichen ascospores. *Revue bryol. lichén.* **36**, 330–332.

GARRETT, R. M. (1971). Studies on some aspects of ascospore liberation and dispersal in lichens. *Lichenologist* **5**, 33–44.

GARRETT, R. M. (1972). Electrostatic charges on freshly discharged lichen ascospores. *Lichenologist* **5**, 311–313.

GERSON, U. (1973). Lichen–arthropod associations. *Lichenologist* **5**, 434–443.

GREGORY, P. H. (1951). Deposition of airborne *Lycopodium* spores on cylinders. *Ann. appl. Biol.* **38**, 357–376.

GREGORY, P. H. (1973). "Microbiology of the Atmosphere" (2nd edition). Leonard Hill, London.

GRESSITT, J. L. (1966). Epizooic symbiosis: the Papuan weevil genus *Gymnophilus* (Leptopiinae) symbiotic with cryptogamic plants, orabatid mites, rotifers and nematodes. *Pacif. Insects* **8**, 221–280.

GRESSIT, J. L., SEDLACEK, J. and SZENT-IVANY, J. J. H. (1965). Flora and fauna on backs of large Papuan Moss-forest weevils. *Science, N.Y.* **150**, 1833–1835.

HALE, M. E. (1955). Phytosociology of corticolous cryptogams in the upland forests of southern Wisconsin. *Ecology* **36**, 45–63.

HALE, M. E. (1961). "Lichen Handbook." Smithsonian Institution, Washington D.C.

HALE, M. E. (1967). "The Biology of Lichens." Arnold, London.

HALE, M. E. (1972). Natural History of Plummers Island, Maryland XXI. Infestation of the lichen *Parmelia baltimorensis* Gyel. & For. by *Hypogastrura packardii* Folsom (Collembola). *Proc. biol. Soc. Wash.* **85**, 287–295.

HAWKSWORTH, D. L. (1969). The lichen flora of Derbyshire. *Lichenologist* **4**, 105–193.

HAWKSWORTH, D. L. (1971). *Lobaria pulmonaria* (L.) Hoffm. transplanted into Dovedale, Derbyshire. *Naturalist, Hull* **1971**, 127–128.

HAYNES, F. N. (1964). Lichens. *Viewpts Biol.* **3**, 64–115.

HEDLUND, T. (1895). Ueber thallusbildung durch pyknokonidien bei *Catillaria denigrata* und *C. prasina. Bot. Zbl.* **63**, 9–16.

HEILMAN, A. S. and SHARP, A. J. (1963). A probable antibiotic effect of some lichens on bryophytes. *Revue bryol. lichén.* **32**, 215.

HENDRICKSON, J. R. and WEBER, W. A. (1964). Lichens on Galapagos giant tortoises. *Science, N.Y.* **144**, 1463.

HENSSEN, A. (1969). An interesting new species of *Lempholemma* from Canada. *Lichenologist* **4**, 99–104.

HENSSEN, A. and JAHNS, H. M. (1973) ["1974"]. "Lichenes. Eine Einführung in die Flechtenkunde." Thieme, Stuttgart.

HIRST, J. M. and STEDMAN, O. J. (1963). Dry liberation of fungus spores by rain drops. *J. gen. Microbiol.* **33**, 335–344.

HOFFMAN, G. R. (1971). Bark samples for use in air pollution—epiphytic cryptogam studies. *Bryologist* **74**, 490–493.

INGOLD, C. T. (1953). "Dispersal in Fungi." Clarendon Press, Oxford.

INGOLD, C. T. (1965). "Spore Liberation." Clarendon Press, Oxford.

INGOLD, C. T. (1971). "Fungal Spores. Their Liberation and Dispersal." Clarendon Press, Oxford.

INGOLD, C. T. and HADLAND, S. A. (1959). The ballistics of *Sordaria*. *New Phytol.* **58**, 46–57.

JAHNS, H. M. (1974) ["1973"]. Anatomy, morphology and development. *In* "The Lichens" (V. Ahmadjian and M. E. Hale, eds), pp. 3–58. Academic Press, New York and London.

KERSHAW, K. A. and MILLBANK, J. W. (1969). A controlled environment lichen growth chamber. *Lichenologist* **4**, 83–87.

KERSHAW, K. A. and MILLBANK, J. W. (1970). Isidia as vegetative propagules in *Peltigera apthosa* var. *variolosa* (Massal.) Thoms. *Lichenologist* **4**, 214–217.

KOFLER, L., JACQUARD, F. and MARTIN, J. F. (1969). Influence des fumées d'usines sur la germination des spores de certains lichens. *Bull. Soc. bot. Fr., Mém.* **1968** *Coll. Lich.*, 219–230.

KRISTINSSON, H. (1970). Report on lichenological work on Surtsey and in Iceland. *Surtsey Res. Prog. Rep.* **5**, 1.

KRISTINSSON, H. (1972). Studies on lichen colonization in Surtsey 1970. *Surtsey Res. Progr. Rep.* **6**, 77.

LAUNDON, J. R. (1967). A study of the lichen flora of London. *Lichenologist* **3**, 277–327.

LETROUIT-GALINOU, M. A. (1974) ["1973"]. Sexual reproduction. *In* "The Lichens" (V. Ahmadjian and M. E. Hale, eds), pp. 59–90. Academic Press, New York and London.

LINDAHL, P. O. (1960). The different types of isidia found in the lichen genus *Peltigera*. *Svensk bot. Tidskr.* **54**, 565–570.

LINDSAY, W. L. (1859). On the spermogones and pycnides of filamentous, fruticulose and foliaceous lichens. *Trans. R. Soc. Edinb.* **22**, 101–303.

LYNGE, B. (1940). Lichens from Iceland collected by Norwegian botanists in 1937 and 1939. 1. Macrolichens. *Skr. Norske Vid.-Akad. Oslo, mat. nat. kl.* **1940** (7), 1–56.

MAAS GEESTERANUS, R. A. (1948) ["1947"]. Revision of the lichens of the Netherlands I. Parmeliaceae. *Blumea* **6**, i–viii, 1–199.

McWHORTER, F. P. (1921). Destruction of mosses by lichens. *Bot. Gaz.* **72**, 321–325.

MARGOT, J. (1973). Experimental study of the effects of sulphur dioxide on the soredia of *Hypogymnia physodes*. *In* "Air Pollution and Lichens" (B. W. Ferry, M. S. Baddeley and D. L. Hawksworth, eds), pp. 314–329. University of London, Athlone Press, London.

MASON, B. J. (1962). "Clouds, Rain and Rainmaking." Cambridge University Press, Cambridge.

MILLER, E. V., GRIFFIN, C. E., SCHAEFFERS, T. and GORDON, M. (1965). Two types of growth inhibitors in extracts from *Umbilicaria papulosa*. *Bot. Gaz.* **126**, 100–107.

MITCHELL, J. C. (1974). Contact allergy from *Frullania* and respiratory allergy from *Thuja*. *J. Can. Med. Assn.* **110**, 653–657.

MIYOSHI, M. (1901). Ueber die sporocarpenevacuation und darauf erfolgendes sporenausstrauen bei einer flechte. *Journ. Coll. Sci. imp. Univ. Tokyo* **15**, 367–370.

MOLLER, A. (1887). Ueber die cultur flechtenbildener ascomyceten ohne algen. *Unters. Bot. Inst. K. Akad. Munster* **1887**, 1–52.

MOLLER, A. (1888). Ueber die sogenannten Spermatien der Ascomyceten. *Bot. Ztg* **66**, 421–425.

NIENBURG, W. (1919). Studien zur Biologie der Flechten. I. II. III. *Z. Bot.* **11**, 1–38.

PEAKE, J. F. and JAMES, P. W. (1967). Lichens and mollusca. *Lichenologist* **3**, 425–428.

I

PEIRCE, G. J. (1898). On the mode of dissemination and on the reticulations of *Ramalina reticulata*. *Bot. Gaz.* **25**, 404–417.

PETTERSSON, B. (1940). Experimentelle untersuchingen über die euanenomochore verbreitung der sporenpflanzen. *Acta bot. fenn.* **25**, 1–103.

POELT, J. (1965). Uber einige artengruppen der flechtengattungen *Caloplaca* und *Fulgensia*. *Mitt. bot. Staatssamml., Münch.* **5**, 571–607.

POELT, J. (1970). Das Konzept der Artenpaare bei den Flechten. *Vortr. bot. Ges. [Dtsch. bot. Ges.]*, *n.f.* **4**, 187–198.

POELT, J. (1972). Die taxomische Behandlung von Artenpaaren bei den Flechten. *Bot. Notiser* **125**, 77–81.

POELT, J. (1974a) ["1973"]. Systematic evaluation of morphological characters. *In* "The Lichens" (V. Ahmadjian and M. E. Hale, eds), pp. 91–115. Academic Press, New York and London.

POELT, J. (1974b) ["1973"]. Classification. *In* "The Lichens" (V. Ahmadjian and M. E. Hale, eds), pp. 599–632. Academic Press, New York and London.

PYATT, F. B. (1967). The inhibitory influence of *Peltigera canina* on the germination of graminaceous seeds and the subsequent growth of the seedlings. *Bryologist* **70**, 326–329.

PYATT, F. B. (1968a). The occurrence of a rotifer on the surfaces of apothecia of *Xanthoria parietina*. *Lichenologist* **4**, 74–75.

PYATT, F. B. (1968b). An investigation into conditions influencing ascospore discharge and germination in lichens. *Revue bryol. lichén.* **36**, 323–329.

PYATT, F. B. (1969). Studies on the periodicity of spore discharge and germination in lichens. *Bryologist* **72**, 48–53.

PYATT, F. B. (1973). A note on the discharge of ascospores with accompanying algal cells in *Pertusaria pertusa*. *Revue bryol. lichén.* **39**, 345–347.

PYATT, F. B. (1974) ["1973"]. Lichen propagules. *In* "The Lichens" (V. Ahmadjian and M. E. Hale, eds), pp. 117–145. Academic Press, New York and London.

RHODES, P. G. M. (1931). The lichen-flora of Hartlebury Common. *Proc. Bgham. nat. Hist. Soc.* **16**, 39–43.

RICHARDSON, D. H. S. (1967). The transplantation of lichen thalli to solve some taxonomic problems in *Xanthoria parietina* (L.) Th. Fr. *Lichenologist* **3**, 386–391.

RICHARDSON, D. H. S. (1975). "The Vanishing Lichens." David and Charles, Newton Abbott, London, and Vancouver.

RIDLEY, H. N. (1930). "The Dispersal of Plants Throughout the World." Reeve, Ashford, Kent.

RONDON, Y. (1966). Action inhibitrice de l'extrait du lichen *Roccella fucoides* (Dicks.) Vain sur la germination. *Bull. Soc. bot. Fr.* **113**, 1–2.

ROSE, F. (1973). A mobile lichen—in which grid square? *Bull. Br. Lichen Soc.* **2** (33), 13.

RUDOLPH, E. D. (1970). Local dissemination of plant propagules in Antarctica. *In* "Antarctic Ecology" (M. W. Holdgate, ed.) **2**, 812–817. Academic Press, London and New York.

SATO, M. (1963). Mixture ratio of the lichen genus *Thamnolia*. *Nova Hedwigia* **5**, 149–155.

SAUER, E. G. F. (1962). Ethology and ecology of golden plovers on St Lawrence Island, Bering Sea. *Psychol. Forsch.* **26**, 399–470.

SAUNDERS, P. W. J. (1966). The toxicity of sulphur dioxide to *Diplocarpon rosae* causing black spot on roses. *Ann. appl. Biol.* **58**, 103–114.

VON SCHRENK, H. (1898). On the mode of dissemination of *Usnea barbata*. *Trans. Acad. Sci. St Louis* **8**, 189–198.

SCOTT, G. D. (1959). Observations on spore discharge and germination in *Peltigera praetextata* (Flk.) Vain. *Lichenologist* **1**, 109–111.

SCOTT, G. D. (1964). Studies of the lichen symbiosis: 2. Ascospore germination in the genus *Peltigera*. *Z. allg. Mikrobiol.* **4**, 326–336.

SCOTT, G. D. (1974) ["1973"]. Evolutionary aspects of symbiosis. *In* "The Lichens" (V. Ahmadjian and M. E. Hale, eds), pp. 581–598. Academic Press, New York and London.

SERNANDER, R. (1918). Subfossile flechten. *Flora, Jena* **112**, 703–724.

SHEARD, J. W. (1967). A revision of the lichen genus *Rinodina* (Ach.) Gray in the British Isles. *Lichenologist* **3**, 328–367.

SKOREPA, A. C. and SHARP, A. J. (1971). Lichens in "packets" of Lacewing larvae (Chrysopidae). *Bryologist* **74**, 363–364.

SMITH, A. L. (1921). "Lichens." Cambridge University Press, Cambridge.

SMITH, D. C. (1962). The biology of lichen thalli. *Biol. Rev.* **37**, 537–570.

STEWART, K. W. and SCHLICHTING, H. E. (1966). Dispersal of algae and protozoa by selected aquatic insects. *J. Ecol.* **54**, 551–562.

THOMSON, J. W. (1948). Experiments upon the regeneration of certain species of *Peltigera*; and their relationship to the taxonomy of this genus. *Bull. Torrey bot. Club* **75**, 486–491.

THOMSON, J. W. (1950). The species of *Peltigera* of North America north of Mexico. *Am. Midl. Nat.* **44**, 1–68.

THOMSON, J. W. (1963). The lichen genus *Physcia* in North America. *Beih. Nova Hedwigia* **7**, 1–172.

THOMSON, J. W. (1972). Distribution patterns of American Arctic Lichens. *Can. J. Bot.* **50**, 1135–1156.

TIBELL, L. (1971). The genus *Cyphelium* in Europe. *Svensk bot. Tidskr.* **65**, 138–164.

TOBLER, F. (1911). Zur Biologie von Flechten und Flechtenpilzen. *Jb. wiss. Bot.* **49**, 389–417.

TOBLER, F. (1925). "Biologie der Flechten." Borntraeger, Berlin.

TREUB, M. (1888). Notice sur la nouvelle flore de Krakatau. *Ann. Jard. bot. Buitenz.* **7**, 213–223.

VERSEGHY, K. (1965). Effect of dry periods on the spore production of lichens. *Acta Biol. Hung.* **16**, 85–104.

WADE, A. E. (1965). The genus *Caloplaca* Th. Fr. in the British Isles. *Lichenologist* **3**, 1–28.

WALKEY, D. G. A. and HARVEY, R. (1966). Studies on the ballistics of ascospores. *New Phytol.* **65**, 59–74.

WEBER, W. A. (1974). Two lichen–arthropod associations in Australia and New Guinea. *Lichenologist* **6**, 168.

WERNER, R.-G. (1927). "Recherches Biologiques et Expérimentale sur les Ascomycètes de Lichens." Thesis, Braun, Paris.

WESTMAN, L. (1973). Notes on the taxonomy and ecology of an arctic lichen *Lecanora symmicta* var. *sorediosa*. *Lichenologist* **5**, 457–460.

10 | Distribution Patterns shown by Epiphytic Lichens in the British Isles

B. J. COPPINS

Royal Botanic Garden, Edinburgh, Scotland

Abstract: Most epiphytic lichens can be described as "widespread" although they may vary somewhat in abundance in different regions and may have widely differing distributions outside Britain. A smaller number of species exhibit marked distributional tendencies which are apparently determined, for the most part, by climatic factors. A working classification is presented of the patterns shown by these species in Britain. This is intended to be of temporary value only, leading, it is hoped, to a more comprehensive analysis of the British lichen flora based on European distributions. The classification is supplemented by suggested climatic correlations, some ecological observations and comments as to how the British distribution of some species reflects their European ranges. Difficulties in interpreting distribution patterns are stressed, particularly with regard to the influence of man-made factors and the limitations of available climatic data.

INTRODUCTION

Until recent years information on the distribution of British lichens was only available from the often rather scanty data given in the floras of, for example, Mudd (1861), Leighton (1879) and Smith (1918, 1926) or by extensive searches of local lists and herbaria.

The first major attempt to compile the available data on lichen distributions in the British Isles was that of Watson (1953), who presented records according to the vice-county system; although a useful source of reference, this work contains numerous erroneous and dubious reports.

Following the formation of the British Lichen Society in 1958, interest in lichenology was greatly increased and in 1964 the Society initiated a Distribution Maps Scheme. The recording is based on the 10 km square units of the Ordnance Survey's National Grid and by 1973 the results obtained were very encouraging

Systematics Association Special Volume No. 8, "Lichenology: Progress and Problems", edited by D. H. Brown, D. L. Hawksworth and R. H. Bailey, 1976, pp. 249–278. Academic Press, London and New York.

(Seaward, 1973). Due to a shortage of man-power, our knowledge of lichen distributions is not yet as profound as that for the vascular plants (Perring and Walters, 1962), although it is now sufficiently advanced for us to begin to make reasonably confident statements regarding the distribution of many species. This is particularly true for the species that are exclusively or predominantly epiphytes.

The majority of epiphytic lichens have fairly widespread distributions, although their frequency of occurrence in different areas may vary and their distribution patterns outside Britain may differ widely. However, there is a smaller, yet significant, number of species which exhibit marked distributional tendencies even within the comparatively small area of the British Isles. It is the aim of this paper to demonstrate the different types of distribution that are apparent, and to note some of the climatic factors which show correlations with these types.

PROBLEMS IN COMPILING A CLASSIFICATION OF DISTRIBUTION PATTERNS

1. Unevenness of Recording

Seaward (1973; Fig. 1) showed that most of Britain, apart from Ireland, had received a moderately good degree of attention, and many additional data have been gathered subsequently. Information is still scanty for much of southern and north-eastern Scotland, although it is hoped that these areas will be surveyed in the near future.

2. Influence of Man

The distribution of many lichens, especially lowland species, has been much modified as a result of the impact of man on the environment. For many centuries forest species have been adversely affected by both deforestation and the intensive management of existing woodlands. Several characteristically "old forest" species, which appear to require a continuity of humid microclimatic conditions in order to survive or reproduce, are apparently unable to recolonize planted woodlands, at least in the less humid districts of most of England and eastern Scotland (Rose, 1974; Chapter 11). Generally speaking these are also the areas that have suffered most from the effects of pollution of their air by sulphur dioxide and also from the effects of agricultural chemicals.

Undoubtedly, epiphytes of open habitats, and especially those favouring hypertrophicated (eutrophicated) bark, benefited by forest clearance and the spread of agriculture. These species (e.g. *Anaptychia ciliaris*, *Parmelia acetabulum*, *Physconia* spp., *Physcia clementei*) were presumably able to migrate into areas where suitable natural habitats would scarcely have been available in the primeval landscape. Their distribution thus became more directly controlled by climatic conditions, rather than indirectly by the barrier of forest cover. However, most of these species favoured by the events of the past have since undergone a decline owing to the devastating effects of both air pollution by sulphur dioxide and the introduction of modern agricultural techniques, involving the extensive use of artificial fertilizers, pesticides, stubble burning and the removal of hedgerow trees. Hawksworth *et al.* (1973, 1974) discussed the impact of man on the British lichen flora in more detail.

It is evident that when discussing the distribution shown by the great majority of species we are not dealing with natural situations but ones markedly modified by man. To compensate, in part, for man's effects in more recent times we are fortunate in being able to use rich resources of herbarium and literature data compiled over the last 200 years. Many herbaria still await detailed study and there are several that have not yet been traced but may still exist.

3. Uniqueness of Species Distributions

Even if the above problems could be overcome there is still a major drawback arising from the individuality of the distribution shown by each particular species. The implications of this fact in the production of a phytogeographical classification are well summarized by Ratcliffe (1968) in his consideration of Atlantic bryophytes:

"Any student is faced with the awkward fact that, while it is possible to arrange distribution patterns of species into separate and distinctive groups, no two species have exactly the same geographical scatter or continuity of distribution within their total range and there is such a wide range of variation in these patterns that close links between any two related groups can nearly always be found. The problem is basically the taxonomic one of descriptive analysis of a multidimensional continuum, and geographical groupings are merely convenient approximations describing particular patterns of distribution recognizable within a large and complex field of variation; they are selected reference points—equivalent to 'noda' in vegetation analysis—and should not be regarded as representing real discontinuities."

PROBLEMS IN CORRELATING PATTERNS OF CLIMATE AND LICHEN DISTRIBUTION

(1) If there is a close similarity between the geographical limit of a species and a climatic isopleth it is reasonable to assume that the particular climatic factor probably has a major controlling influence on the distribution of the species. However, a general correlation of this type is not proof of a direct cause and effect relationship and such a supposition needs to be supplemented by detailed autecological studies.

Although there are many works which provide standard meteorological information (e.g. des Abbayes, 1934; Degelius, 1935; Ahlner, 1948; Almborn, 1948; Barkman, 1958; Mitchell, 1961; Schauer, 1965) little has been published in the form of detailed monitoring of microclimatic conditions experienced by particular species. The comparison of continuously monitored microclimatic data between sites where a species appears to be enjoying conditions for optimal growth and reproduction, and sites where it is at its geographical limit, would be most valuable.

(2) The relationship between a particular climatic factor and the distribution of a species is not necessarily the same throughout the range of the species, and the relationship between the two may change in order to compensate for changes in other climatic factors.

(3) There is good evidence to show that some species occur in the form of different physiological races which have become adapted to different sets of climatic conditions (Harris, 1971; Kershaw, 1972). However, whether these are genotypic or phenotypic adaptations is not yet known (Hawksworth, 1973a).

(4) In this paper indications of climatic correlations with distribution patterns are given in a very generalized way by using only a few climatic parameters in the form of isopleth maps obtained from standard meteorological data (Figs 1–5). These maps only serve to present the general climatic trends over the country and they do not take into account local variations in climate resulting from topographical features or vegetation.

A TENTATIVE CLASSIFICATION OF BRITISH DISTRIBUTION TYPES

The classification of distribution types presented below is based solely on British distributions and its purpose should only be regarded as temporary. It is hoped that a better understanding of lichen distributions in Europe as a whole will eventually lead to a more comprehensive and meaningful phytogeographical analysis of the British lichen flora.

Group 1. Western

This group contains the species confined to western Britain and is taken to include a few species which extend eastwards in the southernmost counties of England, as far as the New Forest in Hampshire. All are Eu-Atlantic or Sub-Atlantic species and there appear to be few epiphytic lichens with a marked western distribution in Britain that do not have Atlantic distributions in Europe as a whole.

(a) *General Western.* The distributions within the British Isles of the species grouped here strongly indicates the requirement of a wet climate providing a more or less continual dampness of the atmosphere. The areas in which these species are found characteristically have an annual rainfall of at least 40 in (*c.* 1000 mm) per year, distributed over at least 160 rain days (Fig. 1). Most of the

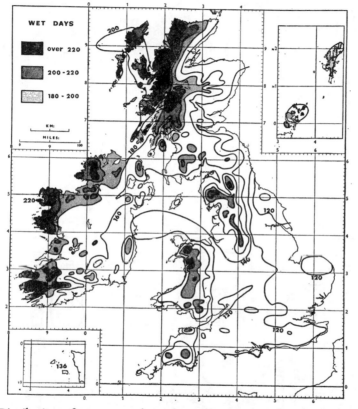

Fig. 1. Distribution of mean annual number of wet days in the British Isles for the period 1951–60. (After Ratcliffe, 1968.) A "wet day" being a period of 24 h in which 0·04 in. (1 mm) of rain is recorded.

species are especially abundant and luxuriant in areas experiencing over 180–200 rain days. When found in relatively low rainfall districts (e.g. the New Forest) the species are restricted to ancient woodland sites (though not necessarily on old trees) where a more or less constantly high humidity has been maintained by the continuous cover of a tree canopy for many centuries. In this district they are also found mainly in woodlands where the water table is near the surface for most of (see the year Rose and James, 1974).

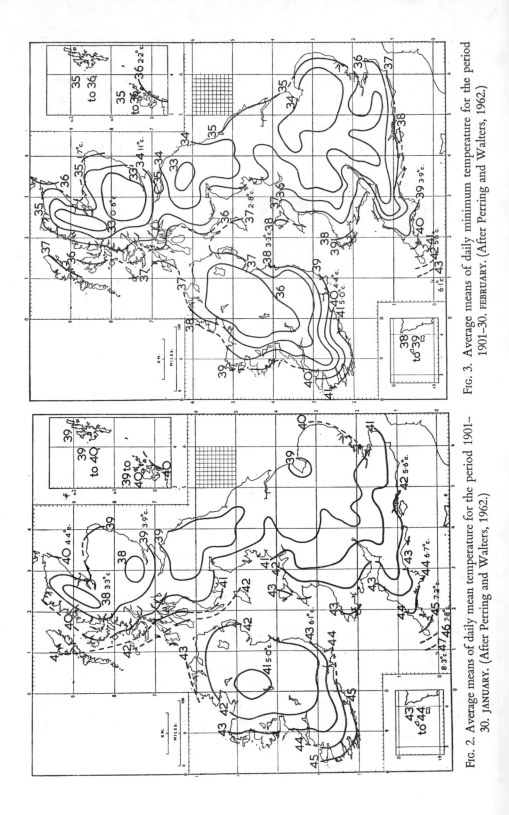

FIG. 3. Average means of daily minimum temperature for the period 1901-30. FEBRUARY. (After Perring and Walters, 1962.)

FIG. 2. Average means of daily mean temperature for the period 1901-30. JANUARY. (After Perring and Walters, 1962.)

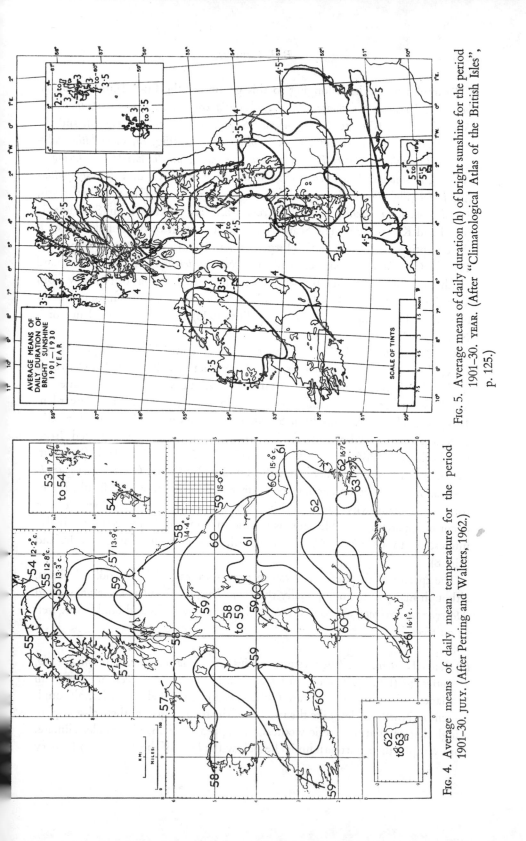

Fig. 5. Average means of daily duration (h) of bright sunshine for the period 1901–30. YEAR. (After "Climatological Atlas of the British Isles", p. 125.)

Fig. 4. Average means of daily mean temperature for the period 1901–30. JULY. (After Perring and Walters, 1962.)

Examples of species belonging to this group are the Eu-Atlantic *Leptogium burgessii*, *Parmelia endochlora*, *Pseudocyphellaria crocata*, *P. lacerata*, *P. thouarsii*, *Sticta canariensis* (Chapter 3, Fig. 4) and *S. dufourii* (Chapter 3, Fig. 3), and the Sub-Atlantic *Alectoria smithii*, *Cetrelia olivetorum* s.l. (Fig. 8), *Gomphillus calycioides*, *Leptogium brebissonii*, *Menegazzia terebrata*, *Parmelia arnoldii*, *P. laevigata* (Fig. 6), *P. sinuosa*, *P. taylorensis* and *Parmeliella atlantica*. Of these (?) *Gomphillus calycioides* (Fig. 7) and *Leptogium brebissonii*, *Parmelia endochlora*, *P. taylorensis*, *Pseudocyphellaria lacerata* and *Sticta canariensis* are unknown in Scandinavia (Dahl and Krog, 1973). The Scandinavian distribution of the remainder, with the exception of *Alectoria smithii* (Jørgensen and Ryvarden, 1970; Hawksworth, 1972a), *Cetrelia olivetorum* s.l. and *Menegazzia terebrata*, is eu-oceanic and southern being predominantly confined to the coastal regions of Norway from Vest-Agder to Sogn og Fjordane. *Pseudocyphellaria crocata*, which has a northerly bias to its European distribution (Degelius, 1935) reaches Nordland in north Norway.

Cetrelia olivetorum s.l. (i.e. *C. olivetorum* and *C. cetrarioides*, see Jørgensen and Ryvarden, 1970; *C. olivetorum* s.s. may be extinct in Britain, see Hawksworth et al., 1974) is very southern in coastal Norway, but is found also in numerous localities further north in inner eastern Norway (Jørgensen and Ryvarden, 1970). It is also known rarely in Sweden, more frequently in Finland, and in eastern Poland (Rydzak, 1961). In south-east Europe it is reported from Czechoslovakia (Pišút, 1970), Romania (Pišút, 1971) and Yugoslavia (Ramaut, 1961), and it is also recorded from the Ukraine by Oxner (1937); *Menegazzia terebrata* has a rather similar European distribution. It becomes difficult, therefore, to explain why species with such extra-British distributions do not occur in apparently suitable habitats in eastern Scotland. In this respect it is also interesting to note that *C. olivetorum* s.l. is so far unrecorded in Ireland (Fig. 8).

The species of the General Western group are essentially woodland plants, although many of them (especially the *Parmelia* spp., *Cetrelia olivetorum* s.l. and *Menegazzia terebrata*) are found frequently on mossy rocks and boulders as well as on trees.

(b) *South Western*. The five western species, *Graphina ruiziana*, *Parmelia dissecta*, *P. horrescens* (Fig. 9), *Phaeographis lyellii* and *Rinodina isidioides*, are absent in Scotland and absent or very rare (*R. isidioides*) in the Lake District, although present in North Wales. Like the species of the former group, these also require a high degree of atmospheric wetness, but appear to demand a milder climate. For example, their northern limits lie below the lines of the 41°F (5°C) January

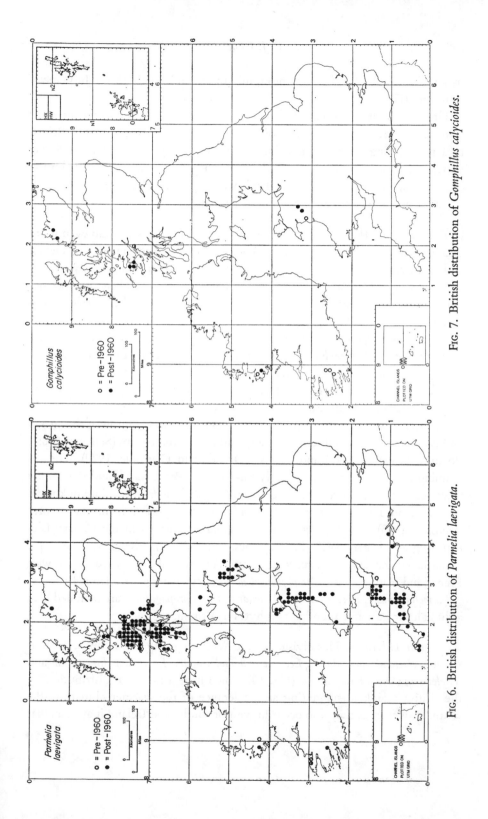

FIG. 7. British distribution of *Gomphillus calycioides*.

FIG. 6. British distribution of *Parmelia laevigata*.

mean, 36°F (2·2°C) February mean minimum and 59°F (15°C) July mean isotherms (Figs 2, 3, 4).

The British distributions of these species are strongly indicative of their European distributions which are Eu-Atlantic and southern, with the British Isles representing their northern limit. The one exception is *Parmelia dissecta* which is also found in the southern Alps (Poelt, 1969). The European distribution of *Phaeographis lyellii* is given by Hawksworth (1972b: Fig. 8). All the species included here are confined to woodland habitats.

(*c*) *Extreme South Western.* This group represents the species with an even more extreme south-westerly distribution. They are restricted to the areas of Devon and Cornwall, the Scilly Isles and south-west Ireland, and apparently require warmer winter temperatures than those of the last group. Their distributions correlate with a January mean of over 43–45°F (6·1–6·7°C) and a February mean minimum of over 38–39°F (3·3–3·9°C).

Haematomma leprarioides (Mitchell, 1970), *Parmelia robusta*, *P. stuppea*, *Parmentaria chilensis*, *Polychidium dendriscum* and *Porina nucula* are only known from south-west Ireland, although *Parmelia robusta* is also recorded from Jersey in the Channel Isles (Mitchell, 1970). *Anaptychia dendritica* var. *propagulifera* is only recorded from the Scilly Isles.

Parmelia stuppea has the most widespread European distribution of this group and is reported from southern Germany and the northern Alps (Schauer, 1965), Portugal (Tavares, 1945) and France (Ozenda and Clauzade, 1970). The remaining species appear to be strongly Eu-Atlantic. *Porina nucula* and *P. hibernica* (Swinscow, 1962), *Parmentaria chilensis* and *Polychidium dendriscum* are unknown on mainland Europe, although *P. dendriscum* is reported from the Azores (Henssen, 1963) and *P. chilensis* and *P. nucula* are known in the Canary Isles (Imshaug and Harris, 1969; P. W. James, unpublished). Elsewhere in Europe, *Parmelia robusta* and *Haematomma leprarioides* are only recorded from west France and Portugal, and *Anaptychia dendritica* var. *propagulifera* from Brittany.

Parmelia robusta, *P. stuppea* and *A. dendritica* var. *propagulifera* are included here although they are not found as epiphytes in Britain.

(*d*) *Scottish–Hibernian.* This division of the Western group so far consists of four species: *Arctomia delicatula*★, *Lecanactis homalotropum*, *Leptogium hibernicum* and *Thelotrema* (*Ocellularia*) *subtile* (Fig. 10) which have a northern distribution on the British mainland suggesting an intolerance of the warmer climate in the south. However, as they also occur in west or south-west Ireland it appears that

---

★ Not an epiphyte.

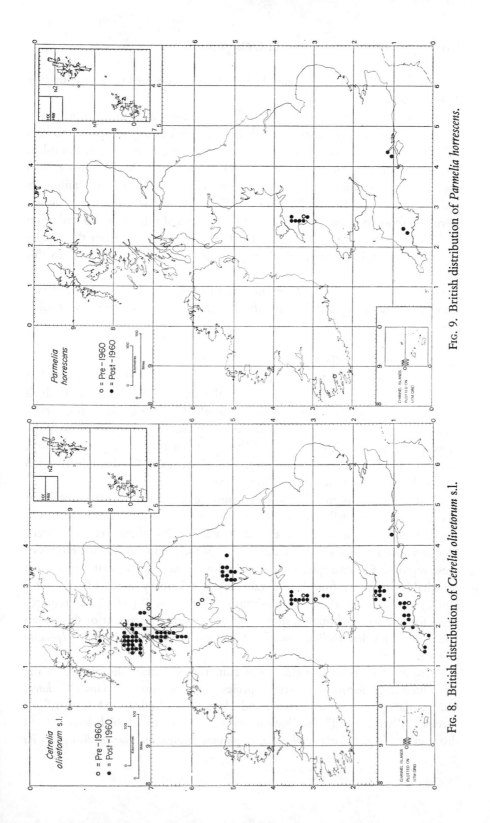

Parmelia
horrescens

o = Pre – 1960
= Post – 1960

FIG. 9. British distribution of *Parmelia horrescens.*

Cetrelia
olivetorum s.l.

o = Pre – 1960
= Post – 1960

FIG. 8. British distribution of *Cetrelia olivetorum* s.l.

they are dependent on a very humid climate, being confined to areas with over
200 rain days.

Leptogium hibernicum is also known from south-west Norway, north-west
France and Portugal (Jørgensen, 1973). *Lecanactis homalotropum* appears to be a
British endemic. *T. subtile* has only one other European locality, in Bohuslän,
Sweden, from where it was described as *Ocellularia suecica* by Magnusson. It is
also found in the south-eastern U.S.A., Australasia, Indonesia, the Philippines
and the Hawaiian Islands (Salisbury, 1972). *Thelotrema* (*Leptotrema*) *monosporum* (unknown elsewhere in Europe but with a similar world distribution
to *T. subtile*) has been collected by Dr T. D. V. Swinscow from the Ardna-
murchan peninsula in Argyll (Salisbury, 1975), and may be referable to this
group.

This distinctive Scottish–Hibernian pattern is also shown by a few Atlantic
liverworts, including *Acrobolbus wilsonii*, *Mastigiophora woodsii* and *Radula
carringtonii* (Ratcliffe, 1968).

Group 2. Western and Southern

This group may be regarded as an extension of the General Western Group (1a)
in that it comprises species that have a similar western distribution but which
extend eastwards of the New Forest in the south. The distinction is rather arbitrary for many species, especially those with only one or two localities east of the
New Forest (e.g. *Arthonia stellaris*, *Biatorella ochrophora*, *Anaptychia obscurata*,
Lecanora jamesii and *Ochrolechia inversa*). The Western and Southern distribution
is probably best exemplified by *Dimerella lutea* (Fig. 11).

The distributions of members of this group are correlated with areas receiving
over 30 in of rain distributed over 120 rain days. The absence of most species in
areas of suitable rainfall in north-east Yorkshire, Northumbria and eastern
Scotland appears to be due to an intolerance of the cold winter temperatures of
these regions. For example, their distributions generally correlate with areas
of suitable humidity (number of rain days), a January mean of over 39°F
(4·0°C) and a February mean minimum of over 34°F (1·1°C). A requirement for a warmer climate is shown by a southern component of this group
which includes *Arthothelium ruanum*, *Parmelia reddenda* (Fig. 12) and *Polyblastia
allobata*.

There appears to be an influence of climate, particularly humidity, on the
reproductive performance of some species in this group (e.g. *Dimerella lutea*,
Ochrolechia inversa, *Parmelia crinita* and *Parmeliella plumbea*) and some of the
species of "widespread" distribution referred to Group 7 which have related
distributions (e.g. *Lobaria amplissima*, *Nephroma laevigatum*, *Normandina pulchella*

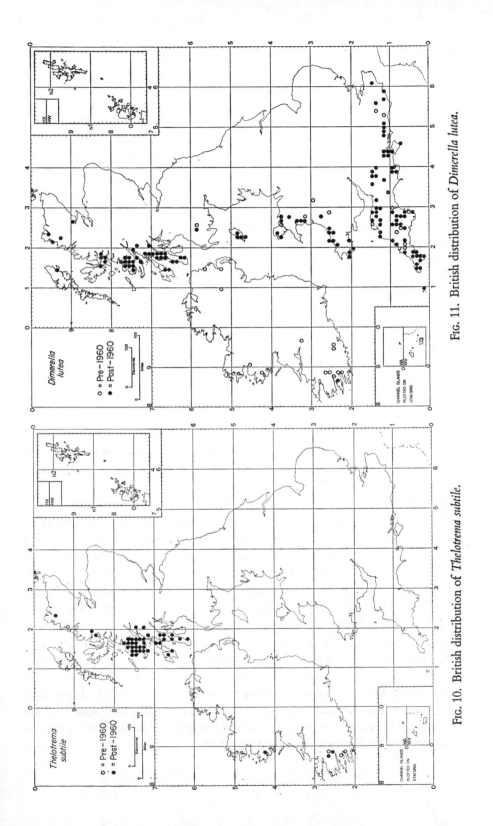

Fig. 11. British distribution of *Dimerella lutea*.

Fig. 10. British distribution of *Thelotrema subtile*.

and *Pannaria pityrea*). In the drier parts of southern England, which experience below about 140 rain days, these species require woodland or semi-woodland parkland sites of great age and continuity and are also confined to the more ancient trees. However, in the higher rainfall and more humid areas of the west they are again typically found in old woodland and parkland, but their greater reproductive capacity is evidenced by their occurrence on younger trees. *Ochrolechia inversa* is confined to old trees of *Quercus* and *Fagus* in the New Forest in Hampshire and St Leonard's Forest in Sussex, but is found on much younger and smaller trees, especially *Betula*, in the west. *Dimerella lutea* is restricted to old woodland sites, and generally to old trees, in Kent, Sussex and Hampshire, whereas in the west it may often be found on quite young trees of, for example, *Acer pseudoplatanus*, in plantations. However, it should be stated that this apparent reduction in reproductive performance in southern England may in some areas be accentuated by low levels of sulphur dioxide pollution (see Hawksworth *et al.*, 1973).

As mentioned above, there is a close similarity between the distribution exhibited by members of this group (especially *Catillaria atropurpurea*, *Pannaria rubiginosa* and *Parmeliella plumbea*) and some placed in the miscellaneous assemblage of Group 7 (e.g. *Catinaria grossa* (Fig. 22), *Collema furfuraceum*, *C. subfurvum*, *Lobaria amplissima*, *Nephroma laevigatum*, *Normandina pulchella*, *Pannaria pityrea*, *Phaeographis dendritica* and *Sticta limbata*) and their separation here is very arbitrary and artificial. The Western and Southern Group forms a link between the Western Group, 1a, and the members of Group 7 with a high humidity requirement which are mostly Sub-Atlantic species.

Perhaps in this group *Arthonia stellaris*, *Arthothelium ruanum*, *Anaptychia obscurata*, *Pannaria rubiginosa*, *Parmeliella plumbea* and *Thelopsis rubella* can be regarded as Sub-Atlantic. *Arthothelium ilicinum* is Eu-Atlantic, being known outside Britain only from Brittany and the Pyrénées–Altlantiques region of France (Coppins, 1971; Vězda and Vivant, 1972) and Norway (Du Rietz, 1919). *Ochrolechia inversa* may be Eu-Atlantic but it has been much overlooked. It is known outside Britain only from Brittany, where it is a common forest species (Coppins, 1971). The same was true for *Lecanora jamesii*, but it has recently been collected in the Austrian Alps from near Steinach, Innsbruck (P. W. James, *in litt.*). The recently described *Porina coralloidea*, recorded only from the British Isles and Brittany, may well occur elsewhere as it is easily overlooked. It would appear from Poelt (1969) that *Parmelia reddenda* has a Eu-Atlantic distribution restricted to South Scandinavia, Britain and France. *Arthothelium ruanum* is quite widespread in central, western and southern Europe, especially if it is regarded as conspecific with *A. ruanideum* (Almborn, 1948).

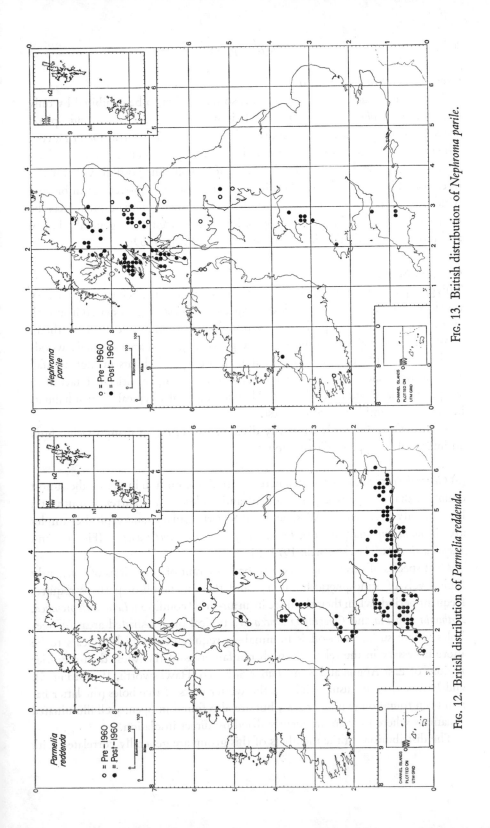

Fig. 13. British distribution of *Nephroma parile*.

Fig. 12. British distribution of *Parmelia reddenda*.

Group 3. Western and Northern

This seemingly anomalous ditribution is so far known to be exhibited by only one species, *Nephroma parile* (Fig. 13). It is widespread in the west and upland areas of Scotland and extends southwards in the west to Dartmoor in Devon. The notable feature of this distribution is the complete absence of records from eastern England and southern England east of Devon. The possibility that it was overlooked in the past by workers in these areas, such as Borrer, Crombie and Mudd, is very remote. The warm and dry summers of these areas may be unsuitable for *N. parile*, and this supposition is supported by its generally boreal and montane distribution in the rest of Europe (Poelt, 1969).

Group 4. Southern

This group comprises species having a strong southern bias, being absent or very rare in Scotland. It does not include those species with a strong western (Groups 1b and 1c) or eastern (Group 5b) bias, although many affinities exist with them.

Whereas the majority of species in the previous groups are essentially woodland plants or plants which can occur in more open habitats in high rainfall areas, most species of this group are characteristic of open, drier habitats, such as open parkland and roadside trees. They are especially typical of such habitats in the more humid areas of the west.

The Southern Group is arbitrarily divided into two and many species show an intermediate type of distribution.

(*a*) *General Southern*. Represented here are species, common in the southernmost counties, that are (or were) of frequent and widespread occurrence south of a line roughly between the Wash and the Severn Estuary, and typical examples are *Arthonia impolita* (Fig. 14), *Lecanactis premnea*, *Phlyctis agelaea* (Fig. 15) and *Usnea articulata* (Hawksworth, 1973b: Fig. 4).

Most species of this group appear to be intolerant of conditions of constantly high levels of atmospheric humidity, and their metabolism perhaps requires frequent periods when their thalli are in an air-dried condition. *Lecanactis premnea*, *Arthonia impolita* and *Opegrapha lyncea* are typical of dry-shaded areas of tree-boles where surface run-off is minimal—a requirement which appears to be more necessary in the relatively high rainfall areas of the West than the drier regions of East Anglia and Kent. *Parmelia borreri* (Hawksworth, 1973b: Fig. 5). and *Physcia clementei* usually inhabit the wetter parts of tree boles (the latter has also been found on exposed rocks) but are also found in open, well illuminated situations where they are frequently dried by direct insolation.

The distributions of the members of this group are generally correlated with

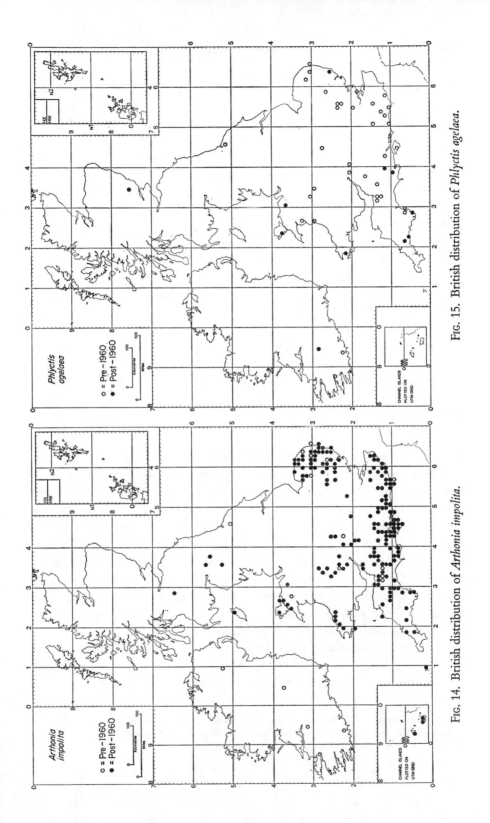

FIG. 15. British distribution of *Phlyctis agelaea*.

FIG. 14. British distribution of *Arthonia impolita*.

areas experiencing less than 160 rain days, and also with warm summer temperatures. Their distributions lie south of the lines of the 59°F (15°C) or 60°F (15·6°C) July mean isotherms (Fig. 4). There is also a correlation with sunshine levels of over about 3·5 h per day (Fig. 5). It would appear that they are inhibited from occurring in the dry, moderately sunny areas of eastern Scotland by the low summer and winter temperatures of that region.

As is expected, all have a southern distribution in Europe and only *Arthonia impolita*, known from east Denmark and southern-most Sweden (Almborn, 1948), *Opegrapha lyncea* and *Phylctis agelaea* are found in Scandinavia. Almborn (1948) placed *P. agelaea* in his *Parmelia acetabulum* group, whose northern limit roughly coincides with that of *Quercus*. *O. lyncea* (L Map 4)* is eastern in the northern part of its range in Britain and its distribution provides a link between the General Southern Group and the South-Eastern Group (5b).

Caloplaca herbidella which may belong here is also recorded from southern Sweden (Magnusson, 1932).

A number of species referred to Group 7 show southern tendencies in that they become much scarcer in Scotland, particularly in the east, which suggests an aversion for the colder climate there. Examples of such species include *Enterographa crassa*, *Parmelia caperata* (Hawksworth *et al.*, 1973: Fig. 6), *P. revoluta* (Hawksworth, 1973a: Fig. 2), *P. subrudecta* (Fig. 20) and *P. tiliacea*.

(*b*) *Extreme Southern*. The distribution characteristic of this group is one concentrated below a line roughly drawn between the estuaries of the River Severn and the River Thames. Some have just a few localities further north, for example, *Anaptychia leucomelaena* (Anglesey), *Opegrapha prosodea* (L Map 5), *Parmelia soredians* (Fig. 16) and *Schismatomma niveum* (South Wales). *Mycoporellum sparsellum* has one station in West Donegal, northern Eire. *Gyalectina carneolutea* has two pre-1960 records for the Isle of Man (Hartley and Wheldon, 1927) and Silverdale in Lancashire, which have been confirmed since its distribution was mapped by Hawksworth (1972b: Fig. 8; 1973b: Fig. 1). *Lithographa dendrographa* (Hawksworth, 1972b: Fig. 7) probably also belongs here even though it has two localities in west Scotland.

The distributions of the species in this group correlate with high summer temperatures, e.g. July mean of over 61°F (16°C), and mild winter temperatures, e.g. January mean of over 40–41°F (4·4–5·0°C) and February mean minimum

---

* The abbreviation "L Map" refers to the "Distribution Maps of Lichens in Britain" series currently appearing in the *Lichenologist*; for Maps 1–7 see **5**, 464–480 (1973) and for Maps 8–17 see **6**, 169–199 (1974).

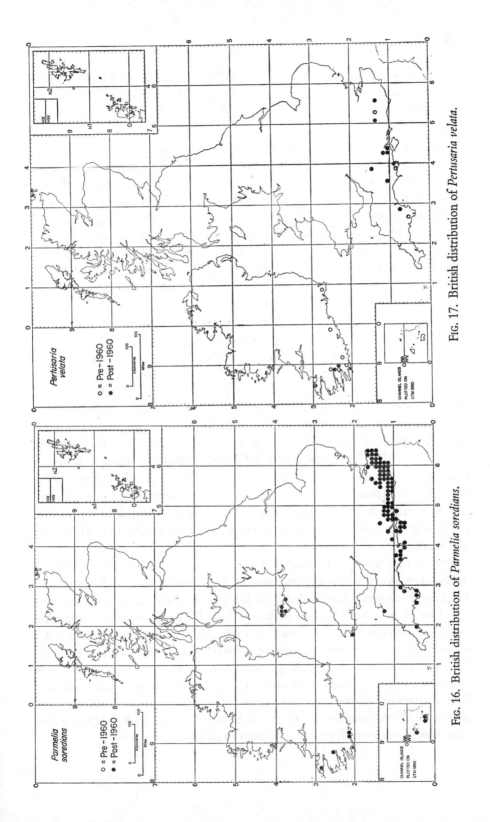

FIG. 17. British distribution of *Pertusaria velata*.

FIG. 16. British distribution of *Parmelia soredians*.

Pertusaria velata

o = Pre-1960
● = Post-1960

Parmelia soredians

o = Pre-1960
● = Post-1960

CHANNEL ISLANDS
PLOTTED ON
UTM GRID

CHANNEL ISLANDS
PLOTTED ON
UTM GRID

of over 35°F (1·7°C). They also correspond to areas receiving high sunshine levels of over 4–4·5 h per day. The distribution of *Parmelia soredians* shows a remarkably close correlation with the isopleth for 4·5 h of sunshine per day (Figs 5 and 16).

Catinaria intermixta, *Mycoporellum sparsellum*, *Pertusaria velata* (Fig. 17) and *Schismatomma niveum* which are woodland species, and *Anaptychia leucomelaena* and *Pseudocyphellaria aurata* which may also occur in open habitats near the sea, appear to require a humid microclimate together with the warm temperature regime found in the south. Therefore, they are perhaps better regarded as an extreme southern component of the Western and Southern Group.

The species favouring more open habitats, such as *Parmelia carporrhizans* (Hawksworth, 1972b: Fig. 10), *P. soredians* (Fig. 16), *Physcia tribacioides* (Hawksworth, 1972b: Fig. 11) and *Teloschistes chrysophthalmus* (Hawksworth *et al.*, 1974: Fig. 6), appear to require frequent periods in a dried-out condition which are possible owing to the high sunshine levels in the southern counties.

All the species in this group are southern in Europe and only two are known from Scandinavia. Almborn (1948) showed *Pertusaria velata* to be very rare in Denmark and southernmost Sweden, and further stated that it is restricted to oceanic and maritime districts, being only known from Schleswig-Holstein and Rügen in North Germany, Portugal and Dalmatia. Many early reports of this species refer to other species, especially *P. hemisphaerica* and *Ochrolechia yasudae*. *Catinaria intermixta* (Nyl.) P. James, which is only correctly known in Britain from the New Forest and the Killarney district of County Kerry, is confined in Scandinavia to southern Sweden and Denmark (Almborn, 1948). Almborn stated that it was distributed in sub-oceanic districts of Europe, that is, South Scandinavia, Germany, France, Britain and Portugal. The possibly conspecific *Catillaria dispersa* is recorded from Germany, Austria, Yugoslavia and Romania.

Physcia tribacioides and *Pseudocyphellaria aurata* are confined to warmer oceanic regions of south-western Europe. *Gyalectina carneolutea* is a southern Eu-Atlantic species (Hawksworth, 1972b: Fig. 8). *Parmelia carporrhizans* was stated by Hawksworth (1972b) to have a broadly mediterranean distribution extending into central and western France, the Alps and the southern U.S.S.R.; its distribution may perhaps be obscured by confusion with the closely related but only doubtfully distinct *P. quercina*. *Tornabenia atlantica*, which is now extinct in Britain (Hawksworth *et al.*, 1974), reached its northern limit here from its mediterranean-coastal distribution. *Opegrapha prosodea* is known from Brittany and southern France, Portugal, Spain, Italy and Malta. *Mycoporellum sparsellum* is unknown from the European mainland.

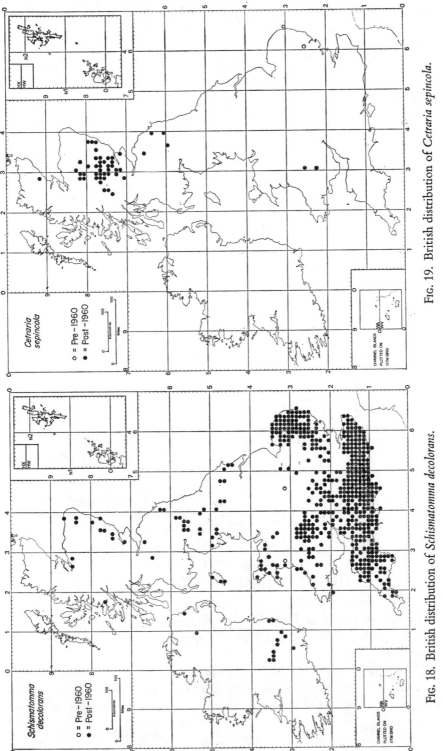

FIG. 19. British distribution of *Cetraria sepincola*.

FIG. 18. British distribution of *Schismatomma decolorans*.

　　　　　　　　　　　B. J. Coppins

Group 5. Eastern

(a) *General Eastern*. Species in this category show a distinct eastern bias in their British distributions and include *Anaptychia ciliaris*, *Caloplaca luteoalba*, *Parmelia acetabulum* and *Pertusaria coccodes*. Distribution maps of the first three species have recently been published (L Maps 2, 3, 6). These species are very rare west

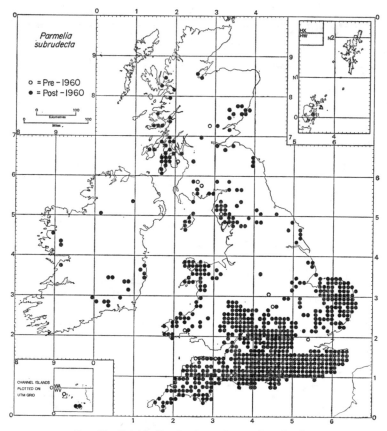

Fig. 20. British distribution of *Parmelia subrudecta*.

of the upland region of Dartmoor in southern England and they are rarely found in Wales. *A. ciliaris* and *C. luteoalba* are also known, though very rarely, in southern Ireland.

This group has affinities with some members of the General Southern Group (e.g. *Arthonia impolita*, *Lecanactis premnea*) in favouring a climate with a relatively low rainfall and its species are mostly confined to areas experiencing less than 140 rain days. However, they differ in being able to tolerate the cooler climate of eastern Scotland.

The distribution of *Schismatomma decolorans* (Fig. 18) shows an intermediate form between that of the General Southern and General Eastern groups.

Almborn (1948: Fig. 4) gave a detailed map of the Scandinavian distribution of *Parmelia acetabulum*, which has a northern limit approximately corresponding to that of *Quercus* except for its absence in south-west Norway. This western limit coincides with the 1000 mm (*c.* 40 in) isohyet for annual rainfall and this is in close agreement with the western limit of this species in Britain. *Anaptychia ciliaris* shows a similar eastern pattern in Scandinavia (Lynge, 1921), although it extends further north.

Both *A. ciliaris* and *P. acetabulum* extend into eastern Europe and south-west Asia, whereas *C. luteoalba* is only recorded from southern and south-western Europe. The distribution of *P. coccodes* is unclear owing to confusion with other taxa, especially *P. coronata*.

The four species included in this group are all typical of moderately eutrophicated and well illuminated bark, although *P. coccodes* is occasionally found in woodlands. Quite a different habitat, namely shaded acid bark or lignum, is favoured by several of the Caliciales (e.g. *Coniocybe hyalinella*, *C. sulphurea*, *Chaenotheca brunneola*, *C. chrysocephala*, *C. trichialis*) that appear to have an eastern distribution in Britain. *Toninia caradocensis* is another species of acid bark and lignum with an apparently marked eastern distribution.

(*b*) *South-Eastern*. This group represents a southern element of the last and includes species mainly confined to England south of Yorkshire. It is characterized by *Cyphelium inquinans* (L Map 8) and in a more extreme form by *Cyphelium notarisii* (L Map 9) and *Catillaria graniformis*. A distribution of the South-Eastern type is also shown by the saxicolous *Caloplaca teicholyta* (Hawksworth *et al.*, 1974, Fig. 10).

The southern distribution of these species correlates with areas of higher summer temperatures with a July mean of over 59°F (15°C) and are mostly concentrated south of the 61°F (16·1°C) isotherm. Their distributions also coincide with areas receiving long periods of sunshine.

European distributions for the two *Cyphelium* species are given by Tibell (1969, 1971). Both almost always occur on lignum and are widespread throughout much of Europe, especially at low altitudes. Both are eastern in Scandinavia: *C. inquinans* is rare in Norway and *C. notarisii* is unknown there.

Two species now extinct in Britain, *Lecania fuscella* and *Lecanora populicola*, may also be referable to this group.

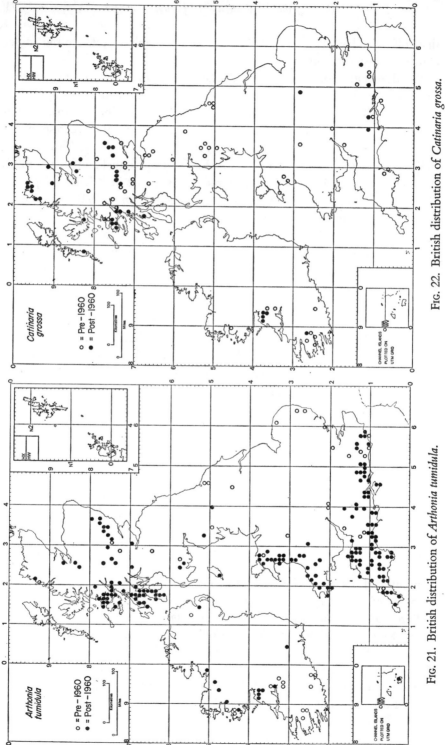

FIG. 22. British distribution of *Catinaria grossa*.

FIG. 21. British distribution of *Arthonia tumidula*.

Group 6. Northern

Apart from some species referred to Group 7, this group contains the boreal, boreal-alpine and generally montane element of our epiphytic lichen flora, although some species (e.g. *Alectoria fuscescens* and *Pseudevernia furfuracea*) are commonly found in lowland areas, especially the rather more "continental" areas of Britain and Europe. Unfortunately the British distributions of many northern species are not so well known as that for the species of the preceding groups. Consequently, it is not yet possible to discuss them as fully.

Most species of the group show an eastern bias in Britain, even those that are fairly widespread, such as *A. fuscescens* (Hawksworth, 1972a: Fig. 9) and *Pseudevernia furfuracea* (Hawksworth and Chapman, 1971). When occurring in western Britain they are usually found at higher altitudes and often in more exposed situations than most species of the Western, and Western and Southern Groups (1 and 2). Other examples of species that have their headquarters in the north, but which are found locally in suitable habitats in other parts of Britain, are *Cetraria sepincola* (Fig. 19). *Lecanora piniperda*, *Lecidea turgidula*, *Mycoblastus sanguinarius*, *Parmeliopsis aleurites* and *P. hyperopta*.

Examples of species confined to Scotland and Northumbria are *Alectoria lanestris*, *A. pseudofuscescens* (?extinct) and *A. vrangiana* (Hawksworth, 1972a: Figs 10, 13, 23), *Cetraria juniperina* (? extinct), *C. pinastri*, *Cladonia botrytes*, *C. cenotea*, *Hypogymnia vittata* (? extinct), *Parmelia septentrionalis* and *Physcia ciliata* (? extinct). *Cavernularia hultenii* is now known from several localities in northern Scotland. Its Scandinavian distribution of West-central Norway and adjacent areas of Sweden is shown by Ahlner (1948: Fig. 9).

Group 7. Widespread

This group contains a large residuum of species which are known, or believed, to occur in most parts of Britain, although not necessarily with equal frequency in each.

The group can be roughly divided on habitat preference:

(*a*) *Ubiquitous species.* These species are found both in woodland and in open situations and include, for example, *Arthonia radiata*, *Arthopyrenia gemmata*, *A. punctiformis*, *Catillaria griffithii*, *Cetraria chlorophylla*, *Cladonia coniocraea*, *Evernia prunastri*, *Hypogymnia physodes*, *Lecanora chlarotera*, *L. expallens*, *Lecidella elaeochroma*, *Lepraria incana*, *Normandina pulchella*, *Opegrapha vulgata*, *O. varia*, *Parmelia caperata*, *P. perlata*, *P. saxatilis*, *P. subrudecta* (Fig. 20), *P. sulcata*, *Pertusaria amara*, *P. pertusa*, *Phlyctis argena*, *Platismatia glauca*, *Ramalina farinacea* and *Usnea subfloridana*.

(b) *General woodland species.* These species are usually associated with deciduous woodlands, both plantations (at least in the more humid areas) and old forest, and include *Arthonia aspersella*, *A. tumidula* (Fig. 21), *Dimerella diluta*, *Enterographa crassa*, *Graphina anguina*, *Graphis elegans*, *G. scripta*, *Gyalideopsis anastomosans*, *Opegrapha sorediifera*, *O. vermicillifera*, *Pertusaria leioplaca*, *Porina chlorotica* var. *carpinea* and *P. leptalea*.

(c) *Old woodland species.* These species are usually associated with old woodland or ancient parkland and include* *Arthonia didyma*, *Catillaria sphaeroides*, *Catinaria grossa* (Fig. 22), *Collema subfurvum*, *Haematomma elatinum*, *Lecidea cinnabarina*, *Lobaria* spp., *Nephroma laevigatum*, *Pachyphiale cornea*, *Pannaria pityrea*, *Peltigera collina*, *Sticta limbata*, *S. sylvatica* and *Thelotrema lepadinum*.

(d) *Species of open habitats.* The species included here prefer trees in well illuminated situations, such as roadsides, hedgerows and open parklands. They include *Bacidea naegelii*, *Caloplaca citrina*, *Candelariella reflexa*, *Lecania cyrtella*, *Parmelia tiliacea* s.l., *Physcia adscendens*, *P. aipolia*, *P. tenella*, *Physconia enteroxantha*, *P. pulverulenta*, *Ramalina fastigiata*, *Rinodina exigua*, *Xanthoria parietina* and *X. polycarpa*.

Many of the "widespread" species show affinities with one or more of Groups 1–6. In many cases these affinities will become clearer following detailed surveys of the little-known areas of northern and south-western Scotland and Ireland.

The affinities between some "widespread" species, with Sub-Atlantic distributions, and the Western and Southern Group (Group 2) has already been mentioned (p. 262).

Affinities with the Western and Southern Group and the General Southern Group (Group 4a) are shown by *Enterographa crassa*, *Parmelia caperata*, *P. perlata* (Hawksworth *et al.*, 1973: Fig. 5), *P. revoluta*, *P. subrudecta* and *P. tiliacea*. In Scotland these species become scarcer northwards, particularly in the east, where they become restricted to sheltered valley sites in old woodland.

Three brown *Parmelia* species, *P. elegantula*, *P. exasperatula* and *P. laciniatula* have eastern tendencies and could be placed in Group 5a. Only *P. exasperatula* is so far known from Ireland (L Maps 12–14). Eastern and perhaps northern tencencies are evident in the distributions of *Cetraria chlorophylla*, *Lecanora chlarona*, *L. intumescens* and *Platismatia glauca*.

The nineteenth-century distribution of *Parmeliopsis ambigua* would place it in the Northern Group (Group 6). However, this species has since become

---

* See also Chapter 11, Tables III–V.

widespread in many parts of Britain as an indirect result of pollution of the air by sulphur dioxide (Hawksworth *et al.*, 1973, 1974).

Probably the most abundant lichen in the British flora is *Lecanora conizaeoides* which is found throughout most of the country, although it is still rare or seemingly absent from some areas of western Britain and the Scottish Highlands where the level of sulphur dioxide in the air is very low (Hawksworth *et al.*, 1973, 1974). The recent records of this species on fence posts and twigs of conifers and birch from these "clean" areas may indicate somewhat higher levels of sulphur dioxide in these regions. However, what is more likely is that some of the vast quantities of diaspores of *L. conizaeoides* produced in moderately polluted areas (with mean winter sulphur dioxide values in the range 55–150 $\mu g/m^3$) are being transported by agencies such as wind, man or migratory birds, and becoming established in habitats where competition from other species is low.

CONCLUSIONS

As previously admitted, the classification given here is a gross simplification of a multidimensional continuum. However, it is hoped that the bringing together of similar distributions into tentative groupings will contribute to our understanding of the factors that control lichen distributions, even if only to enable us to know better what questions to ask.

With our present state of knowledge, the distribution patterns shown by epiphytic lichens in Britain clearly indicate that, all other factors being equal (e.g. distribution of substrates, air pollution, deforestation), it is atmospheric humidity and temperature, and the quantitative and temporal combinations in which they occur, that are the prime factors in determining these patterns.

By comparing the British distributions of lichens with those of angiosperms (Perring and Walters, 1962) it can generally be concluded that, although temperature plays a more or less equal rôle in determining the distributions in both groups, the rôle of atmospheric humidity is perhaps of far greater importance in the case of the lichens. This is not surprising as lichens are far more directly dependent on the atmosphere for their water supply and lack proficient means for controlling their water content.

ACKNOWLEDGEMENTS

I am much indebted to Dr F. Rose and Mr P. W. James for invaluable discussions on lichen distributions and for the use of much unpublished data; to Dr M. R. D. Seaward for providing data from the British Lichen Society's Distribution Map Scheme; to Mr D. M. Henderson for general comments on the manuscript; and to Mr J. Heath and Miss

C. Allen of the Biological Records Centre, Monks Wood, for preparing Figs 6–22. I also wish to thank the following for allowing me to reproduce figures: Dr F. H. Perring and Dr S. M. Walters (Figs 2–4), Dr D. A. Ratcliffe (Fig. 1), and the Controller of Her Majesty's Stationery Office (Fig. 5).

Part of the research in preparation of this paper was undertaken during the tenure of a grant from the Natural Environment Research Council.

REFERENCES

DES ABBAYES, H. (1934). La végétation lichénique du Massif Armoricain, Étude chorologique. *Bull. Soc. Sci. nat. Ouest Fr.*, *sér.* 5, **3** 1–267.

AHLNER, S. (1948). Utbredningstyper bland nordiska barrträdslavar. *Acta phytogeogr. suec.* **22**, i–ix, 1–257.

ALMBORN, O. (1948). Distribution and ecology of some south Scandinavian lichens. *Bot. Notiser, Suppl.* **1** (2), 1–252.

BARKMAN, J. J. (1958). "Phytosociology and Ecology of Cryptogamic Epiphytes." Van Gorcum. Assen.

COPPINS, B. J. (1971). Field Meeting in Brittany. *Lichenologist* **5**, 149–169.

DAHL, E. and KROG, H. (1973). "Macrolichens of Denmark, Finland, Norway and Sweden." Universitetsforlaget, Oslo, Bergen and Tromsø.

DEGELIUS, G. (1935). Das ozeanische Element der Strauch- und Laubflechtenflora von Skandinavien. *Acta phytogeogr. suec.* **7**, i–xii, 1–411.

DU RIETZ, G. E. (1919). Några lavar från det 16:e Skandinavska naturforskarmötets exkursion i Bergens skägård. *Bergens Mus. Årb.* **1918-1919**, 26–31.

HARRIS, G. P. (1971). The ecology of corticolous lichens II. The relationship between physiology and the environment. *J. Ecol.* **59**, 441–452.

HARTLEY, J. W. and WHELDON, J. A. (1927). The lichens of the Isle of Man. *NWest. Nat.* **2**, *Suppl.* 1–38.

HAWKSWORTH, D. L. (1972a). Regional Studies in *Alectoria* (Lichenes) II. The British species. *Lichenologist* **5**, 181–261.

HAWKSWORTH, D. L. (1972b). The natural history of Slapton Ley Nature Reserve IV. Lichens. *Fld Stud.* **3**, 535–578.

HAWKSWORTH, D. L. (1973a). Ecological factors and species delimitation in the lichens. *In* "Taxonomy and Ecology" (V. H. Heywood, ed.), pp. 31–69. Academic Press, London and New York.

HAWKSWORTH, D. L. (1973b). Some advances in the study of lichens since the time of E. M. Holmes. *J. Linn. Soc. (Bot.)* **67**, 3–31.

HAWKSWORTH, D. L. and CHAPMAN, D. S. (1971). *Pseudevernia furfuracea* (L.) Zopf and its chemical races in the British Isles. *Lichenologist* **5**, 51–58.

HAWKSWORTH, D. L., COPPINS, B. J. and ROSE, F. (1974). Changes in the British lichen flora. *In* "The Changing Flora and Fauna of Britain" (D. L. Hawksworth, ed.), pp. 47–78. Academic Press, London and New York.

HAWKSWORTH, D. L., ROSE, F. and COPPINS, B. J. (1973). Changes in the lichen flora of England and Wales attributable to pollution of the air by sulphur dioxide. *In* "Air Pollution and Lichens" (B. W. Ferry, M. S. Baddeley and D. L. Hawksworth, eds), pp. 330–367. University of London, Athlone Press, London.

HENSSEN, A. (1963). Eine Revision der Flechtenfamilien Lichinaceae und Ephebaceae. *Symb. bot. upsal.* **18** (1), 1–123.

IMSHAUG, H. A. and HARRIS, R. C. (1969). *Parmentaria chilensis* Fée. *Lichenologist* **4**, 77–82.

JØRGENSEN, P. M. (1973). On some *Leptogium* species with short *Mallotium* hairs. *Svensk bot. Tidskr.* **67**, 53–58.

JØRGENSEN, P. M. and RYVARDEN, L. (1970). Contribution to the lichen flora of Norway. *Årbok Univ. Bergen, mat.-nat. ser.* **1969** (10), 1–24.

KERSHAW, K. A. (1972). The relationship between moisture content and net assimilation rate of lichen thalli and its ecological significance. *Can. J. Bot.* **50**, 543–555.

LEIGHTON, W. A. (1879). "The Lichen Flora of Great Britain, Ireland and the Channel Islands", (3rd edition.). Privately published, Shrewsbury.

LYNGE, B. (1921). Studies on the lichen flora of Norway. *Skr. Norske Vid.-Akad. Oslo, mat.-nat. kl.* **1921** (7), 1–252.

MAGNUSSON, A. H. (1932). New or interesting Swedish lichens VII. *Bot. Notiser* **1932**, 417–444.

MITCHELL, M. E. (1961). L'Elément eu-océanic dans la flore lichénique du sud-ouest de l'Irlande. *Revta Biol.* **2**, 177–256.

MITCHELL, M. E. (1970). Contribution à la lichénologie irlandaise. III. *Bull. Soc. scient. Bretagne* **45**, 81–84.

MUDD, W. (1861). "A Manual of British Lichens." Privately published, Darlington.

OXNER, A. N. (1937). "Viznačnik lišajnikiv URSR." Akademii Nauk USSR, Kiev.

OZENDA, P. and CLAUZADE, G. (1970). "Les Lichens. Étude Biologique et Flore Illustrée." Masson, Paris.

PERRING, F. H. and WALTERS, S. M. (1962). "Atlas of the British Flora." Nelson, London and Edinburgh.

PIŠÚT, I. (1970). Doplnky k poznania lišajníkov Slovenska 6. *Acta Fac. Rerum. nat. Mus. nat. Slov., Bratisl.* **16**, 31–40.

PIŠÚT, I. (1971). Interessante Flechtenfunde aus Mittel-und südosteuropa. *Fragm. balcan.* **8**, 165–169.

POELT, J. (1969). "Bestimmungsschlüssel europäischer Flechten." Cramer, Lehre.

RAMAUT, J. L. (1961). Contribution à l'étude chromatographique de quelques *Parmelia* de la section *Amphigymnia*, sous-section *Subglaucescens* Vain. *Revue bryol. lichén.* **30**, 131–134.

RATCLIFFE, D. A. (1968). An ecological account of Atlantic bryophytes in the British Isles. *New Phytol.* **67**, 365–439.

ROSE, F. (1974). The Epiphytes of Oak. *In* "The British Oak; its History and Natural History" (M. G. Morris and F. H. Perring, eds), pp. 250–273. Classey, Faringdon.

ROSE, F. and JAMES, P. W. (1974). Regional studies on the British lichen flora I. The corticolous and lignicolous species of the New Forest, Hampshire. *Lichenologist* **6**, 1–72.

RYDZAK, J. (1961). Tree lichens in the forest communities of the Białowieża National Park. *Annls Univ. Mariae Curie-Skłodowska,* **C, 16**, 17–48.

SALISBURY, G. (1972). *Thelotrema* Ach. sect *Thelotrema*. 1. The *T. lepadinum* group. *Lichenologist* **5**, 262–274.

SALISBURY, G. (1975). *Thelotrema monosporum* Nyl. in Britain. *Lichenologist* **7**, 59–61.

SCHAUER, T. (1965). Ozeanische Flechten im Nordalpenraum. *Port. Acta Biol.,* B **8**, 17–226.

K

Seaward, M. R. D. (1973). Distribution maps of lichens in Britain. *Lichenologist* **5**, 464–466.

Smith, A. L. (1918, 1926). "A Monograph of the British Lichens" (2nd edition), 2 vols. British Museum (Natural History), London.

Swinscow, T. D. V. (1962). Pyrenocarpus lichens: 3. The genus *Porina* in the British Isles. *Lichenologist* **2**, 6–56.

Tavares, C. N. (1945). Contribuiçao para o estudo das Parmeliáceas Portuguesas. *Port. Acta Biol.*, **B**, **1**, 1–211.

Tibell, L. (1969). The genus *Cyphelium* in northern Europe. *Svensk bot. Tidskr.* **63**, 465–485.

Tibell, L. (1971). The genus *Cyphelium* in Europe. *Svensk bot. Tidskr.* **65**, 138–164.

Vězda, A. and Vivant, J. (1972). Lichens des Pyrénées occidentales nouveaux pour la flore française. *Bull. Soc. bot. Fr.* **119**, 243–252.

Watson, W. (1953). "Census Catalogue of British Lichens." British Mycological Society, London.

Note added in proof

During 1974 and 1975 a number of records have been made which extend the ranges of some of the species discussed in this Chapter. Brief details are given here together with their 10 km square national grid references.

Biatorella ochrophora: 37/51 (Fife: St Andrews).

Catinaria grossa: 20/87 (S. Devon: Chudleigh Caves).

Cetrelia cetrarioides: 92/88 (S. Tipperary: Glengara Wood).

Cyphelium inquinans: 28/21 (E. Inverness: Glen Moriston), 28/22 (E. Inverness: Guisachan Forest); these records are from relict native pinewoods.

Graphina ruiziana: 18/80 (W. Inverness: Barrisdale).

Parmelia soredians: 36/68 (E. Lothian: Tynninghame).

Phlyctis agelaea: 20/76 (S. Devon: Hembury Woods), 20/87 (S. Devon: Chudleigh Caves), 28/41 (E. Inverness: Glen Moriston), 28/95 (Moray: Darnaway Forest).

Porina hibernica: 41/21 (S. Hampshire: New Forest), 41/30 (S. Hampshire: New Forest).

11 | Lichenological Indicators of Age and Environmental Continuity in Woodlands

F. ROSE

Department of Geography, Kings College, University of London, England

Abstract: An analysis of the corticolous and terricolous lichen flora of a large number of British deciduous woodlands has indicated a wide variation in the number of taxa present per square kilometre, even in relatively unpolluted areas. The data from woodlands known to be very old, such as those of the New Forest, Hampshire, give totals of the order of 120–150 taxa per square kilometre, while woodlands known to be of recent origin often have totals of 40 or less.

It is suggested that the primeval mixed oak forests of the British Lowlands of Atlantic times probably had of the order of 150 taxa per square kilometre.

A comparison of the species present in British woodlands suggests that a number of lichens are relict species of ancient woodland areas which have been little disturbed over a very long period. A number of the species of this type have been used in a simple formula to calculate a Revised Index of Ecological Continuity (RIEC), on the basis of the percentage of these taxa occurring in each site. By reference to historical data where these are available, an attempt is made to show that this Index can give a useful method of calculating, approximately, the continuity of high forest cover in a woodland site.

INTRODUCTION

Two questions can be posed about British woodlands: (*a*) do any truly primary woodland relics (that is continuously existing since prehistoric times) remain anywhere in Britain, and (*b*) if so, how might one be able to recognize them?

For about 7 years the author has studied the corticolous, lignicolous and terricolous lichen floras of a wide range of forests and woodlands all over

Systematics Association Special Volume No. 8, "Lichenology: Progress and Problems", edited by D. H. Brown, D. L. Hawksworth and R. H. Bailey, 1976, pp. 279–307. Academic Press, London and New York.

Britain and also, for comparative purposes, those of forests in various parts of France. The result of these studies has been to make it evident that there is a wide variation in the lichen floras of British woodlands, particularly in the following respects: (a) species density, i.e. the number of species per unit area of woodland; (b) species abundance, in terms of both frequency of occurrence per total number of trees and of cover values of individual species; and (c) the particular species present in a woodland.

For some woodland sites historical documentation, of varying reliability, is available on the degree of continuity of tree cover and on the type of past management; for others this information is either difficult to obtain or does not exist. It has become clear that sites with good evidence of long continuity of woodland cover, particularly of the canopy of mature standard trees, differ much, with respect to the variables (a) and (c) above, from those woodlands that are known to be of relatively recent origin or with a history of major disturbance of canopy continuity.

Obviously there are many other factors that may affect the lichen vegetation of a wood. These include (a) air pollution levels; (b) regional climatic differences; (c) the nature of the phorophytes; and (d) site topography (which in turn will influence microclimate). There is now an extensive literature of the effects of air pollution on lichens, which is usefully summarized by Ferry et al. (1973). Clearly it is unwise, and likely to be misleading, to attempt to study any relationship between past management and present lichen populations in woodlands in areas where air pollution levels are high unless the effect of this latter factor is given full consideration. However, even in heavily polluted areas, such as Epping Forest (Essex), Hawksworth et al. (1973: p. 331) indicated that documentation of the lichen flora *before* air pollution became severe may provide information relevant to the present topic. Regional climatic differences affect to some extent the species present even in lowland woodlands in different parts of Great Britain but the major effect appears to be on the abundance of particular species rather than on what is present or absent. The question of phorophyte species appears to have only marginal significance if only deciduous woodlands dominated by oak (*Quercus*) and beech (*Fagus*), with some elm (*Ulmus*) and ash (*Fraxinus*) present, are considered; other tree species tend to bring in other lichen epiphytes. In areas of low air pollution, the topography of the site seems to be of minor importance within lowland areas, though, where pollution levels are moderate, it must be borne in mind that sheltered valleys or ravines tend to have richer lichen floras than exposed sites. Upland woodlands with very high rainfall have been excluded from this analysis, as their floras show considerable differences from lowland sites (see p. 304).

SPECIES DENSITY PER UNIT AREA OF WOODLAND SITES

1. Choice of Sites

In this section, the data collected in a large number of British lowland woodland sites are examined in terms of the number of lichen taxa per unit area, with a working unit of approximately 1 km^2. The object of this is to ascertain if any pattern emerges from such an analysis in terms of a relationship between the density of taxa per unit area and the age, and past continuity of the canopy, of woodlands in cases where information is available. Such an analysis may provide some information on the probable content and density of primeval British hardwood forests.

Table I gives a list of 102 lowland hardwood woodland sites in various parts of Great Britain. Except where stated otherwise, the sites are all of the order of

TABLE I (see text for explanation)

Site (Fig. 1)	Species density per km^2 (or less)	Index of Ecological Continuity (Rose, 1974)	Revised Index of Ecological Continuity
Section (a) Sites containing old high forest of oak with glades; either medieval parklands or former Royal Forests or Chases: many (indicated by ^{a}) are definitely known to have contained woodland in medieval times			
(i) South-west England			
1. Boconnoc Park, Cornwall^a	(188) 184	100	100
2. Trebartha Park, Cornwall	153	75	100
3. The Dizzard, Cornwall	128	55	85
4. Brownsham Woods, Devon	98	70	100
5. Arlington Park, Devon	121	60	75
6. Walkham Valley, Devon	116	60	75
7. Holne Chase, Devon^a	106	70	100
8. Becka Falls, Devon	78	65	95
9. Horner Combe, Somerset^a	(150) 146	70	100
10. Mells Park, Somerset	(115) 101	50	55
11. Melbury Park, Dorset^a	(173) 162	85	100
12. Lulworth Park, Dorset	(115) 105	45	50
13. Cranborne Chase, Dorset and Wiltshire^a	(148) 133	60	85
14. Longleat Park, Wiltshire^a	(142) 133	75	95
15. Great Ridge Wood, Wiltshire	98	45	55
16. Savernake Forest, Wiltshire^a	112 (c. 6 km^2)	50	60

TABLE I—*continued*

Site (Fig. 1)	Species density per km² (or less)	Index of Ecological Continuity (Rose, 1974)	Revised Index of Ecological Continuity
(ii) New Forest, Hampshire[a]	259	100	100
	(*c.* 36 km²)		
17. Mark Ash Wood	160	95	100
18. Vinney Ridge	116	80	100
19. Busketts Wood	159	90	100
20. Shave Wood	121	90	100
21. Wood Crates	135	70	80
22. Rushpole Wood	123	80	90
23. Great Wood	113	85	100
24. Bramshaw Wood	122	95	100
25. Bignell Wood	117	95	100
26. Lucas Castle Wood	132	95	100
(iii) South-east England			
27. Up Park, Sussex[a]	131	45	65
28. Parham Park, Sussex[a]	(162) 149	50	60
29. St Leonards Forest, Sussex[a]			
(i) 1974	107	55	65
(ii) pre-1805	143	95	100
30. Eridge Park, Sussex[a]			
(i) 1975	(186) 177	70	85
(ii) 1842	(200) 191	85	100
31. Ashburnham Park, Sussex[a]	(151) 145	60	70
(iv) East Anglia			
32. Staverton Park, Suffolk[a]	65	30	35
33. Sotterley Park, Suffolk[a]	(92) 83	15	20
(v) Midland England			
34. Wychwood Forest, Oxfordshire[a]	86	30	35
35. Brampton Bryan Park, Herefordshire[a]	(104) 94	50	50
36. Downton Castle Park, Herefordshire	67	35	40
(vi) Wales			
37. Gwaun Valley, Pembrokeshire	117	55	65
38. Coedmore Woods, Cardigan	(117) 111	55	65

Tᴀʙʟᴇ I—*continued*

Site (Fig. 1)	Species density per km² (or less)	Index of Ecological Continuity (Rose, 1974)	Revised Index of Ecological Continuity
39. Coed Crafnant, Merionethshire	101	60	100
40. Swallow Falls, Bettws-y-Coed, Caernarvonshire	74	60	85
41. Coed Hafod, Denbighshire	112	50	65
42. Glan Conwy, Denbighshire	62	50	70
43. Coed Felinrhyd, Merionethshire	71	50	75
44. Coedydd Aber, Caernarvonshire	90	55	70
(vii) Northern England			
45. Shipley Wood, Durham			
(i) 1974	108	50	55
(ii) 1805	114	60	80
46. Great Wood, Borrowdale, Cumberland	102	65	95
47. Low Stile Wood, Borrowdale, Cumberland[a]	100	65	100
48. Gowbarrow, Cumberland	82	45	70
49. Naddle Low Forest, Westmorland[a]	73	25	40
50. Witherslack Woods, Westmorland	56	40	55
(viii) Apparently natural oakwoods in western Scotland			
51. Camasine Woods, Loch Sunart	177	70	100
52. Ellary Woods, Loch Caolisport	127	65	100
53. Taynish Woods	98	40	65
54. Avinagillan, Loch Tarbert	91	55	90
55. Coire Buidhe, Loch Sunart	87	40	60
56. Loch na Droma Buidhe	77	40	70
57. Dorlin, Loch Moidart	71	40	70
58. Meall à Bhroin, Loch Sunart	100	50	80
59. Rubha aird Druimnich, L. Sunart	92	40	70
60. Glasdrum, Loch Creran	85	30	65
61. Ard Trilleachan, Loch Etive	45	25	50
62. Bonawe Woods, Loch Etive	87	55	90
63. Dalnasheen, Appin	67	45	75
64. Kinauchdrach, Jura	61	40	55
65. Doire Dhonn, Jura	73	30	65

TABLE I—*continued*

Site (Fig. 1)	Species density per km² (or less)	Index of Ecological Continuity (Rose, 1974)	Revised Index of Ecological Continuity
66. Ardmore, Islay	60	30	55
67. Coill'a'Chorra Ghiortein, Islay	55	30	45

Section (b) Sites with naturally regenerated mature forest on former common grazing land

68. Ebernoe Common, Sussex	70	35	40
69. The Mens, Sussex	75	25	40

Section (c) Mature hardwood plantations within the New Forest, Hampshire

70. Pitts Wood Enclosure	54	30	40
71. Brockishill Enclosure	80	35	50
72. Pond Head Enclosure	31	20	30
73. South Bentley Enclosure	48	30	40

Section (d) Mature high forests of oak, etc., known to have been clear-felled and replanted within the last 200 years

74. Dalegarth Woods, Eskdale, Cumberland	68	15	20
75. Nagshead Enclosure, Forest of Dean, Gloucestershire	16	0	0
76. Manesty Park, Derwent Water, Cumberland	20	0	0
77. Banneriggs, Grasmere, Westmorland	15	0	0
78. Rob Ragg Wood, Elterwater, Westmorland	12	0	0
79. Skelwith Force Wood, Lancashire	18	0	0
80. Claife Woods, Windermere, Lancashire	13	0	0
81. Low Coppice Wood, Grizedale, Lancashire	16	0	0
82. Anna's Wood, Coniston Water, Lancashire	13	0	0
83. Holme Wood, Loweswater, Cumberland	36	0	0
84. St Michael Penkevil, Cornwall	54	10	20
85. Cranford Cross, North Devon	24	5	5
86. Five Lords Wood, Quantocks, Somerset	29	10	10

TABLE I—*continued*

Site (Fig. 1)	Species density per km² (or less)	Index of Ecological Continuity (Rose, 1974)	Revised Index of Ecological Continuity
87. Holford Combe, Quantocks, Somerset	50	20	25
88. Pamber Forest, Silchester, Hampshire	29	0	0

Section (e) Old coppice woodlands with oak or ash standards

89. East Dean Park Wood, Sussex[ab]	70	40	40
90. Combwell Wood, Goudhurst,[ab] Kent[a]	63	25	25
91. Brenchley Wood, Kent	44	5	5
92. Ham Street Woods, Kent	45	5	5
93. Near Loxwood, Sussex	33	0	0
94. Napwood, Frant, Sussex	45	5	5
95. Maplehurst Wood, Westfield, Sussex	54	5	10
96. Marline Wood, Hastings, Sussex	54	15	15
97. Foxley Wood, Norfolk[a]	15	5	5
98. Hayley Wood, Cambridgeshire[a]	35	5	5
99. Hintlesham Great Wood, Suffolk[a]	10	0	0
100. Felsham Hall Wood, Suffolk[a]	13	0	0
101. Park Coppice, Coniston, Lancashire	16	0	5

Section (f) Old Royal forest, little altered in structure but now much affected by air pollution

102. Epping Forest, Essex			
(i) 1974	38	5	10
(ii) early 19th century	(130) 109	25	30

[b] These sites in section (e) contain areas of old trees in open high forest, and site 89 is known to have been a medieval deer park.

1 km², or sometimes rather less, in area. Some sites within parklands contain free-standing trees with bark hypertrophicated (eutrophicated) by the excreta of grazing animals. Such trees tend to carry lichens which are not characteristic of woodland or forest habitats, but are species of the Xanthorion alliance, particularly species of the genera *Buellia*, *Caloplaca*, *Physcia*, *Ramalina* and *Xanthoria*. In these sites the total number of species is given in parentheses, followed by a corrected figure, without parentheses, obtained by subtracting the total of species present that are normally confined to hypertrophicated bark

(e.g. *Anaptychia cilaris, Caloplaca citrina, Lecanora dispersa* and the species of *Physcia, Physconia,* and *Physciopsis*). The object of this correction is to make the data from the sites more comparable ecologically.

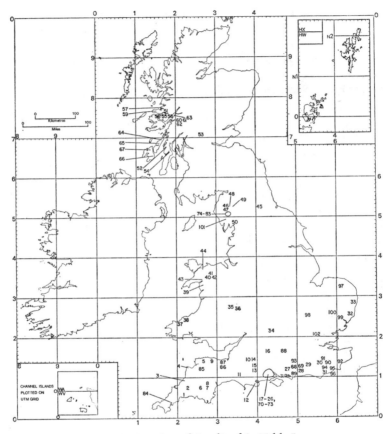

Fig. 1. Location of sites listed in Table I.

The sites in Table I section (a) are all mature to old stands of high forest, often within known ancient deer parklands or Royal Forests of medieval origin, and in many cases (ª) with a long documented history of existence as woodland. They have not been managed as coppice woodland either now or within the last few centuries. Table I section (b) includes two Sussex woodlands that today have a rather similar structure to the sites in section (a) (unmanaged irregular high forest type). These are known to be on old common land which has almost certainly always had some woodland cover. Historical evidence, however, indicates that these two sites have had at least one phase in the last few centuries when they were largely open pastureland which has since regenerated naturally to woodland (R. Tittensor, unpublished data). Table I section (c) consists of mature hardwood plantations within the New Forest. Table I section (d) contains other woodlands known to have been clear felled and replanted within the last 200 years and in several cases, within the last 100 years. Table I section (e) consists of old woodlands that have long been managed as coppice with standards.

Sites with significant air pollution levels (i.e. mean winter levels above *c*. 50 μg SO_2/m^3) have been excluded from the table, as far as available data allow, in order to eliminate this variable. One exception (Epping Forest, Essex), however, is given in Table I section (f) because this site is unique in having very extensive documentation of its past lichen flora (see Hawksworth *et al.*, 1973). Present and past data are given and it is clear that in the early nineteenth century this forest had a species density similar to several sites in section (a). For four other sites (nos 29, 30, 45 and 102) where documentation proved available past data are also given.

2. Discussion of Species Density Data

Comparison of the total numbers of species in the majority of sites in Table I section (a) with those in sections (b), (c), (d) and (e), which represent woodland sites of varied levels of human management or modification (increasing from b to e), prove very significant. Nearly all the sites in section (a), except for those in eastern and central England and for some of those in Scotland, have over 100 lichen species per km² and many have considerably more than this number.

It is now difficult to find any areas of old woodland in eastern and central England, even in old parkland sites, that have not suffered some degree of drying out or "opening up". General environmental modification has been so severe that few extensive hardwood areas with high lichen species density now remain.

In northern England and Scotland a number of species common to southern English forests become rarer or absent through what appear to be major climatic factors. In Table I sections (b) and (c) species density falls considerably to between 30 and 80 species per km², while in (d) and (e), though there is still a range up to 70 species, most sites have very low density figures within the 13–50 range in whatever part of the country they lie.

These data appear to suggest (*a*) that there is a strong negative correlation between lichen species density and the degree of human modification, if one excludes the "invasive" species of hypertrophicated bark that have colonized sites formerly heavily pastured; and (*b*) that the density of the lichen flora of the *primeval* British Lowland hardwood forests of Atlantic times was of the order of at least 120–150 species/km². If the evidence from this very comprehensive sample of what are thought (in section (a)) to be the least altered relics of our ancient forests has any value, this is certainly what it seems to suggest. Indeed, allowing for the various changes that are known to have occurred in all British woodlands (Hawksworth *et al.*, 1974), the original density figures may have been even greater than this, but no further evidence is available.

A similar analysis of bryophyte epiphytes in British woodlands was also attempted but, partly because there are far fewer bryophyte epiphytes overall in the lowland deciduous woodlands of Britain (except in extremely Atlantic sites) and partly because comparatively few of these bryophytes show any real correlation with continuity of forest environments, a comparison between lichens and this group of plants is omitted from this paper, though it has been considered by Rose (1974).

THE CONCEPT OF "OLD-FOREST INDICATOR" LICHEN SPECIES AND THE REVISED INDEX OF ECOLOGICAL CONTINUITY

1. Indicator and Relict Species

Purely numerical comparisons of total epiphytic lichen floras and their density are useful but they tell us little of the character of the flora itself. A very wide range of ecological groupings of lichens can occur in even a small woodland. Different species may be present for very different reasons. The question is whether any group, or groups, of indicator species can be detected that are particularly sensitive to changes in the forest environment with time and whose presence or absence may indicate continuity or disturbance of this environment, thereby providing evidence as to whether a woodland that *looks* old really *is* old and has been relatively little altered with time.

Study of several hundred lowland deciduous woodlands in Britain and France in areas of low air pollution has revealed two facts clearly.

1. Certain species of lichen epiphytes occur in all, or nearly all, woodlands containing standard hardwood trees, whether these are old high forest, coppice-with-standards, or areas of mature oak plantations. Table II lists a number of examples of such species which are also, in many cases, found equally commonly on oaks (or ash trees) in open parkland, pasture, or minor-road verge situations, though nowadays less often by pollution-producing major roads.

2. A number of other lichen species are normally only found in mature, or old, stands of oak or mixed oak high forest.

In many cases it is very difficult to establish the past history of such sites in detail but wherever it has been possible to do so, it has become clear that they are very old, probably primary, woodlands, with strong evidence of some

TABLE II. Some lichen epiphytes common and general on *Quercus* (and on *Fraxinus*) in both woodland and in more open situations in *unpolluted* areas of Lowland Britain

Calicium viride	*P. glabratula* subsp. *glabratula*
Catillaria griffithii	*P. perlata*
Cetraria chlorophylla	*P. revoluta*[a]
Evernia prunastri	*P. saxatilis*
Graphis elegans	*P. subrudecta*[a]
G. scripta	*P. sulcata*
Hypogymnia physodes	*Pertusaria amara*
Lecanora chlarotera	*P. hemisphaerica*
L. expallens	*P. hymenea*
Lecidea quernea	*P. pertusa*
L. scalaris	*Phlyctis argena*
Lepraria incana	*Platismatia glauca*
Ochrolechia androgyna	*Ramalina farinacea*
O. yasudae	*Schismatomma decolorans*[a]
Parmelia caperata[a]	*Usnea subfloridana*

[a] Species becoming rarer or absent in northern and eastern Scotland.

continuity of a high tree canopy (as opposed to coppice) since at least medieval times. Table III lists a number of examples of these species.

Such sites are found in the following types of terrain:

(a) in the "Ancient and Ornamental woodlands" of the New Forest; the only old Royal hunting forest that has remained, in part, open and free from active silvicultural management for some hundreds of years (see Tubbs, 1964; Rose and James, 1974). The New Forest, today, is unique in character in western Europe, and is probably closer in its ecological features to the primeval Atlantic forest than any other area of comparable size west of Poland;

(b) in more fragmentary form in the relics of other old Royal forests, or chases owned by the nobility, such as Savernake Forest, Exmoor Forest, Wychwood Forest, Cranborne Chase, and (until last century) St Leonard's Forest;

(c) in the wooded parts of deer parks of medieval origin;

(d) in remoter parts of western and northern Britain on escarpments and on steep-sided gorges where active forestry has been minimal for reasons of remoteness or difficulty of the terrain.

Such sites are not found in parks created in the eighteenth-century phase of "landscape gardening" or later.

There is good evidence (Brandon, 1963) that many medieval deer parks and chases were formed from relict areas of the primeval wilderness still existing at

TABLE III. Some lichen epiphytes that appear to be "faithful" to old hardwood forests in lowland Britain

Arthonia didyma	*Pannaria mediterranea*
A. stellaris	*P. pityrea*
Arthopyrenia cinereopruinosa	*P. rubiginosa*
Biatorella ochrophora	*P. sampaiana*
Bombyliospora pachycarpa	*Parmelia arnoldii*
Caloplaca herbidella	*P. crinita*
Catillaria atropurpurea	*P. horrescens*
C. pulverea	*P. reddenda*
C. sphaeroides	*Parmeliella corallinoides*
Chaenotheca brunneola	*P. plumbea*
Dimerella lutea	*Peltigera collina*
Enterographa crassa	*P. horizontalis*
Haematomma elatinum	*Pertusaria velata*
Lecanactis premnea	*Phyllopsora* sp.
Lecidea cinnabarina	*Porina coralloidea*
Leptogium burgessii	*P. hibernica*
L. teretiusculum	*P. leptalea*
Lobaria amplissima	*Pseudocyphellaria crocata*
L. laetevirens	*P. thouarsii*
L. pulmonaria	*Pyrenula nitida*
L. scrobiculata	*P. nitidella*
Lopadium pezizoideum	*Rinodina isidioides*
Nephroma laevigatum	*Schismatomma niveum*
N. parile	*Stenocybe septata*
Ochrolechia inversa	*Sticta limbata*
Opegrapha lyncea	*S. sylvatica*
Pachyphiale cornea	*Thelopsis rubella*
Thelotrema lepadinum	

the time of their formation, which contained at least some fragments of primary forest. Most coppice-with-standards woodlands, although they may be primary in the sense that some sort of woodland cover has always existed on the site, have undergone continual drastic environmental modification in the coppicing cycle.

The species in Table III can therefore be regarded as "faithful" to the type of woodland sites listed in (*a*) to (*d*) above, though by no means all of them are of constant occurrence. Some of them show a degree of geographical restriction within Britain, presumably due to climatic tolerance limits; others are very widespread though of varied frequency.

It would appear that such species may in fact be relics of the ancient forest epiphyte flora (similar phenomena can be seen amongst the Coleoptera and Hemiptera in the forest fauna) and it is suggested that the presence of a sufficient number of them in a site could be taken as evidence of continuity of the ancient forest canopy at that site. They could, therefore, be regarded as "indicator species" in two senses: (*a*) ecological indicators of the existence of a particular type of forest environment at the present time; (*b*) historical indicators of lack of environmental change, within certain critical limits, over a long period of time. As relict species, it is suggested that they can still maintain themselves in their present, now isolated, habitats but are no longer able to colonize new habitats at a distance.

Within the more or less continuous blanket of primeval forest cover that existed in many parts of Britain even into early medieval times, the dispersal and colonization of epiphytic lichens, and bryophytes, was probably relatively straightforward, and they were probably more or less ubiquitous in their appropriate ecological niches. From early medieval times to the present day, however, the old oak woodlands became more and more fragmented and most of those that remained became increasingly modified by various forms of management. The isolated scattered fragments that remained least modified in such sites as those discussed above would have provided habitats where many forest epiphytes could have survived but, as they were now surrounded by unfavourable terrain for colonization, re-establishment of such species in new plantations, or in felled woodlands showing regeneration, would have become increasingly difficult. The general drying-out of the landscape due to agricultural practice in the last few hundred years must have played a part in this. It is noteworthy in this connection that some epiphytic lichens and bryophytes that were known to fruit freely up to the early or mid nineteenth century in many parts of lowland Britain (there is abundant evidence of this from both the literature and old herbarium specimens) are now very rarely found fertile in lowland Britain, even in relatively unaltered old woodlands, though some still fruit freely in areas such as the humid west of Scotland (see Hawksworth *et al.*, 1973, 1974).

Thus we have the concept of many of the forest epiphytes today being not only "indicator species" but also of being "relict species".

2. Revised Index of Ecological Continuity

A short list of 20 of the species, listed in Table III, have been previously used
(Rose, 1974) to calculate an "Index of Ecological Continuity", by calculating
the percentage of these 20 selected species that occur in a number of British
woodland sites (see Table IV). In the present paper a larger number of these

TABLE IV. Lichens used in Table I to calculate the
original Index of Ecological Continuity (Rose,
1974)

Arthonia didyma[a]	Ochrolechia inversa[a]
Catillaria atropurpurea[a]	Opegrapha lyncea[b]
C. sphaeroides[a]	Pachyphiale cornea[a]
Dimerella lutea[a]	Pannaria pityrea[a]
Enterographa crassa[b]	Parmelia crinita[a]
Haemotomma elatinum	P. reddenda[a]
Lecanactis premnea[b]	Peltigera horizontalis[a]
Lecidea cinnabarina	Sticta limbata[a]
Lobaria laetevirens[a] or	Thelopsis rubella[a]
L. amplissima[a] or	Thelotrema lepadinum
L. scrobiculata[a]	
L. pulmonaria[a]	

[a] No longer present in midland and eastern England
as a result of both environmental modification
and air pollution.
[b] Very rare, or absent, in the northern half of Britain.

species (30) have been taken to calculate a Revised Index of Ecological Continuity (RIEC) for a larger number (102) of British deciduous woodland sites.
There are problems of selection in devising such an Index. It is desirable to keep
it as simple as possible. Some of the species in Table III are much more localized
geographically than others, as stated above (p. 287); they also vary much in
their frequency. It was decided, for the revised Index, to try to select 30 species
that are of reasonably frequent occurrence in those woodland sites that are
known to be ancient in England and Wales. This meant that some species that
perhaps may be very significant in Scotland, such as *Parmeliella plumbea*,
Pseudocyphellaria spp., and *Pannaria rubiginosa*, have not been included because
they are too rare now in England and Wales to be of practical value in calculation of an Index. On the other hand, in an attempt to provide an Index comprehensive enough to be of some value in central and eastern England, it was
necessary to include such species as *Enterographa crassa*★, *Opegrapha lyncea*† and
Lecanactis premnea† that are either rare in Scotland (★) or so far unknown there

(†). Ideally perhaps separate regional Indices should be used for both Scotland and for eastern England, but this procedure at once introduces complexity and also the difficulty of drawing necessarily arbitrary regional boundaries.

The Revised Index has been based upon the percentage occurrence of up to a maximum of 20 species, out of a total list of 30 species. Occurrences at any one site of more than 20 of the species are ignored. In this way it was hoped to provide a large enough sample of species to give sufficient representation of what

TABLE V. Species used in Table I to calculate the Revised
Index of Ecological Continuity

Arthonia didyma	*Pachyphiale cornea*
Arthopyrenia cinereopruinosa	*Pannaria pityrea*
Catillaria atropurpurea	*Parmelia crinita*
C. sphaeroides	*P. reddenda*
Dimerella lutea	*Parmeliella corallinoides*
Enterographa crassa[a]	*Peltigera collina*
Haematomma elatinum[a]	*P. horizontalis*
Lecanactis premnea[a]	*Porina leptalea*
Lecidea cinnabarina[a]	*Pyrenula nitida*[a]
Lobaria amplissima	*Rinodina isidioides*
L. laetevirens	*Stenocybe septata*[a]
L. pulmonaria	*Sticta limbata*
L. scrobiculata	*S. sylvatica*
Nephroma laevigatum	*Thelopsis rubella*
Opegrapha lyncea[a]	*Thelotrema lepadinum*[a]

[a] Still existing in eastern England.

appear to be indicator species over different regions of Britain. Table V lists these 30 species; only one site contains all of them (Boconnoc Park, Cornwall).

The Revised Index of Ecological Continuity was calculated, for a given site, as follows:

$$\text{RIEC} = \frac{n}{20} \times 100$$

where *n* is the number of species present (up to a maximum of 20) at the site out of the list of 30 species in Table V.

Since the Revised Index is based on a larger list of species, it should tend to give a higher value than the original Index in areas such as East Anglia where species of the Lobarion, for example, no longer occur and perhaps indicate the less disturbed, older, East Anglian woodland sites more clearly than did the original Index.

The results of the RIEC analysis for the list of woodland sites in Table I is given in the third column of figures in that Table. The original IEC data, calculated on a list of 20 species only, are given in the second column of figures in Table I.

3. Discussion of the RIEC Data

The great majority (61 out of 67) of the sites in section (a) in Table I prove to have an RIEC of 50 or over; 43 (64%) have an Index of 70 or over; and 21 (33%) have an Index of 100.

The majority of the sites with an Index of high value (80–100) in Table I section (a) lie in southern and south-western England from the New Forest to Cornwall, with others in Sussex, central Wales, Lakeland and on the west Scottish coast. If high rainfall were the primary factor controlling lichen species density and the occurrence of exacting forest species, one would perhaps expect to find the highest values for both species density and the RIEC in Table I in western Scotland, but this is not the case. Admittedly the Revised Index has been based primarily and necessarily on the old woodlands of the New Forest because they have a better documented history of long continuity of environment than the great majority of British woodlands (see Tubbs, 1964; Rose and James, 1974). The species list used, however, consists of lichens which all, except two, occur widely distributed in west Scotland.

The RIEC values in Table I sections (b)–(e) show a similar trend to the density figures: only one site (71) reaches an Index value of 50, only five others reach 40 (sites 68, 69, 70, 73 and 89), and in sections (d) and (e) 13 sites out of 28 (nearly 50%) have an RIEC value of zero.

The two sites 68 and 69 in Table I section (b) are much closer in RIEC values (though not in species density) to the majority of the sites in (a) than to those of sections (d) and (e). Some degree of continuity of forest cover, at least locally, since early times is suggested by the data and, in the case of 69, The Mens, which is now well documented historically, the Index correlates well with what is known of its history.

The sites in Table I section (c), though all planted since 1750 (site 73) or since 1800 (sites 70–72), form a special case. They lie within the New Forest and ancient woodlands either adjoin them today (71, 73) or did until early last century (70, 72). Recolonization of "old forest" species is likely to have been easier than in the case of the woodlands of section (d) which are now more isolated (and in many cases already were at the time of their planting) from ancient relict forest areas. The low Index values for many sites in Lakeland, in an area generally favourable climatically for lichen growth, are remarkable. At Dalegarth wood, Cumberland (site 74) there is a mature oak forest which looks like a category (a) site superficially but the owner stated that most of it was clear-felled and replanted in about 1770; hence the low Index value is not surprising. Elements of the Lobarion are entirely lacking there. As Eskdale was already largely deforested by 1770, recolonization would have been very difficult.

Sites in Table I section (e) (old coppice woodlands) have also in general low Index values. In some cases, however (for example sites 89, 90, 92, 94, 98 and 100), there is evidence that they have been under *some* kind of woodland cover since early medieval times. Clearly the nature of coppice management, with heavy shading of the boles of the standard trees during the mature phase of the coppice cycle, and drastic desiccation after cutting of the coppice poles, is not favourable to the survival of "old forest" species or to a rich lichen flora generally. Many such sites, today, also lack large standard trees. Site 89 (East Dean Park Wood, West Sussex) is, however, a special case and so also to a lesser degree is site 90 (Combwell Wood, Kent). There is documentary evidence (M. Collins, pers. comm.) that site 89 was in part wooded in early medieval times, and it appears to have been formerly a deer park. Even today a number of large specimens of *Quercus* and *Fraxinus* remain and it is on these that species such as *Lobaria pulmonaria* and *Sticta limbata* persist. Thus this site, though largely a coppice-with-standards wood, has much more the character in flora, structure and history of a section (a) site.

Epping Forest (site 102) in Table I section (f) has already been discussed (p. 287); air pollution has exterminated most of its epiphytic lichens. The relatively low values for both the IEC and the RIEC early last century, probably reflect the lack of understanding of many small crustose lichens by the lichenologists who worked there, rather than their absence, because W. Borrer (who worked in Sussex at that period and was the most astute lichenologist of his day) recorded many of them in Sussex sites at that period.

To conclude this examination of the data, some sites in section (a) will be briefly considered individually. In southern and south-western England (apart from the New Forest itself which has already been considered p. 289), sites 1, 9, 11 and 14 are outstanding in their species density and high Index values, particularly if the RIEC is considered. There is good evidence that all these sites contained forest (enclosed as deer park) at the time of the Domesday survey. All still contain extensive areas of very old oak or oak–beech forest, of sufficient density to retain humid conditions between the trunk spaces but yet open enough to let in adequate light. Further east in Sussex, site 30 (Eridge Park) is also outstanding. It is known to have been enclosed as a deer park about 800 years ago, and was probably formed from a relic of the primeval wilderness at that time. Records from early last century indicate that it was even richer then in "old forest" species than today. The slight deterioration in the RIEC may well be due to two causes: (a) a slight degree of air pollution affecting hypersensitive species such as *Lobaria scrobiculata* and (b) the reclamation of nearly half of the park in 1958 for agriculture.

Staverton Park (site 32), though it still has a higher IEC and RIEC than other East Anglian sites, nevertheless has a low figure compared with most Table I section (a) sites in southern and western England. This site has been studied in detail by Peterken (1969) who has given strong historical evidence for the continuity of naturally regenerating, uneven-aged forest there until recent times. Today, however, the site has become much more open, due to lack of regeneration, and, in the low-rainfall area of East Anglia, this opening-up may well have caused drying out of the formerly humid forest environment. The larger foliose "old forest" species, such as *Lobaria pulmonaria*, are not there now, though this species was recorded in the early nineteenth century at three nearby stations in East Suffolk (Henslow and Skepper, 1860). On the other hand, species such as *Lecanactis premnea* and *Opegrapha lyncea* survive well there today (and in other old East Anglian park-like woodlands). These are certainly "old forest" species, but they can survive well on relatively isolated ancient trees where the humidity is low. The low Index values indicate the ecological realities of the situation well, as they do also at Sotterley Park, Suffolk (site 33) and in other East Anglian sites not quoted in Table I.

At St Leonards Forest (site 29) environmental deterioration has gone much further in recent years than at Eridge Park (site 30), due to extensive felling and replanting. Since it was well documented by Borrer (see Turner and Dillwyn 1805) at the opening of the nineteenth century, it has been possible to produce both modern and pre-1805 values for the Indices.

Table VI gives full species lists for 16 of the sites listed in Table I. It includes several of the richest sites in Table I section (a) and also examples from sections (b), (c), (d) and (e).

Table I section (a) sites in northern England, Wales and Scotland have lower total floras in many cases than comparable sites in southern England, and in most cases have lower Index values. The two facts that appear to account for this are (i) the absence of many crustose species of dry bark in wetter parts of Britain and (ii) the absence, already discussed (p. 292), of many species of more southern European distribution further north. It is true that a number of hyper-atlantic species come in further northwards and westwards but (a) these are not numerous enough to compensate for the loss of southern-continental species, and (b) they have not been used in the calculation of the Indices.

TABLE VI. Species lists for selected sites

	1	9	11	14	17	30	32	45	47	69	70	83	86	88	90	92
						Sites (see Table I)										
Alectoria fuscescens	.	.	.	+	.	+	.	+	+	.	.	+	.	.	.	.
A. subcana	.	.	.	.	.	.	.	.	.	.	.	+	.	.	.	.
Anaptychia ciliaris	.	.	+	+	.	+	.	.	.	.	.	.	.	.	.	.
Arthonia aspersella	.	+	.	.	.	.	.	.	.	.	.	.	.	.	+	.
A. didyma	+	+	+	+	+	+	.	+	.	+	+	.	.	.	+	.
A. impolita	+	.	+	+	.	.	.	.	.	.	.	.	.	.	.	.
A. leucopellaea	+	.	.	.	.	.	.	.	.	.	.	.	.	.	.	.
A. lurida	+	.	.	+	.	.	.	.	.	.	.	.	.	.	.	.
A. punctiformis	+	+	+	+	+	+	.	+	.	+	.	.	.	.	.	.
A. radiata	+	+	+	+	+	+	+	+	+	+	.	.	.	.	+	+
A. spadicea	+	+	+	+	+	+	+	+	.	+	.	.	+	+	+	+
A. stellaris	+	+	.	.	+	.	.	.	.	.	.	.	.	.	.	.
A. tumidula	+	+	.	+	+	.	.	.	.	.	+	.	.	.	.	.
Arthopyrenia antecellans	+	.	.	+	+	.	+	+	.	.	.	.	.	.	.	.
A. biformis	+	+	+	.	+	+	.	+	.	.	.	.	.	.	.	.
A. cinereopruinosa	+	+	+	+	+	.	.	+	+	.	+	.	.	.	.	.
A. fallax	+	+	.	.	+	+	+	+	+	.	.	.	+	.	.	.
A. gemmata	+	+	+	+	+	+	.	+	.	.	.	.	.	.	.	.
A. punctiformis	+	+	+	+	+	+	+	+	.	.	+	+	.	.	+	.
Arthothelium illicinum	+	.	.	.	+	.	.	.	+	.	.	.	.	.	.	.
Bacidia affinis	.	.	.	.	.	.	.	+	+	.	.	.	.	.	.	.
B. chlorococca	.	.	.	.	+	+	.	.	.	.	.	.	.	.	.	.
B. endoleuca agg.	+	+	+	+	+	+	.	+	+	+	+	.	.	.	.	.
B. friesiana	.	.	.	.	.	+	.	.	.	.	.	.	.	.	.	.
B. incompta	.	.	+	.	.	+	.	+	.	.	.	.	.	.	.	.
B. lignaria★	.	.	.	.	.	.	.	+	+	.	.	.	.	.	.	.
B. melaena	.	.	.	.	.	+	.	+	+	.	.	.	.	.	.	.
B. naegelii	.	.	+	.	.	.	.	.	.	.	.	.	.	.	.	.
B. phacodes	+	+	+	+	.	+	.	.	.	.	.	.	.	.	.	.
B. pruinosa	.	.	.	+	.	+	.	.	.	.	+	.	.	.	.	.
B. rubella	+	+	+	+	.	+	.	+	+	.	.	.	.	.	.	.
B. sphaeroides agg.	.	+	.	.	.	.	.	.	+	.	.	.	.	.	.	.
B. sp. (undescribed)	.	.	.	.	+	+	.	.	.	.	+	.	.	.	.	.
B. umbrina	+	+	.	.	.	.	.	.	.	.	.	.	.	.	.	.
Baeomyces roseus	.	.	.	.	+	+	.	.	.	.	.	.	.	.	.	.
B. rufus	.	+	+	.	+	+	.	+	.	.	+	.	.	.	.	.
Biatorella ochrophora	+	+	.	.	.	.	.	.	.	.	.	.	.	.	.	.
Bombyliospora pachycarpa	.	.	.	.	+	.	.	.	.	.	.	.	.	.	.	.

★ *Micarea liguaria*

TABLE VI.—*continued*

	\multicolumn{16}{c}{Sites (see Table I)}														
	1	9	11	14	17	30	32	45	47	69	70	83	86	88	90	92
"*Botrydina vulgaris*"	.	.	.	.	+	.	.	.	.	.	.	.	.	.	.	.
Buellia alboatra	+	.	+	+	.	+	+	.	.	.	.	.	.	.	.	.
B. canescens	+	+	+	+	.	+	+	.	.	.	.	.	.	.	.	+
B. disciformis	+	+	.	.	+	.	.	+	.	.	.	.	.	.	.	.
B. griseovirens	.	.	+	.	+	+	.	.	.	.	.	.	.	.	.	.
B. punctata	+	+	+	+	.	+	+	.	.	.	.	.	.	.	.	.
B. schaereri	.	.	.	.	.	+	.	.	.	.	.	.	.	.	.	.
Calicium abietinum	+	.	+	.	+	+	+	+	.	.	.	.	.	.	.	.
C. salicinum	.	.	+	+	.	+	.	.	.	.	.	.	.	.	.	.
C. viride	+	+	+	+	.	+	+	+	.	+	.	.	.	.	.	.
Caloplaca cerina	.	.	+	+	.	+	.	.	.	.	.	.	.	.	.	.
C. cerinella	.	.	+	.	.	.	.	.	.	.	.	.	.	.	.	.
C. citrina	.	.	+	+	.	+	.	.	.	.	.	.	.	.	.	.
C. ferruginea	.	.	.	.	.	+	+	.	.	.	.	.	.	.	.	.
C. herbidella	.	.	.	.	.	+	.	.	.	.	.	.	.	.	.	.
C. holocarpa	+	.	+	.	.	.	.	.	.	.	.	.	.	.	.	.
Candelariella reflexa	+	.	+	+	.	+	.	.	.	.	.	.	.	.	.	.
C. vitellina	.	.	+	.	.	+	.	.	.	.	.	.	.	.	.	.
C. xanthostigma	.	+	.	+	.	.	.	.	.	.	.	.	.	.	.	.
Catillaria atropurpurea	+	+	+	+	+	+	.	+	+	+	+	.	.	.	+	.
C. griffithii	+	+	+	+	+	+	+	+	.	+	+	.	.	.	+	+
C. lightfootii	+	.	+	.	+	+	+	+	.	.	.	.	.	.	+	.
C. pulverea	+	+	.	.	+	.	.	.	.	.	.	.	.	.	.	.
C. sphaeroides	+	+	.	+	+	.	.	+	+	.	.	.	.	.	.	.
Catinaria grossa	.	.	.	.	.	+	.	.	.	.	.	.	.	.	.	.
C. intermixta	.	.	.	.	+	.	.	.	.	.	.	.	.	.	.	.
Cetraria chlorophylla	.	+	+	+	.	+	.	+	+	.	.	+	.	.	+	.
Cetrelia cetrariodes	+	+	.	.	+	.	.	.	+	.	.	.	.	.	.	.
Chaenotheca aeruginosa	.	.	.	.	+	.	.	.	.	.	.	.	.	.	.	.
C. brunneola	.	+	.	.	.	+	.	+	.	.	+	.	.	.	.	.
C. chrysocephala	.	.	.	.	+	.	.	.	.	.	.	.	.	.	.	.
C. ferruginea	.	.	+	+	+	+	+	+	.	.	.	.	.	+	+	.
Cladonia bacillaris	.	.	.	.	+	.	+	.	.	.	.	.	.	.	.	.
C. caespiticia	+	+	.	.	+	+	.	.	+	+	.	.	.	.	.	.
C. chlorophaea	+	+	+	+	+	+	+	+	+	+	+	+	.	.	+	+
C. coccifera	.	.	+	.	.	.	.	+	.	.	.	.	.	.	.	.
C. coniocraea	+	+	+	+	+	+	+	+	+	+	+	+	+	+	+	+
C. crispata	.	.	.	.	+	+	.	.	.	.	.	.	.	.	.	.
C. digitata	.	+	.	+	+	+	+	+	+	.	.	.	.	.	.	.
C. fimbriata	.	.	+	+	+	+	+	+	.	+	.	.	.	.	+	.

TABLE VI.—*continued*

	Sites (see Table I)															
	1	9	11	14	17	30	32	45	47	69	70	83	86	88	90	92
C. floerkeana	+	.	+	.	+	.	.	.	.	.	.	.	.	.	.	.
C. furcata	+	.	+	+	+	+	.	+	+	.	.	.	.	.	+	.
C. impexa	+	.	+	+	+	+	.	+	+	.	.	.	.	.	.	.
C. macilenta	+	+	+	+	+	+	+	+	+	+	.	.	.	.	.	+
C. ochrochlora	+	+	+	+	+	+	.	.	.	+	+	.	.	.	.	.
C. parasitica	+	+	+	.	+	+	+	+	.	.	.	.	.	.	+	+
C. pityrea	+	.	.	.	.	+	.	.	.	.	.	.	.	.	.	.
C. polydactyla	+	+	+	.	+	+	.	+	+	.	+	+	.	.	.	.
C. pyxidata	+	.	.	+	+	+	.	.	.	.	.	.	.	.	.	.
C. squamosa	+	+	+	.	+	+	.	.	+	.	+	+	.	.	.	.
var. *allosquamosa*	.	.	.	+	+	.	.	+	.	.	.	.	.	.	.	.
C. subulata	.	.	.	.	+	+	.	.	.	.	.	.	.	.	.	.
C. tenuis	+	.	+	.	.	+	.	+	.	.	.	.	.	.	.	.
Collema fragrans	.	.	+	.	+	.	.	.	.	.	.	.	.	.	.	.
C. furfuraceum	.	.	.	.	.	.	.	.	.	+	.	.	.	.	.	.
Coniocybe furfuracea	.	.	+	.	.	.	.	.	.	.	.	.	.	.	.	.
Coriscium viride	.	.	.	.	+	.	.	.	.	.	.	.	.	.	.	.
Cyphelium inquinans	.	.	.	.	.	+	.	.	.	.	.	.	.	.	+	.
C. sessile	.	.	+	+	.	+	.	.	.	.	.	.	.	.	+	.
Cystocoleus niger	.	.	.	.	.	.	.	.	.	+	.	.	.	.	.	.
Dermatina quercus	+	.	+	.	.	+	.	+	.	+	.	.	.	.	.	.
Dimerella diluta	+	+	+	+	+	+	.	+	+	+	.	.	.	.	.	.
D. lutea	+	+	.	+	+	+	.	.	.	.	.	.	.	.	.	.
Enterographa crassa	+	+	+	+	+	+	+	+	+	+	+	.	.	.	.	.
Evernia prunastri	+	+	+	+	+	+	+	+	+	+	+	+	+	+	+	+
Graphina anguina	+	+	+	.	.	.	.	+	+	.	.	.	.	.	.	.
Graphis elegans	+	+	+	+	+	+	+	+	+	+	+	+	.	+	.	+
G. scripta	+	+	+	+	+	+	+	+	+	+	+	.	+	.	+	+
Gyalecta flotowii	.	.	+	.	.	+	.	.	.	.	.	.	.	.	.	.
G. truncigena	+	+	+	+	+	+	.	.	.	.	+	.	.	.	.	.
Gyalectina carneolutea	+	.	+	.	.	.	.	.	.	.	.	.	.	.	.	.
Gyalideopsis anastomosans	+	.	.	.	+	+	.	.	.	.	.	.	.	.	.	.
Haematomma elatinum	+	.	+	.	+	+	+	.	+	.	.	.	+	.	+	.
H. ochroleucum var.																
porphyrium	.	.	+	+	.	+	+	.	.	.	.	.	.	.	.	.
Hypogymnia physodes	+	+	+	+	+	+	+	+	+	+	+	+	+	+	+	+
H. tubulosa	+	+	+	+	+	+	.	+	+	+	.	.	.	.	+	.
Lecanactis abietina	+	+	+	+	+	+	+	+	.	+	+	.	+	+	.	.
L. corticola	+	.	.	.	.	+	.	.	.	.	.	.	.	.	.	.
L. premnea	+	+	+	+	+	+	+	.	.	.	.	.	.	.	.	.

Table VI.—*continued*

	1	9	11	14	17	30	32	45	47	69	70	83	86	88	90	92
Sites (see Table I)																
Lecania cyrtella	+	.	+	+	.	+	.	.	.	.	.	.	.	.	.	.
Lecanora carpinea	.	.	+	.	.	.	.	+	.	.	.	.	.	.	.	.
L. chlarona	.	+	.	.	.	+	.	.	.	.	.	.	.	.	.	.
L. chlarotera	+	+	+	+	+	+	+	+	+	+	+	+	.	+	+	+
L. confusa	+	+	+	+	+	+	.	.	.	.	.	.	.	.	+	.
L. conizaeoides	+	+	+	+	+	+	+	+	+	+	+	+	+	+	+	+
L. dispersa	.	.	+	+	.	+	+	.	.	.	.	.	.	.	.	.
L. expallens	+	+	+	+	+	+	+	+	+	+	+	+	+	+	+	+
L. intumescens	.	.	.	.	+	.	.	+	.	.	.	.	.	.	.	.
L. jamesii	+	.	+	+	+	.	.	.	+	.	.	.	.	.	.	.
L. pallida	+	.	.	.	+	+	.	+	.	+	+	.	.	.	+	+
L. piniperda	.	.	.	.	.	+	.	.	.	.	.	.	.	.	.	.
L. sambuci	.	.	+	.	.	.	.	.	.	.	.	.	.	.	.	.
L. varia	.	.	.	.	.	+	.	.	.	.	.	.	.	.	.	.
Lecidea berengeriana	.	.	.	.	.	.	.	.	+	.	.	.	.	.	.	.
L. cinnabarina	+	.	.	.	+	+	+	+	.	.	.	.	+	.	.	.
L. granulosa	+	+	+	+	+	+	+	+	+	+	+	.	.	+	+	+
L. quernea	+	+	+	+	+	+	+	+	+	+	+	+	+	+	+	+
L. scalaris	.	.	+	.	+	+	+	+	.	.	.	.	.	.	+	.
L. templetonii	.	+	.	.	.	.	.	.	+	.	.	.	.	.	.	.
L. sublivescens	.	.	.	+	.	.	.	.	.	.	.	.	.	.	.	.
L. symmicta	.	.	+	+	.	+	.	.	+	.	.	.	.	.	.	.
L. tenebricosa	+	.	.	.	.	.	.	.	.	.	.	.	.	.	.	.
L. turgidula	.	.	.	.	+	.	.	.	.	.	.	.	.	.	.	.
L. uliginosa	+	+	+	+	+	+	.	+	+	+	+	.	.	.	.	+
L. cf. vernalis	.	.	+	+	.	+	.	.	.	.	.	.	.	.	.	.
Lecidella elaeochroma	+	+	+	+	+	+	.	+	.	+	.	+	.	.	+	+
Lepraria candelaris	+	+	+	+	+	+	+	+	.	+	+	.	.	.	+	+
L. incana	+	+	+	+	+	+	+	+	+	+	+	+	+	+	+	+
L. membranacea	.	+	+	.	.	.	.	.	+	.	.	.	.	.	.	.
Leprocaulon microscopicum	.	.	.	.	.	.	.	.	+	.	.	.	.	.	.	.
Leptogium burgessii	.	.	.	.	.	.	.	.	+	.	.	.	.	.	.	.
L. cyanescens	.	+	.	.	.	.	.	.	.	.	.	.	.	.	.	.
L. lichenoides	+	+	+	+	+	+	.	+	+	.	.	.	.	.	.	.
L. teretiusculum	+	+	+	+	.	+	.	.	.	.	.	.	.	.	.	.
L. tremelloides	+	.	.	.	.	.	.	.	.	.	.	.	.	.	.	.
Leptorhaphis epidermidis	+	.	.	.	.	+	.	.	.	.	.	.	.	.	.	.
Lithographa dendrographa	+	+	.	.	.	.	.	.	.	.	.	.	.	.	.	.
Lobaria amplissima	+	.	+	+	+	.	.	.	.	.	.	.	.	.	.	.
L. laetevirens	+	+	+	.	+	+	.	+	+	.	.	.	.	.	.	.

TABLE VI.—*continued*

	Sites (see Table I)															
	1	9	11	14	17	30	32	45	47	69	70	83	86	88	90	92
L. pulmonaria	+	+	+	+	+	+	.	+	+	.	.	.	.	.	.	.
L. scrobiculata	+	+	.	.	.	.	.	.	.	.	.	.	.	.	.	.
Lopadium pezizoideum	.	.	.	.	.	.	.	.	+	.	.	.	.	.	.	.
Macentina sp.	.	.	.	.	.	+	.	.	.	.	.	.	.	.	.	.
Melaspilea lentiginosa	+	.	.	.	.	.	.	.	.	.	.	.	.	.	.	.
M. ochrothalamia	.	+	.	.	+	.	.	.	.	.	.	.	.	.	.	.
M. proximella	.	.	.	.	+	.	.	.	.	.	.	.	.	.	.	.
Micarea denigrata	+	.	.	.	+	+	.	+	.	.	.	.	.	.	.	.
M. prasina	+	+	+	+	+	+	.	+	+	+	+	.	.	.	.	.
M. violacea	.	.	.	.	+	.	.	.	+	.	.	.	.	.	.	.
Mycoblastus sanguinarius	.	.	.	.	.	.	.	.	+	.	.	.	.	.	.	.
Mycoporellum sparsellum	.	.	.	.	+	.	.	.	.	.	.	.	.	.	.	.
Nephroma laevigatum	+	+	.	.	.	+	.	.	+	.	.	.	.	.	.	.
N. parile	.	+	.	.	.	.	.	.	.	.	.	.	.	.	.	.
Normandina pulchella	+	+	+	+	+	+	.	.	+	+	.	.	.	.	.	.
Ochrolechia androgyna	+	+	+	+	+	+	+	+	+	+	+	+	+	.	+	.
O. inversa	+	.	+	.	+	.	.	.	.	.	.	.	.	.	.	.
O. parella	.	.	.	.	.	+	.	.	.	.	.	.	.	.	.	.
O. tartarea	.	.	.	.	.	.	.	.	+	.	.	.	.	.	.	.
O. turneri	+	+	+	+	+	+	+	+	+	+	.	.	.	.	.	.
O. yasudae	+	+	+	+	+	+	+	+	+	+	+	.	.	.	+	.
Opegrapha atra	+	+	+	+	+	+	+	+	.	.	+	.	.	.	.	.
O. gyrocarpa	+	.	.	.	.	.	.	.	.	.	.	.	.	.	.	.
O. herbarum	+	+	+	.	.	.	.	+	.	.	.	.	.	.	.	.
O. lyncea	+	.	+	+	.	.	+	.	.	.	.	.	.	.	.	.
O. niveoatra	.	.	.	.	.	+	.	.	.	.	.	.	.	.	.	.
O. ochrocheila	+	+	.	.	.	+	.	.	.	.	.	.	.	.	.	.
O. rufescens	.	.	+	.	.	.	.	.	.	+	.	.	.	.	.	.
O. sorediifera	+	+	.	.	+	+	.	.	.	+	.	.	.	.	.	.
O. varia	+	+	+	+	+	+	.	+	.	.	.	.	.	.	.	.
O. vermicellifera	+	+	+	+	.	+	.	+	.	+	.	.	.	.	.	.
O. vulgata	+	+	+	+	+	+	+	+	.	+	.	.	.	.	.	+
Pachyphiale cornea	+	.	+	+	+	+	.	+	+	+	+	.	.	.	+	.
Pannaria mediterranea	+	+	.	.	.	.	.	.	.	.	.	.	.	.	.	.
P. pityrea	+	+	+	.	+	.	.	.	+	.	.	.	.	.	.	.
Parmelia acetabulum	.	.	.	.	.	+	.	.	.	.	.	.	.	.	.	.
P. arnoldii	+	.	.	.	.	.	.	.	.	.	.	.	.	.	.	.
P. borreri	+	.	.	.	.	.	.	.	.	.	.	.	.	.	.	.
P. caperata	+	+	+	+	+	+	+	+	+	+	+	+	+	+	+	+
P. carporrhizans	.	+	.	.	.	.	.	.	.	.	.	.	.	.	.	.

TABLE VI.—*continued*

	Sites (see Table I)															
	1	9	11	14	17	30	32	45	47	69	70	83	86	88	90	92
P. crinita	+	.	+	+	+	+	.	.	+	.	.	.	.	.	.	.
P. dissecta	.	.	.	.	+	.	.	.	.	.	.	.	.	.	.	.
P. elegantula	.	.	+	+	.	+	.	.	.	.	.	.	.	.	.	.
P. endochlora	+	.	.	.	.	.	.	.	.	.	.	.	.	.	.	.
P. exasperatula	.	+	.	.	.	.	.	.	.	.	.	.	.	.	.	.
P. glabratula																
subsp. *glabratula*	+	+	+	+	+	+	+	+	+	+	+	+	+	+	+	+
subsp. *fuliginosa*	+	.	.	.	.	.	.	.	.	.	.	+	.	.	.	.
P. laciniatula	.	+	+	.	.	+	.	.	.	.	.	.	.	.	.	.
P. laevigata	.	+	.	.	.	.	.	.	+	.	.	.	.	.	.	.
P. perlata	+	+	+	+	+	+	.	.	+	+	+	+	.	.	+	.
P. reddenda	+	+	+	+	+	+	.	.	.	.	.	.	.	.	.	.
P. reticulata	+	.	+	+	+	+	.	.	.	+	+	.	.	.	+	+
P. revoluta	+	+	+	+	+	+	+	.	+	+	+	.	.	+	+	+
P. saxatilis	+	+	+	+	+	+	+	+	+	+	+	+	+	+	+	+
P. sinuosa	+	.	.	.	.	.	.	.	.	.	.	.	.	.	.	.
P. soredians	.	.	.	.	.	+	.	.	.	.	.	.	.	.	.	.
P. subaurifera	+	+	+	+	+	+	.	+	.	.	.	.	.	.	.	+
P. subrudecta	+	+	+	+	+	+	+	.	+	+	.	.	.	+	+	+
P. sulcata	+	+	+	+	+	+	+	+	+	+	+	+	+	+	+	+
P. taylorensis	.	+	.	.	.	.	.	.	+	.	.	.	.	.	.	.
P. tiliacea	.	+	+	.	.	+	.	.	.	.	.	.	.	.	.	.
Parmeliella corallinoides	+	+	.	.	.	.	.	.	+	.	.	.	.	.	.	.
P. plumbea	.	.	.	.	.	+	.	.	+	.	.	.	.	.	.	.
Parmeliopsis aleurites	.	.	.	.	.	+	.	.	.	.	.	.	.	.	.	.
P. ambigua	.	.	.	+	.	+	.	+	.	.	.	.	.	.	+	.
P. hyperopta	.	.	.	.	+	.	.	.	.	.	.	.	.	.	.	.
Peltigera canina	+	+	+	+	+	+	.	+	.	.	.	.	.	.	.	.
P. collina	+	+	+	.	.	.	.	.	+	.	.	.	.	.	.	.
P. horizontalis	+	+	+	+	+	.	.	+	+	.	.	.	.	.	.	.
P. polydactyla	+	+	.	.	+	+	.	.	.	.	.	.	.	.	.	.
P. praetextata	+	+	+	+	+	+	.	+	+	.	.	.	.	.	.	.
P. rufescens	+	.	.	.	.	+	.	+	.	.	.	.	.	.	.	.
Pertusaria albescens	+	+	+	+	+	+	.	+	.	.	.	.	.	.	+	+
var. *corallina*	+	+	+	+	+	+	.	+	+	.	.	+	.	+	+	.
P. amara	+	+	+	+	+	+	+	+	+	+	+	+	+	+	+	+
P. coccodes	+	.	+	+	+	+	.	+	.	+	.	.	.	+	+	+
P. coronata	+	.	.	.	.	+	.	.	.	.	.	.	.	.	.	.
P. flavida	.	.	.	+	.	+	.	+	.	+	.	.	.	.	.	.
P. hemisphaerica	+	+	+	+	+	+	+	+	.	+	+	+	+	+	+	.

TABLE VI.—*continued*

	1	9	11	14	17	30	32	45	47	69	70	83	86	88	90	92
P. hymenea	+	+	+	+	+	+	+	+	+	+	+	+	.	+	+	+
P. leioplaca	+	+	+	+	+	+	.	+	+	+	+	.	+	.	+	+
P. multipuncta	+	+	+	+	+	+	.	+	.	+	+	.	+	.	.	+
P. pertusa	+	+	+	+	+	+	+	+	+	+	+	.	+	+	+	+
P. pustulata	.	.	.	.	+	.	.	.	.	.	.	.	.	.	.	.
P. velata	.	.	+	+	+	+	.	.	.	.	.	.	.	.	.	.
Phaeographis dendritica	+	+	+	.	+	+	.	.	.	+	.	+	.	.	.	
P. lyelli	+	.	.	.	.	.	.	.	.	.	.	.	.	.	.	
P. ramificans	+	.	+	.	+	.	+	.	.	.	.	.	.	.	.	.
Phlyctis agelaea	+	.	.	.	.	.	+	.	.	.	.	.	.	.	.	.
P. argena	+	+	+	+	+	+	+	+	+	+	+	+	+	+	+	+
Phyllopsora sp.	+	+	.	.	.	.	.	.	+	.	.	.	.	.	.	
Physcia adscendens	.	+	+	+	+	+	+	.	.	.	.	.	.	.	.	.
P. aipolia	+	+	+	+	.	+	.	.	.	.	.	.	.	.	.	.
P. labrata	+	.	+	.	.	+	.	.	.	.	.	.	.	.	.	.
P. orbicularis	+	+	+	+	.	+	.	.	.	.	.	.	.	.	.	.
P. tenella	+	.	+	.	+	+	.	.	.	.	.	.	.	.	.	.
Physciopsis adglutinata	+	.	+	+	.	+	.	.	.	.	.	.	.	.	.	.
Physconia grisea	.	.	+	+	.	.	.	.	.	.	.	.	.	.	.	.
P. farrea	.	.	+	.	.	+	.	.	.	.	.	.	.	.	.	.
P. pulverulenta	.	+	+	+	.	+	.	.	.	.	.	.	.	.	.	.
Platismatia glauca	+	+	+	+	+	+	.	+	+	+	+	+	+	+	+	+
Porina chlorotica var.																
carpinea	+	+	+	+	+	+	.	+	+	+	.	.	.	.	.	.
P. coralloidea	+	.	+	.	+	+	.	.	.	.	.	.	.	.	.	.
P. hibernica	+	.	.	.	.	.	.	.	.	.	.	.	.	.	.	.
P. leptalea	+	.	.	.	+	.	.	.	+	+	.	.	.	.	.	.
P. olivacea	+	.	.	.	.	.	.	.	.	.	.	.	.	.	.	.
Pseudevernia furfuracea	.	.	+	.	.	+	.	+	.	.	.	.	.	.	+	.
Pyrenula nitida	+	+	+	+	+	+	.	.	+	.	.	.	.	.	.	.
P. nitidella	+	+	+	+	+	+	.	.	.	+	.	.	.	.	.	.
Ramalina calicaris	+	.	+	.	.	.	.	.	.	.	.	.	.	.	.	.
R. farinacea	+	+	+	+	+	+	+	+	+	+	+	+	+	+	+	+
R. fastigiata	+	.	+	+	.	.	.	.	.	.	.	.	.	.	.	.
R. fraxinea	.	.	.	+	.	.	.	.	.	.	.	.	.	.	.	.
R. obtusata	.	+	+	+	.	+	.	.	.	.	.	.	.	.	.	.
R. pollinaria	+	+	+	+	.	.	.	.	.	.	.	.	.	.	.	.
Rinodina exigua	.	.	.	+	.	+	.	.	.	.	.	.	.	.	.	.
R. isidioides	+	.	.	+	.	.	.	.	.	+	.	.	.	.	.	.
R. roboris	+	.	+	+	+	+	+	.	.	.	.	.	.	.	.	.

TABLE VI.—*continued*

	Sites (see Table I)															
	1	9	11	14	17	30	32	45	47	69	70	83	86	88	90	92
Schismatomma decolorans	+	+	+	+	+	+	+	.	.	+	+	.	.	+	+	+
S. niveum	+	.	+	+	+	.	.	.	.	+	+	.	.	.	.	.
Sphaerophorus globosus	.	+	.	.	.	.	.	.	+	.	.	.	.	.	.	.
Stenocybe pullatula	+	+	+	.	+	+	.	+	.	.	.	.	.	.	.	.
S. septata	+	+	+	+	+	+	+	.	+	+	+	.	.	.	.	.
Sticta fuliginosa	+	+	.	.	.	.	.	.	+	.	.	.	.	.	.	.
S. limbata	+	+	+	.	+	.	.	.	+	.	.	.	.	.	.	.
S. sylvatica	+	+	.	.	.	.	.	.	+	.	.	.	.	.	.	.
Teloschistes flavicans	.	.	+	.	.	.	.	.	.	.	.	.	.	.	.	.
Thelopsis rubella	+	+	+	+	+	+	.	.	.	.	+	.	.	.	.	.
Thelotrema lepadinum	+	+	+	+	+	+	+	+	+	+	+	.	.	.	+	+
Tomasellia gelatinosa	.	.	.	.	+	.	.	+	+	.	.	.	.	.	.	.
T. ischnobela	+	.	.	.	.	.	.	.	.	.	.	.	.	.	.	.
Usnea articulata	+	+	+	.	.	.	.	.	.	.	.	.	.	.	.	.
U. ceratina	+	.	+	+	+	+	.	.	.	.	+	+	.	.	+	.
U. florida	+	+	+	+	.	.	.	.	.	.	.	.	.	.	.	.
U. fragilescens	+	+	+	.	.	.	.	.	.	.	.	.	.	.	.	.
U. fulvoreagens	+	.	+	.	.	.	.	.	.	.	.	.	.	.	.	.
U. rubiginea	+	+	+	+	.	+	.	.	.	.	.	.	.	.	+	.
U. subfloridana	+	+	+	+	+	+	+	+	+	+	+	+	+	+	+	+
U. intexta	+	+	.	+	+	.	.	.	.	.	.	+	.	.	+	+
Xanthoria candelaria	.	.	.	+	.	.	.	+	.	.	.	.	.	.	.	.
X. parietina	.	+	+	+	+	+	.	.	.	.	.	.	.	.	.	.
X. polycarpa	.	.	+	.	.	+	.	.	.	.	.	.	.	.	.	.
Xylographa vitiligo	.	.	.	.	.	+	.	.	.	.	.	.	.	.	.	.
Total number of taxa	188*	150*	173	142	160	186	65	108	100	75	54	36	29	29	63	45

4. Upland Woodlands in High Rainfall Areas

The lichen flora of upland deciduous woodlands in western Britain (where there is much leaching of nutrients from the bark due to heavy rainfall in exposed situations) differs in many respects from that of lowland woodlands. Hence upland or exposed western sites have been excluded from consideration in this paper. Such woodlands, however, appear to have a number of "old forest" species of their own, among which appear to be *Alectoria smithii*,

---

* See also p. 307.

Menegazzia terebrata, Ochrolechia tartarea and *Parmelia laevigata,* and it may be possible to devise an Index for them in due course.

5. Western Europe

In lowland France and in mountain forests in the Pyrenees, the Indices used in this paper appear to work reasonably well. Degelius (1965) recorded 127 epiphytic lichen taxa in an old oak forest, Hald Egeskov, of some 2 km² near Viborg in Jutland, Denmark. The RIEC for this site is 35, but it is more continental in climate than any British forest.

6. Coniferous Forests

So far no real attempt has been made to apply similar techniques of analysis to the ancient coniferous forests of the Scottish Highlands, but there is every reason to believe that they will prove susceptible to similar treatment. Coniferous plantations (even mature ones) and subspontaneous pinewoods in lowland England have very limited lichen epiphyte floras. However, some of the Scottish pine forests that Stevens and Carlisle (1959) indicated as being of primary

TABLE VII. Possible "old forest" lichen epiphytes of Scottish Boreal Forests (growing on *Pinus, Sorbus* and *Betula*)

Alectoria capillaris	*Haematomma elatinum*
A. lanestris	*Lecanora ochrococca*
A. sarmentosa subsp. *sarmentosa*	*Lecidea cinnabarina*
Arthonia leucopellaea	*Lecidea friesii*
Buellia schaereri	*Platismatia norvegica*
Cavernularia hultenii	*Usnea extensa*
Chaenotheca spp.	

origin have proved recently to have very rich epiphyte lichen floras, with many species that may lend themselves to use in the preparation of a Boreal Forest Index in the future.

In Table VII, some species are listed that appear to be old Boreal forest lichen epiphytes on the evidence to hand so far.

CONCLUSIONS

It is evident that valuable information on the degree of continuity of the ecological environment in British lowland deciduous woodlands can be obtained, from a combination of (*a*) species density data and (*b*) the proportion of "old forest" species present in a site. These old forest species are essentially environmental indicators, now largely relict in nature.

The major problems lie in the selection of suitable "indicator" species, and in how many of them to use. Those utilized here are both widespread (at least in England south of the Lake District and Yorkshire) and show a high degree of correlation with old woodland areas. One might argue for the exclusion of species of more southern distributions such as *Opegrapha lyncea* and *Lecanactis premnea*. These two species are found in quite dry open parkland situations as well as in forests, and today perhaps are better correlated with the presence of ancient oaks rather than with ancient forest areas. There is little doubt, however, that they are relics of the old forest flora as they are so characteristic of such forests areas as remain intact. Their inclusion does enable some correlation to be made between relict lichen floras and *known* ancient woodland relics in such areas as East Anglia and much of midland England, where the general dryness of the old woodland relics today, due to drainage and opening up of the former canopy in a dry climate, has led to the disappearance of most other epiphytic and humicolous forest lichens that can be used in other areas. In the Sussex Weald, for example, the topography, the higher rainfall and the persistence of much more general woodland cover appear to have permitted the survival of a more representative forest lichen flora at a number of sites. In addition, many lichen epiphytes that are still able to colonize roadside and pasture trees freely in more humid parts of southern and south-western Britain (e.g. *Graphis* spp., *Normandina pulchella*, *Parmelia perlata*, *Rinodina roboris*) show in East Anglia a relict distribution pattern in old woodlands and old, sheltered, parklands. It is not possible to include such species in the construction of a *national* Index. *Enterographa crassa*, though included in the species list for both Indices, also gives rise to similar problems in that in Devon and Cornwall it becomes of quite general distribution as an epiphyte on mature trees, while in northern England and in Scotland it becomes progressively rarer northwards. even in known ancient woodlands. Regional versions of the Index could perhaps be prepared, but, as already mentioned (p. 292), this in itself leads to further problems of definition. The list of species used in the Index could be confined to the members of the Lobarion pulmonariae alliance (see Barkman, 1958), comprising species of the genera *Lobaria*, *Nephroma*, *Pannaria*, *Parmeliella*, *Sticta*, etc. These all certainly behave as "old forest" indicators except in a few situations in west Scotland where they seem able still to colonize freely plantations of hardwood trees. If this were done, comparatively few sites in southern lowland Britain would, today, achieve significant values of such an Index, because nearly all woodlands in southern Britain have lost a proportion of the species of this group.

The species list used here, at least for the Revised Index, represents what appears to be a compromise, that offers some possibility of grading woodlands on a scale of increasing or decreasing levels of past disturbance.

ACKNOWLEDGEMENTS

I would like to thank particularly Mr P. W. James, Mr B. J. Coppins and Dr D. L. Hawksworth for assistance in the field and in the identification of critical species during the survey; and also all the landowners, too numerous to mention individually, who so kindly and willingly granted access to their woodlands.

REFERENCES

BARKMAN, J. J. (1958). "Phytosociology and Ecology of Cryptogamic Epiphytes." Van Gorcum Assen.

BRANDON, P. F. (1963). "The Common Lands and Wastes of Sussex." Ph.D. thesis, University of London.

DEGELIUS, G. (1965). Lavfloran i Hald Egeskov Jylland. Ett bidrag till de danska ekskogsresternas naturhistoria. *Bot. Tidsskr.* **61**, 1–21.

FERRY, B. W., BADDELEY, M. S. and HAWKSWORTH, D. L. (eds) (1973), "Air Pollution and Lichens." University of London, Athlone Press, London.

HAWKSWORTH, D. L., COPPINS, B. J. and ROSE, F. (1974). Changes in the British Lichen Flora. *In* "The Changing Flora and Fauna of Britain" (D. L. Hawksworth, ed.), pp. 47–78. Academic Press, London and New York.

HAWKSWORTH, D. L., ROSE, F. and COPPINS, B. J. (1973). Changes in the lichen flora of England and Wales attributable to pollution of the air by sulphur dioxide. *In* "Air Pollution and Lichens" (B. W. Ferry, M. S. Baddeley and D. L. Hawksworth, eds), pp. 330–367. University of London, Athlone Press, London.

HENSLOW, J. S. and SKEPPER, E. (1860). "Flora of Suffolk." London and Bury St Edmunds.

PETERKEN, G. F. (1969). Development of vegetation in Staverton Park, Suffolk. *Fld Stud.* **3**, 1–39.

ROSE, F. (1974). The Epiphytes of Oak. *In* "The British Oak; its History and Natural History" (M. G. Morris and F. H. Perring, eds), pp. 250–273. Classey, Faringdon.

ROSE, F. and JAMES, P. W. (1974). Regional studies on the British Lichen Flora I. The Corticolous and Lignicolous Species of the New Forest, Hampshire. *Lichenologist* **6**, 1–72.

STEVENS, H. M. and CARLISLE, A. (1959). "The Natural Pinewoods of Scotland." Oliver and Boyd, Edinburgh and London.

TUBBS, C. R. (1964). Early encoppicement in the New Forest. *Forestry* **37**, 95–105.

TURNER, D. and DILLWYN, L. W. (1805). "The Botanist's Guide through England and Wales", 2 vols. Phillips and Fardon, London.

Note added in proof

The following additions to Table VI should be made: *Anaptychia obscurata* (site 1), *Arthopyrenia cerasi* (site 9) and *Micarea cinerea* (site 9).

12 | Studies on the Growth Rates of Lichens

R. A. ARMSTRONG

Botany School, University of Oxford, England

Abstract: Progress in the field of lichen growth rate studies is briefly reviewed. The application of a new method of measuring growth rate to thalli of different size has led to the conclusion that there are changes in the radial growth rate during the life of a lichen thallus. For most of the life of a lichen thallus the radial growth rate is constant and the thallus radius increases linearly. Preceding the linear phase the radial growth rate increases with time and the thallus radius increases logarithmically. There is no evidence for a postlinear phase in the radial growth of a lichen thallus. Studies on the growth rate of lichens are applied both to the problems of determining the age of a lichen thallus on an undated substratum and to an ecological investigation in the field.

INTRODUCTION

The purpose of this chapter is to review, briefly, progress in the field of lichen growth rate studies and to describe some applications of growth rate measurements to other fields of study in lichenology. Over the last 25 years many lichenologists have been interested in the problem of measuring lichen growth rates (e.g. des Abbayes, 1951; Platt and Amsler, 1955; Rydzak, 1956, 1961; Frey, 1959; Hale, 1959, 1970; Brodo, 1964, 1965; Phillips, 1969). These studies have investigated the growth rates of different species and also the changes in growth rate which occur during the life of a lichen thallus.

Progress in the study of lichen growth rates will have important consequences for two fields of study in lichenology. First, data on lichen growth rates can enable estimates of both the age of a thallus and the period of exposure of a substratum to be made. This technique has been applied to the dating of glacial moraines (Beschel, 1961) and more recently to the dating of some archaeological remains. In addition, the age structure of a lichen population can be determined and the population dynamics of lichens studied. Second, the growth rate of a

Systematics Association Special Volume No. 8, "Lichenology: Progress and Problems", edited by D. H. Brown, D. L. Hawksworth and R. H. Bailey, 1975, pp. 309–322. Academic Press, London and New York.

lichen can be used as a measurement of performance after experimental treatment. Although physiological investigation of lichens and methods of measuring and analysing lichen distributions have become sophisticated, the gap between these studies has also widened. A problem for future lichenologists will be to link these two fields of study. It is possible that ecological experiments in the field which use growth rate as a measure of performance will enable lichen distribution to be interpreted in physiological terms.

Most of the work on lichen growth rates to be described in this chapter has been carried out on foliose saxicolous lichens in an area of Ordovician rock at the mouth of the River Dovey in South Merionethshire (Nat. Grid Ref. 22 (SN)/61–96–).

UNITS OF LICHEN GROWTH RATE

Studies on the growth rates of lichens have almost exclusively been confined to measurements of the increase in radius or area of a thallus in unit time. Three units have been used to express this radial growth of a lichen thallus. First, the radial growth rate which is expressed in units of mm/year has been used by Hale (1959) and is the increase in radius of the thallus which occurs each year. Second, growth has been expressed as the percentage increase in surface area of the thallus which occurs each year by Rydzak (1961). Third, Woolhouse (1968) has interpreted the work of Blackman (1919) in a lichenological context and has stated that the biologically significant growth is the growth achieved by a thallus in a given time interval in relation to the area of thallus which is available for the interception of light. This unit of growth can be derived as follows from the "compound interest law" of growth of higher plants where

$$W_1 = W_0 e^{rt}.$$

W_1 is the final weight of the plant, W_0 is the initial weight of the plant, r is the rate at which material already present is used to make new material (i.e. the relative growth rate) and t is time. For a lichen

$$A_2 = A_1 e^{r(t_2 - t_1)},$$

where A_1 and A_2 are the thallus areas at the beginning and end of a time interval $t_2 - t_1$. Taking natural logarithms

$$\log \frac{A^2}{A_1} = r(t_2 - t_1).$$

By rearrangement of this equation the mean relative growth rate r can be found from

$$r = \frac{\log A_2 - \log A_1}{t_2 - t_1}.$$

The units of r (the relative growth rate) are cm²/cm²/unit time.

For the relative growth rate to be a valid measure of the growth of the whole thallus it has to be assumed that all carbon that is fixed by the thallus is potentially available for radial growth. This assumption presupposes that there is translocation of fixed carbon from the centre to the edge of the thallus. However, there are few data to support this assumption and hence there is some doubt of the validity of applying the relative growth rate to the whole thallus. In addition, Armstrong (1974b) has shown that in large thalli of *Parmelia glabratula* subsp. *fuliginosa*, the radial growth rate remains relatively constant after fragmentation of the centre of the thallus. This result suggests that the centre of the thallus does not contribute to the radial growth of large thalli. It is possible that a more accurate measurement of the relative growth rate of large thalli could be made by expressing growth in relation to the peripheral area of the thallus rather than to the area of the whole thallus. More work on the transport of fixed carbon within the lichen thallus is needed to clarify this problem.

METHODS OF MEASURING THE RADIAL GROWTH RATE OF A LICHEN

An ideal method of measuring the radial growth rate of a lichen thallus should be quick, easy, inexpensive, accurate and applicable to lichen thalli on all substrata. If the method is quick, easy and inexpensive, then large numbers of thalli can be measured, while if the method is accurate the growth rate of thalli can be measured over short intervals of time. Neither of the two main methods that have been used in the past fulfils these requirements. Some lichenologists have measured growth rate by tracing the outline of thalli on plastic sheets and then retracing at a later date (Rydzak, 1956, 1961; Hale, 1959; Brodo, 1964, 1965) but this method is inaccurate over short intervals of time. Close-up photography of lichen thalli (Phillips, 1969; Hale, 1970) is a very accurate method but is too expensive and time consuming to be used in a large scale project which measures many thalli. Armstrong (1973), working in South Merionethshire, has described a rapid method of measuring the radial growth of foliose thalli on a slate substratum which is accurate to 0·1 mm per monthly period. The mean rate of advance of a sample of 10 lobes per thallus over graduated millimetres marked on the rock is measured under a magnification of × 10. Thalli growing on isolated slate fragments collected from rock surfaces are placed on horizontal boards in the field. A fully marked thallus can be set up in about 10 min while re-measuring growth after a time interval takes only a few minutes. Growth measurements can be recorded each month for foliose lichens and every 2 or 3 months for crustose lichens. The main disadvantage of the method is that measurements can only be made successfully on hard, smooth substrata (e.g. Ordovician slate) and hence this method is of limited application.

GROWTH PHASES IN THE LIFE OF A LICHEN THALLUS

It has been observed that a lichen thallus does not grow uniformly throughout its life. The growth phases of *Parmelia conspersa* have been studied by Hale (1967) and it was found that the radial growth rate was low until a thallus diameter of 1·0–1·5 cm was reached; the growth rate was then constant until about 12 cm after which growth slowed. A second pattern of growth has been described by Beschel (1958) when lichen growth rates were obtained from dated substrata. The radial growth rate was found to be most rapid in small thalli and reached a peak in thalli between 4 and 8 years old before growth became linear through to maturity. Hale's work suggests that there are three growth phases during the life of a thallus of *P. conspersa*: (*a*) a linear phase of growth where the radial growth rate is constant in thalli from 1·5 cm to 12 cm in diameter; (*b*) a prelinear phase where the radial growth rate in thalli smaller than 1·5 cm in diameter is slower than the constant rate; and (*c*) a postlinear phase where the radial growth rate declines in thalli bigger than 12 cm in diameter. Further studies by Armstrong (1973, 1974c) on the radial growth rate and relative growth rate expressed on a whole thallus basis, of *P. glabratula* subsp. *fuliginosa* thalli from 1 mm to 6 cm in diameter has also led to the proposal of different growth phases during the life of a lichen thallus. A summary of the changes in growth rate with time and the corresponding phases of growth is shown in Fig. 1. For most of the life of a lichen thallus (between about 1·5 cm and 6 cm thallus diameter in *P. glabratula* subsp. *fuliginosa*) the radial growth rate (mm/year) is const nt. During this period there is a constant increase in radius each year and this phase is called the linear phase. In the prelinear phase (thalli smaller than about 1·5 cm in diameter) the radial growth rate increases with time and the thallus radius increases logarithmically. There is no evidence to suggest that the radial growth rate declines either in the largest thalli examined or in the senescent phase after fragmentation of the centre of the thallus in *P. glabratula* subsp. *fuliginosa* and the results are not consistent with a postlinear phase in the radial growth of a lichen thallus (Armstrong, 1974b). If the relative growth rate (cm²/cm²/year) expressed in relation to the total area of the thallus is measured, then the prelinear phase is marked by a rapid rise in the growth rate followed by a dec ne, the linear phase by a negative exponential decline and the onset of the fragmentation of the thallus by a rise in the relative growth rate. It has been suggested that because in large thalli the centre of the thallus may not contribute to radial growth, the relative growth rate should be measured by expressing growth in relation to the peripheral area rather than to the area of the whole thallus. This measurement would not increase after fragmentation of the centre of the thallus. Hence, the data are not consistent with a postlinear phase in the radial growth or the relative growth of a lichen thallus. The negative exponential decline in the relative growth rate during the constant radial growth rate phase may also be attributable to expressing growth on a whole thallus basis. It is possible that the whole of the thallus may contribute to the radial growth of

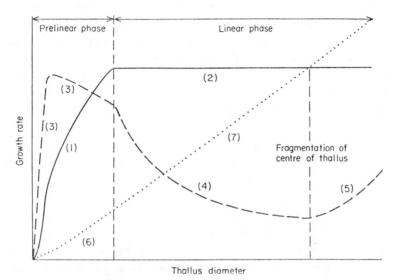

FIG. 1. Summary of the changes in growth rate which occur during the life of a lichen thallus and the corresponding growth phases: (————) = Radial growth rate (mm/year); (— — — —) = Relative growth rate (cm²/cm²/year); (. . . . .) = Thallus radius (cm).

(1) Increasing radial growth rate in prelinear phase.
(2) Constant radial growth in linear phase.
(3) Changes in the relative growth rate in prelinear phase.
(4) Negative exponential decline in relative growth rate in linear phase.
(5) Increasing relative growth rate after fragmentation of the thallus.
(6) Logarithmic increase in radius in prelinear phase.
(7) Constant increase in radius in linear phase.

small thalli so that measurements of the relative growth rate of thalli in the prelinear phase calculated on a whole thallus basis may be valid.

This scheme for the changes in growth rate with thallus size has been shown on a slate substratum for four foliose lichens: *Parmelia conspersa*, *P. glabratula* subsp. *fuliginosa*, *P. saxatilis* and *Physcia orbicularis*; the scheme is consistent in general form with that of Hale (1967) but not consistent with that of Beschel

(1958). Whether the scheme suggested in this chapter is general for all lichens or whether the growth phases differ in different groups of lichens or under different environmental conditions are problems for the future.

If the diameter and the radial growth rate of a lichen thallus are known, then an estimate of the age of a thallus can be made by dividing the diameter by the radial growth rate. However, this method does not take account of the changes in the radial growth rate which occur during the life of a lichen thallus. In addition, if the age structure of a large population of thalli is to be determined, then it is impractical to measure the growth rate of every thallus in the population to be aged. A method of measuring the age structure of a lichen population must take account of these two problems.

To determine the age of an individual plant or animal in the field it is usual to relate age to a biological quantity which is a result of growth. This quantity can be a measure of size or weight or be an internal or external visible effect of growth, e.g. growth rings in trees and in the horns and shells of animals. If the relationship between age and a biological quantity is known for a random sample of individuals, then a calibration curve can be constructed and the age of unknown individuals estimated. In lichen populations greatest thallus diameter is a quick and easy quantity to measure in the field and may be expected to be related to age (Beschel, 1961). Populations of lichen thalli can be quickly sampled for greatest thallus diameter and the results, in size classes, expressed as a size structure for the population. A size structure can be converted to an age structure by using the relationship between radial growth rate and thallus diameter. Once the age structure of a lichen population is known, the population dynamics of lichens can be studied.

The age structure of a population of *Parmelia glabratula* subsp. *fuliginosa* was determined from the radial growth rates of 60 thalli with diameters from 1 mm to 6 cm measured in 1972 and 1973 (Armstrong, 1973, 1974b). The relationship between the radial growth rate (mm/year) and thallus diameter (cm) for *P. glabratula* is shown in Fig. 2. To construct a calibration curve of age against size, thallus diameter was divided into increments of 1·0 cm and the growth rate at the midpoint of each centimetre increment read off a fitted curve (growth rates at the points 0·5, 1·5, 2·5, 3·5 etc.). Each of the centimetre increments was divided by its midpoint growth rate to estimate the time taken to grow each centimetre. The age of a thallus of diameter A cm is then calculated by adding up the times for each centimetre up to A. A calibration curve of age against thallus diameter for *P. glabratula* is shown in Fig. 3. The relationship between

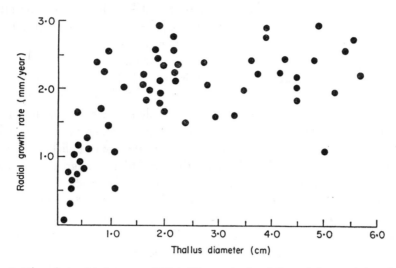

FIG. 2. The relationship between thallus diameter (cm) and the radial growth (mm/year) for 60 thalli of *Parmelia glabratula* subsp. *fuliginosa*.

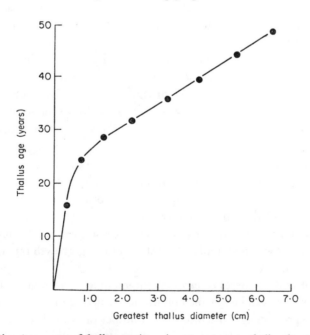

FIG. 3. A calibration curve of thallus age (years) against greatest thallus diameter (cm) for *Parmelia glabratula* subsp. *fuliginosa*.

age and size is linear in thalli above approximately 1·5 cm when the radial growth rate is constant. As a consequence of the low radial growth rates of thalli up to 1·0 cm in diameter, small increments in size represent large increments in age in about the first centimetre of growth.

There are several sources of error in this method of measuring the age of a lichen thallus:

1. Because of the low radial growth rates of thalli less than 1·0 cm in diameter a small error in the measurements of growth rate or diameter of such a thallus will result in a large error in the measurement of the age of a thallus.

2. The method assumes that all thalli in the population have grown either as a result of lichenization or from very small vegetative reproductive structures (soredia or isidia) so that age is zero when thallus diameter is zero. However, if thalli have become established from thallus fragments (which may vary considerably in size) then at the time of establishment (zero age) thallus diameter will be greater than zero. To reduce this error an estimate of the frequency of the different methods of reproduction in a population will be needed.

3. The method assumes that there is a relationship between the size and age of a thallus and implies both that thalli of equal size are the same age and that thalli of different size are of different ages. However, because of the variability in the radial growth rate shown by thalli of equal size in the linear phase of radial growth neither of these assumptions may be valid for all thalli in the population. In a large population it is impractical to measure the growth rate of every thallus to be aged in order to take account of this error. However, the spread of the radial growth rates around the mean from a random sample of thalli in the linear phase of growth from the population will give an estimate of this error in measuring age.

4. The calibration curve does not take account of yearly variation in the radial growth rate of a thallus and some preliminary data by Armstrong (unpublished) on three successive years' growth suggest this error may be large. More data on the growth rates of thalli measured over several successive years will be needed to reduce this error. Consequently, accurate measurements of the age of a lichen thallus which grows on an undated substratum using growth rate data remains a problem for the future.

In spite of many sources of error in the method of measuring the age of a lichen thallus, data can be obtained on the age structure of lichen populations. It is usual to express an age structure as the number of individuals per age class. If the age classes chosen are large (e.g. 5 year age classes could be used) many of the errors due to the method of measuring age will not be significant when large populations are sampled. To obtain information on the recruitment and mortality of thalli in a population from an age structure requires only that the relative numbers of thalli in the age classes be known. Hence, despite the errors in measuring age using present growth rate data the age structure of a population could be determined with sufficient accuracy to obtain information on the population dynamics of lichens (see Armstrong, 1974c).

USE OF LICHEN GROWTH RATES IN AN ECOLOGICAL EXPERIMENT

In 1972 and 1973 a transplant experiment was performed to examine the influence of aspect on the growth of four saxicolous lichens. One of these species, *Parmelia conspersa*, occurs with high frequency on south-east facing rock surfaces and with low frequency on north-west facing rock surfaces in South Merionethshire, Wales (Armstrong, 1974a). Three samples of *P. conspersa* thalli growing on slate fragments were collected from rock surfaces and placed on horizontal boards in the field. The radial growth of each thallus was measured in each month in 1972. Hence, all thalli used in the experiment were grown under similar conditions for a year prior to the experiment. The radial growth of each thallus was measured using the method of Armstrong (1973) and expressed in units of cm²/cm² (Woolhouse, 1968). On 1st January 1973 sample 1, the control (20 thalli) was left on the boards to measure the differences in growth between corresponding months in 1973 and 1972. The rock fragments of sample 2 (12 thalli) were fixed on to a south-east facing rock surface and the rock fragments of sample 3 (12 thalli) were fixed on to a north-west facing rock surface. The rock fragments were placed in "holes" chipped from the rock and were taped to the surface with a strong, water-resistant adhesive tape. The total growth of all thalli in the control sample and in each of the two transplanted samples was measured each month for the first 6 months of 1973. The transplant experiment is illustrated schematically in Fig. 4. The total growth of samples 2 and 3 in 1973 after transplantation was compared with their growth in the corresponding month of 1972 when the samples were under control conditions on the boards and differences in growth which were attributable to differences in years accounted for using sample 1. The result of this transplant experiment for the first six months of 1973 is shown in Fig. 5. The growth of *P. conspersa* thalli was inhibited at the north-west aspect when compared with growth under control conditions, while growth was not significantly different from the control at the south-east aspect when a paired sample *t*-test was applied to the data. Hence, *P. conspersa* occurs with low frequency on north-west facing rock surfaces and growth of the transplanted thalli is inhibited on north-west facing rock surfaces. There are two mechanisms by which the environment may act to influence the distribution of *P. conspersa*: (*a*) the environment may inhibit the establishment of

P. conspersa thalli on north-west rock surfaces; or (*b*) the environment may influence the growth of the thalli after establishment so that when growth conditions are unfavourable thalli of *P. conspersa* may be eliminated by competition. There is evidence to suggest that the establishment of *P. conspersa* thalli is not inhibited on north-west faces and the low frequency of *P. conspersa* on north-west faces may be due to an effect of the environment on the growth of the thalli. The growth rate of a lichen thallus reflects the net carbon assimilation rate of the lichen which in turn reflects the balance between photosyn-

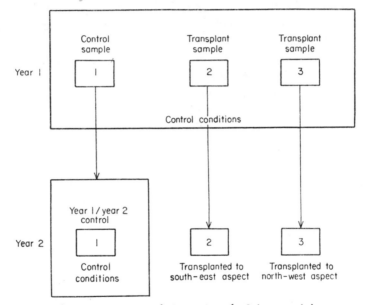

FIG. 4. The transplant experiment technique. Sample 2 in year 1 is compared with sample 2 in year 2 after transplantation. The difference in growth of sample 2 in years 1 and 2 is accounted for using sample 1.

thesis and respiration. The net assimilation rate of the lichen thallus is known to be influenced by light intensity and the water content of the thallus (Harris, 1971) and by the frequency of wetting and drying (Smith and Molesworth, 1973). All of these environmental factors may be expected to differ with aspect. Hence, it is possible that transplant experiments, together with measurements of the radial growth rate, will increase our understanding of lichen ecology by interpreting lichen distributions in physiological terms.

CONCLUSIONS—SOME FUTURE PROBLEMS

The method of measuring the growth rate of lichens described by Armstrong (1973) has been used to investigate the changes in growth rate that occur during

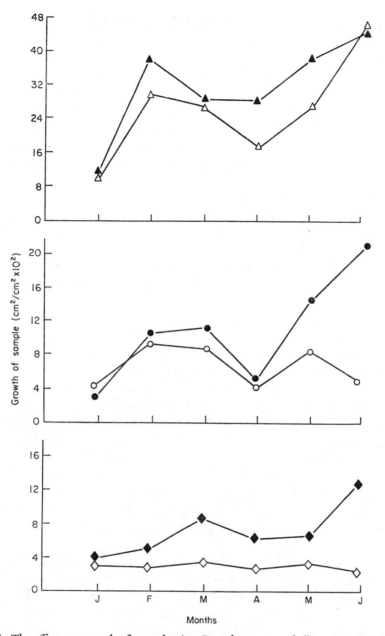

FIG. 5. The effect on growth of transplanting *Parmelia conspersa* thalli to sites of different aspect. Sample 1 ▲ = control year 1, △ = control year 2. Sample 2 ● = control year 1, ○ = growth at south-east aspect year 2. Sample 3 ◆ = control year 1, ◇ = growth at north-west aspect year 2.

the life of thalli of four foliose saxicolous lichens. The delimitation of the growth phases of these lichens is based on growth rate measurements of thalli of different size so that each stage in the life of the thallus is represented by a different thallus. This method of determining the growth phases of a thallus throughout its life is not ideal because the variability in radial growth rate between thalli of equal size when grown under standard conditions is high (Armstrong, 1974c). This variability will obscure the differences in growth rate between thalli which are due to the age of the thallus. The delimitation of the growth phases is also based on measurements of growth rate in one year (1973 for the linear phase and 1972 for the prelinear phase). There is evidence from a preliminary study on three successive years' growth to suggest that the yearly variation in growth rate of a thallus is high (Armstrong, unpublished), and this variability will obscure differences in growth rate which are due to the age of the thallus. An alternative approach to the study of the growth phases of a lichen on an undated substratum could be made by measuring the growth rate of a single thallus over several successive years but such a long-term study is beyond the scope of present investigations.

The radial growth of the thallus has been considered by lichenologists in the past and has been expressed as a radial growth rate (mm/year) or as a relative growth rate ($cm^2/cm^2/year$). Although it has been suggested in this chapter that this measurement of growth can be applied both to ageing a thallus and to ecological experiments in the field, the radial growth rate is only one parameter of growth. There may be changes in the thickness of the thallus with age as well as increases in area which will contribute to the total dry matter production of the thallus. At some stages in thallus development the contribution to the total dry matter production of the thallus by an increase in thickness may be greater than the contribution from radial growth. A diagram of the growth phases of a lichen which is based on changes in thallus thickness as well as radius may be significantly different to the scheme suggested in this chapter. A study of the changes in thallus thickness which occur with time is, again, a problem for the future.

It has been suggested in this chapter that once the problem of measuring lichen growth rate is solved then progress in lichen ecology will be rapid. In a review of lichen physiology Farrar (1973) has described how small but long-term changes in the environment can be important in influencing lichen growth. Some success has been achieved in growing lichens for varying periods in the laboratory (Kershaw and Millbank, 1969; Dibben, 1971), but growth cabinets cannot provide conditions for lichen growth over long periods comparable to those in the field. Some authors have rightly suggested that rapid progress in lichen ecology will come from experiments on whole thalli in growth cabinets in the laboratory but few such experiments have been carried out. However, progress in investigating the ecology of lichens can be made by performing field experiments. There are few measurements of performance that can be made on a thallus in the field after experimental treatment. One of the measurements that can be made is the radial growth rate of the thallus and, consequently, an accurate method of measuring the radial growth rate is needed. Although Armstrong (1973) has described such a method and has performed some field experiments using the method (see text and Armstrong, 1974c), measurements can only be carried out successfully on hard, smooth substrata. The discovery of new methods which can be applied to lichens on other substrata is a major problem for future lichenologists.

ACKNOWLEDGEMENTS

The help of Professors F. R. Whatley and D. C. Smith in discussions of this work and in the preparation of the manuscript is gratefully acknowledged. Most of the work described in this chapter was financed by the Natural Environment Research Council.

REFERENCES

DES ABBAYES, H. (1951). "Traité de Lichenologie." [*Encycl. Biol.* **41**, i–x, 1–217.] Lechevalier, Paris.

ARMSTRONG, R. A. (1973). Seasonal growth and growth rate colony size relationships in six species of saxicolous lichens. *New Phytol.* **72**, 1023–1030.

ARMSTRONG, R. A. (1974a). The descriptive ecology of saxicolous lichens in an area of South Merionethshire, Wales. *J. Ecol.* **62**, 33–45.

ARMSTRONG, R. A. (1974b). Growth phases in the life of a lichen thallus. *New Phytol.* **73**, 913–918.

ARMSTRONG, R. A. (1974c). The Structure and Dynamics of Saxicolous Lichen Communities. D.Phil. thesis, University of Oxford.

BESCHEL, R. E. (1958). Flechtenvereine der Städte, Stadtflechten und ihr Wachstum. *Ber. Naturw.-med. Ver. Innsbruck* **52**, 1–158.

BESCHEL, R. E. (1961). Dating rock surfaces by lichen growth and its application to glaciology and physiography (lichenometry). *In* "Geology of the Arctic" (G. O. Raasch, ed.), **2**, 1044–1062. University of Toronto Press, Toronto.

BLACKMAN, V. H. (1919). The compound interest law and plant growth. *Ann. Bot.* **23**, 353–360.

BRODO, I. M. (1964). Field studies of the effects of ionising radiation on lichens. *Bryologist* **67**, 76–87.

BRODO, I. M. (1965). Studies on growth rates of corticolous lichens on Long Island, New York. *Bryologist* **68**, 451–456.

DIBBEN, M. J. (1971). Whole-lichen culture in a phytotron. *Lichenologist* **5**, 1–10.

FARRAR, J. F. (1973). Lichen Physiology: Progress and Pitfalls. *In* "Air Pollution and Lichens" (B. W. Ferry, M. S. Baddeley and D. L. Hawksworth, eds), pp. 238–282. University of London, Athlone Press, London.

FREY, E. (1959). Die Flechtenflora und -vegetation des National-parks in Unterengardin. II. Die Entwicklung der Flechtenvegetation auf photogrammetrisch kontrollierten Dauerflächen. *Ergebn. wiss. Unters. schweiz Natn Parks* **6**, 241–320.

HALE, M. E. (1959). Studies on lichen growth rate and succession. *Bull. Torrey bot. Club* **66**, 126–129.

HALE, M. E. (1967). "The Biology of Lichens." Arnold, London.

HALE, M. E. (1970). Single-lobe growth rate patterns in the lichen *Parmelia caperata*. *Bryologist* **73**, 72–81.

HARRIS, G. P. (1971). The ecology of corticolous lichens II. The relationship between physiology and the environment. *J. Ecol.* **59**, 441–452.

KERSHAW, K. A. and MILLBANK, J. W. (1969). A controlled environment lichen growth chamber. *Lichenologist* **4**, 83–87.

PHILLIPS, H. C. (1969). Annual growth rates of three species of foliose lichens determined photographically. *Bull. Torrey bot. Club* **96**, 202–206.

PLATT, R. B. and AMSLER, F. P. (1955). A basic method for the immediate study of lichen growth rates and succession. *J. Tenn. Acad. Sci.* **30**, 177–183.

RYDZAK, J. (1956). A method of studying growth in lichens. *Annls Univ. Mariae Curie-Skłodowska* **C, 10**, 87–91.

RYDZAK, J. (1961). Investigations on the growth rates of lichens. *Ann. Univ. Mariae Curie-Skłodowska* **C, 16**, 1–15.

SMITH, D. C. and MOLESWORTH, S. (1973). Lichen physiology XIII. Effects of rewetting dry lichens. *New Phytol.* **72**, 525–533.

WOOLHOUSE, H. W. (1968). The measurement of growth rates in lichens. *Lichenologist* **4**, 32–33.

13 | Performance of *Lecanora muralis* in an Urban Environment

M. R. D. SEAWARD

Postgraduate School of Studies in Environmental Science,
University of Bradford, England

Abstract: Man-made substrates are recommended for use in lichen-based biological scales for the assessment of urban environments. Asbestos-cement, investigated here in terms of its water-holding and buffering capacities, is considered to be particularly valuable. *Lecanora muralis* is able to exploit the unique properties of asbestos-cement and thus extend its ecological range in urban areas in Britain.

The use of biological performance as a biological scale is illustrated by studies on naturally occurring and transplanted material of *L. muralis*. *L. muralis* s.l. is a highly polymorphic species ecologically versatile in natural habitats, but shows a remarkable consistency in its urban facies being treated here as an ecad of var. *muralis*. The morphological, histological and ecological performances of this ecad are compared with material from non-urban environments; particular attention is paid to growth-rate, distributional limits (especially in respect of substrate preference and the change in air pollution levels) and water relations in the West Riding of Yorkshire conurbation centred upon Leeds and Bradford. Transplant material from non-urban sites did not achieve the performance of the naturally occurring urban ecad.

INTRODUCTION

Biological scales for pollution evaluation are particularly valuable tools to the lichen ecologist, where direct measurement of the quality and quantity of the emission, the rainwater, the ground water, or the soil are impracticable. Biological scales are designed for on-the-spot diagnoses, but it is possible to analyse lichen material, in much the same way as the environment, to determine the morphological, histological, ecological and physiological performances for short- or long-duration exposures. The measurements of growth rate, productivity, reproductive capacity, deformity, discoloration, chlorophyll content, respiratory activity and ionic content are all examples of the use of lichens in the construction of biological scales. More research is, however, required for these

Systematics Association Special Volume No. 8, "Lichenology: Progress and Problems", edited by D. H. Brown, D. L. Hawksworth and R. H. Bailey, 1976, pp. 323–357. Academic Press, London and New York.

types of diagnoses; for example, is the increase in size, and number per unit area, of apothecia reflective of an increase in reproductive capacity, or merely of a malformation of the central part of the thallus? This can only be answered by determination of the number and viability of the ascospores produced. Symptom mapping is open to criticism; discoloration and deformity may be naturally associated with general senescence, or injury from insects or diseases, or from localized sprays or coprophilous hypertrophication. A state of rejuvenation, as indicated by the appearance of new thalli and (or) an increase in growth rate, may be taken as an indication that an area is recovering from a previous period of more intense atmospheric pollution.

Biological scales vary from region to region; the lichens used in their construction vary not only according to major geographical differences, such as climate and topography, but also to the availability of substrate(s). The latter depend on major vegetational (for corticolous species), edaphic (for terricolous species), and geological (for saxicolous species) factors operating in the particular region. Biological scales for use in urban environments should therefore be based on widely available man-made substrates, such as mortar, concrete and asbestos-cement, which are a major feature of urban building materials in most latitudes. These substrates have been instrumental in the extension of many lichen distributions into urban areas, in contrast to corticolous substrates which are rapidly disappearing. For a detailed interpretation of a gradient of increasing lichen diversity and luxuriance, it is necessary to have a continuum of a particular substrate; the latter, from a corticolous point of view, is out of the question for many areas, both urban and rural, in the British Isles.

The majority of ecological (including transplantation) and physiological studies on lichens to date have been concerned with corticolous species. Corticolous lichens may sometimes be adapted to higher humidities (and, in many cases, higher rainfall—cf. transplantation work to Gilbert, 1968), and generally their moisture needs are not sustained by the bark substrate in urban environments; the value of corticolous species for transplantation needs careful consideration in these respects. It is ironic that the limitation of trees and other vegetation in urban areas is a major cause of the lowering of humidity there. The growth of the substrate is an added complication for corticolous transplants, and saxicolous transplants suffer damage due to detachment and fixation/ adhesive procedures. Much of the following information is derived from a programme of non-urban to urban, and vice versa, lichen transplantation work. Transplantation of the intact thallus of *Lecanora muralis* together with its accompanying saxicolous substrate (in this case asbestos-cement) proved highly successful and obviated the shortcomings outlined above.

The urban lichen flora depends upon a multiplicity of factors. This line of enquiry has been concerned with a study of the relationship of one species to the urban environment as a whole rather than simply in terms of pollution. The complexity of factors acting directly or indirectly on the lichen in such a way as to produce a gradation of thallial performance along a transect which radiates from the centre of an urban complex and (or) source of pollutant(s). *Lecanora muralis* has proved to be a most suitable lichen for gradation work in respect of its morphological, histological ecological performances.

TAXONOMY

Representatives of the *Lecanora muralis* group are geographically ubiquitous, ecologically versatile and taxonomically polymorphic. The taxon *L. muralis* is by far the commonest member of the complex and there are numerous other described species, subspecies and varieties that are referable to this one taxon. Many present-day North American lichenologists, for example, view this complex with some suspicion, and are apt to aggregate the numerous taxa (several at specific level) described by Fink (1935) under *L. muralis*. An exception is var. *garovaglii*, currently elevated to specific rank. Hale and Culberson (1970) gave specific status to eight members of the complex in the Continental United States and Canada, and Poelt (1958) included the *L. muralis* complex within *Lecanora (Placodium)* sect. *Placodium* s.s. containing 14 species. In both cases, several species would be best described as closely allied to, rather than integral members of, the complex. Seven species are listed within subgen. *Placodium* for the British Isles (James, 1965), of which the following are included in, or closely allied to, the *L. muralis* complex: *L. achariana*, *L. muralis* var. *muralis*, var. *albomarginata*, var. *diffracta* and var. *versicolor*, and *L. straminea*.

From the plethora of synonymy and taxonomic disarray, it is necessary to organize the information relating to the British material named as *L. muralis*. The four taxa belonging to *L. muralis* s.l., and recognized by James as occurring in the British Isles, are described below. The descriptions have been derived from examination of large quantities of field material, collected from a wide range of British habitats, and of herbarium material from the following sources: Bankfield Museum, Halifax (HFX); British Museum (Natural History), London (BM); Borough Museum, Keighley (KGY); Department of Botany, University of Leeds (LDS); and the author's herbarium. For better interpretation of the British varieties, comparative foreign herbarium material has been consulted from the following sources: British Museum (Natural History), London (BM); University of Colorado, Boulder (COLO); Botanical Museum, University of Helsinki (H); Borough Museum, Keighley (KGY); University of

Minnesota, Minneapolis (MIN); National Museum of Canada, Ottawa (CANL); National Museum of Wales (NMW); and the author's herbarium.

Distributional data for the British Isles have been derived from an analysis of completed cards made for the British Lichen Society's Distribution Maps Scheme and from literature sources far too numerous to itemize. The world distribution of *L. muralis* s.l. (Fig. 1) has been constructed from the examination of material from the herbaria listed above, from personal communications (T. Ahti, I. M. Brodo, C. W. Dodge, K. G. Foote, M. Galun, J. Poelt and C. M. Wetmore), and from the following literature sources: Awasthi (1965), Awasthi and Singh (1970), Barkman (1958), Bird (1966), Degelius (1945), Fink (1935), Kleinig (1966), Klement (1965a, b), Krause and Klement (1962), Macoun (1902), Magnusson and Zahlbruckner (1944), Otto and Ahti (1967), Schubert and Klement (1966), Suza (1925), Thomson, Scotter and Ahti (1969), Trass (1963), Weber and Viereck (1967) and Werner (1955).

The synonymy indicated for the four taxa described below has been derived from Crombie (1894), Fink (1935), Grummann (1963), James (1965), Poelt (1958), Smith (1918), Zahlbruckner (1927–28) and other sources; many taxa described in these papers have been omitted here. It must be emphasized that the author has not endeavoured to typify and check the identities of these names and that a thorough revision of the nomenclature of the group is still required.

Lecanora muralis
var. muralis*

Thallus squamulose (perhaps best described as crustose at the centre and foliose at the margins; placodioid); cartilaginous when wet, tartareous when dry; glaucous greyish-green

---

* *Lecanora muralis* (Schreb.) Rabenh., *Deut. Kryptfl.* **2**, 42, 1845 (Basionym: *Lichen muralis* Schreb., *Spic. fl. lips.*, 130, 1771; selected synonyms: *Lobaria muralis* (Schreb.) Hoffm., *Parmelia saxicola* var. *muralis* (Schreb.) Fr., *Patellaria muralis* (Schreb.) Trev., *Placodium murale* (Schreb.) Frege, *Psora muralis* (Schreb.) Hoffm., *Squamaria muralis* (Schreb.) Elenk., *Placolecanora muralis* (Schreb.) Räs.; *Parmelia muralis* var. *albopulverulenta* Schaer., *Lecanora muralis* var. *albopulverulenta* (Schaer.) Rabenh., *L. saxicola* var. *albopulverulenta* (Schaer.) Jatta, *Parmelia saxicola* var. *albopulverulenta* (Schaer.) Trev., *Patellaria albopulverulenta* (Schaer.) Trev., *Placodium albopulverulentum* (Schaer.) Massal., *P. saxicola* var. *albopulverulentum* (Schaer.) Körb., *Squamaria saxicola* var. *albopulverulenta* (Schaer.) Nyl.; *Lichen lingulatus* Vill.; *L. ochroleucus* Wulf., *Parmelia muralis* var. *ochroleuca* (Wulf.) Schaer., *Placodium ochroleucum* (Wulf.) DC., *Lichen luteolus* Gmel.; *Placodium saxicola* var. *compactum* Körb., *Placolecanora muralis* f. *compacta* (Körb.) Kopacz.; *Placodium saxicola* var. *vulgare* Körb., *Lecanora saxicola* var. *vulgaris* (Körb.) Th. Fr., *Psoroma saxicola* var. *vulgare* (Körb.) Flag.; *Lichen saxicola* Poll., *Parmelia saxicola* (Poll.) Ach., *Lecanora saxicola* (Poll.) Ach., *Placodium saxicola* (Poll.) Frege, *Squamaria saxicola* (Poll.) Howitt, *Lecanora muralis* var. *saxicola* (Poll.) Lilj., *Pannularia muralis* var. *saxicola* (Poll.) Räs., *Squamaria muralis* var. *saxicola* (Poll.) Elenk.; note that "*Gasparrinia saxicola* Tornab." listed by Zahlbruckner under *L. muralis* may be *Caloplaca saxicola* (Hoffm.) Nordin—see Nordin, 1972).

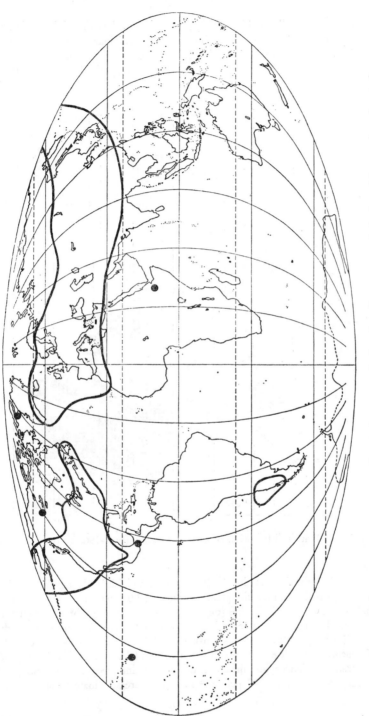

Fig. 1. World distribution of *Lecanora muralis* s.l., being mainly found in the temperate latitudes of the northern hemisphere.

to brown when wet, dull greyish-yellow/green/brown when dry; lobes paler, sometimes more yellowish; lower surface white, with no rhizinae; in young rosettes the
greenish colour of thallus predominates but in older rosettes the colour, with the exception of the marginal lobes, is dictated by the colour of the apothecia; very thin thallus at

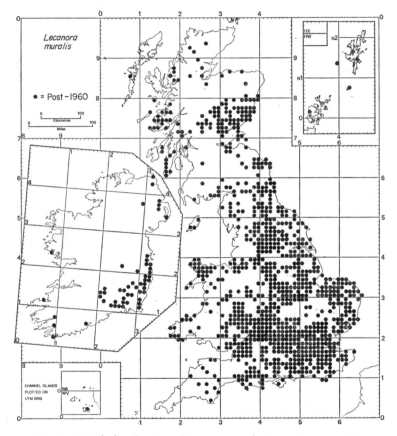

FIG. 2. British distribution of *Lecanora muralis* (post-1960 records).

the centre to relatively thick at the margin; soredia and isidia absent; marginal lobes
simple, sinuately divided or multifid, plane or concave, 2·0–3·5 mm long and 0·5–1·5 mm
broad; thallus next to marginal lobes (or when relatively few apothecia present) in the
form of crowded crenate scales; central area more or less areolate and closely adnate;
cortex K— (? faintly yellow on occasions), C—, KC—; medulla P—(P— or P+ red;
contains fumarprotocetraric acid in North America *fide* Hale, 1969); other lichen products
recorded; atranorin, leucotylin, lichenin, usnic acid and zeorin (Culberson, 1969);
apothecia very common, 70–140 per cm² in central areas; adnate to subsessile, thin

exciple; 0·6–1·8 mm diam; flat to concave disc (the latter perhaps due to crowding), yellow-brown (pale buff) to tawny-red-brown (differing greatly in colour on the same thallus); margin *c.* 0·1 mm wide, slightly raised; entire and smooth at first, becoming crenate, flexuose and deformed later due to crowding; hymenium 50–80 μm; asci 30–40 × 8–12 μm; paraphyses 1–1·5 μm in width, straight and coherent; spores simple, ellipsoid, 9–14 × 4–6 μm.

Many poor forms exist in natural situations which are lacking in some or many of the above characteristics.

Chiefly saxicolous, being widespread on a variety of rocks and building materials, but occasionally corticolous, lignicolous or muscicolous; thermophilic/photophilic, being found in sunny and exposed situations, usually limited to level areas or gentle slopes and rare on steep slopes; nitrophilous/ornithocoprophilous, preferring stonework frequented and manured by birds; hydrophilous (when substrate has a low water-holding capacity then generally found in areas with a high annual rainfall); found on calcareous and non-calcareous substrates in natural situations but limited to the former in more central urban areas; substrate pH range of non-urban material, excluding varieties named below, 4·75–7·80 (mode 6·9–7·1) = neutrophil (cf. Mattick, 1932: range 4·35–5·15, mode 4·4–4·5); a natural component of many saxicolous lichen assemblages, especially in hilly districts, throughout the world, undoubtedly on the increase—synanthropic (associated with man) and spreading throughout industrial, urban and lowland areas of the British Isles (Fig. 2); described as rare in many lichen floras of English counties at the beginning of this century (e.g. indicated as very rare in south Lancashire by Wheldon and Travis, 1915) and although 50 years later the vice-county distribution appears extensive (Watson, 1953) the records used in its compilation are meagre and represent in many cases only one or two records from an entire county; post-1960 10 km grid square mapping has shown this species to be present in approximately 37% of the squares which have received coverage.

var. albomarginata*

This variety is distinguished from var. *muralis* in the following: both central and marginal areas thicker; less adpressed; squamules very frequent, with some or all having white margins; marginal lobes inflexed, ± imbricate, ± crenate and slightly broader; K−, C−; apothecia larger, with brown to deep brown epithecium; asci 30–40× 7–9 μm; paraphyses 70–80× 1 μm; spores 10–12× 6–7 μm.

Substrate pH range 5·65–8·35 (mode 7·7–7·8); frequent in upland areas of Britain, particularly Scotland, but uncommon over most of England; on rocks, walls and cement.

---

* *Lecanora muralis* var. *albomarginata* (Nyl.) Tomin, *Opr. kork. lish. Eur. SSSR*, 424, 1956 (Basionym: *Squamaria saxicola* var. *albomarginata* Nyl., *Not. Sällsk. fauna fl. fenn.* **11**, 181, 1871; selected synonyms: *Lecanora albomarginata* (Nyl.) Stein., *L. saxicola* var. *albomarginata* (Nyl.) Th. Fr., *Squamaria saxicola* subsp. *albomarginata* (Nyl.) Cromb., *Placolecanora albomarginata* (Nyl.) Räs. ex Hakul., *P. muralis* var. *albomarginata* (Nyl.) B. de Lesd., *P. muralis* f. *albomarginata* (Nyl.) Kopacz.).

var. diffracta*

Differs from var. *muralis* in the following: much thinner thallus, almost entirely areolate (i.e. diffract); highly crustaceous and very difficult to separate from substrate; lobes absent, or occasionally short and undeveloped; K−, C−; darker thallus yellowish brown to reddish; squamules limited and black-edged; areoles angular and black-edged; apothecia limited, adnate and 0·5–1·5 mm diam; epithecium brown to dark reddish-brown, flat to slightly convex; margin± crenate, often obscure; spores 8–12× 4–6 μm.

Substrate pH range 4·15–4·55 (mode 4·4–4·5); frequent on siliceous rocks in upland areas of Britain; adaptable to both smooth and rough substrates.

var. versicolor†

Differs from var. *muralis* in the following: thallus small, smooth, thin and very areolate; lobes absent or short and plicate; centre often lacking continuous thallus, but rejuvenating through squamules and/or lobes in older parts; yellowish white to pale buff, ± powdery white; K−, C−; apothecia± immersed, small (0·4–0·9 mm) but prominent due to white pulverulent margins; disc fawn to dark brown, flat to concave; margin subcrenate, becoming angulose through crowding; asci 40× 10–12 μm; spores 8–14× 4–7 μm.

Substrate pH range 5·50–7·80 (mode 6·4–6·7); frequent on calcareous and non-calcareous rocks (especially granite) in Britain, also frequent on erratics; adaptable to both smooth and rough substrates; in both upland and maritime areas.

It will be seen from the above descriptions of the four accepted British infraspecific taxa that the *L. muralis* complex is a highly variable group. One can

---

* *Lecanora muralis* var. *diffracta* (Ach.) Rabenh., *Deut. Kryptfl.* **2**, 42, 1845 (Basionym: *Lichen diffractus* Ach., *Lich. suec. Prod.*, 63, 1798; selected synonyms: *Lecanora diffracta* (Ach.) Ach., *Squamaria diffracta* (Ach.) Duby, *Parmelia saxicola* var. *diffracta* (Ach.) Fr., *P. muralis* var. *diffracta* (Ach.) Schaer., *Placodium saxicola* var. *diffractum* (Ach.) Flot., *P. diffractum* (Ach.) Massal., *P. murale* var. *diffractum* (Ach.) Arnold, *Psoroma saxicola* var. *diffracta* (Ach.) Rabenh., *Squamaria muralis* var. *diffracta* (Ach.) Poetsch, *S. saxicola* var. *diffracta* (Ach.) Fr., *Lecanora saxicola* var. *diffracta* (Ach.) Jatta, *Placolecanora muralis* f. *diffracta* (Ach.) Kopacz.; *Lecidea bolcana* Poll., *Lecanora muralis* var. *bolcana* (Poll.) Massal., *L. bolcana* (Poll.) Poelt, *Placolecanora bolcana* (Poll.) Kopacz.; *Squamaria insulata* DC., *S. saxicola* var. *insulata* (DC.) Nyl.; *S. muralis* var. *areolata* Leight., *Lecanora muralis* var. *areolata* (Leight.) Stein.; *Parmularia laatokkaensis* Räs., *Lecanora laatokkaensis* (Räs.) Poelt, *Placolecanora laatokkaensis* (Räs.) Kopacz.).

† *Lecanora muralis* var. *versicolor* (Pers.) Tuck., *Syn. N. Am. Lich.* **1**, 185, 1882 (Basionym: *Lichen versicolor* Pres., *Usteri's Annln Bot.* **11**, 24, 1794; selected synonyms: *Lobaria versicolor* (Pers.) Hoffm., *Parmelia versicolor* (Pers.) Ach., *Lecanora versicolor* (Pers.) Ach., *Placodium versicolor* (Ach.) Frege, *Parmelia saxicola* var. *versicolor* (Pers.) Fr., *Placodium saxicola* var. *versicolor* (Pers.) Flot., *Gasparrinia saxicola* var. *versicolor* (Pers.) Torn., *Squamaria saxicola* var. *versicolor* (Pers.) Nyl., *Placodium murale* var. *versicolor* (Pers.) Rabenh., *Psoroma saxicola* var. *versicolor* (Pers.) Rabenh., *Lecanora saxicola* var. *versicolor* (Pers.) Th. Fr., *Squamaria muralis* var. *versicolor* (Pers.) Poetsch, *S. versicolor* (Pers.) Oliv., *Placolecanora versicolor* (Pers.) Kopacz.; *Parmelia versicolor* var. *microcymatia* Wallr.; *Lecanora sulphurata* Harm.).

attach little taxonomic significance to the majority of the differences observed between the four "varieties". These differences are mainly connected with the ability of the thallus (probably in relation to its thickness) to produce crustose, squamulose and foliose characteristics to varying degrees in central and marginal areas. There is a gradation from foliose to crustose thalli for typical material of the taxa in the following sequence: "*muralis*", "*albomarginata*", "*diffracta*" and "*versicolor*". Associated with this gradation of thallus characters is the phenomenon of increased areolation which follows the same sequence for the taxa. Other characteristics, with slight differences from "variety" to "variety", are the coloration of thallus, coloration, size and shape of apothecial disc and margin, and size of asci and spores. The nature of the apothecial margin is taxonomically unreliable since its shape so often varies according to the pressure exerted by the crowding of apothecia.

There is also considerable ecological overlap between the taxa, although the substrate pH preference may have a bearing on variations of the thallus. The pH ranges have been determined, electrometrically, from the measurement of at least 50 saxicolous substrates collected from immediately beneath the thalli of each of the four "varieties" from a wide range of British localities. The "varieties" *albomarginata*, *muralis*, *versicolor* and *diffracta* can be described, respectively, as being mildly calcicolous, neutrophilous, mildly acidophilous and strongly acidophilous when judged on their modal substrate pH values. The "average" of the pH measurement of approximately 300 substrate samples, from all the four taxa, proved to be 6·65, with a range of 4·15–8·35. With the exception of the "var. *diffracta*", the taxa are found on a wide range of calcareous and non-calcareous substrates. The "var. *diffracta*" has, however, limited pH preferences and is relatively more distinctive taxonomically. There is strong evidence to support the retention of this latter taxon at a varietal level (cf. *L. bolcana* in Poelt, 1958), together with the var. *muralis*. The varieties "*albomarginata*" and "*versicolor*" are probably best described as ecotypes, although experimental work will be needed to confirm their status.

The plasticity of development of the *L. muralis* complex is emphasized by the existence of a phenotype characteristic of the urban environment. This phenotypic modification is best described as an ecad; transplant experiments supporting this conclusion are described later.

PERFORMANCE IN AN URBAN ENVIRONMENT

Lecanora muralis has been described above as a highly polymorphic taxon. It nevertheless shows consistency in its urban facies. Urban material generally belongs to the var. *muralis*; a variety with a wide tolerance range in terms of the

urban environment. The urban ecad of var. *muralis* has been collected throughout the British Isles in recent years (Fig. 2). Var. *muralis* occurs in urban habitats
throughout Europe but, surprisingly, it has not been recorded from urban
situations in North America, despite its wide distribution and frequent occurrence in other habitats on that continent.

There are numerous intermediates between urban and naturally occurring
material from unpolluted and upland areas of the British Isles. The gradation
from non-urban to urban material is associated with the degree of areolation and

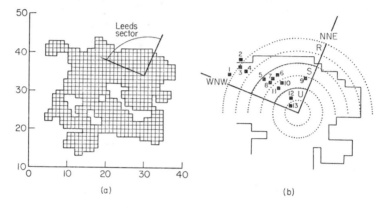

(a) (b)

Fig. 3. (a) West Riding conurbation: urbanized areas (607 km² supporting *c.* 1·7 million
population) expressed as 1 km² recording units of the national grid; west-north-
west to north-north-east sector of Leeds indicated (see also Fig. 11). (b) Location
of sites within west-north-west to north-north-east sector of Leeds used for
morphological, histological, ecological and eco-physiological analyses of the
performance of *Lecanora muralis*. Highly urbanized (U), suburban (S) and semi-
rural (R) zones delimited at 2 mile intervals from the city centre.

development of squamulose/foliose marginal lobes. Since var. *muralis* is syn-
anthropous the intermediates are common in a wide range of man-made and man-
influenced situations that are not necessarily fully urbanized. Typical localities
for such intermediates are farm buildings (where nitrophilous needs are catered for)
and villages. A representative non-urban collection from West Deeping (Lincoln-
shire) proved valuable for transplant and physiological experiments. Large
quantities of rosettes were available and easily dislodged (complete with substrate) owing to the excessive transverse flaking of the flat, predominantly
calcareous (pH 7·35) capstones of the wall. The performance of this material was
matched against that of West Riding conurbation material (transplants and
non-transplants) *in situ* and *in vitro*. Large quantities of *L. muralis* rosettes
(complete with substrate) were also made available for experimental work

owing to the demolition of two factories at suburban sites in North Leeds (site 5, in Fig. 3). These factory buildings had several acres of asbestos-cement sheet roofing (*c.* 30 years old) which supported a relatively diverse lichen flora, mostly containing assemblages dominated by *L. muralis.*

1. Morphological Performance

It is appropriate to describe in some detail the morphology of typical urban material of *Lecanora muralis.* Only the major characteristics in which the urban ecad differs from the non-urban var. *muralis* and the other "varieties" described above are given as follows:

Lecanora muralis var. muralis—urban ecad

Thallus fully squamulose, marginal lobes distinctly foliose; marginal lobes olive grey, often darkening inwardly through olive to olive black when dry; numerous white-edged crenate scales (? = squamules), which are abundant when apothecia are limited, olive grey; apothecial margins whitish olive grey when dry; marginal lobes, scales and apothecial margins olive when wet; the greener coloration of both wet and dry thalli (especially on asbestos-cement roofs) is often the result of a surface covering by algae (non-phycobionts) such as *Pleurococcus*; lower surface greyish white; thallus relatively thick throughout and comparatively easy to remove from its substrate (especially asbestos-cement); thalli regularly orbicular (distortion occurring on asbestos-cement tile junctions and corrugated sheeting); marginal lobes 3·0–4·5 mm long and 0·5–1·5 mm broad; infertile marginal zone 5–15 mm broad, adpressed, and followed inwards by scales/apothecia in ± adpressed zone—ratio of apothecia to scales highly variable; apothecia, in varying stages of development, always present, in numbers which vary from just a few (i.e. less than 40 for the entire thallus) to over 200 per cm^2 (more than 4000 for the entire thallus), the central apothecial/scaling zone representing 55–75% of the total thallial area; 0·4–2·0 mm diam; disc flat, but often concave or convex due to crowding; greyish yellow to brownish yellow or brown (but differing greatly in colour on the same thallus); margin 0·1–0·2 mm wide, slightly raised, highly flexuose and deformed due to crowding of both neighbouring apothecia and abundant scaling; distortion and eruption (non-areolation) of central areas of the thallus is common in older rosettes; asci 40–50× 10–12 μm at the widest point; paraphyses 70–80× 1·5 μm wide, straight to slightly curved, often with club-like tips; spores ovoid 6–11× 3–5 μm (average 8·1× 3·9 μm).

Entirely saxicolous, being widespread on a variety of building materials both calcareous (asbestos-cement tiles and sheeting, concrete), and siliceous (Millstone grit walls); substrate pH range 5·65–11·40 (mode 7·4–7·6). Cf. Mattick (1932) and Klement (1965a) where naturally occurring material is classed as distinctly acidophilous.

An obvious method of assessing the morphological performance of a lichen is to measure its growth rate. This practice has been successfully employed by numerous workers to compare the growth rates of different species; average annual radial rates of 0·01–27·0 mm for foliose species, 1·6–90·0 mm for fruticose species, and 0·57–3·0 mm for crustose species have been summarized by

Hale (1967). As will be appreciated from the ranges presented, the data has been assembled from a variety of sources involving widely different ecological and geographical situations, and numerous unrelated taxa. Brodo (1965), working on corticolous lichens in similar situations, showed uniformity in annual radial growth for different foliose and different crustose species of 1·26–1·78 mm and 0·36–0·86 mm respectively.

There is the necessity for comparisons of growth rates both between species or ecotypes and between *in situ, in vitro* and transplant materials. It is important to consider the most appropriate basis for growth measurement. For the radially symmetrical thalli of *L. muralis* annual radial increases would appear to be the simplest form of measurement. The annual increase of symmetrical thalli, thus measured, can be related to the absolute size of the thallus. Growth measurements, expressing a percentage increase of the entire thallus area (cf. Rydzak, 1961) or mean relative rate of thallus growth (cf. Woolhouse, 1968), may be applicable to the measurement of thalli of widely different sizes. Hale (1967), however, feels that in general the measurement of the rate of increase "is a function of the original size of the thallus" and is "a mathematical rather than a biological phenomenon".

Growth measurements of *L. muralis*, based on annual radial extension, have been confined to thalli within a limited range of diameters. Preliminary growth measurement work on a wide range of suburban material showed that the diameter of the rosette was not directly proportional to its age. Brodo (1965), however, showed no apparent significant difference between growth rates of juvenile thalli (i.e. less than 41 mm diam) and mature thalli of corticolous lichens. The delimitation of juvenile and mature is quite arbitrary and the results questionable in the light of the present investigations.

It can be seen from Fig. 4 that radial extension measurements are possible from the diameter ranges of *L. muralis* rosettes of the widely available lichen assemblages on asbestos-cement roofing. It is difficult to incorporate colonization and germination time factors into such graphical interpretations but a reasonably accurate age of rosette can be deduced. Radial extensions of *L. muralis* were determined over a 2 year period for the preliminary work (and over a 3 year, and in some instances a 4 year, period for the major investigations) from rosettes with an initial diameter within the range of 42·7–69·3 mm (*c.* 8–16 years old); the majority of data, however, was assembled from rosettes measuring 42·7–55·4 mm (*c.* 8–10 years old).

Growth measurements for the present investigation have involved the use of either photography or acetate sheet overlays orientated by markers on the asbestos-cement substrate; the latter involved techniques similar to those used

by Hale (1954) and Brodo (1965) for corticolous lichens. However, there was
no added complication due to the growth of the substrate, and a correction
factor (cf. Brodo, 1965) to account for this was not necessary.

The significant differences Brodo (1965) obtained from one year to another
show the marked effect of the climate on the annual growth rate. Generally,

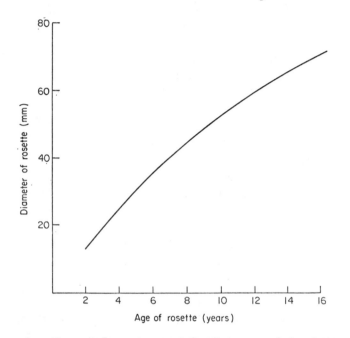

FIG. 4. Growth performance/longevity curve for *Lecanora muralis* based on analyses of
suburban material. (Colonization and germination time factors of the reproductive
propagules have been omitted from this interpretation.)

growth appears to be linearly related to mean annual rainfall. It is noticeable
from the West Riding conurbation studies that there is a greater uniformity in
the growth measurements from year to year (over a 4 year period), for the
urban material, than for the more rural material; the more stable urban climate
may have some bearing on these differences. Furthermore, there is a considerable difference in growth rates between summer and winter in temperate latitudes. Rydzak (1961) found that lichen growth virtually ceases in Poland during
the winter. Mean summer and winter growth measurements have been incorporated here and it is felt that the significant pollution differences between these
two periods may have some bearing on the growth performance of *L. muralis* in
urban areas.

Gilbert (1971) found no significant differences in the annual growth rate of *Parmelia saxatilis* in urban and non-urban sites, which is contrary to the results presented here for *L. muralis*. The annual radial extension (mean of two consecutive years) of naturally occurring *L. muralis*, as measured at nine sites within the west-north-west to north-north-east sector of Leeds (Fig. 3), varied between 2·84 and 6·05 mm (cf. 1·30 mm determined by Hakulinen, 1966), and increases with increasing distance from the city centre (Table I).

TABLE I. Relationship between annual radial extension in *Lecanora muralis* and distance from Leeds city centre

Site number	Distance from city centre (miles)	Annual radial extension (mm)	Number of rosettes measured
11	2·52	2·84	40
10	2·64	3·32	20
9	2·76	2·87	20
8	3·24	3·95	40
7	3·36	3·89	20
6	3·48	3·54	20
5	3·84	3·91	20
4	5·28	4·91	20
2	6·18	6·05	20

Differences in growth between the sites are very highly significant and most of these differences may be attributed to the effect of distance from the city centre. Annual radial extension growth decreases on average by 1·26 mm per mile ($P < 0.1\%$) as the centre is approached. Distance from the city centre, though a good measure of the environment, is not a perfect one, since deviations from the growth to be expected on this basis, though small, are also very highly significant. It may well be that all the rosettes measured were within an area very much under the influence of the urban environment, and therefore may all be classified as ecad material. It may well be that the growth rate here has a direct correlation with the annual rainfall; the significantly lower rainfall in the centre of Leeds needs consideration in this respect (Seaward, 1972, p. 60).

The selection of thalli of *L. muralis* for the purpose of growth measurements on asbestos-cement roofing needs careful attention. The growth form will often be determined by the undulatory nature of the substrate. Corrugated sheeting is particularly rewarding for the abundance of rosettes it supports, but suffers in respect of thallial distortion. Figures 5 and 6 illustrate how two "rosettes", when examined in 1971, could have been similarly interpreted; both thalli have the outward appearance of a more or less orbicular form with an inner zone of degeneration. This is indeed true in the case of thallus 02D but thallus 05A has

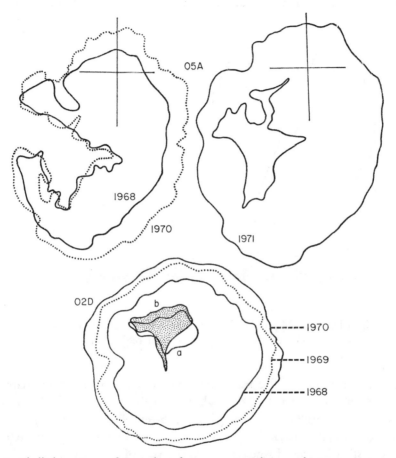

FIG. 5. Thallial tracings of transplanted *Lecanora muralis* on asbestos-cement roofing material; a and b indicate degenerate areas of thallus in March 1970 and April 1971 respectively. A further tracing for specimen 02D in October 1971 is given in Fig. 6.

attained its present shape through substrate restrictions at an earlier stage and the gradual coalescence of two particular outgrowths. The degeneration of the central area of thallus 02D took place after transplantation; its outline prior to transplantation is indicated by the 1968 line in Fig. 5.

A better indication of the nature of this degeneration, involving swelling,

followed by eruption and eventually by the total disappearance of the central area
(a to b in Fig. 5, and c in Fig. 6), is revealed photographically by using an
oblique light and through profile work. The profile of thallus 02D (Fig. 6)
shows in more detail the form assumed through degeneration. In this case

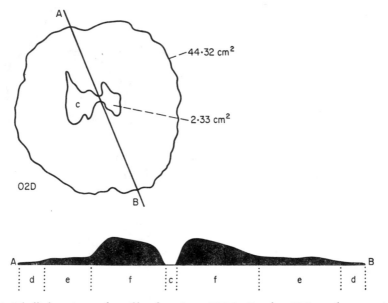

FIG. 6. Thallial tracing and profile of specimen 02D in October 1971, to show continued
degeneration (c) and distortion (f) of central regions of transplanted *Lecanora
muralis*.

degeneration occurred after transplantation from site 5 to site 11 (with, for
example, an approximately 20% higher mean daily SO_2 level). The relative
thicknesses of the marginal lobes (d) and sub-reproductive areas (e) are quite
typical of normal material, but the distortion (f) and total disappearance (c) are
characteristic of hastened senescence in these parts of the thallus.

A balance exists in the case of *L. muralis* between the marginal advance and
the inactivation of the central areas; the thallus in this instance, however, remains relatively intact (cf. crescent formation in *Parmelia saxatilis*, see Gilbert,
1971). At the same time it has been noted above that *L. muralis* produces an
abundance of both apothecia and rejuvenation scales. The reduction in radial
extension in more urban environments would appear to effect a balance between
a marginal advance and an inner reproductive capacity. The apothecial (and to
some extent scale) development, which causes this characteristic eruption, doe
not bring about areolation. The disappearance of the thallus (Fig. 6c) from the

substrate may, to a large extent, be accounted for in the epinastic curling which often follows eruption.

The above explanation of apothecial behaviour in urban environments is substantiated to a large measure by the following transplantation experiment. In March 1968, 40 rosettes of *L. muralis* on asbestos-cement roofing from site 5 were critically measured with the aid of a travelling microscope to determine the proportion of thallus devoted to reproduction and the number and collective surface area of the apothecia. Two batches of the substrate each bearing 20 rosettes were mathematically standardized in terms of these reproductive capacities; one batch remained at site 5 and one batch was transplanted to a similar location at site 11 (1·4 miles nearer the city centre than site 5). Similar measurements were carried out in October 1971 (43 months after the transplantation), and the data presented in Table II were obtained. It will be seen from Table II that the performance of the two batches differed considerably. The transplant material had reductions in total surface and reproductive areas of the thallus and the number and collective surface area of apothecia. However, the number of apothecia per unit area for the transplant material increased such that

TABLE II. Growth and reproductive behaviour of transplanted *Lecanora muralis*

	Site 5 (mean of 40 rosettes) March 1968	Site 5 (mean of 20 rosettes) October 1971	Site 11 (mean of 20 rosettes) October 1971	Significance of difference after transplanting[a]
Area of entire thallus (cm²)	22·87	50·92	42·15	★★★
Annual radial extension (mm)	—	3·72	2·69	★★★
Area of reproductive zone (cm²)	16·19	35·59	29·93	★★★
Total number of apothecia	6589	8046	7695	★
Number of apothecia per cm² of reproductive zone	407	226	257	★★★
Average size of apothecia (mm²)	0·21	0·38	0·39	—
Collective surface area of apothecia (cm²)	13·78	30·39	29·81	—

[a] ★★★, ★: significant at the 0·1%, 5% level respectively; —: not analysed statistically.

99·6% of the reproductive area was apothecial as compared with only 85·4% for the non-transplant material. One can see here the eruptive quality brought about by the higher apothecial activity, not only by an increase per unit area but also by a slight and possibly significant increase in the size of the apothecia, and the slower radial extension of the marginal lobes.

The above experimental work formed part of a more extensive programme of semi-rural to suburban and urban, and suburban to semi-rural lichen transplantation work. As a result of this work, the following additional data on the morphological behaviour of *L. muralis* were assembled. The relationship between the size of the rosette and its annual radial extension has already been discussed but it is clear from the experimental work that graphical interpretation (cf. Fig. 4) is more complex when transplantation and environmental data are taken into consideration.

A generalized picture of growth performance of naturally occurring and transplant material as expressed by annual radial extension per size of rosette measurements (mean of 2 years) categorizes the material into three groups as follows: (1) semi-rural (naturally occurring); (2) suburban (naturally occurring) and semi-rural to suburban (transplant); (3) semi-rural to urban (transplant) and suburban to urban (transplant). The terms "semi-rural", "suburban" and "urban" are utilized for describing material measured within the three zones delimited in Fig. 3. From the data in Tables I and II it is clear that the growth performance of material transplanted from semi-rural to suburban habitats had been reduced to that of the naturally occurring material in those suburban habitats. There appeared to be little difference between both the semi-rural and suburban material transplants to urban habitats, although the material in its entirety was quite clearly responding very differently. The suburban material transplanted to semi-rural sites never truly matched up to the performance of the naturally occurring material, and is best included under category 2. Such transplant material became areolate in the centre in many instances; the marginal lobes showed an increased growth rate during the summer months, but the overall annual radial extension was nullified to a large degree by a decreased growth rate during the winter months.

A synopsis of the transplant growth performance of *L. muralis* in relation to naturally occurring material in the semi-rural/suburban boundary area is given in Fig. 7. This synopsis is based on the measurement of 80, 220 and 40 rosettes at the semi-rural, suburban (including semi-rural/suburban non-transplants) and urban sites respectively. It will be seen from this synopsis that there were significant reductions in radial extension through transplantation to more urban sites in all instances and these reductions were considerably higher during the winter months than the summer months. Transplantation to more rural sites produced a significant increase in radial extension during the summer months but a slight reduction, probably of no significance here, occurred during the winter months.

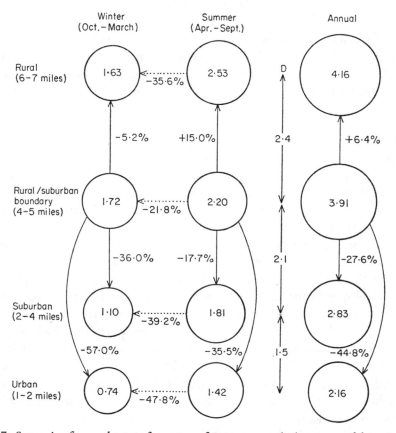

Fig. 7. Synopsis of transplant performance of *Lecanora muralis* (as measured by winter, summer and annual radial extension) at semi-rural, suburban and urban sites. Radial extension values (mm) are the mean of two annual measurements. Percentage differences relative to naturally growing material in the semi-rural/suburban boundary area are indicated. D = average distances (miles) between transplantations.

An important point to be deduced from this work is the variation between the winter and summer growth performances for both transplant and naturally occurring material. These differences, as measured by a percentage reduction in radial extension of the thallus during winter over that of summer, are 39·2 and

M

47·8% for the transplants to suburban and urban sites respectively. These reductions are 79·8% (transplant to suburban site) and 119·2% (transplant to urban site) lower than the naturally occurring (control) material at the semi-rural/suburban boundary site and become less dramatic when compared with the 63·3% reduction experienced by the material transplanted to the semi-rura site.

The general conclusion must be that urban environment factors have a more detrimental effect on growth during the winter months. This would support the view that toxicity (with, for example, higher levels of sulphur dioxide pollution), rather than drought, was the overriding influence on lichen performance.

2. Histological Performance

During the course of the compilation of taxonomic data it became apparent that there were histological differences not only between the several taxa but also between the material collected from the various habitats. There was a particularly noticeable difference between the urban ecad material and all the other material examined.

Histological preparations were obtained through microtome sectioning (setting 7μm) of wax-embedded lichen material previously fixed in Carnoy's fluid (acetic acid/chloroform/100% ethyl alcohol; 1:3:6). Staining techniques were employed in the preliminary work but proved to be unnecessary for histological interpretation. In some instances staining tended to mask the deposition products of urban material and often reacted unfavourably with the substrate (especially asbestos-cement) which was in intimate contact with the thallus section. A wide range of histological preparations made from thalli derived from 203 field sources was used for the following analyses. In many cases the interpretations were made from photographs of the unstained microscopical preparations.

L. muralis is composed essentially of an upper cortex which merges with the algal layer, a medulla and a layer which is in intimate contact with the substrate consisting of fungal hyphae that are unlike those which normally constitute a lower cortex and hence bring the medulla into more direct contact with the substrate. Two tissue layers are recognized here as constituting the major body of the thallus; the phycobiont layer refers collectively to the upper cortex and the algal layer (but excludes epinecral secretions and surface environmental contaminants), and the mycobiont layer refers to the clearly defined medulla (but excludes any less coherent hyphal areas beneath). Measurements were made of the mean vertical thicknesses of these two distinct tissue layers from sections which were cut through lobes in a zone 3–6 mm behind the thallus margin (cf. Hill and Woolhouse, 1966). Figure 8 has been derived from the examination of 125 and 68 thalli from different urban (within the west–north–west to north–north–east sector of Leeds) and rural (widespread in England and Scotland) sites respectively. At least 20 measurements were made from each of five different sections of the same thallus. The mean measurements (in micrometres) for the different tissue layers of urban ecad and rural materials are given in Table III.

TABLE III. Mean thickness of tissue layers (μm) in rural and urban ecad material of *Lecanora muralis*

	Phycobiont (upper cortex and algal layer)	Mycobiont (medulla)	Total thickness	Ratio phycobiont: mycobiont
Urban ecad	45·2	124·3	169·5	1:2·75
Rural ecad	69·4	94·1	163·5	1:1·36

These values indicate that there is no significant difference in the thallus thicknesses of urban ecad and rural material. There is, however, a significant difference in the relative thicknesses of the phycobiont and mycobiont tissue layers clearly indicated by differences in the phycobiont:mycobiont ratios (Table III).

Despite the morphological diversity exhibited by the *L. muralis* complex, it is clear from Fig. 8 that marginal lobes of non–urban material, when present, are histologically more consistent (especially in terms of the thickness of the mycobiont) than the lobes of urban material. Taxonomically the two "forms" would best be described according to these thicknesses as reported in Table IV. Trans-

TABLE IV. Variation in tissue thickness in rural and urban ecad material of *Lecanora muralis*

	Urban ecad (μm)	Rural ecad (μm)
Phycobiont	(15–)25–55(–105)	(35–)45–85(–115)
Mycobiont	(35–)50–150(–315)	(45–)50–155(–175)
Total thickness	(80–)95–205(–360)	(90–)95–205(–345)

plant material (i.e. rural to suburban sites, and suburban to rural sites, within the west–north–west to north–north–east sector of Leeds) showed certain characteristics intermediate between the above measurements. Particularly important in this respect is the ratio of phycobiont to mycobiont which was on average 1:1·89. Measurements for transplant material, however, were based on an analysis of only 20 microscopical preparations and more work is therefore necessary to substantiate what would appear to be a significant ratio.

One of the major anatomical features of *L. muralis* is the arrangement of the phycobiont in relation to the fungal cortex above and the fungal medulla

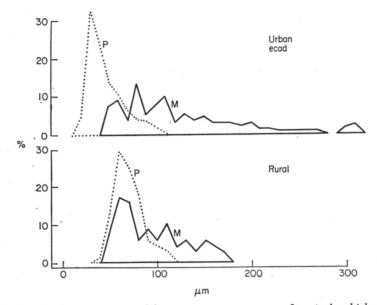

FIG. 8. Graphical interpretation of the percentage occurrence of particular thicknesses of phycobiont (P) and mycobiont (M) layers for urban ecad and rural material of *Lecanora muralis*.

below. In non-urban material the algal cells are arranged in distinctive structures, whereby intrusions of both the upper cortex and the medulla give regular cone-like fungal wedges from above and below (cf. Poelt, 1958). In the case of urban ecad material from less polluted areas, the cone-like appearance disappears and the algal clumps are aggregated into a more continuous layer. Where surface contamination is particularly severe (the algal cells are then lacking), a noticeable wedge-shaped extension of the upper cortex is prominent beneath the contaminant.

Material from a more urban environment has a continuous contamination layer over the surface which is more noticeable from above as a dark discoloration behind the growing area of the marginal lobes. These lobes have a low density of algal cells, with disorderly clumping. The more foliose lobes of the urban ecad receive contamination from above and below the thallus and, under extreme environmental stress, algal cells are limited in number or completely lacking. The thallus areas underlying apothecia are, in urban material, either limited, or lacking, in algal cells, although surface contamination normally appears less severe than for the marginal areas described above.

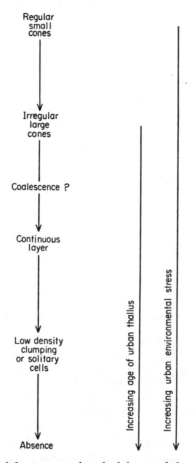

FIG. 9. Summary of modifications to the algal layer of the urban ecad of *Lecanora muralis* with increasing age or urban stress.

It may be concluded that an increasing modification of the algal layer is correlated with increasing urban environmental stress. This conclusion is tentatively summarized in Fig. 9.

3. Ecological Performance
L. muralis has a distinctive zonation within the West Riding conurbation; its main inner limits in 1969 and 1974 are given in Fig. 10. These inner limits

coincide mainly with its distribution on asbestos-cement tile roofing. A more detailed pattern of distribution according to substrate preference in the west-north-west to north-north-east sector of Leeds (Fig. 3) is given in Fig. 11. The mean distance, in miles from the city centre, of the inner limit on the three

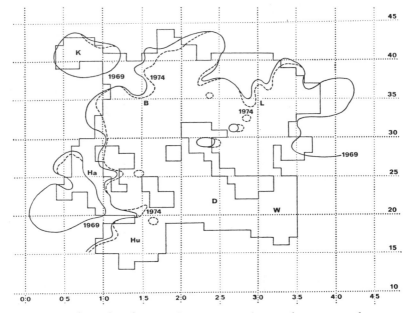

Fig. 10. Map to show the advance of *Lecanora muralis* into the west, north-west, north and north-east of the West Riding Conurbation (B = Bradford, D = Dewsbury, Ha = Halifax, Hu = Huddersfield, K = Keighley, L = Leeds, W = Wakefield) during a 5 year period. Major urbanized areas with national grid lines at 5 km intervals (see also Fig. 3) and tentative major distributional lines of *L. muralis* for 1969 (————) and 1974 (- - - - -) are indicated.

types of substrate in 1970 were: asbestos-cement tile and sheeting roofs, 3·04; cement, concrete and mortar, 3·60; siliceous wall capstones 5·53. The inner limit for the distribution of *L. muralis* for a particular date is so clearly defined that the factors operating in this area at or immediately preceding that time must be critical to the lichen's performance, or even its existence. This inner limit can be equated to a synopsis of factors operating in the two zones between 2·5 and 3·5 miles from the city centre (Seaward, 1972, p. 102), although extreme inner limits of up to 1·5 miles from the city centre have been recorded prior to 1970.

An investigation of the pH of the rainwater falling close to each of these two limits was undertaken during the period October 1969 to March 1970. The percentage of 25 weekly measurements of rainwater pH collected at four sites in each instance, are shown in Table V. Average pH values of 4·75 and 5·20 for the extreme and major inner limits respectively were determined from the above data. Although Gilbert (1970) found *L. muralis* on asbestos-cement roofing

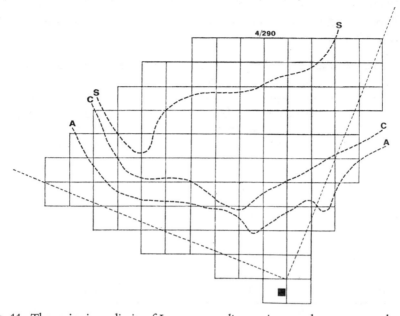

FIG. 11. The major inner limits of *Lecanora muralis* growing on asbestos-cement sheeting and tile roofs (A), on cement, concrete and mortar (C), and on siliceous wall capstones (S) within the west–north–west to north–north–east sector of Leeds (1970). The pH values for the substrates were as follows: new asbestos-cement 9·70–12·70 (av. 11·30), old asbestos-cement 6·50–12·70 (av. 9·80), cement, concrete and mortar 6·95–9·30 (av. 8·35), and siliceous wall capstones 5·20–5·30 (av. 5·25).

at a similar distance from the centre of Newcastle, the material was sterile, only being found with apothecia at a distance of more than 3 miles. Laundon (1970) recorded the species from within 1·25 miles of the centre of London, but gave no account of its reproductive status. With very few exceptions, all material in the West Riding conurbation, even at its inner limit, is highly reproductive.

A subdivision of the pattern of distribution according to substrate preference can be made for the inner limit of *L. muralis* growing on either tiled or sheeting asbestos-cement roofing. From work in the west–north–west to north–north-east sector of Leeds the following delimitations (mean distance in miles from the city centre) have been determined for asbestos tiles and sheets respectively: extreme inner limit, 1·50 and 2·55; mean inner limit, 2·55 and 3·35. Zonational subdivision of the cement, concrete and mortar substrates is not possible due to the varying quantities of sand used in their manufacture. The siliceous substrates are represented mainly by Millstone Grits which for the most part are texturally and chemically consistent, although very occasionally (especially to the west of the West Riding conurbation) calcareous grits are to be found which bear a distinctive lichen flora.

TABLE V. Distribution of rainfall pH values in two urban regions of Leeds city

pH range	Sites up to 0·5 mile outside extreme inner limit (1·50 miles)	Sites up to 0·5 mile inside mean inner limit (3·04 miles)
3·0–3·5	4	0
3·5–4·0	24	7
4·0–4·5	18	16
4·5–5·0	21	14
5·0–5·5	9	27
5·5–6·0	12	14
6·0–6·5	9	7
6·5–7·0	3	11
7·0–7·5	0	4

It is possible from the above analyses to draw up a biological scale (Table VI) based entirely on the use of *L. muralis*, paying particular attention to man-made substrates which are in common use throughout the British Isles (cf. numerous scales which rely upon various corticolous and saxicolous substrates that are often only locally operable). Such a scale will be applicable to urban environents with the full range of substrates indicated.

The distribution patterns, shown in Table VI, were determined during 1970. However, the zonation of *L. muralis*, although clearly defined, is by no means static. The distribution in 1970 showed an inward advance on all substrates since the commencement of the major mapping programme in 1968. Further inward advances have also been recorded over the past 5 years (Fig. 10). Advances of several hundred metres during this period are not uncommon, and already the delimitations of substrate preference prepared in 1970 (Fig. 11) are out of date. The independent appearance of thalli on purely acidic substrates, where previously there had been a primary establishment on mortar joints and pointing, is indicative of a change in the environment and/or the ecological capabilities of this lichen. Noticeable invasions of less calcareous substrates are common throughout the Leeds suburbs; for example, numerous rosettes were noted for the first time in February 1971 on concrete and on non-asbestos-cement roofing tiles in the 2·5–3·0 mile zone.

TABLE VI. Biological scale of status of *Lecanora muralis* compared with distance from Leeds city centre, SO_2 concentration and rainfall pH

Distance from city centre (miles) Leeds, 1970	Status of *Lecanora muralis*	Mean daily sulphur dioxide concentration ($\mu g/m^3$)	Rainwater pH
0–1·5	Absent	>240	4·4–4·7
1·5–2·5	On asbestos-cement tile roofs	200–240	4·7–4·9
2·5–3·5	On asbestos-cement tile and sheet roofs	170–200	4·9–5·1
3·5–5·5	On asbestos-cement roofs, cement, concrete and mortar	125–170	5·1–5·5
5·5+	On asbestos-cement roofs, cement, concrete, mortar and siliceous wall capstones	<125	5·5+

Isolated communities on asbestos-cement tiles, dominated by *L. muralis*, with unique phytosociological properties are frequently to be found within the 1·5–3·0 mile zone. This zone also contains both tiled and sheeting asbestos-cement roofs supporting communities characterized by conspicuous rosettes of *L. muralis*, and a lack of diversity in lichen associates. The large expanses of asbestos–cement roofing are very quickly colonized by a limited number of species and almost total cover by lichens is a common feature. This coverage is mainly by crustose species such as *Lecanora dispersa* and *Candelariella aurella*. There are usually five to seven rosettes of *Lecanora muralis* per m² of roofing. Detailed examination in October 1971 of 14 asbestos-cement tiled roofs near site 11 (see Fig. 3) showed the average rosette size to be 34·7 cm², the total rosette coverage of roof surface was approximately 2% and the associated lichen flora was totally crustose (usually three or four species). Asbestos-cement sheeting roofs near to site 11 supported a more diverse lichen flora. Not only were there usually six or seven crustose species but *Physcia caesia* and *P. orbicularis* accompanied *L. muralis* and the total cover by foliose species was over 5%. The asbestos-cement crown ridges to these roofs were particularly well colonized by foliose species where cover, especially by *Physcia* species, of over 15% were frequent. The richness of the lichen flora on the ridges is associated with bird activity.

Analysis of the frequency of *L. muralis* and associated lichens in assemblages occurring on asbestos-cement tile and sheet roofs within 20 km of the centre of the West Riding conurbation provides only a general picture, owing to wide differences in frequency from zone to zone: *Candelariella aurella* and *Lecanora dispersa*—abundant; *Caloplaca citrina*, *Lecanora muralis*, *Lecidella stigmatea*, *Physcia caesia* and *P. orbicularis*—common; *Lecania erysibe*, *Lecanora conizaeoides* and *Physcia adscendens*—frequent; *Caloplaca decipiens*, *C. holocarpa*, *Lecanora campestris*, *Physcia tenella*, *Rinodina subexigua*, *Verrucaria muralis*, *Xanthoria aureola* and *X. parietina*—occasional; *Candelariella vitellina*, *Physcia nigricans*, *Verrucaria nigrescens*, *Xanthoria elegans*—uncommon or rare.

The above list matches to a large extent the assemblages described by Brightman (1959), Laundon (1967, 1970) and Gilbert (1968), but differs in the relative frequency of the species. Laundon and Gilbert recorded 11 and 22 species respectively from asbestos-cement roofs. Laundon unfortunately did not examine this substrate in any detail, because of its inaccessibility, but he noted the value of detailed recording from this habitat for the mapping of lichens in relation to air pollution.

A most surprising feature to emerge from a comparison between Gilbert's work in Newcastle upon Tyne and the West Riding conurbation work is the absence of *Lecanora conizaeoides* from asbestos-cement roofs in the former area. This species is locally frequent on asbestos-cement roofs throughout the West Riding conurbation. *Xanthoria elegans* appears to be on the increase throughout the British Isles on asbestos-cement and concrete. *Physcia nigricans* may also be increasing and (or) overlooked in the field due to its similarity to *P. caesia*. Additional *Caloplaca* species (e.g. *C. heppiana* and *C. saxicola* recorded by Gilbert, 1968) and also species of *Bacidia*, *Catillaria*, *Protoblastenia* and *Verrucaria* may be present on asbestos-cement roofing in outer suburbs and beyond the conurbation boundary.

It may be concluded that the wide and increasing use of asbestos-cement for roofs has enabled *L. muralis* and many of the other species listed to extend their ecological and geographical ranges. The favourable buffering and water-holding capacities of this substrate have been instrumental in their spread in urban areas. It may well be that the water-holding capacity and surface texture, and perhaps buffering capacity to acidic air pollution, of this substrate have allowed a similar extension of their range in non-urban areas. It may be predicted that man-made substrates, such as asbestos-cement, will be the determining factor for the future success of *L. muralis* and many other British lichens.

4. Eco-Physiological Performance

The water content of the lichen thallus is one of the most important factors governing metabolic processes. There are no specialized parts of the thallus concerned with the uptake or loss of water, nor does there appear to be any water conservation mechanism. The water regime of the thallus is governed by physical, rather than metabolic, processes. Frequent alternations between near desiccation and near saturation of the thallus occur naturally. Toleration of cycles of wetting and drying is a feature of lichens (Farrar, 1973) and alternations may well be necessary for lichen survival.

Saturated water content measurements of up to 3900% of the dry weight of the thallus have been recorded, but the general range for foliose, fruticose, and to some extent crustose, lichens is 100–300% (Smith, 1962). The saturated water content of the majority of *L. muralis* material examined was in the range of 150–180%. The water content of freshly gathered material from asbestos-cement roofing was mainly in the range of 23–55% and had a water availability in the range of 0·05–0·07 g/cm² of thallus. Water content measurements of less than 6% were not uncommon for *L. muralis* exposed to laboratory conditions and such material generally revived when inundated with water.

Different areas of a saturated thallus may contain different amounts of water. Smith (1960) showed that the medulla of *Peltigera polydactyla* contained 25% more water than the algal layer and upper cortex. The apothecial areas of the urban material of *L. muralis* contained on average approximately 15% more water than the marginal areas; this may be explained by the predominance of the fungal, rather than the algal, components here. These data may help to explain the results from the following experiment on the water intake of *L. muralis*. Ten rosettes were inundated with tap water at a constant temperature of 21°C. The material was periodically removed, carefully dried with absorbent paper, and weighed. The mean gain in weight was expressed as a percentage of the mean oven-dry weight of the thalli (determined at the conclusion of the experiment). The water-holding capacity was determined as a percentage of that attained after 31 h of inundation (i.e. saturation), since there was a negligible increase in water uptake over the last 7 h of the experiment. The data obtained are presented in Table VII.

Water absorption by dry thalli is a rapid process, and it must be appreciated that the thalli could not be subjected to a lower than air-dry (21°C) water content for fear of some cellular damage. The air-dry rosettes had a water content, on average, of 21·46% of their dry weight at the commencement of the inundation and therefore the 76·7% uptake is not a true reflection of uptake for this particular type of experiment, but it is indicative of the total water content of the thalli. It will be seen that nearly 75% of the water necessary for eventual saturation had been absorbed within 5 min and that this figure had risen to over 85% after 30 min. Smith (1962) reviewed the time taken for air-dry thalli to

TABLE VII. Water uptake by *Lecanora muralis* (see text for experimental details)

Time (min)	% Water content of dry weight of thallus	% Water-holding capacity attained
1	76·7	50·43
2	94·4	62·22
3	100·7	66·19
4	104·7	68·84
5	111·9	73·55
7	118·0	77·59
9	124·3	81·73
13	127·7	83·97
37	135·1	88·82
75	141·5	93·01
116	144·4	94·95
235	147·7	97·09
1440	151·9	99·86
1860	152·1	100·00

become completely saturated, the figures presented varying from 1–2 min to 90 min. In the light of the above work with *L. muralis*, one must view the term "completely saturated" with some suspicion. *L. muralis* continued to absorb water over a 24 h period. It is possible that a rapid intake of water by the medulla was followed by a slower permeation into the upper cortex (cf. Smith, 1960). Certainly the different phycobiont:mycobiont ratio and thickness of thallus of the urban ecad of *L. muralis* may have some bearing on the longer time needed for saturation. Possibly the surface contamination of urban material of *L. muralis* may inhibit water entry (and indeed control water loss) in the same manner as the non-wettable surface of those crustose species which obtain their water from vapour rather than liquid sources.

The effectiveness of an asbestos-cement substrate as a reservoir of water to *L.*

muralis after 4 h of dehydration (Fig. 12) may be contributory to the success of this lichen in urban environments. The efficiency of this lichen as an urban colonizer may depend very much on the evapotranspiratory gradient set up from substrate to lichen to atmosphere (Seaward, 1972, p. 177), which involves both input and output of water. This reliance on the substrate for water replenishment is certainly put to the test in the very exposed habitats favoured by *L. muralis* rosettes.

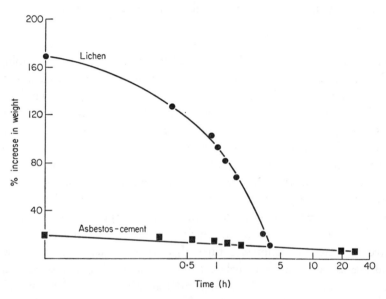

FIG. 12. Graph to show the water loss (expressed as percentage increase over oven-dry weight, 24 h at 100°C) from saturated asbestos-cement and *Lecanora muralis*.

In conditions of severe drought, the water content of most lichens is within the range of 2–14·5% of the dry weight (Smith, 1962). Material of *L. muralis* which was subjected to a 2 month desiccation (average water content = 4·96% of the dry weight), resumed a respectable level of respiratory performance (70% of untreated material) when inundated with water, but failed to respond after sulphur dioxide fumigation (Seaward, 1972, pp. 270–271). Survival of lichens subjected to drought depends more on the duration of the condition than its intensity (cf. Rogers, 1971), and there is a broad correlation between the habitat of the species and drought tolerance. Both these factors are contrary to the "drought" theory for the absence/reduction of lichens in urban areas. Long periods of severe drought are not a feature of urban environments, although it is appreciated that measurable humidity deficits and higher temperatures are commonplace. Lichens from warm, dry habitats certainly show more tolerance to drought than those from cool, moist habitats. From this point of view, the use of lichens favouring the latter type of habitat for transplantation into urban environments (cf. Gilbert, 1968) is unsatisfactory. A cool, moist environment is necessary for the metabolism of *L. muralis*. It must, therefore, be concluded that its ecological requirements are satisfied by the unique properties of asbestos-cement, a substrate which maintains a water regime compatible with the lichen's existence and which also mitigates the harshness of urban conditions.

DISCUSSION

Since lichen classification is based almost entirely on anatomical, morphological and chemical characters, these features of a taxon are singled out as being reflections of genetic relationships. This conception of species is theoretical rather than practical since breeding populations of lichens are unknown. The problem of the lichenologist is to make a judgement as to what is a stable genetic phenomenon and what is a phenotypic response to a varying environmental condition.

The ecad concept rarely seems to have been applied to lichenology. For the most part, lichen facies have attained the taxonomic rank of form, variety, species or even genus (e.g. *Lecanora albomarginata*, *L. versicolor*). Clements (1928) defined an ecad as being a direct and demonstrable adaptation to a habitat. This is probably a description of an emergent ecotype; the latter has inheritable-adaptive characteristics impressed through environmental selection. One line of evidence as to the status of urban *L. muralis* stems from transplant experiments. These might show the ability of non-urban material to modify its subsequent growth and physiological characteristics on transplantation to an urban environment and vice versa (i.e. the ability of thalli from polarized sources ultimately to thrive in each other's habitats). From the ecological performance data presented above, there are some reasonable grounds for acknowledging the status of ecotype for the urban material of *L. muralis*. The morphological, histological, ecological and physiological responses of the transplant material from non-urban sites did not match up to the naturally occurring urban ecad, at least during the period of this investigation. Much more needs to be done on transplant work (e.g. Richardson, 1967), which gives due consideration to the substrate (cf. use of differing substrates in *L. muralis* transplantations).

How "reversible" is the particular phenotype of urban material of *L. muralis*, and what responses are observed in its transplantation to non-urban environments? Most lichen taxa show a remarkable range of form which cannot be catered for by narrow concepts imposed by lichen taxonomists working on limited herbarium material. Statistical comparisons are essential for meaningful taxonomy. The apparently indeterminate growth of most lichens makes this approach difficult and reliance has been placed upon the few structures, such as asci and ascospores, which have attained their ultimate size. All too often taxonomic differentiation is based on dubious criteria such as spore dimensions (cf. variability of spore size in the *L. muralis* complex). Can one justify a new taxon based on a difference in cortical thickness? The concept of variation has played a very minor part in lichen taxonomy. The plethora of taxa described within the *L. muralis* complex bears witness to this fact. Future progress in lichen taxonomy lies to a large extent in the hands of the cytologist, geneticist and physiologist. The field lichenologist should be aware of the part played by all of the interacting dynamic environmental factors in moulding the lichen phenotype.

ACKNOWLEDGEMENTS

Much of this paper is based on part of the work (Seaward, 1972) accepted for the degree of Ph.D. by the University of Bradford. I am indebted to Trinity and All Saints' Colleges, Horsforth, Leeds, for a grant towards the cost of this research programme; to Dr D. J. Hambler for his encouragement, advice and invaluable supervision; to Miss V. A Hinton and Dr D. L. Hawksworth for reading and commenting on the original manuscript; to Dr P. B. Topham for her assistance with the statistical analyses; and to Professor E. Lees under whose direction the work was made possible.

REFERENCES

Awasthi, D. D. (1965). Catalogue of the lichens from India, Nepal, Pakistan, and Ceylon. *Beih. Nova Hedwigia* **17**, 1–137.

Awasthi, D. D. and Singh, K. P. (1970). A note on lichens from Kashmir, India. *Curr. Sci.* **39**, 441–442.

Barkman, J. J. (1958). "Phytosociology and Ecology of Cryptogamic Epiphytes." Van Gorcum, Assen.

Bird, C. D. (1966). "A Catalogue of the Lichens Reported from Alberta, Saskatchewan and Manitoba." University of Calgary, Department of Biology, Calgary. [Mimeographed.]

Brightman, F. H. (1959). Some factors influencing lichen growth in towns. *Lichenologist* **1**, 104–108.

Brodo, I. M. (1965). Studies of growth rates of corticolous lichens on Long Island, New York. *Bryologist* **68**, 451–456.

Clements, F. E. (1928). "Plant Succession and Indicators." Wilson, New York.

Crombie, J. M. (1894). "A Monograph of Lichens found in Britain", Vol. 1. British Museum (National History), London.

Culberson, C. F. (1969). Chemical and Botanical Guide to Lichen Products." University of North Carolina Press, Chapel Hill, N.C.

DEGELIUS, G. (1945). Ett sydberg i Kebnekaise-området och dess lavflora. *Bot. Notiser* **1945**, 390–412.

FARRAR, J. (1973). Lichen physiology: progress and pitfalls. *In* "Air Pollution and Lichens" (B. W. Ferry, M. S. Baddeley and D. L. Hawksworth, eds), pp. 238–282. University of London, Athlone Press, London.

FINK, B. (1935). "The Lichen Flora of the United States." University of Michigan Press, Ann Arbor.

GILBERT, O. L. (1968). "Biological Indicators of Air Pollution." Ph.D. thesis, University of Newcastle upon Tyne.

GILBERT, O. L. (1970). Further studies on the effect of sulphur dioxide on lichens and bryophytes. *New Phytol.* **69**, 605–627.

GILBERT, O. L. (1971). Studies along the edge of a lichen desert. *Lichenologist* **5**, 11–17.

GRUMMANN, V. J. (1963). "Catalogus lichenum Germaniae." Gustav Fischer, Stuttgart.

HAKULINEN, R. (1966). Über die Wachstumgeschwindigkeit einiger Laubflechten. *Annls bot. fenn.*, –, 167–179.

HALE, M. E. (1954). First report on lichen growth rate and succession at Aton Forest, Connecticut. *Bryologist* **57**, 244–247.

HALE, M. E. (1967). "The Biology of Lichens." Arnold, London.

HALE, M. E. (1969). "How to Know the Lichens." Brown, Dubuque, Iowa.

HALE, M. E. and CULBERSON, W. L. (1970). A fourth checklist of the lichens of the continental United States and Canada. *Bryologist* **73**, 499–543.

HILL, D. J. and WOOLHOUSE, H. W. (1966). Aspects of the autecology of *Xanthoria parietina* agg. *Lichenologist* **3**, 207–214.

JAMES, P. W. (1965). A new check-list of British lichens. *Lichenologist* **3**, 95–153.

KLEINIG, H. (1966). Beitrag zur Kenntnis der Flechtenflora von Kreta. *Nova Hedwigia* **11**, 513–626.

KLEMENT, O. (1965a). Flechtenflora und Flechtenvegetation der Pityusen. *Nova Hedwigia* **9**, 435–501.

KLEMENT, O. (1965b). Zur Kenntnis der Flechtenvegetation der Kanarischen Inseln. *Nova Hedwigia* **9**, 503–582.

KRAUSE, W. and KLEMENT, O. (1962). Zur Kenntnis der Flora und Vegetation auf Serpentinstandorten des Balkans: 5. Flechten und Flechtengesellschaften auf Nord-Euböa (Griechenland). *Nova Hedwigia* **4**, 189–262.

LAUNDON, J. R. (1967). A study of the lichen flora of London. *Lichenologist* **3**, 277–327.

LAUNDON, J. R. (1970). London's lichens. *Lond. Nat.* **49**, 20–69.

MACOUN, J. (1902). "Catalogue of Canadian Plants: Part VII—Lichenes and Hepaticae." Queen's Printer, Ottawa.

MAGNUSSON, A. H. and ZAHLBRUCKNER, A. (1944). Hawaiian lichens: II. The families Lecideaceae to Parmeliaceae. *Ark. Bot.* **31**, 1–109.

MATTICK, F. (1932). Bodenreaktion und Flechtenverbreitung. *Beih. bot. Zbl.* **49**, 241–271.

NORDIN, I. (1972). "*Caloplaca* sect. *Gasparrinia* i Nordeuropa." Skriv Service AB, Uppsala.

OTTO, G. F. and AHTI, T. (1967). "Lichens of British Columbia: Preliminary Checklist." University of British Columbia, Vancouver. [Mimeographed.]

POELT, J. (1958). Die Lobarten Arten der Flechtengattung *Lecanora* Ach. sensu ampl. in der Holarktis. *Mitt. Bot. Staatssamml., Münch.* **1958**, 411–589.

RICHARDSON, D. H. S. (1967). The transplantation of lichen thalli to solve some taxonomic problems in *Xanthoria parietina* (L.) Th. Fr. *Lichenologist* **3**, 386–391.

ROGERS, R. W. (1971). Distribution of the lichen *Chondropsis semiviridis* in relation to its heat and drought resistance. *New Phytol.* **70**, 1069–1077.

RYDZAK, J. (1961). Investigations on the growth rate of lichens. *Annls Univ. Mariae Curie-Skłodowska*, **C, 16,** 1–15.

SCHUBERT, R. and KLEMENT, O. (1966). Beitrag zur Flechtenflora von Nord- und Mittelindien. *Nova Hedwigia* **11**, 1–73.

SEAWARD, M. R. D. (1972). "Aspects of Urban Lichen Ecology." Ph.D. thesis, University of Bradford.

SMITH, A. L. (1918). "A Monograph of the British Lichens", Second edition, Vol. 1. British Museum (Natural History), London.

SMITH, D. C. (1960). Studies in the physiology of lichens, 1–3. *Ann. Bot.* **24**, 52–62, 172–185, 186–199.

SMITH, D. C. (1962). The biology of lichen thalli. *Biol. Rev.* **37**, 537–570.

SUZA, J. (1925). Nástin zeměpisného rozšíření lišejníků na Moravě vzhledem k foměrum Evropským. *Přírodov. fak. Masarykova univ. Brno* **55**, 1–152.

THOMSON, J. W., SCOTTER, G. W. and AHTI, T. (1969). Lichens of the Great Slave Lake Region, Northwest Territories, Canada. *Bryologist* **72**, 137–177.

TRASS, H. (1963). On the lichen-flora of Kamchatka I. *Acad. Sc. R.S.S. Estonia* **1963**, 170–220.

WATSON, W. (1953). "Census Catalogue of British Lichens." British Mycological Society, London.

WEBER, W. A. and VIERECK, L. A. (1967). Lichens of Mt. McKinley National Park, Alaska. *Bryologist* **70**, 227–235.

WERNER, R.-G. (1955). Synthèse phytogéographique de la flore lichénique de l'Afrique du Nord française d'après des données récentes et essai de paléogéographie lichénique. *Bull. Soc. bot. Fr.* **102**, 35–50.

WHELDON, J. A. and TRAVIS, W. G. (1915). The lichens of South Lancashire. *J. Linn. Soc., Bot.* **43**, 87–136.

WOOLHOUSE, H. W. (1968). The measurement of growth rates in lichens. *Lichenologist* **4**, 32–33.

ZAHLBRUCKNER, A. (1927–28). "Catalogus Lichenum Universalis", Vol. 5. Borntraeger, Leipzig.

14 | Nutritional Aspects of Marine and Maritime Lichen Ecology

A. FLETCHER

Department of Marine Biology, Marine Science Laboratories,
University College of North Wales, Anglesey, Wales

Abstract: Field observations are reviewed which suggest the importance of nutritional factors in the zonation of lichens on rocky shores. Consideration is given to the differences in species composition and distribution on siliceous and calcareous substrates, the rôle of salinity in estuaries, pH differences and the problem of ornithocoprophilous species.

Results from some recent experiments are presented, testing the hypothesis that seashore lichens are distributed according to the amount of seawater or freshwater in the environment. Littoral thalli are the least affected by continuous immersion in seawater or freshwater, while supralittoral thalli either die or the bionts separate, regardless of salinity. Tolerance towards repeated drying and wetting shows a similar gradient; littoral thalli die, while supralittoral thalli die more slowly or the symboisis breaks down. Salinity is of secondary importance of wetness.

Thalli accumulate inorganic cations to concentrations much greater than are found in the environment. However, this generally depends on the amounts of cations in the environment. Xeric supralittoral species have the lowest concentrations, consistent with their dominance in a drought-prone, hence nutrient-deficient, habitat. Only potassium and calcium appear to be concentrated independently of environmental levels.

Calcium is important to *Xanthoria parietina* in decreasing potassium loss to freshwater but no effect is seen in *Parmelia saxatilis*. The effects vary in the dark and light. The relationship of these results to field data is discussed.

INTRODUCTION

The zonation of lichens on rocky shores in northern Europe is a well-known phenomenon and is frequently remarked upon by even the most casual observers. Equally characteristic is the general conclusion that this zonation is related to the influence of seawater. In this paper it is intended to review the very limited

Systematics Association Special Volume No. 8, "Lichenology: Progress and Problems", edited by D. H. Brown, D. L. Hawksworth and R. H. Bailey, 1976, pp. 359–384. Academic Press, London and New York.

evidence available from published sources which supports this conclusion and also to present some new evidence.

The term "nutrition" is applied here in a rather wide sense. It is intended to cover effects arising from the ionic content of water, usually termed salinity, which involves two components: first the osmotic effect exerted by the concentration of dissolved solids, and secondly the effect of ions, individually or in combination in a purely chemical sense. Naturally it is hard to separate these two influences, but due consideration will be given to them in the following treatment. "Nutrition" is usually considered to apply only to growth, but it is accorded a slightly wider treatment in this paper by including tolerance effects as well as presumed requirements. In the present state of lichen-physiological knowledge where growth requirements of lichen thalli are still unknown, owing to the absence of any standardized lichen culture techniques, it is almost impossible to separate tolerance from requirement criteria in both laboratory experiments and field data.

FIELD EVIDENCE IMPLICATING NUTRITIONAL FACTORS

The literature relating to the distribution of lichens on rocky shores in the British Isles has been reviewed by Fletcher (1973a, b), who also gave the most recent classification of seashore lichen communities. The floras of seashores vary markedly according to their substrate, whether siliceous or calcareous, and these will be discussed in turn.

1. Siliceous Shores

Some 250–300 lichen species occur on British seashores of siliceous rocks (Fig. 1). About 100 of these are found only on the seashore, while the rest may be found inland as well. Particular lichen species reach their peaks of abundance in different regions of the shore. The degree of overlap between the zones is shown in Fig. 1 (right-hand side). Clearly littoral (*Verrucaria*, *Lichina*, etc.) and mesic supralittoral thalli (*Caloplaca marina*, *Lecanora helicopis*, etc.) are restricted to their individual zones. Many of the xeric supralittoral species are also restricted to their zone (*Ramalina siliquosa*, *Lecidella subincongrua*, etc.) but a large proportion also spread into, or have spread from, the terrestrial region (*Lecanora atra*, etc.). Thus there are more species in zones close to the sea that are restricted to such regions than there are restricted species in zones farther away from the sea. A number of terrestrial species can penetrate downshore (e.g. *Lecania erysibe* penetrating into the mesic supralittoral) and are thus occasionally subjected to seawater submersion by the tides. The relatively low number of species in the littoral and mesic supralittoral, compared with the high number in upshore zones, suggests that rather few species, in proportion to the total, are especially restricted in their seawater tolerance or requirements.

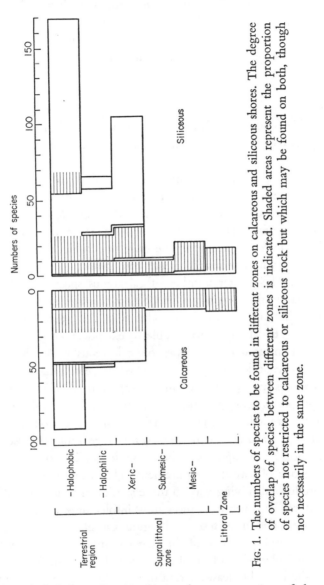

Fig. 1. The numbers of species to be found in different zones on calcareous and siliceous shores. The degree of overlap of species between different zones is indicated. Shaded areas represent the proportion of species not restricted to calcareous or siliceous rock but which may be found on both, though not necessarily in the same zone.

It can be concluded that siliceous shores show a zonation of those species apparently needing maximal amounts of seawater (littoral and mesic supralittoral), to species which are perhaps only tolerant towards seawater (terrestrial halophilic), since the latter also occur inland, far away from even salt spray. An intermediate group of species, confined to the xeric supralittoral, are only found on the seashore in an environment farthest away from both sea and terrestrial influences. It is hard to say at this stage whether such species require seawater specifically or, being merely seawater tolerant, are occupying a niche which seawater-requiring species find nutrient deficient, and terrestrial species find too saline.

2. Calcareous Shores

Published data concerning lichens on limestone seashores are very scarce. Fletcher (1972) gave some quantitative data, while Knowles (1913) gave a purely subjective account. The zonation of species is much less obvious to the casual observer than that on siliceous shores, only two zones being obvious. These are (*a*) a littoral zone, which is practically identical in species composition (Fig. 1, left-hand side) to that on siliceous shores, and (*b*) a more or less homogeneous supralittoral and terrestrial region, which is composed mainly of species restricted to calcareous substrata. The only noticeable zonation in the upper shore is that many new, terrestrial species come in as the shore is ascended, while none die out. The greater number of calcareous species are therefore those which are terrestrial in origin but can perhaps tolerate seawater to a greater or lesser degree. A complication is introduced in that species of the mesic and submesic supralittoral appear to be able to extend much farther upshore into terrestrially dominated communities on calcareous rocks than they do on siliceous rocks, regardless of aspect, wave exposure, etc. It may be concluded that calcareous shores have (*a*) no exclusively supralittoral species, only terrestrial, salt-tolerant species being present, and (*b*) exclusively littoral and mesic supralittoral species are able to penetrate farther upshore than on siliceous rocks of comparable aspect and wave exposure. At this stage it may be postulated that on calcareous shores there appears to be a simple gradient of seawater tolerance overlapping one of seawater requirement. Thus those species requiring most seawater are littoral, giving way to those species most tolerant of seawater in the supralittoral, and eventually those most intolerant of seawater and requiring freshwater are found in the terrestrial region. On siliceous shores this scheme is much more complex because of the larger number of species which are exclusive to particular zones of the shore. A simple gradient from most seawater tolerant in the littoral to least tolerant or requiring freshwater in the terrestrial region could explain the distribution of about half the species on the shore, but the exclusively mesic and xeric supralittoral species seem to tolerate or require an exact amount of seawater in the environment.

The ability of littoral and mesic supralittoral species to penetrate upshore on calcareous rocks could perhaps, at first sight, be accounted for by the higher water-retaining properties of such rocks. This is considered to be unlikely, however, because these upshore penetrating-thalli may often be found along-side terrestrial thalli, presumably experiencing the same conditions. Also the phenomenon is not exhibited on porous siliceous schists. It is tempting to speculate that the chemical composition of the limestone, perhaps particularly its calcium content, enables seawater-loving species to live farther away from the sea.

An alternative explanation of the calcareous/siliceous problem may be that seawater overcomes the chemical nature of the substrate, enabling limestone species to live on siliceous shores, particularly in the littoral and mesic supra-littoral (*Xanthoria parietina, Lecanora dispersa, Rinodina subexigua, Lecania erysibe, Dermatocarpon miniatum* and many others). Above this zone the floras diverge because the chemical nature of the substrate predominates over that of seawater and determines the species to be present.

3. Correlations between Seashore Lichen Floras and Environmental Nutrients

There is no direct evidence that seawater deposition varies up the shore, but relevant publications have been reviewed by Goldsmith (1967), Waisel (1972) and Fletcher (1973b). Brown and Di Meo (1972) found that the position on tree trunks of *Parmelia caperata* and *P. perlata* correlated with direction and distance from the sea and with chloride content in the thalli. This was thought to be derived from seawater spray.

Santesson (1939) noted an interesting correlation between the geographical location of marine *Verrucaria* species in the entrance to the Baltic Sea and their position on the shore. A salinity gradient exists in these regions, ranging from seawater in the western Baltic to almost freshwater in the east. *Verrucaria mucosa* was present in all regions where the seawater salinity exceeded 20–30 g/l, while *V. maura* persisted into low salinity water of 4–6 g/l. *V. striatula* which occupies an intermediate position on the shore, between upper *V. maura* and lower *V. mucoss,* was present in waters of intermediate salinity and above. Fischer-Piette (1931) found a similar correlation with other species in an estuary in northern France. While Santesson (1939) implied that salinity was the limiting factor in the distribution of *Verrucaria* species, Fischer-Piette acknowledged that other factors may play a part. These may include temperature gradients, espec-ially pronounced in the Baltic and estuaries, and the tendency of many species to be photo- or sciaphilic, especially *Verrucaria maura*. The salinity of seawater is accompanied by many other factors, such as differences in pH and carbon dioxide and oxygen availability, so that without evidence to segregate these factors it may be premature to imply salinity as being the sole limiting factor.

The pH factor has received some attention. Fletcher (1972, 1973b) measured the pH of soil at the tops of cliffs subject to varying degrees of wave action, and also in crevices ascending the shore. Lichen species most tolerant towards seawater were found nearest to soils of high pH (7–8), which were low on the shore or on cliff tops most exposed to wave action. Less seawater-tolerant species correlated with soils of lower pH (5–6). The soil pH on limestone cliffs was always 7–8 regardless of wave action or position (Fletcher, 1972). The high pH of soil most subject to seawater spray was attributed to the high pH of seawater; after drought or rain the soil pH was always lower.

A more direct approach was adopted by Du Rietz (1932) who determined the pH of the water in which lichens had been immersed. The species from lowest on the shore (*Verrucaria maura*) gave a pH of 6·6, while *Caloplaca marina* higher upshore gave a lower pH of 6·1. *Anaptychia fusca* gave a value of pH 4·9 and was the highest species on the shore studied. Values of pH for *Lecanora helicopis* varied according to the position of the thalli; those from low on the shore gave higher pH values than those from higher up. Limestone species all gave values of 7·5–7·7 regardless of height on the shore.

Kärenlampi (1966), working on the archipelago of southern Finland, found that shores with a dense tree cover in the upper regions had no zonation of lichens, but rocks farther out to sea had a typical Baltic zonation. This lack of zonation was attributed to acid water run-off from the tree-clad soil, which neutralized seawater and inhibited the normal flora. This acidity, as well as the natural acidity of the substrate, could be neutralized by seawater (wave action) and bird manure, to allow a zonation pattern to become established. Fletcher (1973b) considered that the development of a terrestrial flora was part of the natural sequence on seashores. The lack of a xeric supralittoral in situations analogous to those studied by Kärenlampi (sheltered) was thought to be a result of slower drying out and, therefore, generally wetter conditions. In addition, acidic soil water could affect a larger amount of shore, at such sites, owing to the reduced incidence of neutralizing seaspray. The consequent reduction in the pH of the water was considered to favour the establishment of a terrestrial flora. On exposed shores, similar to the rocks farther out to sea mentioned by Kärenlampi, increased salt spray, higher pH and tendency to rapid drying out left a niche unfavourable for terrestrial and littoral lichens, but one which could be occupied by xeric supralittoral species.

Since seawater has a high pH of 8–8·4 (Harvey, 1955), pH measurements have been used to estimate the amount of seawater influence to which lichens are subjected. Such measurements must be made with caution, however. Since usually a volume of water is added to a sample of soil, rock or lichen thallus, it must be ensured that the proportion of water to sample is constant or at least comparable. In addition, the physical properties of soils may vary, particularly the ratio of organic to inorganic matter, so that comparative weights or volumes of soil do not necessarily afford an adequate basis for comparison. Determining the pH of living thalli is particularly hazardous because pH changes in the medium are induced as a result of gas exchange arising from photosynthesis and respiration.

There is no published field evidence which separates nutritional factors from osmotic effects.

4. Miscellaneous Field Correlations Implying Nutritional Control of Zonation

The resurgence of maritime lichen communities well inland has been mentioned by Laundon (1966, 1970). Xeric supralittoral species in particular may often be found on wind-exposed rocks at distances ranging from several to 100 km from the nearest seashore. Such wind-exposed sites presumably act as deposition centres for salts, and this has been discussed by Edwards and Claxton (1964). The simple salt deposition theory as an explanation for the recurrence of maritime floras inland is somewhat complicated by the common observation that such wind-exposed sites are often used as bird perches, and are thus subject to bird manuring. Many Scandinavian authors have listed ornithocoprophilous seashore lichens, while Kärenlampi (1966), Ferry and Sheard (1969) and Fletcher (1973b), using quantitative techniques, have shown the types of modifications to be expected on shores subject to bird manure. Hakulinen (1962) and Massé (1966) determined the pH and nitrogen contents of ornithocoprophilous lichens and found that high pH and high nitrogen contents were associated with such thalli. Presumably a high calcium and high phosphate level would also be implicated. A common factor which may be discerned between maritime floras and ornithocoprophilous floras is the occurrence of high pH and probably also high calcium contents.

It may be concluded from this review of field evidence that many species require seawater, while others merely tolerate it. In addition, the influence of particular ions is suggested, especially calcium. The principal difficulty with evidence of this kind is that correlations of field evidence with chemical factors are strictly circumstantial; there is no evidence that chemical factors are directly involved in the maintenance of lichen thalli. However, field evidence does allow hypotheses to be made which can be tested under laboratory conditions. Such laboratory evidence will be dealt with in the next section.

LABORATORY EVIDENCE RELATING TO THE NUTRITIONAL PHYSIOLOGY OF SEASHORE
LICHENS

Physiological experiments range from investigations of the total cations in
lichen thalli and their correlation with environmental position, to seawater and
freshwater tolerance, and to the effects of individual ions on metabolism.

1. Total Cation Contents

Typical total cation contents from lichens on a single shore are given in Figs 2, 3
and 4. Recalculated on a volume basis (1 g dry wt = 1–10 ml of thallus volume),

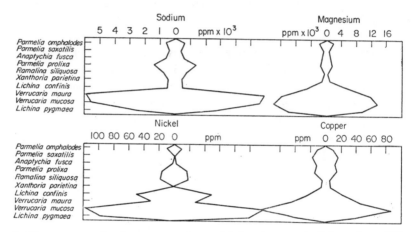

FIG. 2. The contents of sodium, magnesium, nickel and copper in lichens from different
zones of the shore; littoral below, terrestrial above. (Experimental details: thalli
washed, oven dried, ashed at 450°C for 6 h, ash dissolved in hot HCl/HNO$_3$,
diluted with water. Ion analysis by atomic-absorption spectrophotometry.)

the concentrations of cations are very much higher than in seawater. For example,
the concentration of nickel in littoral thalli is 10^6 times that in seawater.

Most cations are found in their highest concentrations in littoral thalli and
show a progressive decline in successive upshore lichens. This is obviously
related to the incidence of seawater, and thalli are thus merely reflecting the
amounts of cations available in the environment. The parallels with the hydrogen ion results (previous section) are apparent.

Many elements show an attenuated distribution in thalli of submesic and
xeric supralittoral lichens (e.g. *Xanthoria parietina*, *Ramalina siliquosa*, *Parmelia
prolixa*), their cation contents being generally lower than in littoral, upper
supralittoral and terrestrial species. This could perhaps be accounted for by a less
efficient retention mechanism in supralittoral thalli. More likely, however, is the

probability that xeric supralittoral thalli are receiving a lower concentration of cations. This is because the xeric supralittoral is thought to be drought prone (Fletcher, 1973b), receiving a minimum of both seawater and freshwater compared with the littoral and terrestrial zones. It is, consequently, relatively mineral-deficient, since minerals are largely conveyed in solution.

Ions less common in acid soil (zinc, iron, manganese) are abundant in ter-

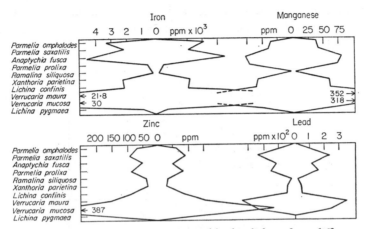

FIG. 3. The contents of iron, manganese, zinc and lead in lichens from different zones of the shore. (Experimental details: see Fig. 2.)

restrial thalli, while those cations easily leached from soil or less abundant in soil than seawater (e.g. sodium, magnesium) are not especially abundant in terrestrial thalli.

This evidence suggests that two gradients of ion abundance may be present on the shore; one at the seawater end, and one at the terrestrial end. These two gradients decline in the xeric supralittoral.

Some elements, especially potassium and calcium, show anomalous distributions (Fig. 4). Potassium is rather less variable than most other elements, showing a relatively low value in littoral species and a relatively high value in supralittoral and terrestrial species. The latter values are only 2–3 times lower than in littoral species. This relative constancy in potassium levels in all seashore lichen thalli suggests perhaps a preferential accumulation of this element. Calcium is very variable and apparently random in its accumulation, and does not correlate with environmental position. It is especially high in certain littoral and some xeric supralittoral species, and may possibly be a structural component (thus, calcium pectates may possibly be involved in maintaining the structural rigidity of *Ramalina siliquosa*). *Xanthoria parietina* has hardly any calcium (1–300 p.p.m.), and since this is the only supralittoral species examined which is also found on limestone inland, it may perhaps suggest that this species needs large amounts of calcium but is unable to retain it.

As with the pH data it appears that thalli from high-seawater environments are richer in seawater-derived ions, while those from environments low in seawater are poor in such elements but are richer in soil-water-derived ions. Except in the cases of potassium and calcium, there is little evidence that thalli accumulate any particular ions. Accumulation appears to be related to environmental position.

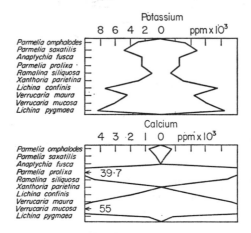

Fig. 4. The contents of potassium and calcium in lichens from different zones of the shore. (Experimental details: see Fig. 2.)

2. Tolerance Experiments Involving Seawater and Freshwater

Tolerance towards submersion of freshwater lichen thalli was investigated with some success by Ried (1960), who found that the duration of submersion, coupled with the length of time taken to recover from such treatment, were directly correlated with the position of the species in the environment. Extending this approach to seashore species becomes complicated by the osmotic-salinity components of seawater. Work already performed by Feige (1972, 1973) involved submerging *Lichina pygmaea* in seawater after a period in distilled water. The total soluble carbohydrates increased while the insoluble fraction decreased and this was considered to be a mechanism which increased the osmolarity of the cell contents to match that of the surrounding medium. This work is a promising line of research, but needs to be extended to a range of seashore species before its establishment as a general phenomenon.

On the seashore all species are subject to greater or lesser amounts of seawater in times of storm or freshwater in times of rain. Only littoral thalli are regularly submerged, the frequency of this submergence depending on the tidal cycle. The upper littoral thalli are only submerged at spring tides for a few days per month on sheltered shores, while lower littoral thalli may be submerged daily. On exposed shores none of the littoral species needs to be submerged at all and wetting by seawater from spray appears to be sufficient for their requirements. Long periods of drought are rare occurrences on the shore and it has been found possible to keep littoral and supralittoral thalli alive in the laboratory at 60% relative humidity, 10–15°C, 4000 lx light period, for several weeks without damage to photosynthesis or respiration.

(a) *Effects of constant submersion.* The results of keeping thalli constantly submerged in aerated fresh (distilled) water or aged seawater and monitoring photosynthesis and respiration are shown in Figs 5 and 6. Considering the seawater treatment first, *Verrucaria mucosa*, the most marine of the species studied, was more or less unimpaired in its ability to photosynthesize, but showed a slight decline after 30 days' submersion. *V. maura*, the next species in the gradient towards drier conditions, showed a slight increase after 30 days. *Xanthoria parietina* showed a marked increase, while the xeric supralittoral *Ramalina siliquosa* showed a relatively rapid cessation of photosynthesis. The increases in photosynthesis in *V. maura* and more noticeably in *Xanthoria parietina*, were correlated with an increase in phycobiont numbers and hence breakdown of the symbiosis. In the latter species, phycobiont cells erupted through the lower cortex and produced zoospores which were shed into the medium and discarded daily. Thus a phycobiont growth curve was established. This overgrowth of phycobiont cells was confirmed by parallel experiments performed in the dark. Under these conditions, no growth of algae took place. Instead, a steady decline in photosynthetic capacity was seen over the experimental period, but by only 20% of the initial value after 14 days.

In freshwater a remarkably similar result can be seen (Fig. 6), the major difference from seawater being the time of the responses. Thus the increase in photosynthetic capacity of *Verrucaria maura* is more rapid than in seawater. *Xanthoria parietina* shows a well defined phycobiont growth curve, while the photosynthesis of *Ramalina siliquosa* declines more slowly. The major difference between submersion in freshwater and seawater is that freshwater slightly favours the less marine thalli while seawater slightly favours the more marine thalli. *Verrucaria mucosa* is relatively unaffected by both freshwater and seawater treatments.

Respiration rates in these experiments are hard to interpret since the curves are composites of respiration derived from mycobiont, phycobiont, bacterial and foreign fungal material. In addition the possibility of wound-enhanced respiration must not be overlooked. During the course of the experiments,

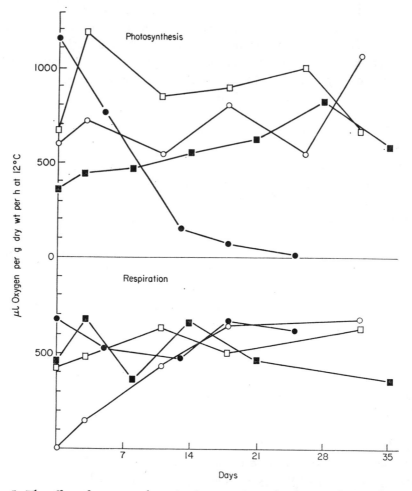

FIG. 5. The effect of constant submersion in seawater on the photosynthesis and respiration of lichens from different zones of the shore. □ = *Verrucaria mucosa,* ○ = *V. maura,* ■ = *Xanthoria parietina,* ● = *Ramalina siliquosa.* (Experimental details: thalli submerged in aerated seawater which was changed daily, 16 h days 6500 lx, 15°C. Oxygen exchange measured by Gilson photo-respirometry at 12°C, 6500 lx, using 1 ml of external buffer for CO_2 (300 p.p.m. = atmospheric) based on diethanolamine.)

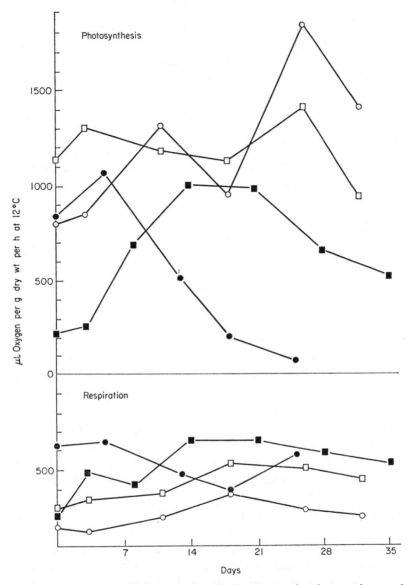

FIG. 6. The effect of constant submersion in freshwater on the photosynthesis and respiration of lichens from different zones of the shore. (Experimental details and symbols: see Fig. 5).

Ramalina siliquosa and *Xanthoria parietina* lost their usual texture, becoming soft and white with overgrown fungal hyphae. The experiments were terminated when the thalli fragmented during manipulation. Similar effects were not observed in *Verrucaria* species. The cessation of photosynthesis by *Ramalina siliquosa* was accompanied by plasmolysis of the phycobiont, while in seawater treatments additional damage was noted in strains containing salazinic acid in that the thalli turned red–brown after 2–3 h of submersion, presumably by breakdown of this chemical constituent. It is tempting to speculate that since the same colour is obtained by potassium hydroxide in the familiar lichenological "K" test, the seawater effect may be a consequence of its alkalinity.

It is important to note that the general features of tolerance to, or requirement for, constant submersion are the same in both seawater and fresh water for an individual species. The salinity aspect has a lesser part to play than the effect of the water itself.

(*b*) *Effects of alternating wetting and drying.* A more usual state of affairs for seashore lichens is that of periods of submersion or wetting by seawater and freshwater followed by periods of drying out. The relative frequency of these conditions could determine the position of the species in the environment (cf. Ried, 1960). In designing experiments to test this hypothesis it must be borne in mind that the results will involve two components: (*a*) the effect of submersion and the consequent recovery from it; and (*b*) the effects of drying out and subsequent recovery. A further complication is introduced when seawater is used because, while a constant salinity is maintained in the submersion period, the thallus is subjected to a gradually increasing salinity as the water evaporates on drying out, until eventually salt crystals appear on the thallus surface. Osmotic stresses to which lichens are subject on the seashore must be considerable.

A daily cycle of 3 h wetting followed by 21 h drought gave the results seen in Figs 7 and 8. When seawater was used, two general responses could be seen. The first was of species showing a general decline in photosynthesis (*Verrucaria maura*, *V. mucosa*), the other was of a general recovery or stabilization after an initial period of decline (*Caloplaca marina*, *Ramalina siliquosa*, *Parmelia prolixa*). Photosynthesis in *V. mucosa* ceased after 9 days but in *V. maura* continued for much longer. Mesic and xeric supralittoral species, after an initial decline, started to recover, so that the net decline over about 2 weeks in these experimental conditions was most serious in the thalli most subject to seawater in nature. The response of *V. mucosa* in freshwater was identical to that in seawater, but the eventual decline of *V. maura* was much more marked. The supralittoral thalli showed similar results to the seawater treatment but the response was slower. In the case of *Parmelia prolixa* in freshwater, the apparent recovery was due to overgrowth of the phycobiont cells and breakdown of the symbiosis. This was probably also the case with the other supralittoral thalli studied.

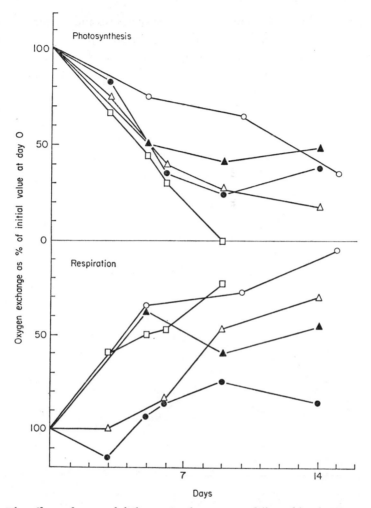

Fig. 7. The effects of repeated daily wetting by seawater followed by drying-out on the photosynthesis and respiration of lichens from different zones of the shore. □ = *Verrucaria mucosa*, ○ = *V. maura*, ▲ = *Caloplaca marina*, ● = *Ramalina siliquosa*, △ = *Parmelia prolixa*. (Experimental details: thalli submerged 3 h at start of light period of 16 h, 6500 lx, 15°C, left to dry for rest of light and whole of dark periods. Oxygen exchange measured as in Fig. 5.)

The overall conclusion is that littoral thalli are more seriously affected than supralittoral thalli. Comparison with the results in the previous section showed that littoral thalli are the most tolerant towards submersion, so that it must be the drought period which is having the greater effect. In the relatively submersion-intolerant supralittoral species, it is the submersion period which exerts the

N

maximum effect by causing breakdown of the symbiosis and overgrowth of the phycobiont. Since the length of the drought period appeared, in preliminary experiments, to have little effect on photosynthesis and respiration, the mechanism of these responses almost certainly involves the ability of a thallus to recover

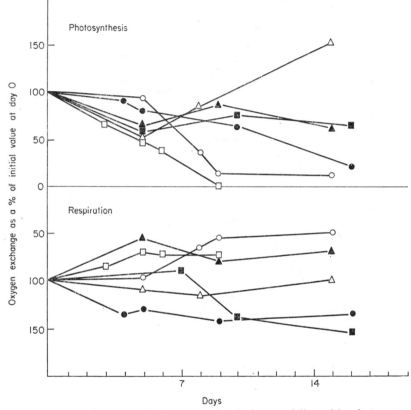

FIG. 8. The effects of repeated daily wetting by freshwater followed by drying-out on the photosynthesis and respiration of lichens from different zones of the shore. □ = *Verrucaria mucosa*, ○ = *V. maura*, ▲ = *Caloplaca marina*, ▓ = *Xanthoria parietina*, ● = *Ramalina siliquosa*, △ = *Parmelia prolixa*. (Experimental details: see Fig. 7.)

from drought or submersion. In thalli which are intolerant towards drought, the wetting period was not long enough to allow complete recovery, so that, if droughts follow each other too closely, a net decline in the health of the thallus results. A similar argument can be applied to the submersion-intolerant supralittoral thalli. These results and hypotheses are very similar to those of Ried (1960) with freshwater lichens.

The remarkable similarity between seawater and freshwater treatments argues again that salinity is of less importance than the relative effects of wetting and drying.

3. Effects of Individual Ions

Ion uptake and exchange has received little attention in lichenological literature, but a moderate amount of information is available relating to heavy metal accumulation (see James, 1973). Ion uptake by fungi has also been the subject of little attention except in unicellular yeasts (see Rothstein, 1965). Generally, however, it is held that most ions may be taken up by ion exchange (Tuominen, 1967; Puckett *et al.*, 1973), but certain ions (e.g. potassium and zinc) may be partly taken up by active metabolic processes (Brown and Slingsby, 1972).

(*a*) *Ion uptake.* The major seawater ions, sodium, potassium, calcium, magnesium and phosphate are taken up by lichen thalli selected from two extremes of the seawater gradient (*Xanthoria parietina* and *Parmelia saxatilis*) as shown in Table I. Only with phosphate has it been possible to demonstrate a temperature-dependent and hence active-metabolic uptake, and then only in the light (Fig. 9). It is possible that the amount of active uptake of cations is small compared with passive and purely physical uptake.

Parmelia saxatilis takes up rather more sodium than *Xanthoria parietina* (Table I). In the dark, sodium appears actually to be excreted by *X. parietina*. Excretion of sodium, often by special glands, is a common feature of halophytes (Waisel, 1972). Both species take up similar amounts of potassium, independent of light and dark conditions. The effects of light and dark are considerable on the uptake of calcium and magnesium, since more is taken up in the dark than in the light. It is possible that some kind of specificity is operating in the light relative to the dark, but (and equally possible) uptake may be linked with hydrogen-ion exchange resulting from carbon dioxide uptake and loss via photosynthesis and respiration.

(*b*) *Ion displacement.* (i) *Sodium* seems to effect slightly more displacement of equivalents of other ions in both species. In the dark sodium is lost by *Xanthoria parietina*, but other ions are still lost. No explanation can be given for this. Slightly more magnesium is lost by *Xanthoria parietina* than by *Parmelia saxatilis*, but more calcium is lost by *P. saxatilis* especially in the dark. With potassium the reverse occurs, more being lost in the light than in the dark. The effect for potassium with *Xanthoria parietina* is independent of light.

(ii) *Potassium* causes less displacement of other ions in both species, but slightly

TABLE I. Ions exchanged in media containing major cations and in distilled water, in μg of ion exchanged per g dry weight (top row), and microequivalents of H+c bottom row). Cations supplied as follows: sodium 0·435mM, potassium 0·256mM, calcium 0·250 mM, magnesium 0·4111 mM. 20 ml of solution used to 100 mg dry wt of lichen, ions determined by atomic absorption spectrophotometry after 6 h in light (3500 lx) and dark at 18°C. Results are means of six experiments. Ion *uptake* by thalli denoted by minus sign.

MEDIUM	Xanthoria parietina								Parmelia saxatilis							
	LIGHT				DARK				LIGHT				DARK			
	Na	K	Ca	Mg	Na	K	Ca	Mg	Na	K	Ca	Mg	Na	K	Ca	Mg
Sodium	−50 −2·17	40 1·02	20 1·00	20 1·62	100 4·34	50 1·28	50 2·50	30 2·43	−100 −4·34	170 4·34	40 2·00	5 0·40	−100 −4·34	90 2·30	80 3·99	5 0·40
Potassium	10 0·43	−200 −5·10	50 2·50	15 1·21	10 0·43	−200 −5·10	55 2·74	40 3·24	5 0·22	−200 −5·10	50 2·50	20 1·62	5 0·22	−200 −5·10	50 2·50	20 1·62
Calcium	20 0·87	10 0·26	−200 −9·98	200 16·18	20 0·87	150 3·83	−500 −24·95	200 16·18	10 0·43	100 2·56	−150 −7·49	150 12·14	5 0·22	100 2·56	−500 −24·95	150 12·14
Magnesium	10 0·43	10 0·26	200 9·98	−90 −7·28	20 0·87	10 0·26	190 9·48	−260 −21·03	5 0·22	95 2·42	200 9·98	−60 −4·85	10 0·43	110 2·81	245 12·22	−260 −21·03
Distilled water	800 34·70	800 20·40	0 0	30 2·43	240 10·42	100 2·56	100 4·99	30 2·43	250 10·85	50 1·28	0 0	10 0·81	250 10·85	0 0	0 0	0 0

more magnesium is displaced in *Xanthoria parietina* in the dark. Magnesium and calcium are the principal ions displaced, but sodium is hardly affected, surprisingly in view of the similar chemical natures of sodium and potassium. The generally greater uptake of potassium relative to loss of other ions may argue for selective and perhaps metabolic uptake of this ion.

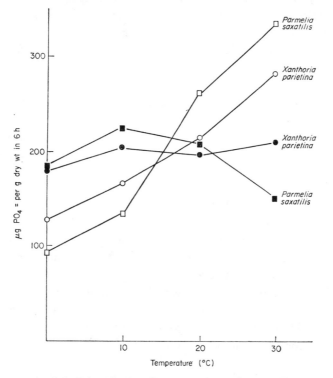

FIG. 9. The uptake of phosphate by *Xanthoria parietina* and *Parmelia saxatilis* at different temperatures in the light (3500 lx) and dark. Phosphate measured by the molybdate method. Closed symbols = dark; open symbols = light.

(iii) *Calcium* causes more equivalents of ions to be lost than are taken up in the light, but the reverse occurs in the dark, especially for *Parmelia saxatilis*. Magnesium is the principal ion displaced, as might be expected, but this is less apparent for *P. saxatilis*. Potassium is also displaced, with considerable differences in the light and dark for *Xanthoria parietina*, where little is lost in the light, but much more in the dark; *Parmelia saxatilis* loses an appreciable amount in both light and dark. Sodium is hardly affected.

(iv) *Magnesium* causes effects similar to calcium in that the similar ion, calcium, is displaced. Sodium and potassium are hardly affected in *Xanthoria parietina*, but potassium is displaced in *Parmelia saxatilis*.

The general trends may be summarized as follows: *Parmelia saxatilis* loses more potassium as a result of ion exchange by base ions (calcium and magnesium), than *Xanthoria parietina*; in *X. parietina*, the effect of calcium can be minimized by light conditions; sodium has a greater effect on *Parmelia saxatilis* than on *Xanthoria parietina*, which can also excrete sodium in the dark. The results thus generally agree with their habitat preferences since *Parmelia saxatilis* lives in environments poor in sodium, calcium and magnesium, while *Xanthoria parietina* inhabits situations rich in one or more of these ions.

It is difficult to obtain results for ion uptake and exchange in concentrations of elements similar to those in seawater because the amounts exchanged are small relative to the high background levels in seawater. It appears from these experiments using dilute solutions of ions, however, that lichens may be able to exert some effect over their ion uptake and are not merely dependent on ambient levels, particularly when the effects of light and dark are considered.

(c) *Effects of distilled water*. Although originally intended as a control for the previous experiments, Table I shows that distilled water (approximating to freshwater) exerts a significant effect in its own right. The results represent cation loss after a 24 h starvation period. Large differences can be seen in the loss of individual ions, in the relative loss between species, and between the light and dark treatments. It is unlikely that loss is due to simple diffusion of seawater or environmental water from water-filled spaces within the thallus, since the levels bear no resemblance to seawater levels (in *Xanthoria parietina*). In addition, cations in this species would be quickly taken up by the cells or cell walls (see p. 375).

Considerable amounts of sodium are lost by both species in light and dark, but in the light, *Xanthoria parietina* loses very large amounts. This again may be evidence of a sodium–excreting mechanism which is perhaps suppressed by other ions in the light. Considerable amounts of potassium are also lost by this species in the light, and to a lesser extent in the dark, but the effect is minimal in *Parmelia saxatilis*. Little calcium and magnesium is lost by *Parmelia saxatilis*, but considerable amounts of calcium are lost by *Xanthoria parietina*, especially in the dark. It might be concluded that freshwater exerts a potentially more serious effect on *Xanthoria parietina* mainly because of this loss of potassium.

(d) *Antagonistic effects of individual ions*. Some data have already been discussed with respect to ion displacement in section (b). Table I shows that loss of individual ions to distilled water is frequently much greater than in dilute solutions of other ions. For example, in distilled water large amounts of sodium are lost by both species, but this is minimal when any of the other ions are added. Exceptions are in the cases of calcium and magnesium which cause some replacement of each other. The effect of calcium is particularly interesting because it overcomes the loss of potassium to distilled water by *Xanthoria parietina* (Fig. 10). Calcium appears to exert this effect only in the light, however. In *Parmelia saxatilis* (Fig. 11) calcium causes a delayed loss of potassium in the dark but has no influence in the light. This suggests that *Xanthoria parietina* may prefer calcium-rich environments in order to minimize the loss of potassium to freshwater, assuming that this loss, which may be predominantly extracellular, is of consequence to that species. A deleterious effect of calcium is suggested with *Parmelia saxatilis*, but as it is only in the dark, perhaps the photophilic and sciaphilic tendencies of species must be considered further.

DISCUSSION

A common feature of lichenology at the present time is that field evidence is far in advance of that produced by laboratory investigations. Thus there are many hypotheses which remain to be verified or denied by physiological work. The seashore distribution problem is no exception.

The field evidence produced so far, relating to seashore lichen distribution, suggests that on the shore species are distributed according to nutritional or osmotic factors originating in seawater or soil-water. The chemistry of the substrate, whether siliceous or calcareous, and the presence of nutrient enrichment, such as bird manure, also imply nutritional factors. Generally it is held that the zonation is one of seawater-tolerating or seawater-requiring species low on the shore, to seawater-intolerant or freshwater-requiring species on the upper shore. Calcicole and ornithocoprophilous species are often associated with seawater-tolerant species, while calcifuge and ornithocoprophobous species are associated with seawater-intolerant or freshwater-tolerant species. A common factor here is the presence of calcium. Phosphate and nitrogen may be discounted from the present discussion because they are scarce in seawater.

The two main conclusions to be discussed are thus the role of calcium in the zonation, and the question of tolerance towards salinity.

1. The Role of Calcium

Calcium is known to be involved in maintaining the structure and efficiency of cell membranes, especially with regard to permeability and selective ion absorption (Rains, 1972). The results presented here may well be linked with this. *Xanthoria parietina* is always found in calcium-rich situations: in the

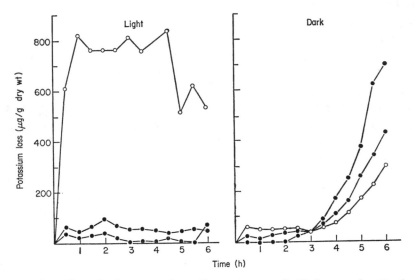

FIG. 10. The effect of calcium on loss of potassium to distilled water by *Xanthoria parietina* in the light and dark. ○ = without calcium, ● = with calcium.

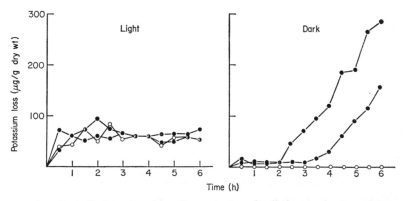

FIG. 11. The effect of calcium on loss of potassium to distilled water by *Parmelia saxatilis* in the light and dark. ○ = without calcium, ● = with calcium.

submesic supralittoral, on limestone and calcareous substrata, on bird perches, etc. The evidence suggests that it does not retain calcium in the thallus very efficiently since it has a very low thallus content of 1–300 p.p.m. Calcium appears to be useful to this species in preventing the loss of potassium to freshwater and, if this proves to be an active-metabolic process, will fit in well with results for the role of calcium in higher plants. The calcifuge species *Parmelia saxatilis* is always found in calcium-deficient situations, but does appear to be efficient in retaining calcium. This high thalline content may be sufficient to prevent excessive potassium loss to freshwater and thus enable it to live in calcium-poor surroundings. Anomalous levels of calcium have been found in xeric supralittoral thalli, especially *Ramalina siliquosa* and *Parmelia prolixa*; it remains to be seen what the role of this element is in such thalli. Possibly a mechanism similar to that found in *Xanthoria parietina*, or structural functions as mentioned earlier in the text, are involved.

2. Salinity Tolerance

The tolerance of thalli to extended periods of wet conditions, whether in seawater or in freshwater demonstrates that species may be distributed according to the amount of wetness in the environment. For example, littoral species are less adversely affected by continuously wet conditions than supralittoral thalli. The mechanism of this effect of wetness appears to vary with the species. Those from drier parts of the shore, such as the xeric supralittoral *Ramalina siliquosa*, simply die off, both bionts being killed. Thalli from the wetter parts of the supralittoral break down because the symbionts separate and become free living. The most interesting observation, however, is that the salinity of the water appears to exert a relatively small effect. Littoral thalli appear to be favoured slightly by seawater conditions, while supralittoral thalli are slightly favoured by freshwater conditions. As these experiments were performed over periods of a few weeks this suggests that extended periods of wet conditions are not particularly deleterious to seashore thalli, either in seawater or freshwater.

The effects of alternating wet and dry periods suggest that littoral thalli are most affected by rapid successions of drought periods, while the supralittoral thalli are most affected by frequent successions of wet periods. Supralittoral thalli appear to be able to adjust to the imposed conditions while littoral thalli die off. Again, however, as with the submersion experiments, the salinity of the medium exerts a lesser effect than the water factor.

Salinity tolerance experiments suggest that in relatively long-term experiments, extending for periods of weeks, the effects of salinity are small, so that thalli are relatively tolerant towards long-term salinity fluctuations.

The more short-term data on ion uptake and exchange, particularly the effects of submersion in distilled water and dilute ion solutions, strongly suggested that thalli are affected. For example, the haline *Xanthoria parietina* loses potassium in distilled water, especially in the light, but not when calcium and, to a lesser extent, other seawater ions are added to the medium. This gives a strong indication that freshwater, or at least calcium-deficient media, would be deleterious to this species. But the long-term experiments on constant sub-mersion do not really confirm this. It remains possible that this loss of potassium is truly serious but that damage to thalli was not apparent from monitoring only photosynthesis and respiration. Perhaps some other metabolic process is im-paired and re-investigation of perhaps carbohydrate activity or protein syn-thesis should be considered when measuring the viability of thalli subjected to potentially harmful experimental conditions. Alternatively, if the loss of potassium takes place largely from cell walls, and the intracellular levels are unaffected, then depletion of this "environmental" potassium may not have a harmful effect, even in experiments as relatively long-term as those described.

The relative effects of dark and light periods on ion displacement and uptake are unusual since generally more ions appear to be absorbed in the dark than in the light. This is contrary to current opinion, based on data on algae and higher plants, which states that ions are taken up to a greater extent in the light (Sutcliffe, 1962; Leggett, 1968; Higinbotham, 1973). However, the situation in higher plants is by no means certain and it has been admitted that results are conflicting and more information is required (Briggs et al., 1961). No precedent may be found in the fungal literature and it is usually presumed that ion uptake by fungi is much the same as that in higher plants and algae (Rothstein, 1965). It is possible that the effects described above are due to interactions between the mycobiont and phycobionts; perhaps light favours uptake by the alga but dark favours translocation to the fungal partner, resulting in an enhanced dark uptake. Alternatively pH changes in the medium induced by respiration in the dark and photosynthesis in the light may be responsible. It is difficult to predict the effects of such changes on ion uptake since a large number of opposing factors are involved (Briggs et al., 1961), and more information is required in this area of lichenology.

CONCLUSION

The experimental work outlined in this paper has been of a preliminary nature, designed to open up promising fields for new research. The tone of this review has, as a consequence, been deliberately slanted to accentuate problems at the expense of progress. It is felt that future research will be most promising if growth criteria are used to evaluate the responses of thalli to experimental stimuli. This is particularly important in the field of nutrition, which often carries the connotation of growth. In addition, a separation of tolerance from requirement criteria would be facilitated. Despite some promising work on artificial culture which has appeared recently, a standardized, inexpensive method of culturing lichen thalli has yet to be adopted and this constitutes an urgent need.

ACKNOWLEDGEMENTS

I wish to acknowledge the help of Miss M. Spear and Mr P. Jonas in obtaining some of the results, the Natural Environment Research Council for financial assistance and to Mrs M. Fletcher for correcting the manuscript.

REFERENCES

BRIGGS, G. E., HOPE, A. B. and ROBERTSON, R. N. (1961). "Electrolytes and Plant Cells." Davis, Philadelphia.

BROWN, D. H. and DI MEO, J. A. (1972). Influence of local maritime conditions on the distribution of two epiphytic lichens. *Lichenologist* **5**, 305–310.

BROWN, D. H. and SLINGSBY, D. R. (1972). The cellular location of lead and potassium in the lichen *Cladonia rangiformis* (L.) Hoffm. *New Phytol.* **71**, 297–305.

DU RIETZ, G. E. (1932). Zur Vegetationsökologie der östschwedische Kustenfelsen. *Beih. bot. Zbl.* **49**, 61–112.

EDWARDS, R. S. and CLAXTON, S. M. (1964). The distribution of air borne salt of marine origin in the Aberystwyth area. *J. appl. Ecol.* **1**, 253–263.

FEIGE, G. B. (1972). Ecophysiological aspects of carbohydrate metabolism in the marine blue green algal lichen *Lichina pygmaea* Ag. *Z. Pflanzenphysiol.* **68**, 121–126.

FEIGE, G. B. (1973). Untersuchungen zur ökologie un physiologie der marinen blaualgen flechten *Lichina pygmaea* Ag. II. Die reversibilitat der osmoregulation. *Z. Pflanzenphysiol.* **68**, 415–421.

FERRY, B. W. and SHEARD, J. W. (1969). Zonation of supralittoral lichens on rocky shores around the Dale Peninsula, Pembrokeshire, *Fld Stud.* **3**, 41–67.

FISCHER-PIETTE, E. (1931). Sur la penetration des diverses espèces marines sessiles dans les estuaries et sa limitation par l'eau douce. *Annls Inst. océanogr.*, Monaco **10**, 217–243.

FLETCHER, A. (1972). "The Ecology of Marine and Maritime Lichens of Anglesey." Ph.D. Thesis, University of Wales.

FLETCHER, A. (1973a). The ecology of marine (littoral) lichens on some rocky shores of Anglesey. *Lichenologist* **5**, 368–400.

FLETCHER, A. (1973b). The ecology of maritime (supralittoral) lichens on some rocky shores of Anglesey. *Lichenologist* **5**, 401–422.

GOLDSMITH, F. B. (1967). "Some Aspects of the Vegetation of Seacliffs." Ph.D. thesis, University of Wales.

HAKULINEN, R. (1962). Ökologische beobachtungen über die flechtenflora der vogel-steine in süd und mittelfinnland. *Arch. bot. Soc. zool.-bot. fenn. Vanamo* **17**, 12–15.

HARVEY, H. W. (1955). "The Chemistry and Fertility of Seawaters." Cambridge University Press, Cambridge.

HIGINBOTHAM, N. (1973). Electropotentials of plant cells. *A. Rev. Pl. Physiol.* **24**, 25–46.

JAMES, P. W. (1973). The effects of air pollutants other than hydrogen fluoride and sulphur dioxide on lichens. *In* "Air Pollution and Lichens" (B. W. Ferry, M. S. Baddeley and D. L. Hawksworth, eds), pp. 143–175. University of London, Athlone Press, London.

KÄRENLAMPI, L. (1966). The succession of the lichen vegetation in the rocky shore geolittoral and adjacent parts of the epilittoral in the south western archipelago of Finland. *Annls bot. fenn.* **3**, 79–85.

KNOWLES, M. C. (1913). The maritime and marine lichens of Howth. *Scient. Proc. R. Dubl. Soc.* **14**, 79–143.

LAUNDON, J. R. (1966). Hudson's *Lichen siliquosus* from Wiltshire. *Lichenologist* **3**, 236–241.

LAUNDON, J. R. (1970). Lichens new to the British flora. *Lichenologist* **4**, 297–308.

LEGGETT, J. E. (1968). Salt absorption by plants. *A. Rev. Pl. Physiol.* **19**, 333–346.

MASSÉ, L. (1966). Étude comparée des teneurs en azote total des lichens et leur substrat: les espéces "ornithocoprophiles". *C. r. hebd. Séanc Acad. Sci., Paris, Sér.* **D, 262**, 1721–1724.

PUCKETT, K. J., NIEBOER, E., GORZYNSKI, M. J. and RICHARDSON, D. H. S. (1973). The uptake of metal ions by lichens: a modified ion exchange process. *New Phytol.* **72**, 329–342.

RAINS, D. W. (1972). Salt transport by plants in relation to salinity. *A. Rev. Pl. Physiol.* **23**, 367–388.

RIED, A. (1960). Stoffwechsel und Verbreitungsgrenzen von Flechten II. Wasser- und Assimilationshaushalt, Entquellungs- und Submersionresistenz von Krustenflechten benachbarter Standorte. *Flora, Jena,* **149**, 345–385.

ROTHSTEIN, A. (1965). Uptake and translocation I. Uptake. *In* "The Fungi" (G. C. Ainsworth and A. E. Sussman, eds), **1**, 429–456. Academic Press, New York and London.

SANTESSON, R. (1939). Amphibious pyrenolichens I. *Ark. Bot.* **29A** (10), 1–67.

SUTCLIFFE, J. F. (1962). "Mineral Salts Absorption by Plants." Pergamon Press, Oxford.

TUOMINEN, Y. (1967). Studies on the strontium uptake of *Cladonia alpestris* thallus. *Annls bot. fenn.* **4**, 1–28.

WAISEL, Y. (1972). "Biology of Halophytes." Academic Press, New York and London.

15 | The Lichen as an Ecosystem: Observation and Experiment

J. F. FARRAR

Department of Botany, University of Dar es Salaam, Dar es Salaam, Tanzania

Abstract: Lichens can be considered as ecosystems with two trophic levels: algal producers and fungal consumers. Other organisms could be included if shown to be regularly associated with lichens. The flow of carbon through thalli is considered in relation to the fluctuating environment of most lichens and it is argued that net productivity is a low percentage of gross primary production. Nutrient flow is considered briefly. Stability of lichens is discussed and related to carbon flow and its control. Some comparisons with other ecosystems are made and the importance of symbiosis discussed. The possibility of co-evolution of the symbionts is raised.

INTRODUCTION

Studies of ecosystems have largely fallen into two categories: either descriptive, the ecosystem being too complex to manipulate experimentally; or experimental, oversimplified "ecosystems" being created in the laboratory or computer. A lichen thallus, along with its immediate environment, can be considered as an ecosystem and has the advantage of being both natural and fairly simple. This paper will attempt to integrate our knowledge of lichen biology with current ecological thinking. Such an attempt is valuable for three reasons. First, it provides a vocabulary which stresses that lichens are not single organisms but associations between two or more very distinct species. Secondly, it forces us to emphasize the relationship between the thallus and its physical and chemical environment—a relationship fundamental to the understanding of lichen biology. Thirdly, perhaps the attempt will indicate that studies of lichens could generate insights relevant to a variety of ecosystems.

Systematics Association Special Volume No. 8, "Lichenology: Progress and Problems", edited by D. H. Brown, D. L. Hawksworth and R. H. Bailey, 1976, pp. 385–406. Academic Press, London and New York.

THE LICHEN AS AN ECOSYSTEM

"Actually the systems we isolate mentally are not only included as parts of larger ones, but they also overlap, interlock and interact with one another. The isolation is partly artificial, but is the only possible way in which we can proceed."

(Tansley, 1935)

Lichen thalli share several properties with larger ecosystems and it is upon these similarities that their claim to be treated as ecosystems rests. The two critical properties are their independence of other organisms, and the occurrence of different trophic levels within the thallus.

Many lichens need only inorganic sources of nutrients, being able to survive without any organic compounds. Thalli have been maintained in growth cabinets for prolonged periods on inorganic growth media (Kershaw and Millbank, 1969, 1970; Pearson, 1970; Harris and Kershaw, 1971). Some lichens may utilize organic carbon and nitrogen sources when available, but in general they need only a purely physico-chemical environment for survival. Animals from gastropods to reindeer feed on lichens (Peake and James, 1967; Gerson, 1973; Richardson, 1975) and some of these associations may be fairly constant (Gilbert, 1971). It has not been shown that such grazing is quantitatively important in the regulation of any lichen species (but see Laundon, 1967). No fungus or bacterium, apart from the fungal biotroph, has been shown to be a regular member of any lichen, although the occurrence of bacteria is disputed between physiologists who do not find them (Bond and Scott, 1955; Scott, 1956) and electron microscopists who do (Brown and Wilson, 1968; Jacobs and Ahmadjian, 1969; Ahmadjian and Jacobs, 1970). The ecosystem concept could readily embrace such grazers and extra consumers and it would be instructive to include them if suitable data were available. Epiphytic lichens are clearly dependent on plants for support, and their hosts must also modify their nutrient supply and light and humidity regimes. The environment of some terricolous lichens is similarly modified by the plants around them. Such lichens can be considered either as components of the larger ecosystem or as ecosystems in their own right. It could also be argued that a lichen community, say one covering a rock face, should be taken as the ecosystem, but for the moment discussion will be restricted to individual thalli. Most of what will be said below is true of communities as well as single thalli.

Virtually all of the accretion of dry matter by lichens is thought to be due to algal photosynthesis, hence the algae can be regarded as primary producers. The fungus is dependent on the alga for organic compounds and can thus be thought of as a primary consumer. The lichen therefore contains two trophic levels. If other organisms were shown to be associated with lichens they could readily be included in such a scheme if their mode of nutrition were known.

Mature communities are characteristically stable over long periods of time (Whittaker, 1970) and lichens share this longevity. Estimates of maximum age of lichens range up to 4500 years (Beschel, 1958) and life-spans are commonly 50–80 years for corticolous lichens (Hale, 1967) and several hundred years for crustose saxicolous lichens (Beschel, 1955). It is not known whether individual cells live this long, but two observations make it likely that the algae at least are continually turned over: the number of algal cells per unit area changes during the year in *Parmelia caperata* (Harris, 1971); and electron microscopy has revealed both dividing algal cells (Ahmadjian and Jacobs, 1970) and recognizably older cells (Jacobs and Ahmadjian, 1973), and also how algae of different ages are stratified in the thallus in such a way as to suggest turnover (Galun *et al.*, 1970).

A fourth characteristic of mature ecosystems is their low growth rate, or low net productivity (Whittaker, 1970). That lichens grow slowly is known from casual and precise observation (Hale, 1967, 1974; Armstrong, 1973) and several ideas have been advanced to account for the low growth rate (Ahmadjian, 1967; Farrar, 1973a).

Lichens thus resemble ecosystems in several ways. As we would be justified in considering them as ecosystems on the basis of their independence of other organisms and their consisting of two trophic levels, it is gratifying to find two other similarities with mature ecosystems. They are, certainly, most atypical ecosystems both in their simplicity and in the close physiological relationship between the components. But it is precisely these characteristics which make them of particular interest.

THE FLOW OF CARBON

"Though the organisms may claim our primary interest, when we are trying to think fundamentally we cannot separate them from their special environment, with which they form one physical system."

(Tansley, 1935)

The growth of an organism can be approximately described by its accumulation and loss of energy, and energy itself can be considered as equivalent to the flow of carbon compounds through the system. This section attempts to relate lichen growth to the flow of carbon compounds through the two species in the thallus.

Although there is considerable variation in the growth rates of different lichens (Kärenlampi, 1971; Armstrong, 1973) all are low compared with those of most higher plants. Relative growth rate falls during the life of the lichen (Kärenlampi, 1970; Armstrong, 1973; Farrar, 1974). Growth rate data can clearly be used as a means of assessing net productivity. The low net productivity of lichens has also been described as a high biomass accumulation ratio, biomass: net production (Forrest, 1971). Measurements of biomass and calorific value of lichens were given by Bliss (1962), Coppins and Shimwell (1971), Forrest (1971) Richardson and Finegan (1972) and Bliss *et al.* (1973).

Lichens are metabolically active only when moist (Smith, 1962; Farrar, 1973a). One reason for low productivity is that lichens spend a large proportion of their life dry and thus metabolically inactive, or nearly so. However, low productivity does not seem to be entirely due to low gross primary production. Indeed, lichen algae appear to photosynthesize very efficiently (Table I) and it

TABLE I. Rates of carbon dioxide fixation in lichens

Species	Rate of CO_2 fixation (μmoles CO_2 per mg chlorophyll per hour)	Conditions	Method
Hypogymnia physodes[a]	160	20°C, 20 000 lx	$H^{14}CO_3^-$ in mM $NaHCO_3$
Hypogymnia physodes[b]	132	*c.* 3·5°C, 60 klx	Infrared gas analysis
Cladonia subtenuis[c]	41·8	23°C, 420 W/m²	Infrared gas analysis

[a] From Farrar (1973b).   [b] Calculated from data of Schulze and Lange (1968). Calculated from data of Rundel (1972).

can be assumed that a large amount of carbon could be acquired photosynthetically (although good measurements are desirable). Rather, several lines of evidence suggest that gross fixation is high, but that a large percentage of the carbon thus assimilated is subsequently lost. In other words, net productivity is a small percentage of gross primary production.

One reason for this high loss of fixed material is simply that the biomass of the producer (alga) is low relative to that of the consumer (fungus). Algae have been estimated to occupy only 3–10% of the thallus by weight (Drew and Smith, 1966; Millbank, 1972 and Chapter 18) and yet supply all or nearly all of the organic carbon required by the fungus. As a result, a high percentage of all the carbon fixed by the alga passes to the fungus, mostly as carbohydrate (Smith *et al.*, 1969; Table II). The problems of measuring the rate of carbohydrate transfer are discussed by Hill and Smith (1972), Farrar (1973a) and Richardson (1974). Estimates based on the "inhibition" technique are likely to be too low, as are those based on estimates of ^{14}C transfer if the radioactivity is diluted by a pool of unlabelled carbohydrate in the alga (for example the ribitol pool in *Trebouxia*). It can safely be assumed that in lichens containing *Nostoc* or *Trebouxia* at least 70–80% of the carbon fixed by the alga passes to the fungus. The producer, then, has a low biomass relative to the consumer but a high throughput of carbon.

That a high percentage of the carbon so partitioned is subsequently lost is a result of the interaction between the lichen and its fluctuating and unpredictable environment (Farrar, 1973a). This loss has three componenents—basal respiration, resaturation respiration and leakage.

Lange *et al.* (1970) have shown elegantly how the desert lichen *Ramalina maciformis* has a carbon surplus much lower than its net fixation. In a typical day, the lichen fixes 1·32 mg CO_2/g in the short period between sunrise and its drying out, then remains dry until rewetted by dewfall at night, when it loses

TABLE II. Proportion of photosynthetically fixed ^{14}C released by the alga

Species	Alga	Carbohydrate released	% passed to known fungal compounds	% released to an inhibition medium	Time (h)
Peltigera polydactyla	*Nostoc*	Glucose	73	—	0·75
			—	33[a]	3–4
Dermatocarpon miniatum	*Hyalococcus*	Sorbitol	—	55	24
Coccocarpia sp.	*Scytonema*	Glucose	—	48	24
Xanthoria aureola	*Trebouxia*	Ribitol	50+	—	24
			—	40	24
Hypogymnia physodes	*Trebouxia*	Ribitol	59[b]	—	24
Sphaerophorus globosus	*Trebouxia*	Ribitol	—	57	24
Peltigera aphthosa	*Coccomyxa*	Ribitol	—	65	24

All but two of these figures are from Hill and Smith (1972), who discuss the inhibition technique at length. The other figures are from [a] Drew and Smith (1967) and [b] Farrar (1973b).

0·78 mg CO_2/g before dawn by respiration. Clearly the exact figures will vary according to conditions, but this indicates that about 59% of fixed carbon is lost within 24 h of its acquisition. This loss is due exclusively to basal respiration which must be a property of both alga and fungus. If a lichen is, like *R. maciformis*, moist in the dark for much longer than it is moist in the light, it must lose a great deal of carbon by basal respiration alone. A simulation model of the growth of *Parmelia caperata* (Harris, 1972) considered basal respiration as the only form of carbon loss; even so, during some weekly periods the simulated thallus suffered a net loss of carbon.

Many temperate lichens are rewetted after dry periods by liquid water (as rain) rather than by absorption of water vapour. Such sudden rewetting has drastic effects on metabolism. A period of high but steadily decreasing respiration follows rewetting in a variety of species, and its intensity and duration are greater the moister the normal habitat of the lichen (Ried, 1953, 1960; Lange, 1969; Smith and Molesworth, 1973; Farrar, 1973b). It can be seen from Table III that the quantities of carbon lost would take an appreciable time to replace by photosynthesis. The source of the lost carbon has not been identified but it seems reasonable to assume that both fungus and alga are involved.

When subjected to a series of wetting and drying cycles, *Hypogymnia physodes* shows such a high loss of polyols and of ^{14}C after preincubation in ^{14}C-bicarbonate, that the loss cannot be accounted for by resaturation respiration and basal respiration alone (Farrar, 1973b). Similarly, in *Peltigera polydactyla* approxi-

TABLE III. Resaturation respiration

Species	Duration (hours)	Total loss of CO_2	Basal respiration: CO_2 loss per hour	Assumed loss due to resaturation respiration	Habitat
Peltigera polydactyla	9·75	0·265 μl per 15 discs	0·012 μl CO_2 per 15 discs	0·144 μl per 15 discs	Moist woodland floor
Hypogymnia physodes	1	0·77 ml/g	0·23 ml/g	0·54 ml/g	Open woodland
Xanthoria aureola	2[a]	0·52 ml/g	0·36 ml/g	0·16 ml/g	Tiled roof

[a] Only the first hour after rewetting was analysed.

All experiments were at 20°C. The assumed loss due to resaturation respiration is the total loss less the basal respiration over the period of resaturation respiration.

Data for *Peltigera* and *Xanthoria* from Smith and Molesworth (1973); data for *Hypogymnia* from Farrar (1973b).

mately 2–3 mg mannitol per 15 discs was lost during the 17 h following re-wetting, whilst over the same period only 0·81 mg CO_2 would be lost by respiration (composed of 0·53 mg CO_2 resaturation respiration in the first 10 h and 7 h basal respiration at 0·04 mg CO_2/h per 15 discs) (Smith and Molesworth, 1973 and calculations from their data). In *H. physodes* the extra loss could be accounted for by a leakage of polyols and sucrose into the medium in which the thallus was rewetted; the amount of carbon lost by leakage approximated to that lost by resaturation respiration (Farrar, 1973b). Compounds unique to each symbiont were found to leak out of *H. physodes* and so it appears that both alga and fungus lose carbon in this way, the alga losing ribitol and sucrose and the fungus losing arabitol and mannitol (Farrar, 1973b).

These phenomena associated with rewetting have been described in the laboratory and it remains to determine their significance in the field. They may be an important source of carbon loss for the many lichens which are frequently wetted and dried in nature. Such wetting and drying may be obligatory for many lichens as it has been argued that many species are intolerant of constant conditions, particularly water saturation (Farrar, 1973a). Thus *H. physodes* retains high photosynthetic activity (Fig. 1) and high polyol contents when subjected

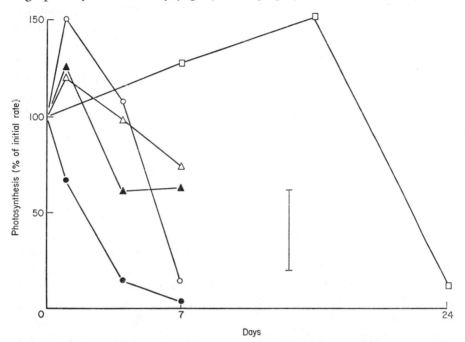

FIG. 1. Photosynthesis of *Hypogymnia physodes* from various environmental regimes. The vertical bar represents the least significant difference for $P = 0.05$. Thalli were subjected to the following regimes: soaking in the light at 20°C (O); soaking in the dark at 20°C (●); soaking in the light at 5° C (△); soaking in the dark at 5°C (▲); and wetting and drying in a 12 h day at 5°C (□). Photosynthesis was assayed by $H^{14}CO_3^-$ fixation at 20°C for 1 h after equilibrating the thalli under assay conditions. Data of Farrar (1973b).

to wetting and drying cycles but not when soaked continuously (Farrar, 1973b). It is probable that the majority of temperate lichens can only survive in environmental regimes which necessarily involve a high loss of fixed carbon.

The loss of such a high percentage of gross production in response to a fluctuating environment could be expected to place a considerable strain on metabolism. The strategy which lichens use to minimize this strain I shall call "physiological buffering". In lichens containing *Trebouxia*, a high percentage of fixed carbon remains in ethanol-soluble carbohydrates and very little enters insoluble material (Richardson and Smith, 1968; Farrar, 1973b). When *H. physodes* is maintained in a variety of regimes (soaking in the dark or light, wetting and drying in the light) the carbon it loses comes entirely from the ethanol-soluble material and there is no evidence of breakdown of insoluble material (Farrar, 1973b). Only a small percentage of the insoluble nitrogen in *Peltigera polydactyla* is mobilized after 6 days' dark starvation (Smith, 1960a). Thus the inevitable disturbances in metabolism are largely confined to changes in the ethanol-soluble fraction, which largely consists of carbohydrates, other compounds being buffered against change to a considerable extent. We can thus consider that soluble carbohydrates in lichens act as physiological buffers and can regard lichen metabolism as broadly divided into growth and repair processes on the one hand and respiratory and stress resistance processes on the other. Photosynthate enters the respiratory substrates and stress resistance compounds (the soluble carbohydrates) which are turned over rapidly, a small amount (which we can estimate at 10–20%) being tapped off into growth and repair processes (see Fig. 4). Such compartmentation of metabolism must occur in both alga and fungus.

The soluble carbohydrates involved in buffering are commonly polyols although polyol glycosides and sucrose are found in many lichens (Culberson, 1969). They are present at very high levels: *H. physodes* commonly contains 10 mg/g ribitol (in the alga), 10 mg/g mannitol and 50 mg/g arabitol (both in the fungus) on a dry weight basis (Farrar, 1973b). If we allow a water content of 2·5 g/g dry weight (Barkman, 1958; B. W. Ferry, pers. comm.) and assume the alga to occupy 10% by weight of the thallus, this implies a concentration of ribitol in the alga of 0·26 M, and of arabitol and mannitol in the fungus of 0·14 M and 0·026 M respectively, even if all the water were intracellular. Obviously these apparent molarities will rise as the lichen dries out. This has two important implications: (*a*) damage to cell contents could be expected at such concentrations of aldoses, e.g. glucose (Scott, 1960); the unreactive polyols are thus most suitable for lichens that air-dry frequently; (*b*) polyols may protect lichens during drying both by acting as osmotica reducing or slowing water loss and by protection of macro-molecules by direct substitution for water molecules in their hydration shells (Webb, 1963; Farrar, 1973b). Thus polyols are active in stress protection directly as well as acting as physiological buffers. That the importance of polyols to lichens is not confined to drought resistance and buffering is indicated by the relationship between levels of ribitol and photosynthesis, and between arabitol and phosphate uptake, in *H. physodes* (Fig. 2).

The salient features of the flow of carbon through *H. physodes* as pictured here (which may be typical of many lichens) are as follows. First, net production is a low percentage of gross primary production. Secondly, a high percentage of gross primary production passes to the primary consumer. Thirdly, as much of the carbon received by the consumer is in the form of ribitol, which it can

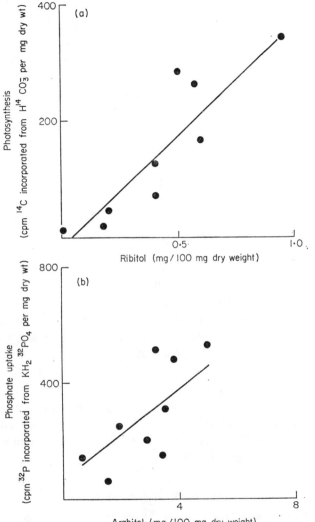

FIG. 2. Polyol levels and assimilative ability. Linear regressions are fitted to the relationships between (a) ribitol levels and photosynthesis ($r = 0.90$) and (b) arabitol levels and phosphate uptake ($r = 0.74$). Thalli which had been subjected to a variety of experimental regimes were assayed for photosynthesis by $H^{14}CO_3^-$ fixation, and polyols were measured by gas–liquid chromatography. From Farrar (1973b).

utilize readily, assimilation efficiency (energy ingested/energy assimilated; see Ricklefs, 1973) will be very high, perhaps 95–100%. The latter two features are both related to the apparent compartmentation of metabolism referred to above. As the primary product of algal photosynthesis is ribitol, this can be lost to the fungus with no lasting damage to the alga and the fungus is receiving an energy source which it can utilize with very high efficiency. The importance of large pools of sugars in the origins of mutualistic symbiosis has been discussed by Lewis (1973a).

A much lower percentage of gross primary production passes to consumers in most ecosystems than the 70–80% estimated here. The examples listed by Wiegert and Owen (1971) range from 1·5–2·5% of net primary production exploited by herbivores in mature deciduous forest, to 30–45% exploited in managed rangeland. Phytoplankton, however, are 60–99% exploited by herbivores and thus marine communities are, in this sense, energetically similar to lichens. The very high assimilation energy of the fungus is approached only in carnivorous animals and cases of endo-parasitism (Welch, 1968; Walkey and Meakins, 1970) and will perhaps be found in other cases of mutualistic symbiosis. The low net production, high throughput of energy and presence of mutualistic relationships are characteristic of mature ecosystems rather than their earlier developmental stages (Odum, 1969), so in this sense the mature lichen thallus resembles mature ecosystems.

It is thus possible to describe the flow of carbon through the lichen ecosystem in some detail. Confirmation and quantification of the proposed pathways in a range of species is badly needed. Much of the carbon flow is concerned with polyol metabolism and it is possible to suggest several functions for these polyols. The intimate relationship between carbon flow and environmental conditions is clearly seen. An attempt to summarize some of the above points is made in Figs 3 and 4, and Table IV, where the flow of carbon through *H. physodes* is presented. This has involved making several assumptions which further experiments would clarify.

THE FLOW OF NUTRIENTS

". . . and there is a constant interchange of the most various kinds within each system, not only between the organisms but between the organic and the inorganic."

(Tansley, 1935)

The dynamics of the many other components of lichens are relatively little studied. Nitrogen and cations have received some attention and phosphorus a little. The considerable body of work implicating sulphur oxides as pollutants of lichens has yielded little of relevance here. Two aspects of nutrient flow are of special interest. Firstly, to what extent are methods of nutrient accumulation adapted to the particular types of nutrient present in a habitat? Secondly, how are nutrients distributed between the alga and fungus; in particular, how do algae get any nutrients at all when they are surrounded by fungal hyphae, which

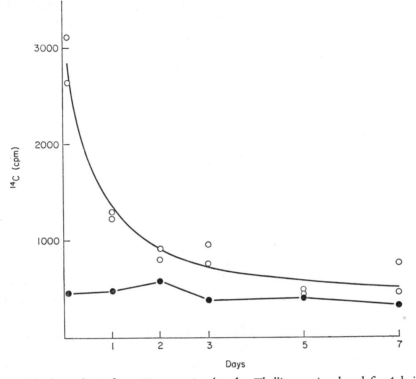

FIG. 3. The loss of ^{14}C from *Hypogymnia physodes*. Thalli were incubated for 1 h in $H^{14}CO_3$ in the light and then subjected to 7 cycles of 3 h soaking and 21 h air-drying, in the light at 20°C. Under these conditions ability to photosynthesize and pools of polyols remain at high levels. Radioactivity remaining in the 80% methanol-insoluble fraction (● c.p.m. per g thallus dry weight) and in total polyols (○, c.p.m. per 10 mg polyol per g thallus dry weight) was assayed. The fall in radioactivity in the polyols is described by the equation $\log x = 3\cdot39 - 0\cdot8 \log y$. Data from Farrar (1973b).

are known to be extremely efficient accumulators of nutrients (Harley, 1971; Smith, 1974)?

Nitrogen commonly limits the growth of plants and animals in nature (Black, 1968; Harley, 1971; Southwood, 1973) and its movement through the lichen ecosystem is therefore of particular interest. Environmental sources of nitrogen encompass molecular nitrogen, ammonium and nitrate in the atmosphere (Burns and Hardy, 1973) and inorganic and organic compounds in run-off and splash from plants and soil. There seems to be little relationship between the sources of nitrogen available to a particular lichen and its method of accumulation. Thus lichens containing *Nostoc*, which can fix atmospheric nitrogen, are not characteristic of habitats poor in combined nitrogen. Smith (1960a, b) found that the habitat of *Peltigera polydactyla* was rich in organic nitrogen compounds and that the lichen could accumulate various amino acids rapidly and ammonium rather

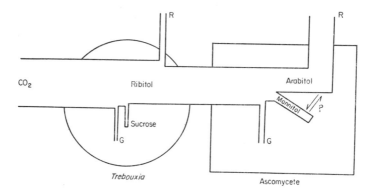

FIG. 4. The flow of carbon in *Hypogymnia physodes* maintained in a favourable environment. The width of the pathways represents the estimated relative rates of carbon flow. R = respiration and leakage, G = growth (synthesis of proteins, etc.). The diagram is based on evidence from Farrar (1973b), which can be summarized as follows:

1. Photosynthetically fixed carbon enters a ribitol pool in the alga and much passes to the fungal compounds arabitol and mannitol.
2. Pools of ribitol, arabitol and mannitol are maintained at roughly constant levels.
3. Thalli preincubated in $NaH^{14}CO_3$ subsequently lose a high percentage of the ^{14}C (Fig. 3.)
4. Loss of ^{14}C is highest from arabitol, ribitol, and mannitol, low from sucrose, and not detected from insoluble material.

less rapidly. Amongst coprophilous lichens, there are some that show high levels of uricase activity but others have levels similar to species not considered coprophilous (Massé, 1969).

The most complete description of nitrogen flow is available for *Peltigera aphthosa* (Millbank and Kershaw, 1974). Atmospheric nitrogen is fixed by *Nostoc* in cephalodia on the thallus and virtually all of this passes to the fungus. The rate of nitrogen fixation is very high, perhaps as the fungus reduces the partial pressure of oxygen around the algae thereby creating anaerobic conditions favourable for fixation (Stewart and Pearson, 1970). The alga *Coccomyxa* receives only 3% of the fixed nitrogen expected on a proportional share basis (Millbank and Kershaw, 1974). Kershaw and Millbank (1970) observed that the lichen was healthier if supplied with dilute potassium nitrate solution and suggested that *Coccomyxa* utilizes this directly.

Although some non-lichenized fungi can utilize nitrate, it is not a preferred nitrogen source; if both ammonium and nitrate are available to the fungus ammonium will be used preferentially (Nicholas, 1965; Burnett, 1968). Thus even if lichen fungi are potentially capable of using nitrate, the presence of

TABLE IV. Some aspects of carbon flow in *Hypogymnia physodes*

Ecological term	Physiological term	Estimated value
1. $\dfrac{\text{gross intake of consumer}}{\text{gross primary production}}$	% gross photosynthesis passing to fungus	70–90%
2. $\dfrac{\text{net production}}{\text{gross primary production}}$	% gross photosynthesis contributing to growth	10–20%
3. assimilation efficiency of consumer	% of carbon received by fungus from alga which is utilizable	95–100%
4. $\dfrac{\text{net production}}{\text{biomass}}$	growth rate, % per year	7–15%

Sources and assumptions:
  1, 2. Based on data of Farrar (1973b) and assumptions mentioned in the text.
    3. Assuming ribitol is the only form of carbon passing from alga to fungus.
    4. Based on published radial growth rates (Bitter, 1901; Ozenda, 1963) and assuming circular thalli of radius 4 cm.

ammonium (Burns and Hardy, 1973) may reduce their ability to utilize it. Smith (1960a, b) showed that *Peltigera polydactyla* accumulates nitrate very slowly, compared with ammonium and amino acids. It is thus possible that green algae in lichens can obtain nitrogen as nitrate because the fungus does not accumulate it efficiently.

The absorption of phosphate and cations has been reviewed recently (Farrar, 1973a; Tuominen and Jaakkola, 1974). The possible role of cell walls in binding potentially toxic cations is worthy of detailed investigation.

Although lichens are good accumulators of a wide variety of small molecules (Smith, 1962) the levels of nutrients reported from thalli are (with the exception of some cations, which may be extracellular) in the order of those found in non-lichenized algae and fungi (see Cochrane, 1958; Round, 1965; Solberg, 1967). Phosphorus is at consistently lower levels in lichens, but the significance of this remains to be determined. There is no indication of storage of these nutrients. Furthermore, the expected rate of nitrogen-fixation in lichens containing *Nostoc* is greatly in excess of the apparent requirement (Hitch and Stewart, 1973). There are two possible explanations for this apparent paradox: either lichen growth is limited by nutrients or there is some loss of nutrients that have been taken up. The former is unlikely to be a complete explanation as it would be expected that one nutrient would be limiting and others would accumulate. Further, there is direct evidence that loss of nutrients can occur. The leakage of soluble carbohydrates from *H. physodes* that occurs on rewetting (see above) is thought to be a physical process in which other compounds can be lost (Simon, 1974) and indeed both inorganic and hexose phosphate were found to leak from *H. physodes* (Farrar, 1973b). Such leakage has three important implications. First, if a nutrient is in short supply it may be advantageous for the lichen to keep it in a non-leaking form. The storage of phosphate as polyphosphate (Simonis and Feige, 1969; Farrar, 1973b) may be explained in this way. Secondly, the compounds leaking from the fungus may represent a source of nutrients for the alga. Thirdly, it may provide a means of excretion.

The possibility of a number of other substances, both organic and inorganic, moving from alga to fungus and fungus to alga cannot be excluded. Indeed Lewis (1973a, b, 1974) has suggested that tight nutrient cycling consequent upon the mutualistic relationship between alga and fungus may expedite the efficient utilization of nutrient-poor environments.

Ahmadjian (1967) points out that several lichen algae secrete vitamins into their growth media when cultured, that amongst these are biotin and thiamine, and that many cultured lichen fungi require biotin and thiamine for growth. If such a relationship holds in the intact thallus yet another flow must be considered. A fungal factor inducing leakiness in the alga is unlikely to exist as an exhaustive search for such a factor has proved unsuccessful (Green, 1970).

Consideration of fluxes of nutrients through thalli indicates the shortcomings of our knowledge of nutrient requirements, localization and turnover in lichens. Next to reproduction, this is the area of lichen biology most in need of clarification.

STABILITY

"There is in fact a kind of natural selection of incipient systems, and those which can attain the most stable equilibrium survive the longest."

(Tansley, 1935)

Some physiological aspects of the lichen–environment complex have been described and a few comparisons made with other ecosystems. It remains to discuss the stability of the lichen system. For the purpose of this discussion, lichen stability will be taken to mean the prolonged existence of a thallus in which alga and fungus maintain their relative abundance and mutualistic relationship over the range of environmental conditions to which it would normally be exposed. The distinction between stability and resilience (Clapham, 1973; Holling, 1973) is not drawn. Observation of lichens readily indicates their longevity and the apparent balance between the symbionts. What is the cause of this stability?

Most of the work on stability in ecosystems is concerned with variation in numbers of organisms. Harris (1971) observed that the number of *Trebouxia* cells per unit area in *Parmelia caperata* varied by an order of 1·7 during the year, but beyond this we have no information on changes of organism number within lichens. It is evident that the stability of lichens cannot be explained by changes in abundance of its components. Rather we must seek an explanation which embraces the comparative stability of organism number.

Diversity in ecosystems has long been suggested as a major cause of stability (MacArthur, 1955; Elton, 1958; Kormondy, 1969) but recent theoretical considerations suggest that an increase in diversity can lead to reduced stability (May, 1972, 1973; Hubbell, 1973). Indeed, a lichen is an extremely simple system and quite stable. The absence of grazers may be a major factor in that once they are present in any quantity, still more complexity may be added by introducing predators. Clearly, it is not possible to attribute lichen stability to diversity.

It is instructive to ask why few predators are found on lichens. One possibility is that some lichen acids render the thallus unpalatable (Culberson, 1969) and, if this is a method for reducing predation, it is of considerable interest. More fundamental is a consideration of energy budgets. The rate of grazing which a lichen could tolerate without permanent damage must be considerably less than its rate of growth. It must thus be so low that it is unlikely that a stable relationship between lichen and predator could develop. Conversely, if the lichen grew rapidly, a predator–lichen relationship would develop and the system would become more complex. Perhaps it is no coincidence that the one well established stable relationship between grazer and lichen—caribou and species of *Cladonia* (Richardson, 1975)—involves lichens with a high growth rate. The basis of lichen stability may thus be sought in its carbon balance.

Carbon balance could be achieved by either of two mechanisms. A sufficiency of organisms able to respond rapidly to changing conditions, by growth or reproduction, could be organized in such a way as to keep a system in equilibrium. This mechanism may operate in complex ecosystems and could explain the partial relationship between diversity and stability. However, it also explains why diversity *per se* is not the cause of stability since organisms of a suitable type must be present to ensure stability. This mechanism is clearly not applicable to lichens. The second mechanism is the exercising of control over the flow of carbon through the system. There are three points where we might expect control to be found in lichens, though further research would doubtless reveal more. In higher plants the rate of photosynthesis can be controlled by the plant to some extent (Sweet and Wareing, 1966; Wareing *et al.*, 1968; Habeshaw, 1973). It has been suggested that accumulated carbohydrates exercise this control but there is no evidence of a causal connection between carbohydrate levels and photosynthesis in flowering plants (Neales and Incoll, 1968). It is not possible, therefore, to decide whether control of photosynthesis in lichen algae may be by accumulated photosynthate, in spite of the relationship between ribitol levels and photosynthesis demonstrated above in *H. physodes*. Transfer of carbo-hydrate from alga to fungus is a crucial part of carbon flow in lichens and there are indications that this varies with experimental conditions and pretreatment of the material (Hill and Smith, 1972; Farrar, 1973b; D. C. Smith, unpublished). Perhaps the alga can exercise control over its carbon loss. Controls of the type of carbohydrate moving means that directional flow can be maintained (Smith *et al.*, 1969) and that the efficiency of transfer is very high. The third point of control may be over the synthesis of protein, polysaccharide and nucleic acids within alga and fungus. That such control is operative is suggested by the apparent compartmentation of metabolism referred to above. Critical experi-mentation is necessary to determine whether such control applies under a range of environmental (and particularly nutritional) regimes or is simply a reflection of unfavourable experimental conditions. What is important is that one of these points of control involves close physiological cooperation between taxonomi-cally dissimilar organisms. In ecosystem terms, mutualistic symbiosis increases both the efficiency of carbon transfer between trophic levels and the degree of control over carbon flow. This may be of great importance in maintaining the stability of the lichen.

Pomeroy (1970) has argued that stability can be related to the type of nutrient cycling in an ecosystem. This argument is in some ways analogous to that adopted here except that in lichens carbon fluxes are considered to be of more importance than nutrient fluxes.

Close physiological association in an ecosystem is one extreme of a continuum of types of relationship between organisms which runs from cooperation to antagonism (Whittaker, 1962; Langford and Buell, 1969; Lewis, 1974). It corresponds in part to the concept of integrated species associations suggested by

Langford and Buell (1969). If we enlarge our lichen ecosystem to embrace grazing animals, epiphytic or endophytic bacteria and adjacent thalli, less specific associations with less close physiological relationships will be found. The existence of species associations as closely integrated as those found in lichens implies that the evolution of the component species must be related in some way as integration increases the pressure towards coevolution (Goodall, 1963). The concept of coevolution has been discussed extensively (Allee *et al.*, 1949; Whittaker, 1962; Goodall, 1963; Baker, 1966; Ehrlich and Raven, 1968; Pimentel, 1968; Langford and Buell, 1969) and evidence for coevolution has been presented (Pimentel, 1968). Hypotheses on the evolutionary relationships between biotrophic fungi and their hosts have been reviewed by Harley (1969, 1971) and Lewis (1973a, 1974). It has been suggested that such coevolution leads to a balance between two (or possibly more) species, which in the case of lichens, and other mutualistic symbioses, such as sheathing mycorrhizas (Lewis, 1974), has resulted in physiological integration. If we look at lichen evolution and reproduction in these terms, the joint reproduction of alga and fungus in isidia and soredia acquires a significance beyond that of increasing the frequency of colonization.

ACKNOWLEDGEMENT

Through his willingness to share his considerable knowledge and understanding of lichens, Professor D. C. Smith provided the stimulus for many of the points raised here. He is in no way to blame for my abuse of them.

REFERENCES

AHMADJIAN, V. (1967). "The Lichen Symbiosis." Blaisdell, Waltham, Mass.

AHMADJIAN, V. and JACOBS, J. B. (1970). The ultrastructure of lichens. III. *Endocarpon pusillum*. Lichenologist **4**, 268–270.

ALLEE, W. C., EMERSON, A. E., PARK, O., PARK, T. and SCHMIDT, K. P. (1949). "Principles of Animal Ecology." Saunders, Philadelphia.

ARMSTRONG, R. A. (1973). Seasonal growth and growth-rate colony size relationships in six species of saxicolous lichens. *New Phytol.* **72**, 1023–1030.

BAKER, H. G. (1966). Reasoning about adaptations in ecosystems. *Bioscience* **16**, 35–37.

BARKMAN, J. J. (1958). "Phytosociology and Ecology of Cryptogamic Epiphytes." Van Gorcum, Assen.

BESCHEL, R. E. (1955). Individuum und alter bei flechten. *Phyton* **6**, 60–68.

BESCHEL, R. E. (1958). Lichenometrical studies in West Greenland. *Arctic* **11**, 254.

BITTER, G. (1901). Zur morphologie und systematik von *Parmelia*, untergattung *Hypogymnia*. *Hedwigia* **40**, 171–274.

BLACK, C. A. (1968). "Soil–Plant Relationships." Wiley, New York.

BLISS, L. C. (1962). Caloric and lipid content in alpine tundra plants. *Ecology* **43**, 753–757.

BLISS, L. C., COURTIN, G. M., PATTIE, D. L., RIEWE, R. R., WHITFIELD, D. W. A. and WIDDEN, P. (1973). Arctic tundra ecosystems. *A. Rev. Ecol. Syst.* **4**, 359–400.

BOND, G. and SCOTT, G. D. (1955). An examination of some symbiotic systems for fixation of nitrogen. *Ann. Bot.* **19**, 67–77.

BROWN, R. M. and WILSON, R. (1968). Electron microscopy of the lichen *Physcia aipolia* (Ehrh.) Nyl. *J. Phycol.* **4**, 230–240.

BURNETT, J. H. (1968). "Fundamentals of Mycology." Arnold, London.

BURNS, R. L. and HARDY, R. W. F. (1973). "Nitrogen Fixation in Bacteria and Higher Plants." Springer, New York.

CLAPHAM, W. B. (1973). "Natural Ecosystems." Macmillan, New York.

COCHRANE, V. W. (1958). "Physiology of Fungi." Wiley, New York.

COPPINS, B. J. and SHIMWELL, D. W. (1971). Cryptogam complement and biomass in dry *Calluna* heath of different ages. *Oikos* **22**, 204–209.

CULBERSON, C. F. (1969). "Chemical and Botanical Guide to Lichen Products." University of North Carolina Press, Chapel Hill, N.C.

DREW, E. A. and SMITH, D. C. (1966). The physiology of the symbiosis in *Peltigera polydactyla* (Neck.) Hoffm. *Lichenologist* **3**, 197–201.

DREW, E. A. and SMITH, D. C. (1967). Studies in the physiology of lichens. VIII. Movement of glucose from alga to fungus during photosynthesis in the thallus of *Peltigera polydactyla*. *New Phytol.* **66**, 389–400.

EHRLICH, P. R. and RAVEN, P. H. (1968). Butterflies and plants: a study in coevolution. *Evolution* **18**, 586–608.

ELTON, C. S. (1958). "The Ecology of Invasions by Animals and Plants." Methuen, London.

FARRAR, J. F. (1973a). Lichen physiology: progress and pitfalls. *In* "Air Pollution and Lichens" (B. W. Ferry, M. S. Baddeley and D. L. Hawksworth, eds), pp. 238–282. University of London, Athlone Press, London.

FARRAR. J. F. (1973b). "Physiological Lichen Ecology." D.Phil. thesis, University of Oxford.

FARRAR, J. F. (1974). A method for investigating lichen growth rates and succession. *Lichenologist* **6**, 151–155.

FORREST, G. I. (1971). Structure and production of North Pennine blanket bog vegetation. *J. Ecol.* **59**, 453–479.

GALUN, M., PARAN, N. and BEN-SHAUL, Y. (1970). Structural modifications of the phycobiont in the lichen thallus. *Protoplasma* **69**, 85–96.

GERSON, U. (1973). Lichen–Arthropod associations. *Lichenologist* **5**, 434–443.

GILBERT, O. L. (1971). Some indirect effects of air pollution on bark-living invertebrates. *J. appl. Ecol.* **8**, 77–84.

GOODALL, D. W. (1963). The continuum and the individualistic association. *Vegetatio* **11**, 297–316.

GREEN, T. G. A. (1970). "The Biology of Lichen Symbionts." D.Phil. thesis, University of Oxford.

HABESHAW, D. (1973). Translocation and the control of photosynthesis in sugar beet. *Planta* **110**, 213–226.

HALE, M. E. (1967). "The Biology of Lichens." Arnold, London.

HALE, M. E. (1974) ["1973"]. Growth. *In* "The Lichens" (V. Ahmadjian and M. E. Hale, eds), pp. 473–492. Academic Press, New York and London.

HARLEY, J. L. (1969). "The Biology of Mycorrhiza" (2nd edition). Leonard Hill, London.

HARLEY, J. L. (1971). Fungi in ecosystems. *J. Ecol.* **59**, 653–668.

HARRIS, G. P. (1971). The ecology of corticolous lichens. II. The relation between physiology and the environment. *J. Ecol.* **59**, 441–452.

HARRIS, G. P. (1972). The ecology of corticolous lichens. III. A simulation model of productivity as a function of light intensity and water availability. *J. Ecol.* **60**, 19–40.

HARRIS, G. P. and KERSHAW, K. A. (1971). Thallus growth and the distribution of stored metabolites in the phycobionts of the lichens *Parmelia sulcata* and *P. physodes. Can. J. Bot.* **49**, 1367–1372.

HILL, D. J. and SMITH, D. C. (1972). Lichen physiology XII. The "inhibition" technique. *New Phytol.* **71**, 15–30.

HITCH, C. J. B. and STEWART, W. D. P. (1973). Nitrogen fixation by lichens in Scotland. *New Phytol.* **72**, 509–524.

HOLLING, C. S. (1973). Resilience and stability of ecological systems. *A. Rev. Ecol. Syst.* **4**, 1–24.

HUBBELL, S. (1973). Populations and simple food webs as energy filters. II. Two-species systems. *Am. Nat.* **107**, 122–151.

JACOBS, J. B. and AHMADJIAN, V. (1969). The ultrastructure of lichens. I. A general survey. *J. Phycol.* **5**, 227–240.

JACOBS, J. B. and AHMADJIAN, V. (1973). The ultrastructure of lichens. V. *Hydrothyria venosa*, a freshwater lichen. *New Phytol.* **72**, 155–160.

KÄRENLAMPI, L. (1970). Morphological analysis of the growth and productivity of the lichen *Cladonia alpestris. Rept Kevo Subarct. Res. Stn* **7**, 9–15.

KÄRENLAMPI, L. (1971). Studies on the relative growth rate of some fruticose lichens. *Rept Kevo Subarct. Res. Stn* **7**, 33–39.

KERSHAW, K. A. and MILLBANK, J. W. (1969). A controlled environment lichen growth chamber. *Lichenologist* **4**, 83–87.

KERSHAW, K. A. and MILLBANK, J. W. (1970). Nitrogen metabolism in lichens II. The partition of cephalodial-fixed nitrogen between the mycobiont and phycobionts in *Peltigera aphthosa. New Phytol.* **69**, 75–79.

KORMONDY, E. J. (1969). "Concepts of Ecology." Prentice-Hall, New Jersey.

LANGE, O. L. (1969). Experimentell-okologische untersuchungen an flechten der Negev-Wuste. I. CO$_2$-Gaswechsel von *Ramalina maciformis* (Del.) Bory unter kontrollierten Bedingungen in Laboratorium. *Flora, Jena* **158B**, 324–359.

LANGE, O. L., SCHULZE, E.-D. and KOCH, W. (1970). Experimentell-okologische untersuchungen an flechten der Negev-Wuste. II. CO$_2$-Gaswechsel und Wasserhaushalt von *Ramalina maciformis* (Del.) Bory am naturlichen Standort wahrend der sommerlichen Trockenperiode. *Flora, Jena* **159**, 38–62.

LANGFORD, A. N. and BUELL, M. F. (1969). Integration, identity and stability in the plant association. *Adv. ecol. Res.* **6**, 84–135.

LAUNDON, J. R. (1967). A study of the lichen flora of London. *Lichenologist* **3**, 277–327.

LEWIS, D. H. (1973a). Concepts in fungal nutrition and the origin of biotrophy. *Biol. Rev.* **48**, 261–278.

LEWIS, D. H. (1973b). The relevance of symbiosis to ecology and taxonomy with particular reference to the exploitation of marginal habitats. *In* "Taxonomy and Ecology"(V. H. Heywood, ed.), pp. 151–172. Academic Press, London and New York.

LEWIS, D. H. (1974). Micro-organisms and plants: the evolution of parasitism and mutualism. *Symp. Soc. gen. Microbiol.* **24**, 365–392.

MACARTHUR, R. (1955). Fluctuations of animal populations and a measure of community stability. *Ecology* **36**, 533–536.

MASSÉ, L. (1969). Quelques aspects de l'uricolyse enzymatique chez les lichens. *C. r. hebd. Séanc. Acad. Sci., Paris* **286**, 2896–2898.

MAY, R. M. (1972). Will a large complex system be stable? *Nature, Lond.* **238**, 413–414.

MAY, R. M. (1973). Qualitative stability in model ecosystems. *Ecology* **54**, 638–641.

MILLBANK, J. W. (1972). Nitrogen metabolism in lichens. IV. The nitrogenase activity of the *Nostoc* phycobiont in *Peltigera canina*. *New Phytol.* **71**, 1–10.

MILLBANK, J. W. and KERSHAW, K. A. (1974) ["1973"]. Nitrogen metabolism. *In* "The Lichens" (V. Ahmadjian and M. E. Hale, eds), pp. 289–307. Academic Press, New York and London.

NEALES, T. F. and INCOLL, L. D. (1968). The control of leaf photosynthesis rate by the level of assimilate concentration in the leaf: a review of the hypothesis. *Bot. Rev.* **34**, 107–125.

NICHOLAS, D. J. D. (1965). Utilization of inorganic nitrogen compounds and amino acids by fungi. *In* "The Fungi" (G. C. Ainsworth and A. S. Sussman, eds), vol. 1, pp. 349–376. Academic Press, New York and London.

ODUM, E. P. (1969). The strategy of ecosystem development. *Science N.Y.* **164**, 262–270.

OZENDA, P. (1963). Lichens. *In* "Handbuch der Planzenanatomie" (2nd edition) (W. Zimmermann and P. Ozenda, eds), **6** (9), i–x, 1–199. Borntraeger, Berlin.

PEAKE, J. F. and JAMES, P. W. (1967). Lichens and mollusca. *Lichenologist* **3**, 425–428.

PEARSON, L. C. (1970). Varying environmental factors in order to grow lichens intact under laboratory conditions. *Am. J. Bot.* **57**, 659–664.

PIMENTEL, D. (1968). Population regulation and genetic feedback. *Science* **159**, 1432–1437.

POMEROY, L. R. (1970). The strategy of mineral cycling. *A. Rev. Ecol. System.* **1**, 171–190.

RICHARDSON, D. H. S. (1974). Photosynthesis and carbohydrate movement. *In* "The Lichens" (V. Ahmadjian and M. E. Hale, eds), pp. 249–288. Academic Press, New York and London.

RICHARDSON, D. H. S. (1975). "The Vanishing Lichens." David and Charles, Newton Abbot, London and Vancouver.

RICHARDSON, D. H. S. and FINEGAN, E. J. (1972). Lichen productivity studies, Devon Island. *In* "Devon Island I.B.P. Project High Arctic Ecosystem: Project report 1970 and 1971" (L. C. Bliss, ed.), pp. 197–214.

RICHARDSON, D. H. S. and SMITH, D. C. (1968). Lichen physiology. IX. Carbohydrate movement from the *Trebouxia* symbiont of *Xanthoria aureola* to the fungus. *New Phytol.* **67**, 61–68.

RICKLEFS, R. E. (1973). "Ecology." Nelson, London.

RIED, A. (1953). Photosynthese und Atmung bei xerostabilen und xerolabilen Krustenflechten. *Planta, Jena* **41**, 436–438.

RIED, A. (1960). Nachwirkungen der Entquellung auf den Gaswechsel von Krustenflechten. *Biol. Zbl.* **79**, 659–678.

ROUND, F. E. (1965). "The Biology of the Algae." Arnold, London.

RUNDEL, P. W. (1972). CO_2 exchange in ecological races of *Cladonia subtenuis*. *Photosynthetica* 6, 13–17.

SCHULZE, E.-D. and LANGE, O. L. (1968). CO_2-gaswechsel der Flechten *Hypogymnia physodes* bei teifen temperaturen im Freiland. *Flora, Jena* 158B, 180–184.

SCOTT, G. D. (1956). Further investigations of some lichens for fixation of nitrogen. *New Phytol.* 55, 111–116.

SCOTT, W. J. (1960). A mechanism causing death during storage of dried micro-organisms. *In* "Recent Researches in Freezing and Drying" (A. S. Parkes and A. V. Smith, eds), pp. 188–201. Blackwell, Oxford.

SIMON, E. W. (1974). Phospholipids and plant membrane permeability. *New Phytol.* 73, 377–419.

SIMONIS, W. and FEIGE, B. (1969). Untersuchungen zur Physiologie der flechte *Cladonia convoluta* (Lam.) P. Cout. I. Allgemeines und Methodik der Untersuchungen. *Flora, Jena* 160A, 552–560.

SMITH, D. C. (1960a). Studies in the physiology of lichens. I. The effects of starvation and of ammonia absorption upon the nitrogen content of *Peltigera polydactyla*. *Ann. Bot.* 24, 52–62.

SMITH, D. C. (1960b). Studies in the physiology of lichens. II. Absorption and utilisation of some simple organic nitrogen compounds by *Peltigera polydactyla*. *Ann. Bot.* 24, 172–185.

SMITH, D. C. (1962). The biology of lichen thalli. *Biol. Rev.* 37, 537–570.

SMITH, D. C. (1974). Transport from symbiotic algae and symbiotic chloroplasts to host cells. *Symp. Soc. exp. Biol.* 28, 485–520.

SMITH, D. C. and MOLESWORTH, S. (1973). Lichen physiology. XIII. Effects of rewetting dry lichens. *New Phytol.* 72, 525–533.

SMITH, D. C., MUSCATINE, L. and LEWIS, D. H. (1969). Carbohydrate movement from autotrophs to heterotrophs in parasitic and mutualistic symbiosis. *Biol. Rev.* 44, 17–90.

SOLBERG, Y. J. (1967). Studies on the chemistry of lichens. IV. The chemical composition of some Norwegian lichen species. *Annls bot. fenn.* 4, 29–34.

SOUTHWOOD, T. R. E. (1973). Insect–plant relationships—an evolutionary perspective. *In* "Insect–Plant Relationships" (H. F. Van Emden, ed.), pp. 3–22. Blackwell, Oxford.

STEWART, W. D. P. and PEARSON, H. W. (1970). Effects of aerobic and anaerobic conditions on growth and metabolism of blue-green algae. *Proc. R. Soc. Lond.* B, 175, 293–311.

SWEET, G. B. and WAREING, P. F. (1966). Role of plant growth in regulating photosynthesis. *Nature, Lond.* 210, 77–79.

TANSLEY, A. G. (1935). The use and abuse of vegetational concepts and terms. *Ecology* 16, 284–307.

TUOMINEN, Y. and JAAKKOLA, T. (1974) ["1973"]. Absorption and accumulation of mineral elements and radioactive nuclides. *In* "The Lichens" (V. Ahmadjian and M. E. Hale, eds), pp. 185–223. Academic Press, New York and London.

WALKEY, M. and MEAKINS, R. H. (1970). An attempt to balance the energy budget of a host–parasite system. *J. Fish. Biol.* 2, 361–372.

WAREING, P. F., KHALIFA, M. M. and TREHARNE, K. J. (1968). Rate-limiting processes in photosynthesis at saturating light intensities. *Nature, Lond.* 220, 453–457.

O

WEBB, S. J. (1963). Possible role for water and inositol in the structure of nucleoproteins. *Nature, Lond*. **198**, 785–787.

WELCH, H. E. (1968). Relationship between assimilation efficiencies and growth efficiencies for aquatic consumers. *Ecology* **49**, 755–759.

WHITTAKER, R. H. (1962). Classification of natural communities. *Bot. Rev*. **28**, 1–239.

WHITTAKER, R. H. (1970). "Communities and Ecosystems." Macmillan, London.

WIEGERT, R. G. and OWEN, D. F. (1971). Trophic structure, available resources and population density in terrestrial vs. aquatic ecosystems. *J. theor. Biol*. **30**, 69–81.

Notes added in proof
1. Work in higher plants has shown that the rate of respiration is a function of both total dry matter and of photosynthesis, implying that respiration for repair and maintenance is a function of total dry matter whilst respiration for growth is a function of photosynthesis (McCree, *in* "Prediction and Measurement of Photosynthetic Productivity" (I. Setlik, ed.), pp. 221–230, 1970; Thornley, *Nature, Lond*. **227**, 304–305, 1970; Biscoe *et al.*, *J. appl. Ecol*. **12**, 269–294, 1975). This supports the concept of compartmentalized metabolism suggested here (p. 392).
2. Control of photosynthesis by accumulated carbohydrate (p. 400) has been demonstrated in two recent papers (Natr *et al.*, *Ann. Bot*. **38**, 589–595, 1974; Throne and Koller, *Pl. Physiol*. **54**, 201–207, 1974) and so it is now probable that similar control will be found in lichen algae.

16 | Sulphur Dioxide Uptake in Lichens

B. W. FERRY

Botany Department, Bedford College, London, England

and

M. S. BADDELEY

The Open University, Newcastle upon Tyne, England

Abstract: It is clear from much recent evidence that lichen thalli placed in dilute solutions of sulphur dioxide are markedly inhibited in their respiratory and photosynthetic metabolism. These observations support the view that sulphur dioxide in the air is an important factor in determining lichen distributions in industrial countries. This paper presents data relating to the process of sulphite (sulphur dioxide) uptake from solution by lichens, particularly *Cladonia impexa* and also a range of corticolous species.

There appears to be no clear correlation between the amounts of sulphite taken up by different species and their field sensitivities to air pollution, based on mapping studies.

Under the experimental conditions used, uptake, from both buffered and non-buffered solutions, seems to be partly or largely passive, perhaps involving cell wall sites for adsorption. This passively adsorbed fraction is less for acid bark species than for basic bark species. Total sulphite uptake, in both live and dead tissue, is always greater from buffered solutions. Evidence for some sulphite movement into cells, involving an energy-requiring active process, is presented, based mainly on experiments with metabolic inhibitors. A considerable proportion of the total sulphite taken up can be eluted out with basic solutions at about pH 9 and further uptake can then be obtained following "recharging" of thalli with dilute acid.

The relevance of laboratory experiments with sulphur dioxide to field situations is discussed.

INTRODUCTION

In view of the great interest generated in recent years in the subject of air pollution and lichens (Ferry *et al.*, 1973), particularly in relation to sulphur dioxide toxicity, it is surprising that virtually nothing is known about uptake of this pollutant by lichens. The few shreds of information currently available have been obtained from measurements of total sulphur contents of lichen thalli

Systematics Association Special Volume No. 8, "Lichenology: Progress and Problems", edited by D. H. Brown, D. L. Hawksworth and R. H. Bailey, 1976, pp. 407–418. Academic Press, London and New York.

growing in, or transplanted into, polluted areas (Gilbert, 1965, 1968, 1969; Griffith, 1966; LeBlanc and Rao, 1972). These workers showed that more sulphur accumulated in lichens in polluted areas than in non-polluted areas, and Gilbert (1968, 1969) demonstrated greater accumulation in live *Usnea filipendula* than in dead material. Lack of knowledge extends also to the metabolism and mode(s) of toxicity of sulphur dioxide in lichen tissues (Farrar, 1973), although some recent work on sulphur dioxide, photosynthetic rates and chlorophyll inactivation in lichens throws light on these matters (Puckett *et al.*, 1973).

Our interest in sulphur dioxide uptake by lichens has also received impetus from sources other than those outlined above. Results of laboratory experiments on the effects of sulphur dioxide on lichen respiration (Baddeley *et al.*, 1972, 1973) and photosynthesis (unpublished) do not always correlate with results from mapping studies (Hawksworth and Rose, 1970) when the relative sensitivities of different species are considered. This seems to be especially true for lichens of basic bark (equivalent to species of hypertrophicated or eutrophicated bark as used in the sense of Barkman (1958) and Hawksworth and Rose (1970)). Further, some preliminary results from transplant experiments with this same group of species also give apparently conflicting results (unpublished). In these experiments, lichen thalli were transplanted into "clean" air, slightly polluted and heavily polluted sites and respiratory and photosynthetic rates were monitored over a period of several months. The relative sensitivities of these basic bark lichens, as assessed by laboratory, mapping and transplant techniques, are summarized in Table I. One notable anomaly relates to the position of *Ramalina farinacea*, which appears to be only moderately sensitive in mapping studies, relatively insensitive in the laboratory and most sensitive under transplant conditions. The same pattern is obtained if photosynthesis is followed.

Various arguments may be put forward as to the suitability of these different methods for assessing sensitivity of lichens to sulphur dioxide. Although the recorded level of sulphur dioxide in the air is the key factor in mapping studies (Hawksworth and Rose, 1970; Hawksworth, 1973) the influence of other factors, particularly local (including micro-) climate, is not properly understood. Such factors may be very important but virtually impossible to assess in mapping studies and transplant experiments. In the laboratory the general problem is one of relating experimental conditions to those, unknown, conditions which pertain in the field. Türk *et al.* (1974) have recently shown that, for a small range of species including some of those we have studied, relative laboratory sensitivities to sulphur dioxide relate somewhat more closely to field sensitivities if thalli are exposed to gaseous sulphur dioxide rather than to sulphite in solution. They monitored respiratory and net photosynthetic rates.

We have previously expressed the view that field conditions for uptake of sulphur dioxide and its subsequent fate might be difficult to simulate in controlled laboratory experiments, and that uptake characteristics might prove to be important in determining the relative sensitivities of different lichens to sulphur dioxide (Baddeley *et al.*, 1973).

Before the problem of uptake of sulphur dioxide can be investigated in the field it seems sensible that it should first be studied in the laboratory. The main aim of this investigation was to establish the pattern of uptake of sulphur

TABLE I. Relative sensitivities of basic bark lichens to sulphur dioxide in the laboratory and to air pollution (presumed sulphur dioxide) in the field

Field sensitivities from mapping studies (Hawksworth and Rose, 1970)	Laboratory sensitivities from respiration rates measured at pH 5·2 (unpublished)	Transplant sensitivities from respiration rates measured at pH 5·2 (unpublished)
Ramalina fastigiata	*Parmelia acetabulum*	*Ramalina farinacea*
⎡*Ramalina farinacea* ⎤ ⎣*Parmelia acetabulum*⎦	*Xanthoria parietina*	⎡ *Parmelia sulcata* ⎤ ⎣*Parmelia acetabulum*⎦
⎡*Xanthoria parietina*⎤ ⎣ *Parmelia sulcata* ⎦	⎡ *Parmelia sulcata* ⎤ ⎣*Ramalina fastigiata*⎦ *Ramalina farinacea*	⎡*Xanthoria parietina*⎤ ⎣*Ramalina fastigiata*⎦

Species are listed in each column in order of decreasing sensitivity. The columns are not comparable horizontally and species bracketed together are of similar sensitivities.

dioxide from solution by lichen thalli. Uptake of gaseous sulphur dioxide, which may prove to be more important in relation to the relative sensitivity problem, will be dealt with in subsequent investigations.

EXPERIMENTAL RESULTS AND DISCUSSION

Most of our work to date has been carried out with *Cladonia impexa* because of plentiful supplies on the shingle at Dungeness, Kent. Data on the sensitivity of this species to air pollution are lacking, but laboratory studies with sulphur dioxide in solution suggest that it is a sensitive species compared with other terricolous *Cladonia* species. It remains to be seen whether it is a typical species with regard to sulphur dioxide uptake properties. Some work has also been done with a range of corticolous species, previously used in laboratory pollution studies (Baddeley *et al.*, 1972).

The main technique employed has involved incubating pre-soaked lichens (15 min in distilled water) in solutions of 10 ppm sulphur dioxide (as ^{35}S-sulphite, sodium salt) in stoppered brown glass bottles. Incubations were normally for 4 h, following which samples were washed for 1 h and then digested in "Domestos" before measuring ^{35}S uptake with a liquid scintillation counter. ^{35}S-sulphite was preferred to gaseous ^{35}S-sulphur dioxide in these studies because of its relative economy and ease of handling. The ionic states of sulphur-containing ions in solution, which are identical irrespective of whether sulphite or sulphur dioxide is used, are dependent on pH and therefore the collective term sulphite is used in this paper for convenience.

A second technique, which permits direct measurement of sulphite in solution using a water-insoluble methylene blue/iodine complex (Kato, 1959), has also proved useful to confirm results obtained by the tracer method.

TABLE II. Sulphite uptake by *Cladonia impexa*

	Non-buffered solution	Buffered solution
Live tissue	62·5	135·5
Dead tissue	32·0	136·5

Figures are μg SO_2 absorbed per g wet lichen over 4 h in the dark at 25°C and pH 4·2.

All of the experiments described below were performed using the tracer technique and were carried out at pH 4·2 and 25°C unless otherwise indicated. Duplicate samples were set up for each treatment in any experiment and each set of results is the mean of at least two experiments. Results are expressed as μg sulphur dioxide absorbed per g wet lichen over the 4 h incubation period. Data on wet weight: dry weight ratios for the various species tested indicate that expression of the results on a dry weight or wet weight basis in no way materially affects the discussions that follow. Relative uptake by the various species remains the same.

The uptake of sulphite by *Cladonia impexa* is shown in Table II for live and dead tissue from buffered solutions (McIlvaines phosphate-citrate buffer—0·2 M disodium hydrogen phosphate plus 0·1 M citric acid) and from non-buffered solutions adjusted to the required pH with dilute hydrochloric acid. Dead tissue was obtained by immersing samples in boiling distilled water for 5 s. There is clearly a very marked effect of the buffer on uptake by both live and dead tissue.

Comparable results have been obtained with the range of corticolous species as shown in Table III. There are differences between species in the amounts of

TABLE III. Sulphite uptake by various corticolous lichens

Species	Non-buffered solution				Buffered solution			
	Live	Dead	Live-dead	Live:dead	Live	Dead	Live-dead	Live:dead
ACID BARK SPECIES[a]								
Evernia prunastri	81·5	22·5	59·0	3·6	110·5	117·5	−7·0	0·9
Hypogymnia physodes	106·5	11·5	95·0	9·3	130·0	78·5	51·5	1·7
Parmelia caperata	75·5	27·0	48·5	2·8	106·0	90·5	15·5	1·2
Parmelia saxatilis	125·0	31·0	94·0	4·0	121·5	101·5	20·0	1·2
Usnea intexta	123·0	26·0	97·0	4·7	134·0	101·5	32·5	1·3
BASIC BARK SPECIES[b]								
Parmelia acetabulum	78·0	38·5	39·5	2·0	131·0	123·0	8·0	1·1
Parmelia sulcata	71·0	34·5	36·5	2·1	83·0	90·0	−7·0	0·9
Ramalina farinacea	108·0	75·5	32·5	1·4	144·0	141·5	2·5	1·0
Ramalina fastigiata	97·5	59·5	38·0	1·6	146·0	142·5	3·5	1·0
Xanthoria parietina	86·5	23·0	63·5	3·8	99·0	56·5	42·5	1·8

Figures are μg SO_2 absorbed per g wet lichen over 4 h in the dark at 25°C and pH 4·2.

[a] On oak (*Quercus* spp.), Bedgebury, Kent and Beckley, Sussex.

[b] On hypertrophicated narrow-leaved willow (*Salix* spp.), Appledore, Kent.

sulphite taken up under both buffered and non-buffered conditions. However, no component of uptake (live, dead or excess live over dead) correlates well with either field sensitivities of different species to air pollution, based on mapping studies and transplant experiments, or with sulphur dioxide sensitivities of different species based on laboratory experiments.

There is a significant amount of uptake by dead tissue of all species indicating some passive uptake, perhaps by adsorption onto cell wall or intracellular sites. This dead uptake, under non-buffered conditions, is generally lower for species of acidic bark resulting in generally higher live:dead uptake ratios for these species (Table III). The exceptional species are *Hypogymnia physodes* and *Xan-*

TABLE IV. Temperature and sulphite uptake in *Cladonia impexa*

	Temperature (°C)				
	0	5	15	25	35
Live tissue	29·0	35·0	46·5	63·5	71·0
Dead tissue	12·5	15·0	24·0	32·0	33·5

Figures are μg SO_2 absorbed per g wet lichen over 4 h in the dark at pH 4·2, from non-buffered solution.

thoria parietina, both having excessively high live:dead uptake ratios for their respective ecological groups. Under buffered conditions dead uptake is markedly increased in all species and live:dead uptake ratios generally approach unity, the exceptions again being *Hypogymnia physodes* and *Xanthoria parietina*.

The effect of increased temperature on sulphite uptake from non-buffered solution is interesting in that both live and dead tissue of *Cladonia impexa* respond similarly (Table IV), although live uptake always exceeds dead uptake over the temperature range tested. The Q_{10} for live and dead uptake is between 1·3 and 1·4 over the range 5–25°C. Preliminary results in buffered solutions show a similar temperature response at least up to 25°C.

Varying the pH of non-buffered and buffered solutions has an effect on sulphite uptake in both live and dead tissue of *Cladonia impexa*, the response being greater under buffered conditions (Table V). The optimum pH for uptake in this species is between pH 4·2 and pH 5·2 and preliminary results for the corticolous species, under buffered conditions, show similar peaks for sulphite uptake. There appears to be no clear distinction between acidic and basic bark species in this respect. Both *Cladonia impexa* and the corticolous species show less uptake at pH 3·2 and markedly decreased uptake at pH 6·2.

The increased uptake demonstrated with phosphate-citrate buffer, compared with non-buffered solutions (Table II), is not confined to this buffer alone but also occurs, at least for *Cladonia impexa*, with phthalate, succinate and acetate buffers. It would appear, therefore, that the effect is essentially a buffering one although the use of low concentrations of phosphoric or acetic acid, instead of hydrochloric acid, to adjust the pH of non-buffered solution does produce a slightly enhanced sulphite uptake. This buffer-enhanced uptake, which appears to be greater in dead tissue than in live tissue of all species tested, must presumably involve passive adsorption. If cell wall or intracellular sites for adsorption

TABLE V. pH and sulphite uptake in *Cladonia impexa*

	pH of solution			
	3·2	4·2	5·2	6·2
From non-buffered solution:				
Live tissue	45·0	47·0	47·5	42·5
Dead tissue	17·5	19·0	16·5	12·5
From buffered solution:				
Live tissue	65·0	87·5	82·5	44·0
Dead tissue	66·0	81·5	85·0	41·0

Figures are μg SO_2 absorbed per g wet lichen over 4 h in the dark at 25°C.

are involved it appears that strong buffering conditions somehow increase the affinity of such sites for sulphite ions. The effect could be either on the ions being adsorbed or on the adsorption sites, or both.

Under non-buffered conditions there is some indication that a proportion of the sulphite taken up by thalli of all species is absorbed into cells and that, for most species, a smaller quantity enters cells under buffered conditions. Although in *Cladonia impexa* dead uptake may actually exceed live uptake under buffered conditions this does not necessarily preclude any entry into cells, a point which is discussed further below. If uptake by live tissue is, in part, an energy-requiring, active process it should be susceptible to the action of metabolic inhibitors.

The results of experiments on the effects of metabolic inhibitors on uptake of sulphite by *Cladonia impexa* from buffered solutions are given in Table VI. Only azide and dinitrophenol reduce uptake significantly more in live tissue than in dead tissue. These two inhibitors have a marked effect on respiration, which is to be expected at the concentrations used. It is assumed that dinitrophenol also acts as an uncoupler of oxidative phosphorylation, although this is not demonstrated. It should be noted that the relative respiration rates given in Table VI are those obtained after incubation for 4 h with the inhibitors alone and do not take into account any effect of sulphite itself, which would also reduce respiration by about 20% unless its uptake were impeded by the inhibitor. The results suggest that about 10–15% of the total sulphite uptake by live tissue from buffered solution is actively taken into cells. This is equivalent to 15–20 μg sulphur dioxide per g wet lichen.

In a preliminary experiment with the same four inhibitors, but under non-buffered conditions, azide and dinitrophenol again reduced uptake by live tissue

TABLE VI. The effects of certain metabolic inhibitors on sulphite uptake in *Cladonia impexa*

Inhibitor (10^{-2}M)	Live tissue	Dead tissue	Relative respiration rate of live tissue in presence of inhibitor
From buffered solution:			
Control	121·5	125·5	100
Sodium azide	95·0	120·5	7·5
Sodium fluoride[a]	105·5	106·0	74·6
Potasium cyanide	93·5	99·0	95·6
Dinitrophenol	70·5	89·0	19·1
From non-buffered solution:			
Control	62·5	29·0	100
Sodium azide	45·0	48·0	12·5
Sodium fluoride[a]	76·5	40·0	65·0
Potassium cyanide	14·5	3·0	93·0
Dinitrophenol	53·0	53·5	70·0

Figures are μg SO_2 absorbed per g wet lichen over 4 h in the dark at 25°C and pH 4·2.
[a] Sodium fluoride concentration; $2·5 \times 10^{-2}$M

compared with dead tissue (Table VI). A clear interpretation of these results is difficult because of the very marked effects of the inhibitors on dead uptake. The simplest interpretation is that, in the absence of inhibitors, the excess uptake by live tissue over dead tissue is all active uptake, about 30–35 μg sulphur dioxide per g wet lichen, and this would be in agreement with the data for azide- and dinitrophenol-treated material in this experiment, if allowance is made for the increased uptake by dead material.

The results of the inhibitor experiments suggest that less sulphite is actively absorbed into cells of *Cladonia impexa* under buffered conditions than under non-buffered conditions. Although the impression is strongly reinforced by the results presented in Table II, there is the problem that dead uptake under buffered conditions equals or even slightly exceeds live uptake of sulphite. Why this should be so we are not certain. Perhaps dead tissue, in which membrane barriers are damaged and proteins denatured, displays more sites for passive uptake of sulphite than live tissue. Such additional, perhaps intracellular, sites would not be so exposed to sulphite in live tissue.

It is not possible with the present data to be sure of the relative importance of cell wall versus intracellular sites for sulphite adsorption in dead and live tissue. Preliminary results with somewhat crude preparations of *Cladonia impexa* cell wall material indicate considerable passive adsorption, under both buffered and non-buffered conditions, by cell walls but suggest also the existence of other (intracellular) adsorption sites, at least in dead tissue. If active uptake is actually reduced under buffered conditions, as the inhibitor data suggest, this might be because of increased efficiency for passive adsorption by cell walls under these conditions, resulting in decreased availability of sulphite at the cell membrane level for active uptake.

Finally we have some preliminary data which show that much of the sulphite (in excess of 60%) taken up by live and dead tissue of *Cladonia impexa*, under buffered and non-buffered conditions, can be eluted out with basic solutions at about pH 9. Sodium hydroxide, ammonium hydroxide and sodium bicarbonate, at concentrations ranging from 0·001 to 1 N, all have much the same effect, as also does 0·2 M disodium hydrogen phosphate. We also find that dead tissue can be "recharged" with 0·01 N hydrochloric acid, following elution, such that it will take up a further normal quantity of sulphite. It is interesting to speculate that these processes of elution and "recharging" of adsorption sites might occur under natural conditions.

FURTHER DISCUSSION AND CONCLUSIONS

These studies on sulphite uptake by lichens are still at a fairly early stage but nevertheless enough data are available to permit some tentative suggestions.

1. The lack of correlation between amounts of sulphite taken up by different species (live uptake, dead uptake and excess live over dead uptake, under either buffered or non-buffered conditions) and laboratory or field sensitivities to sulphur dioxide or air pollution might be due to one of two related possibilities: (*a*) The conditions of pH etc. chosen to follow uptake may not be the ones relevant to field situations and in this respect species may differ. (*b*) Sulphite uptake is probably only one of a multiplicity of factors contributing to the overall sensitivity of a lichen, and again species will probably vary with regard to the importance of the uptake factor. We have previously listed a number of factors, associated either directly with the lichen itself (including its capacity to

take up sulphite) or with its immediate habitat, which probably contribute to the overall sensitivity of a species to air pollution (Baddeley and Ferry, 1973). Türk *et al.* (1974) also concluded that sensitivity of lichens to air pollution is controlled by a complex of factors and they referred to the analysis of Levitt (1972) for any organism subjected to environmental stress, which they adapted to suit the lichen/air pollution situation. Amongst the many factors to which they referred was uptake of sulphur dioxide.

2. Under laboratory conditions uptake of sulphite, from both buffered and non-buffered solutions, appears to be partly active and partly passive. The difference in uptake between live and dead tissue, under non-buffered conditions, and the effects of metabolic inhibitors on uptake, under both buffered and non-buffered conditions, are fairly good evidence for this, at least for *Cladonia impexa*. The temperature data for this species also suggest that a significant component of uptake is by a similar, presumably passive, mechanism in live and dead tissue, because the Q_{10} values are similar for the two types of tissue.

It remains to be shown precisely how much sulphite is actively taken into cells under various conditions but in *Cladonia impexa* this amount rarely, if ever, seems to exceed 50%. Under most conditions passive adsorption appears to be the dominant process but the nature of the adsorption sites remains obscure. If, as seems likely, cell walls are important in this respect, then what are the charged sites? Because of the lack of data on the chemical constituents of lichen cell walls it is difficult to comment realistically. Chitin, glucans and protein have been identified as cell wall materials in non-lichenized Ascomycotina and, of these, the last would seem to be a possible candidate for anion adsorption. At low pH values proteins would tend to be positively charged by the generation of NH_3^+ sites, while chitin, which consists of residues of N-acetylglucosamine, and glucans would not become charged. Further, if intracellular adsorption sites prove to be important in dead tissue, these could also be provided by proteins.

The capacity to adsorb sulphite seems to relate to the natural habitat of the lichen, basic bark species tending to adsorb a greater proportion of their total uptake than acid bark species from non-buffered solutions. Externally applied buffer improves this capacity for passive adsorption in all species tested, the effect being more startling in the acid bark species. *Cladonia impexa*, a terricolous species most commonly associated with acid soils, resembles the acid bark species in this respect.

A considerable proportion of the sulphite taken up by *Cladonia impexa* at pH 4·2 can be eluted out at pH 9. Further passive uptake can be obtained if eluted thalli are "recharged" with dilute hydrochloric acid (pH 2–3).

3. The differences in sulphite uptake under buffered compared with non-buffered conditions, observed for all species tested, might be significant with regard to laboratory studies on sulphur dioxide toxicity. Probably all such studies to date have involved the use of buffered solutions and it may be more relevant to natural conditions to use non-buffered solutions.

ACKNOWLEDGEMENTS

We particularly wish to express our thanks to Mrs Elaine Wilson, research assistant to one of us, who has carefully and meticulously carried out all of the experiments reported in this paper. The work has been conducted during the tenure of a grant from the Science Research Council.

REFERENCES

BADDELEY, M. S. and FERRY, B. W. (1973). Summary: Scope and Direction of Future Work. *In* "Air Pollution and Lichens" (B. W. Ferry, M. S. Baddeley and D. L. Hawksworth, eds), pp. 368–377. University of London, Athlone Press, London.

BADDELEY, M. S., FERRY, B. W. and FINEGAN, E. J. (1972). The effects of sulphur dioxide on lichen respiration. *Lichenologist*, **5** 283–291.

BADDELEY, M. S., FERRY, B. W. and FINEGAN, E. J. (1973). Sulphur dioxide and respiration in lichens. *In* "Air Pollution and Lichens" (B. W. Ferry, M. S. Baddeley and D. L. Hawksworth, eds), pp. 299–313. University of London, Athlone Press, London.

BARKMAN, J. J. (1958). "Phytosociology and Ecology of Cryptogamic Epiphytes." Van Gorcum, Assen.

FERRY, B. W., BADDELEY, M. S. and HAWKSWORTH, D. L. (eds) (1973). "Air Pollution and Lichens." University of London, Athlone Press, London.

FARRAR, J. F. (1973). Lichen physiology: progress and pitfalls. *In* "Air Pollution and Lichens" (B. W. Ferry, M. S. Baddeley and D. L. Hawksworth, eds), pp. 238–282. University of London, Athlone Press, London.

GILBERT, O. L. (1965). Lichens as indicators of air pollution in the Tyne Valley. *In* "Ecology and the Industrial Society" (G. T. Goodman, R. W. Edwards and J. M. Lambert, eds), pp. 35–47. Blackwell, Oxford.

GILBERT, O. L. (1968). Bryophytes as indicators of air pollution in the Tyne Valley. *New Phytol.* **67**, 15–30.

GILBERT, O. L. (1969). The effect of SO_2 on lichens and bryophytes around Newcastle upon Tyne. *In* "Air Pollution, Proceedings of the first European Congress on the influence of air pollution on plants and animals, Wageningen 1968", pp. 223–235. Centre for Agricultural Publishing and Documentation, Wageningen.

GRIFFITH, J. L. (1966). "Some Aspects of the Effect of Atmospheric Pollution on the Lichen Flora to the West of Consett, Co. Durham." M.Sc. thesis, University of Durham.

HAWKSWORTH, D. L. (1973). Mapping studies. *In* "Air Pollution and Lichens" (B. W. Ferry, M. S. Baddeley and D. L. Hawksworth, eds), pp. 38–76. University of London, Athlone Press, London.

HAWKSWORTH, D. L. and ROSE, F. (1970). Qualitative scale for estimating sulphur dioxide air pollution in England and Wales using epiphytic lichens. *Nature, Lond.* **227**, 145–148.

KATO, T. (1959). Colorimetric determination of minute amounts of sulphite ion. *Nippon Kagaku Zasshi* **80**, 52–54.

LeBLANC, F. and RAO, D. N. (1972). Effects of sulphur dioxide on lichen and moss transplants. *Ecology*, **54**, 612–617.

LEVITT, J. (1972). "Responses of Plants to Environmental Stresses." Academic Press, New York and London.

PUCKETT, K. J., NIEBOER, E., FLORA, W. and RICHARDSON, D. H. S. (1973). Sulphur dioxide: Its effect on photosynthetic ^{14}C fixation in lichens and suggested mechanisms of phytotoxicity. *New Phytol.* **72**, 141–154.

TÜRK, R., WIRTH, V. and LANGE, O. L. (1974). CO_2-Gaswechsel-Untersuchungen zur SO_2-Resistenz von Flechten. *Oecologia, Berl.* **15**, 33–64.

17 | Mineral Uptake by Lichens

D. H. BROWN

Department of Botany, The University, Bristol, England

Abstract: The substantial absorption of lead, copper, zinc, nickel and cobalt by *Cladonia rangiformis* under laboratory conditions is used to illustrate the cation selectivity and competition which can be achieved by a passive uptake mechanism. The value of such data in suggesting the nature of cation-binding sites is considered and a discussion provided on the categories of compounds which could potentially be involved. A selective solubilization procedure is used to demonstrate the probably cellular location of potassium (intracellular), lead (extracellular) and zinc (intermediate) using naturally enriched material of *Parmelia glabratula* subsp. *fuliginosa*. Failure of transplanted material of this species to accumulate lead or zinc in a highly polluted region is related to lichen morphology, the nature and source of the heavy metals and meteorological conditions. The rôle of specific lichen substances in cation uptake by lichen thalli is considered and experimental evidence provided to suggest that it is of no significance.

INTRODUCTION

In recent years there has been an increasing interest in cation uptake by lichens. This has been due, at least in part, to the high values of radionuclides and heavy metals recorded from lichens, suggesting their possible value as pollution monitors. As two extensive reviews of this subject have appeared recently (James, 1973; Tuominen and Jaakkola, 1974) only brief comment will be made here.

One feature which makes lichens highly suitable organisms for pollution monitoring is the apparent absence of any impermeable cuticular material. It is frequently assumed that this feature, coupled with the lack of a well developed root system analogous to that of higher plants, means that the atmosphere is the major, if not the exclusive, source of cations for lichens. This is supported by data showing poor correlation between lichen and substratum mineral content

Systematics Association Special Volume No. 8, "Lichenology: Progress and Problems", edited by D. H. Brown, D. L. Hawksworth and R. H. Bailey, 1976, pp. 419–439. Academic Press, London and New York.

for saxicolous (Jenkins and Davies, 1966) and epiphytic (Micóvić and Stefanović, 1961; Kuziel, 1973) species and the enrichment of the radionuclide caesium-137 (derived from atmospheric fall-out from nuclear detonation) in the upper portion of lichen thalli (e.g. Nevstrueva et al., 1967; Ritchie et al., 1971). However, the general agreement between lichen and substratum cation content on silicic compared to ultrabasic rocks (Lounamaa, 1956), old mine workings (Lange and Ziegler, 1963; Lambinon et al., 1964; Brown, 1973) and non-mineralized rocks (LeRoy and Koksoy, 1962), coupled with the general observation that certain lichen species are confined to regions of particular mineral content (Brodo, 1974), suggests that the substratum may have a substantial effect in determining the mineral content of lichens. In order to rationalize such opposing views it must be realized that contamination by mineral-rich particles may occur (Nieboer et al., 1972), although specific reports of ion accumulation by lichens above levels found in the substratum rule out contamination in these cases (Lambinon et al., 1964), and also that lichens are capable of altering the proportions of ions incorporated into the thallus (Tuominen, 1967; Puckett et al., 1973). Although strontium-90 and caesium-137 are both artificial radionuclides with similar mechanisms of formation and dispersal to lichens, strontium-90, unlike caesium-137, is generally found uniformly distributed throughout the living thallus of Cladonia species (Nevstrueva et al., 1967). This difference emphasizes that generalizations about cation metabolism must take into account the nature of the cation involved and the physiological state of the lichen at the time. This is illustrated by the reports of Tuominen (1968, 1971) which showed caesium-137 to be more mobile than strontium-90 in dead Cladonia stellaris thalli in closed tubes. However, Nevstrueva et al. (1967) showed with living Cladonia rangiferina that, when the tip or base of the thallus was dipped in solutions, only limited translocation of caesium-137 occurred from the point of application whereas strontium-90 became evenly distributed within the thallus. Hanson et al., (1967) reported similar results to those of Nevstrueva et al. (1967) with field experiments using strontium-85 and caesium-134. The likely explanation for these differences of behaviour is that, in the living lichen, the monovalent caesium is probably incorporated, by an active process, into the cell cytoplasm by the carrier normally used by its chemical homologue, potassium (Tuominen and Jaakkola, 1974). The divalent cation strontium, mimicking calcium, is not so readily incorporated into the cell but is reversibly bound to ionized groups in the cell wall. In dead lichens the active incorporation of caesium is not possible and its low affinity, as a monovalent cation, for the cell wall ionized groups results in weaker binding and hence its greater mobility in the thallus.

The passive and active methods of cation uptake from the environment are reflected in the final cellular location of ions. Passive uptake by ion exchange or chelation results in ions remaining extracellular while active uptake results in ions becoming intracellular. Such compartmentalization has been reported for lead, which is extracellular, and potassium, which is mainly intracellular, in *Cladonia rangiformis* (Brown and Slingsby, 1972). It is currently assumed that extracellular elements, being outside the plasmalemma, are of no immediate metabolic importance. The converse does not necessarily apply as intracellular elements may be either bound to other components or otherwise inactivated. Hence a high thallus content of a toxic mineral does not necessarily mean that the plant possesses some cellular tolerance mechanism (Lange and Ziegler, 1963). Identification of the cellular location of potentially toxic compounds is therefore one important stage in establishing whether such elements affect cellular metabolism.

SELECTIVITY OF ION UPTAKE

Short-term laboratory experiments, such as those of Tuominen (1967) with strontium, Handley and Overstreet (1968) with caesium and Puckett *et al.* (1973) with a range of heavy metals, have tended to stress the dominance of the passive, physical, cation uptake mechanism rather than the substantially slower, active, biological process. Selectivity of cation uptake is possible with both mechanisms. In the case of the active process, the relative affinities of particular ions for specific carriers in the cell membrane system determines the proportion entering the cell. In the passive process, the chemical nature of the ions involved and that of the acceptor sites similarly determines an order of preference for ion binding. The following experiments illustrate the latter process using *C. rangiformis.*

1. *Experimental Data*

C. rangiformis was collected from populations on slag heaps from either a disused lead smelting area (Somerset: Mendip Hills, Charterhouse) or coal mining area (Somerset: Mendip Hills, Dunkerton) (Brown and Slingsby, 1972; Brown, 1973). The terminal 1 cm portions of deionized, washed and air-dried thalli were used in the proportions 0·2 g lichen/10 ml solution containing singly either 10 or 1000 μmol/10 ml of lead, copper, zinc, nickel or cobalt. In single ion solutions chlorides were used, except for lead where, because of the limited solubility of the chloride, nitrate was used. Mixtures of these cations were also made at 10 μmol/10 ml either with chlorides (except for lead nitrate) or, because of precipitation, all as nitrates. These proportions of lichens to heavy metal ion were the same as those used by Puckett *et al.* (1973). Heavy metals were supplied by mechanically shaking the lichen and solution for 1·5 h. This was followed by thorough filtration and washing (including shaking in 20 ml deionized water), air drying of the thallus for at least 15 h at room temperature, and complete discharge of cations from the lichen by shaking for 1·5 h with 10 ml N HNO₃, before filtration. Nitric acid rather than hydrochloric acid was used to avoid possible precipitation of lead. The concentration of individual cations was estimated by atomic absorption spectrophotometry. All treatments were triplicated.

The ability of lichens to influence the uptake of cations is shown in Table I. Although single cations were supplied here at the same initial concentration,

TABLE I. Recovery of ions supplied singly or in mixtures to a lead slag heap population of
Cladonia rangiformis

Supply	Ion recovered (μmol/g lichen)				
	Lead	Copper	Zinc	Nickel	Cobalt
Singly					
1·0 mM	37·6	31·4	25·9	25·2	23·5
0·1 M	84·4	73·7	60·0	57·8	54·1
Mixed					
1·0 mM chlorides	6·9	15·3	6·3	7·2	6·1
1·0 mM nitrates	26·3	11·3	4·6	5·1	4·4

different proportions were taken up. The relative affinity of the lichen for different cations appears to be independent of concentration. With a 100-fold increase in the initial cation concentration, the sequence of decreasing binding efficiency was still lead > copper ≫ zinc > nickel > cobalt. Puckett *et al.* (1973) found, however, that cation selectivity appeared to be concentration-dependent, at least in *Umbilicaria muhlenbergii*. However, as noted before (Tuominen, 1967; Brown and Slingsby, 1972; Puckett *et al.*, 1973) there is not a linear relationship between the initial concentration of the ion supplied and the quantity actually taken up. Here uptake is slightly more than doubled for a 100-fold increase in the cation supplied. Hence, if a similar process is involved in the natural acquisition of heavy metals in the field then high values recorded from the thallus may represent a substantial underestimate of the environmental levels involved.

It is not certain whether, on this occasion, the uptake observed from 0·1 M solutions represents the saturation value for each cation. Puckett *et al.* (1973) reported saturation with copper for *U. muhlenbergii* at about this concentration giving a value of 50·7 μmol/g recovered from the thallus. In a survey of *Umbilicaria* and *Cladonia* species they reported values of from 41·3 to 68·8 μmol/g lichen but unfortunately did not include *C. rangiformis*. Hence the values recorded here, for copper at least, are at the upper end of the capacity range of Puckett *et al.* and suggest that the cation exchange sites were probably saturated.

When all five heavy-metal cations (at 1·0 mM) were supplied together in a mixed solution of nitrates there was a decrease in the uptake of individual ions compared to the amount accumulated from single ion solutions at the same concentration (Table I). The total uptake (41·8 μmol/g) is above that of the most efficiently bound single cation, lead, at 1·0 mM but below that of all cations at 0·1 M. Hence the cations in the mixture are not being taken up in isolation but are competing with the other ions present for the available binding sites. In the mixture of nitrates, lead and copper still dominate, with nickel, zinc and cobalt bound with substantially lower efficiency. This agrees with the report of Puckett *et al.* (1973) who, however, used chlorides (see below). In other experiments using different material (e.g. see Table II) the relative binding efficiency of nickel, zinc and cobalt varied but, as all of these values are very close together, this probably only reflects slight differences in experimental conditions.

When copper, nickel, zinc and cobalt were present in a mixture as chlorides, precipitation of lead occurred. This is reflected in the very low value of lead recovered from the lichen (Table I) and, because of the lower lead concentration in solution creating a decreased competition with the other ions for available binding sites, higher uptake was observed with the remaining ions compared to that observed with the nitrate mixture. The individual uptake values are, however, still below those reported from the single ion experiment.

The data considered so far used a population of *C. rangiformis* from a lead-rich area. The actual quantities of lead and zinc in the sample used in Table I were 1·21 μmol/g (252 p.p.m.) and 1·46 μmol/g (96 p.p.m.) respectively. These values are lower than previously reported from this site (zinc 240 p.p.m., lead 1220 p.p.m., Brown (1973); lead 4375 p.p.m., Brown and Slingsby (1972)) and reflect the variability in the heavy metal content of lichens and soils from different sites within the same general area (Brown, 1973). An area of high environmental lead and zinc might, in cation uptake studies, either show competition between the endogenous heavy metal and the cations supplied in solution, thus reducing their uptake, or an ability to exclude or accumulate those toxic ions to which it has been exposed in the natural environment (see e.g. Antonovics *et al.*, 1971; Peterson, 1971, for such behaviour in higher plants). A comparison was made, therefore, with a population of *C. rangiformis* from a coal tip area containing 44 p.p.m. zinc and 0·5 p.p.m. lead. Total uptake of cations from a mixture of nitrates showed very little difference between the two populations although lead uptake was slightly higher and zinc uptake slightly lower in the lead slag material (Table II). At the present time, because

TABLE II. Recovery of ions supplied in a mixed ion solution to *Cladonia rangiformis*

	Ion recovered (μmol/g lichen)				
Treatment	Lead	Copper	Zinc	Nickel	Cobalt
Untreated					
(1) Lead slag heap population	26·3	11·3	4·6	5·1	4·4
(2) Coal slag heap population	24·4	10·6	5·9	5·4	5·2
Pretreated[a]					
(1) Boiled	24·1	12·4	4·0	4·9	5·4
(2) Acetone extracted	27·6	15·8	6·8	6·5	6·7
(3) Acid discharged	9·9	1·5	15·4	17·9	14·8

[a] Coal slag heap population. See text for experimental details.

no consistent pattern was observed in other experiments, it is probably reasonable to conclude that these differences are insignificant although it is possible that populations from zinc-rich areas may be able to exclude zinc while populations from lead-rich areas may bind lead more efficiently. These ions may well be bound to different sites within the lichen (see below). The similarity between the two populations in terms of total cation incorporation indicates that the heavy metals already present have no competitive effect on the uptake process.

Brown and Slingsby (1972) have already shown that similar lead uptake occurs with living and dead *C. rangiformis*. Although only 1·5 h incubation periods were employed here, some of the uptake could have been metabolically linked. The possible contribution of metabolism was eliminated by three different treatments of the lichen before cation feeding: (*a*) boiling for 0·5 h in 10 ml deionized water, filtration and washing for 1 h in 20 ml deionized water, (*b*) shaking for 1 h in 10 ml acetone, three times, with similar washing and (*c*) shaking with 20 ml N HNO$_3$ and washing. Death was assessed by photosynthetic incorporation of ^{14}C-bicarbonate and/or loss of potassium from the lichen; the latter was only partial in treatment (*b*).

Boiling the lichen hardly altered the total cation uptake and cation selectivity (Table II), indicating that metabolically linked uptake in previous experiments was insignificant. Prior acetone extraction slightly enhanced total uptake but also had little effect on cation selectivity, perhaps indicating the removal of some organic-soluble barrier without exposing any new kind of binding sites. The nitric acid treatment was designed to elute all endogenous cations bound to the lichen, thereby exposing the acceptor sites without the competition afforded by such cations, as well as rupturing the cell membranes. Brown and Slingsby (1972) used N HCl for this purpose with little difference in lead uptake capacity. Using HNO_3 on this occasion resulted in a substantial change in the pattern of cation uptake. The binding of nickel, zinc and cobalt was substantially raised above the level observed in other competition studies and uptake of lead and copper was reduced dramatically from their previous levels. The failure of all five cations to behave in the same way suggests that this is probably not a simple pH phenome-

TABLE III. Elution of ions from a lead slag heap population of *Cladonia rangiformis*[a]

Recovered ion	Solution concentration	Supplied ion				
		Lead	Copper	Zinc	Nickel	Cobalt
Zinc	1·0 mM	0·82	1·29	—	0·62	0·53
	0·1 M	1·20	1·41	—	1·03	1·19
Lead	0·1 M	—	1·16	0·53	1·07	0·94

[a] Figures in μmoles/g lichen.

non but more likely reflects the loss of suitable binding sites for lead and copper following nitric acid treatment. Whether new sites are exposed which preferentially bind the remaining ions or the reduction in lead and copper uptake reduces competition at their normal sites is not clear. This phenomenon does not appear to be an experimental artefact but has been observed in other experiments using similarly treated material.

An alternative approach to ion selectivity studies is to determine the quantities of particular ions already present in the lichen which are released by competing ions in solution (Rühling and Tyler, 1970). Table III reports the quantities of endogenous zinc and lead released from the material of *C. rangiformis* collected from lead slag, to which Table I refers, by the addition of single cation solutions. The values for lead recovered from 1·0 mM solutions were generally below the detection limit for the apparatus, indicating that zinc was more readily replaceable than lead. The sequences of both zinc and lead replacement differ from those reported for cation uptake (Table I). Thus copper is the best replacing agent for zinc whereas zinc is the least effective lead-replacing agent. The implication from this data is that lead and zinc are partially bound to different sites but that copper is capable of attacking both sites.

2. *Nature of the Binding Sites*

Uptake of cations by the reversible passive mechanism in lichens has been interpreted as an ion exchange phenomenon possibly modified by metal complex formation or chelation (Tuominen, 1967; Brown and Slingsby, 1972; Puckett *et al.*, 1973). Tuominen (1967) determined the pK values of the ionogenic groups in a range of lichens and found marked differences between species. Pectic carboxyl groups (pK 2·7–3·3) comprised only part of the total uptake capacity under the buffered conditions used (pH 6·2) ranging from *Cladonia stellaris* 16% and *Umbilicaria deusta* 37% to *Parmelia saxatilis* 82% and *Usnea filipendula* + *U. subfloridana* 74%. Strontium uptake was therefore considered to be only partially based on uronic acid carboxyls. Tuominen also suggested that organophosphorus compounds or side-chain carboxyl groups of proteins might be responsible for the substantial pK 5·0–7·0 values found in certain species. However, a correlation between uronic acid content of the thallus and strontium uptake was reported and the greater uptake capacity and uronic acid content of epiphytic compared to terricolous species noted. Since this work, other investigators have worked exclusively with terricolous species.

Puckett *et al.* (1973) considered that cation uptake consisted of an ion exchange process modified by metal complex formation involving oxygen and nitrogen donors on the basis of experimentally determined binding sequences for a range of heavy metals. The data presented here are comparable to those of Puckett *et al.* and permit a comparison to be made. Unfortunately there are certain discrepancies in the binding sequences observed in the two studies which may be due, partly, to experimental differences (degree of homogenization of the thallus, nature of the eluting acid) and also to use of different species. On the latter point it is noteworthy that although Tuominen (1967) observed differences between the pK values recorded from different species Puckett *et al.* (1973) obtained generally similar uptake sequences with *Cladonia mitis*, *Stereocaulon paschale* and *Umbilicaria muhlenbergii* (with some variation in the position of nickel and zinc) which are also in general agreement with the sequences obtained here. The variations in the actual order of binding nickel, zinc and cobalt observed under various experimental conditions but with a single species probably reflect the experimental limitations of the method and not differences in the actual binding groups involved. It is therefore considered premature to use the cation selectivity method for suggesting possible binding groups within the lichen. However, this approach does have merit in that it emphasizes the varied chemistry of suitable binding sites which, although potentially available for each different cation, have quite different ion affinities and preferences.

Brown and Slingsby (1972) reported that the site of passive cation uptake is extracellular, probably consisting of one of the cell wall components. Much emphasis has been placed on the suggestion that polyuronic acid (pectic) substances may be the site of ion binding in lichens (Tuominen, 1967; Brown and Slingsby, 1972; Puckett *et al.*, 1973). While this may be true for mosses, which show many similarities with lichens in their cation uptake ability (Clymo, 1963; Rühling and Tyler, 1970; Spearing, 1972), the reports of uronic acid content of lichens (Knight *et al.*, 1961; Clymo, 1963; Tuominen, 1967; Quillet and de Lestang-Laisné, 1969) need to be treated with caution due to the accepted inadequacies of the techniques employed (Tuominen and Jaakkola, 1974). The paucity of reliable data on the cell wall composition of the fungal and algal components of lichens (Culberson, 1969, 1970) means that any discussion must make use of reports on the cell wall composition of free-living species. If uronic acids are present in lichens they are most likely to be associated with the algal component (Tuominen and Jaakkola, 1974; Percival and Turvey, 1973) as they are not satisfactorily recorded from ascomycetes (Bartnicki-Garcia, 1968, 1973). Because lichen algae usually comprise less than 10% of the total thallus weight, the contribution of uronic acids to ion uptake can be expected to be slight.

Important fungal cell wall constituents are β-glucans, fats, chitin and proteins. In a representative ascomycete (*Penicillium griseofulvum*) the proportions of these cell wall constituents were 45·0%, 34·5%, 12·0% and 5·5% respectively (Smithies, 1952). Of these, the β-glucans and lipids are unlikely to be of importance as cation-binding agents due to the lack of suitable acceptor groups. Chitin has been reported from the fungal component of a number of lichens on the basis of specific colour reactions, chemical separation and enzymatic digestion (Boissière, 1969; Quillet and de Lestang-Laisné, 1969). Chemically it is a polymer of *N*-acetyl-D-glucosamine residues and it is possible that the *N*-acetyl side groups could be involved in nitrogen–oxygen–cation complex formation as required by Puckett *et al.* (1973). Cations could also be bound to the side groups of protein molecules and most reports of fungal wall proteins indicate an excess of acidic (carboxyl) groups due to aspartic and glutamic acid residues Mitchell and Taylor, 1969). However, the occurrence of amide residues has usually been overlooked because strong acid hydrolysis of the protein has been used prior to identification of the constituent amino acids. Manocha and Colvin (1967) showed in trypsin digests of *Neurospora crassa* cell wall proteins that much of the aspartic and glutamic acid was neutralized with ammonia.

Again it is possible that such proteins would be capable of nitrogen–oxygen–cation complex formation and it is noteworthy that amides would be converted to acidic residues by nitric acid treatment. This could account for the decreased binding of lead and copper noted in Table II following such treatment, if these cations were poorly bound to the resulting carboxyl residues. It is also possible, but rather unlikely (Taylor and Cameron, 1973), that nitric acid treatment might also deacetylate the chitin molecule but this would yield amine side groups which, although good binders of nickel, would also be expected to bind lead and copper successfully. The greater binding efficiency of lead compared to zinc (Tables I and II) and the indication that these cations may be bound to different sites (Table III and below) may be related to the preference of the former for sulphur-containing compounds. Sulphate esters of polysaccharides have been reported (Percival and Turvey, 1973) in Chlorophycean algae, mainly in marine species. If such compounds occur in lichen algae, loss of sulphur might be achieved by acid treatment possibly accounting for the changed cation-binding efficiency reported in Table III. Sulphydryl groups in proteins could also be involved but most reports suggest (e.g. Manocha and Colvin, 1967) mycelial fungi contain few, or no, suitable sulphur amino acids.

From this discussion it is obvious that until more detailed knowledge is available on the chemical composition of fungal and algal cell walls in lichens suggestions as to the nature of cation binding sites derived from studies of pK values or cation binding sequences must remain highly speculative. The approach of specifically modifying the cell wall components prior to such studies appears to be a fruitful field for further research.

ION UPTAKE AND LOCALIZATION IN NATURE

Recently lichens have been used to monitor the heavy metal fall-out from industrial sources by determining the cation content of lichens growing in their immediate vicinity (Burkitt *et al.*, 1972; Nieboer *et al.*, 1972; Seaward, 1973). Goodman and Roberts (1971) and Little and Martin (1974) used mosses imported from uncontaminated areas to measure the rate of heavy metal uptake and to assess the distribution of fall-out in such areas. The latter authors showed that mosses rapidly acquired a substantial burden of heavy metals which might represent the rapid, passive uptake of such cations by the cell wall as discussed above. However, Little (1973) has shown that much of the heavy metal uptake observed with tree leaves (Little and Martin, 1972) is due to heavy metal rich particles of fall-out adhering to the leaf surface rather than incorporation into the cell wall of either cytoplasm or of the leaf cells.

It was therefore considered of interest to study the rate of uptake of heavy metals by lichens imported from non-industrial areas into a heavily polluted region. In addition, an attempt was made to determine the proportion of heavy metal cations in the various cellular fractions of such imported material as well as material which had grown in the polluted region. The detailed work of Burkitt *et al.* (1972) and Little and Martin (1972, 1974) enabled predictions to be made as to the heavy metal fall-out in regions to the north-east of the Avon-mouth (Bristol) industrial complex. Owing to the high levels of sulphur dioxide, carbon black and, more recently, fluoride pollution in the district (zone 3 to 4 on the scale proposed by Hawksworth and Rose, 1970) it was not possible to find suitable and abundant epiphytic lichen species. For this reason a saxicolous species, *Parmelia glabratula* subsp. *fuliginosa*, was chosen for study.

1. Experimental Data

P. glabratula subsp. *fuliginosa* was obtained from siliceous rocks at Easter Compton churchyard and near Priddy, Mendip Hills. These are respectively 5 km north-east and 28 km south of the main smelter complex at Avonmouth. The polluted nature of the former site was indicated by G. Nickless (pers. comm.) using the *Sphagnum* moss bag technique of Little and Martin (1974) who showed uptake of 867 p.p.m. lead and 906 p.p.m. zinc in a 6 week period. Lichen-bearing stones were transferred from Priddy to Easter Compton for the 2 month period between January and March 1974. Analyses were performed on this material as well as fresh material from both Priddy and Easter Compton collected at the end of the exposure period. The lichen was analysed by the following procedure, based on the method of Bates and Brown (1974). Samples of lichens, 0·1 g, were washed three times in 10 ml deionized water for 0·5 h; treated twice, for 1 h, with 20 ml 1000 p.p.m. strontium chloride solution; washed for 20 min in 20 ml deionized water; boiled for 20 min in 5 ml deionized water and washed in 20 ml deionized water; treated twice with 20 ml N HNO_3 for 1 h and washed as above and finally the lichen totally digested by heating with 2 ml concentrated HNO_3. Determinations were made in triplicate.

(a) *Ion location.* Bates and Brown (1974) reported that their procedure, which lacked the present N HNO_3 elution stage, permitted potassium, sodium, calcium and magnesium in mosses to be separated into (a) extracellular and surface contamination, (b) intracellular and (c) exchangeable (i.e. cell wall) fractions. For potassium it is possible to interpret the data presented in Table IV in a similar fashion. Roughly 10% of the total recoverable potassium is accounted for as extracellular contamination in the fresh material from both sites. Strontium eluted only 1–3% of the total, indicating that very little potassium was in a form exchangeable with this cation. The bulk of the potassium (59–76%) was removed by boiling and this has been interpreted as representing material free in the

TABLE IV. Recovery of ions from *Parmelia glabratula* subsp. *fuliginosa* following various pretreatments[a]

| | Pretreatment | | | | | |
| | Easter Compton material | | | | Priddy material | |
Fraction	Heat	Acetone	1% Detergent	H_2O	Fresh	2 months[b]
Potassium						
Water soluble	2791	1823	506	411	319	333
Strontium elutable	174	473	63	130	45	7
Boiling water soluble	12	334	2271	2344	2144	2388
N HNO_3 soluble	56	113	200	256	184	210
Total digestible	807	841	825	831	401	202
Total	3840	3584	3865	3972	3093	3140
Lead						
Water soluble	0	0	0	0	0	0
Strontium elutable	23	12	100	70	11	11
Boiling water soluble	6	6	6	8	3	2
N HNO_3 soluble	539	531	409	479	158	136
Total digestible	29	31	40	25	4	4
Total	597	580	645	582	176	143
Zinc						
Water soluble	35	20	6	2	2	2
Strontium elutable	170	146	54	64	30	33
Boiling water soluble	12	3	47	51	16	22
N HNO_3 soluble	61	97	144	168	57	60
Total digestible	750	789	780	716	50	40
Total	1028	1055	1031	1001	155	157

[a] See text for experimental details. Figures in parts per million.
[b] Transplanted to Easter Compton for 2 months.

cytoplasm of the cell. Strontium was, in part, used to exchange ions bound to the cell wall but was also considered to act as a block to further uptake by the cell wall of ions released from within the cell by the boiling procedure. If this is correct, it would be necessary to interpret the 5–7% recovery of potassium by N HNO$_3$ treatment as representing material bound in a form not exchangeable with strontium. Strontium is known to be complexed with oxygen rather than nitrogen donors (Mellor, 1964) but as potassium shows a similar preference for oxygen donors it is most likely that the nitric acid-elutable material represents intracellular potassium released from within the cell and bound to the cell wall either by replacing the previously bound strontium or by being bound to sites not occupied by strontium ions. The latter possibility exists for the following reason: relatively low levels of strontium were used because Brown and Slingsby (1972) reported that high levels of nickel resulted in potassium loss from within the cell.

That the nitric acid-elutable fraction contains potassium not previously bound to the cell wall is shown by comparison of the values recovered in this fraction from fresh material with those from material which had previously been killed either by adding 2 ml water and drying in an oven at 100°C (equivalent to boiling treatment of Table II) or shaking twice with 10 ml acetone for 0·5 h. Following these treatments, the first water wash contained the intracellular potassium. Reference to Table IV shows that the oven drying of wet material resulted in nearly complete loss of intracellular potassium, only 12 p.p.m. being finally recovered at the boiling water stage. The acetone treatment was less efficient but still resulted in a substantial loss of potassium from within the cells. Following both pretreatments any potassium lost to the cell wall would have been replaced by the strontium treatment. The nitric acid eluate would there-fore be expected to contain less potassium because the boiling water stage did not result in the release of appreciable additional potassium from within the cells.

The potassium in the total digest (6–21% of the total recovered) probably represents material in substratum particles which were not removed by the previous treatments. Pretreatment with a non-ionic detergent solution (1% Brij 35 = polyoxyethylene lauryl ether) did not reduce the total digestible fraction, indicating that such particles were firmly bound to the lichen (see also Nieboer et al., 1972; Puckett et al., 1973).

Consideration of the distribution of lead and zinc in exactly the same fractions (Table IV) shows that zinc behaves like potassium but that lead is different. Thus strontium-elutable zinc increases following killing pretreatments and a certain proportion of the zinc is freely soluble within the cell. This is lost to the washing fraction following death but at the same time a significant quantity of zinc is still elutable by nitric acid following such a pretreatment. This may, as suggested above, represent zinc complexed to a site not subject to attack by strontium. The data for lead support this view. Killing fails to increase the amount of lead replaceable by strontium and hence the very low values recorded in the boiling water fraction reflect either the very low levels of intracellular lead or the capture of such lead by cell wall sites which are not capable of exchange with strontium. Both lead and zinc are capable of forming complexes with oxygen or nitrogen although the latter may show a preference for oxygen donors (Sidgwick, 1941). This would be in agreement with its being mainly bound to strontium-elutable sites (see discussion above) whereas lead is bound to a different, probably nitrogenous or sulphurous site. This agrees with the data presented in the previous section, which indicated the different behaviour of lead and zinc following nitric acid pretreatment of *C. rangiformis* (Table II) and the different ability of other cations to replace these two elements from the same species growing on lead- and zinc-rich soils (Table III).

Both lead and zinc show very low values in the original washings of fresh material. Acidification of these water-washing fractions to a final concentration of 20% nitric acid failed to increase the amount of either ion by more than 10 p.p.m. This is in marked contrast to the work of Little (1973) in which he showed that the water washings of tree leaves contained particles from which lead and zinc could be discharged by acidification. Up to 40% of the zinc and 80% of the lead recovered by total digestion of leaf material was in this form. The values of lead and zinc in the total digest fraction (Table IV) did not change appreciably with killing or detergent treatment which suggests that if these values correspond to substratum particles, perhaps equivalent to the particles in Little's washing, then they are firmly bound to the lichen.

The data reported here for *P. glabratula* subsp. *fuliginosa* suggest that most of the potassium is intracellular and soluble, most of the lead is bound to the cell wall in an exchangeable form and zinc is intermediate between these extremes. This is in agreement with the work of Brown and Slingsby (1972) who suggested a similar compartmentalization of ions in *C. rangiformis* using a different technique of analysis. Noeske *et al.* (1970) have reported zinc to be uniformly distributed throughout the thallus of *Acarospora smaragdula* using electron microprobe analysis. The same authors demonstrated that iron was deposited on the surface of the thallus, probably as an oxide layer, which is another way in which heavy metals may be associated with a lichen thallus. From the present data and the reports of Brown and Slingsby (1972) and Brown and Bates (1972) it appears reasonable to conclude that very little lead penetrates into the cells of lichens or bryophytes and hence may be of little metabolic significance. Skaar *et al.* (1973) have, however, reported the occurrence of lead within the nucleus of moss cells, even in plants with relatively low total levels of lead. Thus the present method of analysis may well underestimate the quantity of lead within the cell.

(*b*) *Ion uptake.* The total lead and zinc contents, as well as the quantities in each of the various fractions, were obviously lower in material from the non-polluted site at Priddy than in the highly polluted material from Easter Compton (Table IV). Potassium contents on the other hand were very similar at both sites. Comparison of the fresh material from Priddy with that transplanted to Easter Compton for 2 months shows virtually no change in the quantities or distribution of potassium, lead or zinc. Because the intracellular potassium level remained constant, it is apparent that the transplanted material is neither dead nor ionically unstable as a result of uptake of toxic cations. The slight increase in intracellular zinc is probably within the limits of accuracy of the analytical technique and therefore shows that the system is apparently capable of excluding the high levels of this potentially toxic cation present in the environment.

Little (1973) showed that particles adhering to leaf surfaces required treatment with dilute acid before all of the lead and zinc contained in them was analysable; washing with water was insufficient. It is not clear how far the high levels of heavy metals taken up by moss bags (Little and Martin, 1974; G. Nickless, pers. comm. above) could be available to the plant. The reported decline in heavy metal content of lichens away from the Avonmouth smelter (Burkitt *et al.*, 1972) may, in part, only represent ions held on the exterior of the lichen. If such particles were associated with the *Parmelia glabratula* used here then they would have been expected to have been found both in the water washings following acidification and in the N HNO_3 and total digest fractions. Because Little and Martin (1974) reported a general ratio of zinc to lead of 2·98, in particles it is unlikely that the high lead values in the N HNO_3 fraction are derived from such particles; one would then have expected even higher values for zinc. Similarly, it is unlikely that the total digest fraction represents only such particles.

It is perhaps surprising that if heavy metal-rich particles were deposited on the lichen during the test period no trace of them remained. One possible explanation for this is that the shiny upper surface of the thallus represents a poor surface for particles to become attached to (compared with the highly fructicose morphology of a *Sphagnum* plant) and this, coupled with high rainfall during the test period, means that virtually no particles remained on the surface. The shiny surface may also act as a barrier to penetration of cations in solution and thus reduce the rate of uptake of heavy metals produced by solution of these particles. The isidiate parts of the thallus do not result in greater retention of particles; these portions of the thallus were included in the analysis.

Despite the above explanation, material grown under these polluted conditions does become enriched with heavy metals (Table IV). It has often been suggested that the longevity of lichens is one of the reasons why they may acquire high levels of many cations. The present data appear to support this idea but, as noted in the previous section, if much of the cation content of a lichen is in an exchangeable form it can be acquired very rapidly. The important point is to be able to assess what is the available concentration of a particular cation, what is the rate at which it is supplied to the lichen and the nature of the competitive cations. The cation selectivity noted in the previous section means that low concentrations of an ion avidly bound by the lichen will be preferentially bound in competition with a more abundant but less efficiently bound ion. Estimates of the total ion content of a particular source (whether it is soil or air-borne particles) do not necessarily represent the quantity which is available to the lichen and thus discussion of relative enrichments or exclusions of cations may be meaningless (Lounamaa, 1956; Lambinon et al., 1964; Brown, 1973; Tuominen and Jaakkola, 1974). In the present case time of year may also be an important consideration as, following a dry spell, dew formation or the first rain may be responsible for dissolving cations from the particles which have rested on the thallus for a much longer period than is possible during wet conditions.

THE ROLE OF LICHEN SUBSTANCES IN ION UPTAKE

The ability of lichen substances (particularly depsides and depsidones with —OH, —COOH and —CHO donor groups in the *ortho* position) to form chelates with divalent cations has been considered of possible importance in the biogeochemical weathering of rocks by lichens (Syers and Iskandar, 1974). Release of iron, aluminium, calcium and magnesium from silicate rocks by lichen substances was shown by Iskandar and Syers (1972) but other organic acids, such as citric and salicylic, released greater quantities of these cations due to their greater water solubility compared to lichen substances. Such information has led certain workers to suggest that lichen substances might be of importance in ion binding within the lichen thallus (Nieboer et al., 1972; Seaward, 1973). Recorded values for lichen-substance contents of lichen thalli usually range from 1 to 10% of the dry weight (Hale, 1967) but may be as high as 25·6% (lecanoric acid in *Parmelia tinctorum fide* Ahmann and Mathey, 1967).

The greater solubility of lichen substances in organic solvents (e.g. acetone) compared to other possible cation-binding substances permits the hypothesis that they may be of importance in cation binding to be tested experimentally. The removal of fumarprotocetraric acid from *C. rangiformis* and lecanoric acid from *P. glabratula* by the acetone treatment previously mentioned was verified by the loss of *p*-phenylenediamine or calcium hypochlorite-reactive material from the respective thalli after treatment with acetone. Table II therefore records the uptake of cations by thalli with and without the potential cation-binding fumarprotocetraric acid. Acetone treatment, instead of decreasing the total uptake of cations as might be expected if a binding substance were being removed, actually increases it, probably by removing lipophilic molecules which impeded uptake to the actual binding sites. The selectivity sequence is not significantly altered by such treatment. The data in Table IV show that acetone treatment similarly failed to alter either the total content or distribution of lead and zinc recovered from *P. glabratula*. Acidification of the acetone extract itself, to a final concentration of 20% nitric acid, failed to demonstrate more than trace quantities of either of these cations. Initial inspection of the data in Table IV might suggest that potassium was associated with acetone soluble material in that the total quantity recovered is lower than any other treatment of the Easter Compton material. If this was the case then such material represented only 10% of the total potassium as the boiling water soluble fraction is reduced and fumarprotocetraric acid is unlikely to be substantially soluble in this fraction. The more likely explanation is that acetone, being capable of disrupting the cell membrane systems, merely effected the release of some of the already soluble potassium from within the cell. Thus the data presented here show that in two species with lichen substances suggested as being suitable for cation chelation these compounds are not responsible for detectable cation binding either of material naturally incorporated into the thallus or of material supplied in the laboratory.

CONCLUDING OBSERVATIONS

Despite the numerous reports of the cation content of lichens, particularly of radionuclides (see Tuominen and Jaakkola, 1974), it is fair to say that our knowledge of the mechanism of ion uptake by lichens is rudimentary. Much emphasis has been placed on the passive, physical process of ion uptake which, although responsible for a substantial proportion of the total content of many ions, is of uncertain significance in the life of the lichen. Information on the active uptake process for ions entering the cell cytoplasm and the compartmentalization achieved by a combination of the two mechanisms is a necessary step in our understanding of the effect particular cations have on the metabolism of lichens. How far extracellular cation binding acts as a buffer to the ionic extremes of the environment, as implied by Brown and Slingsby (1972), or as a source of essential nutrients, as has commonly and vaguely been implied, is as yet unknown. Lambinon *et al.* (1964) categorized lichens according to their apparent ability to accumulate or exclude heavy metals but, as mentioned previously, until we know what fraction of the total environmental cation concentration is actually available to a particular lichen species, suggestions about the ability to exclude cations appear premature. Whether lichens from extreme ionic environments have particular tolerance mechanisms is also unclear because, although Brown and Slingsby (1972) and the present data suggest that there is no difference in the ability of *C. rangiformis* from different habitats to take up heavy metals, no information is available on the metabolic consequences of such uptake.

Experimentally, cation uptake in lichens has been largely treated as the response of a single, rather than a composite, organism. Although the fungal and algal components under normal conditions behave as an integrated association it is important to remember that physiologically they are extremely dissimilar and their response to cations may be quite different. As the proportions of the two components may differ in the same species under different environmental conditions or with age, comparisons between material from dissimilar habitats may reflect this type of variation as much as any possible physiological adaptations to the different conditions.

ACKNOWLEDGEMENTS
I am very grateful to Dr J. W. Bates, Dr G. Nickless and Dr R. M. Brown for helpful discussion and comment throughout this work.

REFERENCES
AHMANN, G. B. and MATHEY, A. (1967). Lecanoric acid and some constituents of *Parmelia tinctorum* and *Pseudevernia intensa*. *Bryologist* **70**, 93–97.

ANTONOVICS, T., BRADSHAW, A. D. and TURNER, R. G. (1971). Heavy metal tolerance in plants. *Adv. ecol. Res.* **7**, 2–85.

BARTNICKI-GARCIA, S. (1968). Cell wall chemistry, morphogenesis and taxonomy in fungi. *A. Rev. Microbiol.* **22**, 87–108.

BARTNICKI-GARCIA, S. (1973). Fungal cell wall composition. *In* "Handbook of Microbiology. Vol. 2. Microbial composition" (A. I. Laskin and H. A. Lechevalier, eds), pp 201–214. Chemical Rubber Co., Ohio.

BATES, J. W. and BROWN, D. H. (1974). The control of cation levels in seashore and inland mosses. *New Phytol.* **73**, 483–495.

BOISSIÈRE, J.-C. (1969). La chitine chez quelques Lichenes: mise en évidence, localisation. *Bull. Soc. bot. Fr., Mém.* **1968**, Coll. Lich., 141–150.

BRODO, I. M. (1974) ["1973"]. Substrate ecology. *In* "The Lichens" (V. Ahmadjian and M. E. Hale, eds), pp. 401–441. Academic Press, New York and London.

BROWN, D. H. (1973). The lichen flora of the lead mines at Charterhouse, Mendip Hills. *Proc. Bristol Nat. Soc.* **32**, 267–274.

BROWN, D. H. and BATES, J. W. (1972). Uptake of lead by two populations of *Grimmia doniana*. *J. Bryol.* **7**, 181–193.

BROWN, D. H. and SLINGSBY, D. R. (1972). The cellular location of lead and potassium in the lichen *Cladonia rangiformis* (L.) Hoffm. *New Phytol.* **71**, 297–305.

BURKITT, A., LESTER, P. and NICKLESS, G. (1972). Distribution of heavy metals in the vicinity of an industrial complex. *Nature, Lond.* **238**, 327–328.

CLYMO, R. S. (1963). Ion exchange in *Sphagnum* and its relation to bog ecology. *Ann. Bot.*, N.S. **27**, 309–324.

CULBERSON, C. F. (1969). "Chemical and Botanical Guide to Lichen Products." University of North Carolina Press, Chapel Hill, N.C.

CULBERSON, C. F. (1970). Supplement to "Chemical and Botanical Guide to Lichen Products". *Bryologist* **73**, 177–377.

GOODMAN, G. T. and ROBERTS, T. M. (1971). Plants and soil as indicators of metal in the air. *Nature, Lond.* **231**, 287–292.

HALE, M. E. (1967). "The Biology of Lichens." Arnold, London.

HANDLEY, R. and OVERSTREET, R. (1968). Uptake of carrier-free ^{137}Cs by *Ramalina reticulata*. *Pl. Physiol., Lancaster* **43**, 1401–1405.

HANSON, W. C., WATSON, D. G. and PERKINS, R. W. (1967). Concentration and retention of fallout radionuclides in Alaska arctic ecosystems. *In* "Radio-Ecological Concentration Processes" (B. Åberg and F. P. Hungate, eds), pp. 233–245. Pergamon Press, Oxford.

HAWKSWORTH, D. L. and ROSE, F. (1970). Qualitative scale for estimating sulphur dioxide air pollution in England and Wales using epiphytic lichens. *Nature, Lond.* **227**, 145–148.

ISKANDAR, I. K. and SYERS, J. K. (1972). Metal complex formation by lichen compounds. *J. Soil Sci.* **23**, 255–265.

JAMES, P. W. (1973). The effect of air pollutants other than hydrogen fluoride and sulphur dioxide on lichens. *In* "Air Pollution and Lichens" (B. W. Ferry, M. S. Baddeley and D. L. Hawksworth, eds), pp. 143–175. University of London, Athlone Press, London.

JENKINS, D. A. and DAVIES, R. I. (1966). Trace element content of organic accumulations. *Nature, Lond.* **210**, 1296–1297.

KNIGHT, A. H., CROOKE, W. M. and INKSON, R. N. E. (1961). Cation exchange capacities of tissues of higher and lower plants and their relation to uronic acid contents. *Nature, Lond.* **192**, 142–143.

KUZIEL, S. (1973). The ratio of K to Ca in thalli of several species of lichens occurring on various trees. *Acta Soc. Bot. Pol.* **42**, 63–71.

LAMBINON, T., MAQUINAY, A. and RAMAUT, J. L. (1964). La teneur en zinc de quelques lichens des terrains calaminaires Belges. *Bull. Jard. bot. État Brux.* **34**, 273–282.

LANGE, O. L. and ZIEGLER, H. (1963). Der schwermetallgehalt von Flechten aus dem *Acarosporetum sinopicae* auf Erzschlackenhalden des Harzes. I. Eisen und Kupfer. *Mitt. flor.-soz. Arb. Gemein.* n.f. **10**, 156–183.

LE ROY, L. W. and KOKSOY, M. (1962). The lichen—a possible plant medium for mineral exploration. *Econ. Geol.* **57**, 107–111.

LITTLE, P. (1973). A study of heavy metal contamination of leaf surfaces. *Environ. Pollut.* **5**, 159–172.

P

LITTLE, P. and MARTIN, M. H. (1972). A survey of zinc, lead and cadmium in soil and natural vegetation around a smelter complex. *Environ. Pollut.* **3**, 241–254.

LITTLE, P. and MARTIN, M. H. (1974). Biological monitoring of heavy metal pollution. *Environ. Pollut.* **6**, 1–19.

LOUNAMAA, K. J. (1956). Trace elements in plants growing wild on different rocks in Finland. A semi-quantitative spectrographic survey. *Annls bot. Soc. zool.-bot. fenn.* '*Vanamo*' **29**, 1–196.

MANOCHA, M. S. and COLVIN, T. R. (1967). Structure and composition of the cell wall of *Neurospora crassa. J. Bact.* **94**, 202–212.

MELLOR, D. P. (1964). Historical background and fundamental concepts. *In* "Chelating agents and metal chelates" (F. P. Dwyer and D. P. Mellor, eds), pp. 1–50. Academic Press, New York and London.

MICÓVIĆ, V. M. and STEFANOVIĆ, V. D. (1961). Studies on the chemical composition of Yugoslav lichens. I. Parallel studies on the chemical composition of the ash of some Yugoslav lichens and of the ash of oak bark. *Bull. Acad. serbe Sci. Cl. Sci. math.-nat.* **26**, 113–117.

MITCHELL, A. D. and TAYLOR, I. E. P. (1969). Cell-wall protein of *Aspergillus niger* and *Chaetomium globosum. J. gen. Microbiol.* **59**, 103–109.

NEVSTRUEVA, M. A., RAMZAEV, P. V., MOISEER, A. A., IBATULLIN, M. S. and TEPLYKH, L. A. (1967). The nature of ^{137}Cs and ^{90}Sr transport over the lichen-reindeer-man food chain. *In* "Radio-Ecological Concentration Processes" (B. Åberg and F. P. Hungate, eds), pp. 209–215. Pergamon Press, Oxford.

NIEBOER, E., AHMED, H. M., PUCKETT, K. J. and RICHARDSON, D. H. S. (1972). Heavy metal content of lichens in relation to distance from a nickel smelter in Sudbury, Ontario. *Lichenologist* **5**, 292–304.

NOESKE, O., LÄUCHLI, A., LANGE, O. L., VIEWEG, G. H. and ZIEGLER, H. (1970). Konzentration und Lokalisierung von Schwermetallen in Flechten der Erzschlackenhalden des Harzes. *Vortr. bot. Ges. [Dtsch. bot. Ges.] n.f.* **4**, 67–79.

PERCIVAL, E. E. and TURVEY, J. R. (1973). Polysaccharides of algae. *In* "Handbook of Microbiology". Vol. 2, Microbial Composition (A. I. Larkin and H. A. Lechevalier, eds), pp. 147–165. Chemical Rubber Co., Ohio.

PETERSON, P. J. (1971). Unusual accumulation of elements by plants and animals. *Sci. Prog., Lond.* **59**, 505–526.

PUCKETT, K. J., NIEBOER, E., GORZYNSKI, M. J. and RICHARDSON, D. H. S. (1973). The uptake of metal ions by lichens: A modified ion-exchange process. *New Phytol.* **72**, 329–342.

QUILLET, M. and DE LESTANG-LAISNÉ, G. (1969). Convenance glucidique de l'association symbiotique Cyanophycée- Ascomycète. *Bull. Soc. bot. Fr., Mém.* **1968**, *Coll. Lich.* 135–140.

RITCHIE, C. A., RITCHIE, J. C. and PLUMMER, G. L. (1971). Distribution of fallout caesium-137 in *Cladonia* mounds in Georgia. *Bryologist* **74**, 359–362.

RÜHLING, A. and TYLER, G. (1970). Sorption and retention of heavy metals in the woodland moss *Hylocomium splendens* (Hedw.) Br. et Sch. *Oikos* **21**, 92–117.

SEAWARD, M. R. D. (1973). Lichen ecology of the Scunthorpe heathlands. I. Mineral accumulation. *Lichenologist* **5**, 423–433.

SIDGWICK, N. V. (1941). Complex formation. *J. chem. Soc.* 433–443.

SKAAR, H., OPHUS, E. and GULLVÅG, B. M. (1973). Lead accumulation within nuclei of moss leaf cells. *Nature, Lond.* **241**, 215–216.

SMITHIES, W. R. (1952). Chemical composition of a sample of mycelium of *Penicillium griseofulvum* Dierchx. *Biochem. J.* **51**, 259–264.

SPEARING, A. M. (1972). Cation exchange capacity and galacturonic acid content on several species of *Sphagnum* in sandy Ridge Bog, Central New York State. *Bryologist* **75**, 154–158.

SYERS, J. K. and ISKANDAR, I. K. (1974) ["1973"]. Pedogenetic significance of lichens. *In* "The Lichens" (V. Ahmadjian and M. E. Hale, eds), pp. 225–248. Academic Press, New York and London.

TAYLOR, I. E. P. and CAMERON, D. S. (1973). Preparation and quantitative analysis of fungal cell walls: Strategy and tactics. *A. Rev. Microbiol.* **27**, 243–259.

TUOMINEN, Y. (1967). Studies on the strontium uptake of the *Cladonia alpestris* thallus. *Annls. bot. fenn.* **4**, 1–28.

TUOMINEN, Y. (1968). Studies in the translocation of caesium and strontium ions in the thallus of *Cladonia alpestris*. *Annls bot. fenn.* **5**, 102–111.

TUOMINEN, Y. (1971). Studies on some concentration–distance curves of the diffusion of ^{137}Cs and ^{90}Sr ions in columns composed of the thallus of *Cladonia alpestris*. *Annls bot. fenn.* **8**, 245–253.

TUOMINEN, Y. and JAAKKOLA, T. (1974) ["1973"]. Absorption and accumulation of mineral elements and radioactive nuclides. *In* "The Lichens" (V. Ahmadjian and M. E. Hale, eds), pp. 185–223. Academic Press, New York and London.

18 | Aspects of Nitrogen Metabolism in Lichens

J. W. MILLBANK

Botany Department, Imperial College, University of London, England

Abstract: The lichens which have been shown to fix nitrogen by means of the acetylene reduction technique, or the use of $^{15}N_2$, are listed. They number some 50 species of 18 genera. It can be assumed that any lichen with *Nostoc, Calothrix* or *Scytonema* as phycobiont is capable of nitrogen fixation.

Results using more refined techniques of cell estimation have shown that the rate of nitrogen fixation by *Nostoc* in the thallus of *Peltigera* species is similar to that of the free-living organism, rather than considerably greater as previously suggested. In symbiotic *Nostoc*, heterocyst frequencies are of the order of 3% of the total blue-green algal cells. In *Nostoc* present in cephalodia much higher heterocyst frequencies occur, of the order of 25%, with correspondingly more rapid rates of nitrogen fixation.

Dissection experiments have shown that a significant fraction of the fixed nitrogen released to the mycobiont by the *Nostoc* phycobiont is in the form of peptides, with molecular weights of the order of 1100. They have similar properties and composition to those released by the algae when free living.

Under conditions conducive to high rates of nitrogen fixation (high light intensity and enhanced P_{CO_2}) a net evolution of oxygen by lichen thalli has been demonstrated. Preliminary trials have also indicated that when nitrogen fixation is taking place the conditions within lichen thalli are unlikely to be anaerobic.

The implications of these findings are briefly discussed.

NITROGEN FIXATION IN LICHENS

1. Nitrogen-fixing Lichen Species

Reports of nitrogen fixation in or upon lichen thalli go back more than 50 years. However, no definitive results were published until those of Bond and Scott (1955) from experiments on *Collema auriculatum* and *Leptogium lichenoides* using the heavy isotope ^{15}N. This work was extended by Scott (1956) using *Peltigera praetextata*, and Watanabe and Kiyohara (1963) with *Peltigera virescens*.

Systematic Association Special Volume No. 8, "Lichenology: Progress and Problems", edited by D. H. Brown, D. L. Hawksworth and R. H. Bailey, 1976, pp. 441–455. Academic Press, London and New York.

Since then a large number of lichens have been investigated by means of $^{15}N_2$ and, although the techniques required are cumbersome and the equipment expensive, force of necessity stimulated the development of procedures in which $^{15}N_2$ studies could be carried out under field conditions (Stewart, 1967). Using such techniques, Fogg and Stewart (1968) reported on the fixation of nitrogen by *Collema pulposum* and *Stereocaulon* sp. in the Antarctic. Subsequently an entirely different technique for the demonstration and assay of nitrogenase activity was developed, in which acetylene is reduced to ethylene by the enzyme. This technique (see e.g. Postgate, 1972) is overwhelmingly more convenient, rapid and sensitive and has been used to establish the nitrogen-fixing ability of a wide range of organisms including a variety of lichens. The lichen species that have been definitively shown to fix nitrogen by the use of these techniques now number around 50, and they are listed in Table I. It will be noted that the blue-green phycobiont is commonly a species of *Nostoc* with

TABLE I. Nitrogen-fixing lichens

Lichen	Phycobiont	Technique	Source
Collema auriculatum	*Nostoc*	C_2H_2	Hitch and Millbank (1975b)
		$^{15}N_2$	Bond and Scott (1955)
C. coccophorus	*Nostoc*	$^{15}N_2$	Rogers, Lange and Nicholas (1966)
C. crispum	*Nostoc*	C_2H_2	Hitch and Stewart (1973)
C. fluviatile	*Nostoc*	C_2H_2	Hitch and Millbank (1975b)
C. furfuraceum	*Nostoc*	C_2H_2	Hitch and Millbank (1975b)
C. pulposum	*Nostoc*	$^{15}N_2$	Fogg and Stewart (1968)
C. subfurvum	*Nostoc*	C_2H_2	Hitch and Millbank (1975b)
C. tuniforme	*Nostoc*	C_2H_2	Henriksson and Simu (1971)
Ephebe lanata	*Stigonema*	C_2H_2	Hitch and Millbank (1975b)
Leptogium burgessii	*Nostoc*	C_2H_2	Hitch and Millbank (1975b)
L. lichenoides	*Nostoc*	$^{15}N_2$	Bond and Scott (1955)
L. sinuatum	*Nostoc*	C_2H_2	Hitch and Millbank (1975b)
L. teretiusculum	*Nostoc*	C_2H_2	Hitch and Millbank (1975b)
Lichina confinis	*Calothrix*	$^{15}N_2$	Stewart (1970)
L. pygmaea	*Calothrix*	$^{15}N_2$	Stewart (1970)
Lobaria amplissima	n.d. (internal cephalodia)	C_2H_2	Hitch and Millbank (1975b)
	(external cephalodia★)	C_2H_2	Hitch and Millbank (1975b)
L. laetevirens	n.d. (internal cephalodia)	C_2H_2	Hitch and Millbank (1975b)
L. pulmonaria	*Nostoc* (cephalodia)	C_2H_2	Millbank and Kershaw (1970)
Massalongia carnosa	*Scytonema*	C_2H_2	Hitch and Millbank (1975b)

★ *Dendriscocaulon*-morphotype.

Lichen	Phycobiont	Technique	Source
Nephroma arcticum	*Nostoc*	C_2H_2	Kallio, Suhonen and Kallio (1972)
N. laevigatum	*Nostoc*	C_2H_2	Hitch and Millbank (1975b)
Pannaria pezizoides	*Nostoc*	C_2H_2	Hitch and Millbank (1975b)
P. pityrea	*Nostoc*	C_2H_2	Hitch and Millbank (1975b)
P. rubiginosa	*Nostoc*	C_2H_2	Hitch and Millbank (1975b)
Parmeliella atlantica	*Nostoc*	C_2H_2	Hitch and Millbank (1975b)
P. plumbea	*Nostoc*	C_2H_2	Hitch and Millbank (1975b)
Peltigera aphthosa	*Nostoc* (cephalodia)	C_2H_2	Hitch (1971)
var. *variolosa*	*Nostoc* (cephalodia)	$^{15}N_2$	Millbank and Kershaw (1970)
P. canina	*Nostoc*	C_2H_2	Millbank and Kershaw (1970)
P. polydactyla	*Nostoc*	$^{15}N_2$	Watanabe and Kiyohara (1963)
P. praetextata	*Nostoc*	$^{15}N_2$	Scott (1956)
P. pruinosa	*Nostoc*	$^{15}N_2$	Watanabe and Kiyohara (1963)
P. rufescens	*Nostoc*	C_2H_2	Henriksson and Simu (1971)
Placopsis gelida	n.d. (cephalodia)	C_2H_2	Hitch and Stewart (1973)
Placynthium nigrum	*Dichothrix*	C_2H_2	Hitch and Millbank (1975b)
P. pannariellum	*Dichothrix*	C_2H_2	Hitch and Millbank (1975b)
Polychidium muscicola	*Scytonema*	C_2H_2	Hitch and Millbank (1975b)
Pseudocyphellaria thouarsii	*Nostoc*	C_2H_2	Hitch and Millbank (1975b)
Solorina crocea	*Nostoc* (internal layer)	C_2H_2	Kallio, Suhonen and Kallio (1972)
S. saccata	*Nostoc* (internal layer)	C_2H_2	Hitch and Millbank (1975b)
S. spongiosa	*Nostoc* (cephalodia)	C_2H_2	Hitch and Millbank (1975b)
Stereocaulon sp.	*Nostoc* (cephalodia)	$^{15}N_2$	Fogg and Stewart (1968)
Sticta fuliginosa	*Nostoc*	C_2H_2	Hitch and Millbank (1975b)
S. limbata	*Nostoc*	C_2H_2	Hitch and Millbank (1975b)

occasional examples of *Calothrix* and *Scytonema*. As yet there are no definitive reports of nitrogen fixation by lichens containing species of *Gloeocapsa* as phycobiont, but this state of affairs may soon be remedied as a result of studies on specimens of *Peccania* which have been kindly made available to the author by Professor M. Galun. It can now be confidently assumed that if a lichen contains *Nostoc*, *Calothrix* or *Scytonema* as a phycobiont, it will be able to fix nitrogen.

2. The Content of Blue-green Phycobiont in Lichens

Work on the nitrogen-fixing abilities of lichens soon revealed the necessity for a good method of estimating the blue-green algal fraction of the total cell population in a thallus sample. The methods presenting themselves were direct counting in homogenized thallus samples, estimates of photosynthetic pigment content, and separation by differential centrifuging followed by chemical estimation of nitrogen or some other characteristic feature.

Early studies (Millbank and Kershaw, 1969; Kershaw and Millbank, 1970) used differential centrifugation on sucrose density gradients to separate the *Nostoc* cells from dissected *Peltigera aphthosa* cephalodia, a technique which was a refinement and development of one described by Drew and Smith (1967). A figure of about 2·5% was obtained for the proportion of *Nostoc* nitrogen in the cephalodia, and this resulted in an apparently very rapid rate of nitrogen fixation per unit weight of *Nostoc*. A generation time of about 12 h at 25°C would follow if all the nitrogen fixed was utilized by the alga for growth. In fact, a very large proportion of the total nitrogen fixed was released from the algal cells and made available to the mycobiont. When this method of estimation of the phycobiont was later compared with total cell number estimates, the latter obtained both directly by counting and indirectly by pigment extraction, large discrepancies were found (Millbank, 1972). In the case of *Peltigera canina*, it appeared that over 50% of the *Nostoc* cells were being disrupted when the thallus was homogenized, even though relatively gentle techniques were used. A revised estimate for the number of *Nostoc* cells in a typical specimen was needed and a figure of 3×10^6 cells/cm² thallus was quoted. Following from this, increased figures for the proportion of *Nostoc* nitrogen in the cephalodia of *P. aphthosa* (5·5%) and the estimated theoretical generation time (18 h) were quoted by Millbank (1972).

From the finding of apparently rapid rates of nitrogen fixation by the *Nostoc* in *Peltigera*, estimates of the heterocyst populations were made. The first method employed was to examine electron micrographs of ultra-thin sections since it had emerged that, in lichenized algae, cell rupture during homogenization destroyed both vegetative cells and heterocysts indiscriminately. The method, although laborious, was definitive and a figure of 3·3% was reported for *P. canina* (Griffiths *et al.*, 1972). Subsequently Hitch and Millbank (1975b) have investigated the frequency of heterocysts in the *Nostoc* filaments of a range of lichens, using optical microscopy. The thalli were treated with CrO_3 solution at concentrations up to 10% (w/v) for periods from 30 min to 16 h, depending on the lichen. In this way the structure was softened and the mycobiont hyphae partially dissolved, leaving the algal filaments intact. The existence of very long filaments was revealed and the cell:heterocyst ratio easily estimated (Plate I). Results obtained by the use of this technique, based on the method of Hill and Woolhouse (1966), are given in Table II. It is clear that, with the exception of the external cephalodia of *Lobaria amplissima*, the proportion of heterocysts are very much higher in the filaments of algae found in cephalodia than those found in the cortex of the thallus. The exception just mentioned may not in fact be a real one, as these structures on *Lobaria amplissima* are viewed by some authorities as being capable of independent existence and the generic name *Dendriscocaulon* has been applied (see Chapter 3).

TABLE II. The percentage of heterocysts in blue-green phycobiont filaments

Lichen	Cephalodia		Thallus
	Internal	External	
Leptogium burgessii	—	—	4·5
Lobaria amplissima	21·6	3[a]	—
L. laetevirens	30·4	—	—
L. pulmonaria	35·6	—	—
L. scrobiculata	—	—	3·9
Nephroma laevigatum	—	—	4·1
Peltigera aphthosa var. *variolosa*	21·1	—	—
P. canina	—	—	4·9
P. polydactyla	—	—	5·8
P. praetextata	—	—	4·4
P. venosa	—	—	7·8
Sticta limbata	—	—	4·9
S. sylvatica	—	—	3·5

[a] *Dendriscocaulon*-like morphotype.

There is no obvious explanation for this large variation in heterocyst:cell ratio. Kulasooriya *et al.* (1972) proposed that the C:N ratio of the algal cells was a critical factor in heterocyst differentiation, the environmental oxygen tension also playing a part, a low P_{O_2} promoting differentiation. It would seem on the face of it rather unlikely that there is a very wide variation of C:N ratio in algae in the different thalli but only delicate and detailed studies can throw light on this. Other results, to be referred to later, indicate that the cephalodia of *Peltigera aphthosa* (whose algae are very rich in heterocysts) are at an oxygen tension equal to the atmospheric level.

These experiments, in which the intact filaments of blue-green phycobionts from various lichens were examined, also enabled much more reliable estimates to be made of the algal cell population per unit of thallus. The results of these estimates are presented in Table III. It is clear that the population of *Nostoc* cells per square centimetre of thallus is very much higher even than had been re-estimated (see above). An intensive investigation into the cause of this discrepancy revealed that the errors were occurring in the centrifugal separation steps after homogenizing the thallus. Sediments considered to comprise whole cells of algae and fungus mycelial fragments, also contained subcellular particles and membranes from the algae which were expected to have remained in suspension. When the algal cells and fungus mycelium were separated on a sucrose density gradient, the fungus fraction was in fact found to be heavily contaminated with these algal membrane fragments. The pigments associated with these were, however, very resistant to extraction by methanol and thus

TABLE III. Blue-green phycobiont population of lichen thalli (from Hitch and Millbank 1975a)

Lichen	Nostoc
Peltigera canina	9×10^6 cells/cm^2 thallus
P. polydactyla	$10 \cdot 3 \times 10^6$ cells/cm^2 thallus
P. aphthosa[a]	4×10^6 cells/mg dry wt cephalodia
Sticta limbata	$9 \cdot 7 \times 10^6$ cells/cm^2 thallus

[a] 1 cm^2 thallus contains approximately 0·25 mg dry weight of cephalodia.

the erroneous conclusion was drawn, when the technique was originally developed, that the sediments were entirely fungal. The *Nostoc* population of the cephalodia of *Peltigera aphthosa*, estimated by this direct counting technique, was revealed as representing very nearly 50% of the total cephalodial nitrogen. Thus, *Peltigera aphthosa* was again exemplified as an exceptionally useful lichen for study, the blue-green phycobiont being confined to easily dissectable growths in which these very active cells are especially abundant and continue active nitrogen fixation after separation from the vast bulk of the mycobiont. It is not, however, possible to demonstrate nitrogen-fixing activity after homogenizing the cephalodia themselves.

3. The Rate of Nitrogen Fixation

In the light of these new findings, the estimates of the rate of fixation by the blue-green phycobiont of lichens needed to be revised. In *Peltigera canina* the mean rate is 3 nmol C_2H_4/min/mg protein, with a maximum rate of 5. Such a rate is still remarkably rapid, especially in view of the fact that cells are in a non-proliferating state and in view of a markedly sub-optimal nutritional status. The rates of fixation per unit of *Nostoc* protein in *Peltigera polydactyla* and *Sticta limbata* are closely similar. The rate for the *Nostoc* cells in the cephalodia of *Peltigera aphthosa* was, however, of a different order. This emerged from a trial in which 48 discs of thallus were assayed for acetylene reduction, dry weight, total nitrogen, *Nostoc* cell numbers and the ratio of cephalodia to total thallus dry weight. The mean rate of acetylene reduction was 11 nmol/min/mg *Nostoc* protein, but rates of up to 20 were recorded, and other trials confirmed this on several occasions. This extremely rapid rate of acetylene reduction is associated with a high proportion of heterocysts (see Table II), and there is evidently a correlation here. The percentage of heterocysts in *Peltigera aphthosa* is eight times that in *Peltigera canina*, and the rate of acetylene reduction is seen to be up to five times greater. Nitrogen fixation by the *Nostoc* in *Peltigera aphthosa* cephalodia would seem to be exclusively in heterocysts but the rate in *Peltigera canina* and *Peltigera polydactyla* may leave grounds for the possibility of vegetative cell fixation; this matter is still very open (see Hitch and Millbank, 1975a).

THE NATURE OF TRANSLOCATED NITROGEN

That much of the nitrogen fixed by blue-green algae is released into the surroundings has been known for many years (see Fogg *et al.*, 1973, p. 208 *et seq.*). In free-living organisms, release is proportionately greatest in the lag and stationary phases of growth and when conditions are adverse or changing. In lichens the proportion of nitrogen fixed that is subsequently released is high (see Scott, 1956; Millbank and Kershaw, 1969).

1. Electrophoretic Studies

Separation by high voltage paper electrophoresis of the substances released from the cephalodia of *Peltigera aphthosa*, *Placopsis gelida* and *Lobaria amplissima*, and the cortex of *Peltigera polydactyla* has been reported (Millbank, 1974). All of these lichens were chosen for the facility with which structures or layers containing the blue-green phycobiont could be dissected off and thus separated from the bulk of the mycobiont without excessive disturbance of the algal cells. An artificial substratum for absorbing the released nitrogenous products was provided in the form of sterilized moist filter paper. After an arbitrary, but necessarily rather long, period of time the products were eluted from the paper and analysed. Two week's exposure were normally allowed before analysis. The location of the separate components of the mixture of materials released was carried out by means of u.v. fluorescence after treatment with phenylacetalde-hyde–ninhydrin. This technique gave a 10-fold or greater improvement in sensitivity over conventional ninhydrin and was particularly suitable for peptide

substances. The materials demonstrated were compared with those released by
the same strain of *Nostoc* grown under nitrogen-fixing conditions in pure
culture. The separation patterns obtained from eluates from the various lichen
structures are shown in Fig. 1 and those from the concentrated media after
growth of various *Nostoc* isolates, in Fig. 2. This also shows the patterns

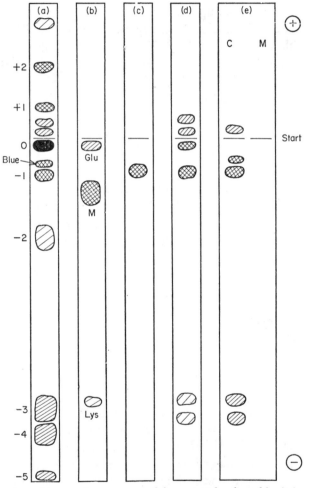

Fig. 1. Electrophoretic separation patterns of the materials released by lichens containing
blue-green phycobionts. (after Millbank, 1974). (a) From the cephalodia of
Peltigera aphthosa. (b) Hydrolysed eluate from (a). (c) From the cephalodia of
Lobaria amplissima. (d) From the cephalodia of *Placopsis gelida*. (e) From the
dissected cortex (C) and medulla (M) of *Peltigera polydactyla*. For explanation of
symbols, see Fig. 2.

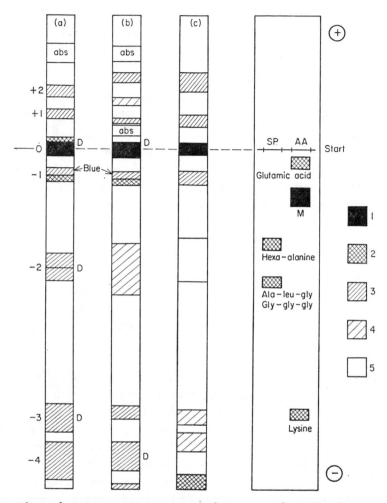

Fɪɢ. 2. Electrophoretic separation patterns of concentrated media after growth of isolated *Nostoc* phycobionts of *Peltigera* species compared with pure amino acids (AA) and synthetic peptides (SP). (a) From *Peltigera aphthosa*; (b) from *P. canina*; (c) from *P. polydactyla*. (after Millbank, 1974.)

Key to intensities—1: Extremely intense→5: Just detectable.

    M = Migration zone of free amino acids; composed of a mixture of alanine, aspartic acid, glycine, glutamine, leucine, phenylalanine, serine, threonine and valine.

    D = indicates two components probably present.

    abs = ultraviolet absorption zone.

obtained from some synthetic peptides and pure amino acids. The various *Nostoc* strains evidently released a generally similar range of substances when in the free-living state and, when these were compared with the separation patterns obtained from dissected lichen material rich in algae, it appeared that substances with very similar electrophoretic mobilities were being released when the algae were living symbiotically. *Peltigera aphthosa* and *Peltigera polydactyla* were the most convenient lichens for study, *Placopsis gelida* being smaller and less easily available and *Lobaria amplissima* having dendroid cephalodia with only a limited cut area for release of material. Stewart and Rowlands (pers. comm.) in a similar study using *Peltigera aphthosa* have found that substantial amounts of nitrogen in the form of ammonia appear to be released from the symbiotic algae. These findings indicate that the nitrogen metabolism of symbiotic *Nostoc* is modified from that of free-living cells in a manner only partly similar to that applying to the carbon metabolism as demonstrated by Smith *et al.* (1969). Thus, glucose and ammonia are released only by symbiotic *Nostoc*; both are simple compounds. Organic nitrogen, on the other hand, is released, in similar forms, by both free-living and symbiotic cells, implying little modification to the metabolism of the alga. Thus the nitrogenous release mechanism appears to be modified and augmented rather than created *de novo*.

2. Chromatography of Hydrolysates

This was carried out to establish the amino acid composition of the peptide substances released by the algae. It was only practicable to study the components found in growth media, the quantities released by dissected lichen material being sufficient only to demonstrate their presence qualitatively and to demonstrate the similarity of their electrophoretic behaviour to that of the compounds released by free-living cells. The detailed composition of the more prominent components is shown in Table IV. The constituent amino acids do not, in general, differ qualitatively from those reported as present in the extracellular peptides of other algae (see Millbank, 1974). Among the materials released by the lichen algae, however, the simplicity of band − 2 and the generally low level of serine is noticeable. The mirror image abundances of aspartic acid compared to component X in bands − 3 and − 4 are also apparent; the significance of these differences is however unknown.

3. Molecular Weight of the Substances Released

Peptides released by *Nostoc* from *Peltigera aphthosa*, *Peltigera canina* and *Peltigera polydactyla* were investigated by means of gel filtration on "Sephadex" columns. The materials released by all three strains were in all cases broadly similar, with molecular weights of the order of 1100, but with appreciable, though lesser, quantities of materials with molecular weights of about 300. Thus, octa- to decapeptides predominate but tri- or tetrapeptides are also significant.

TABLE IV. Amino acid composition of the major electrophoretically separable peptides in concentrated culture media after growth of *Nostoc* isolated from *Peltigera aphthosa* (columns A) and *P. canina* (columns B) (From Millbank, 1974)

| Amino acid | Electrophoresis Zone (see Fig. 1) | | | | | | | | | | |
| | +1 | 0 | | −1 | | −2 | | −3 | | −4 | |
	A	A	B	A	B	A	B	A	B	A	B
Alanine	3	8	3	10	5	7	4	1	1	1	1
Arginine	–	–	–	2	–	–	–	1	–	1	–
Aspartic acid	2	2	1	1	–	–	–	3	5	7	4
Citrulline	1	3	2	1	3	–	–	–	–	1	tr
Glutamic acid	4	6	2	6	2	–	1	1	1	1	1
Glycine	2	6	2	7	3	1	1	1	1	1	1
Leucine	2	6	2	8	4	1	1	1	1	–	1
Ornithine	–	–	–	2	–	–	–	1	–	1	–
Phenylalanine	1	4	tr	4	2	–	–	–	–	–	–
Serine	tr	4	tr	2	1	–	1	1	1	1	1
Threonine	1	3	1	1	3	–	–	–	–	–	–
Valine	2	4	1	4	3	1	1	1	tr	–	1
B	–	4	–	6	1	–	–	–	–	–	–
X	–	1	–	–	–	–	–	7	–	3	–
Y	–	1	–	–	–	–	–	1	–	1	–
Z	–	1	–	–	–	–	–	–	–	1	–

The figures give a guide to the relative abundance of the components of the individual peptides and do not indicate the relative abundance of the peptides themselves.

B, X, Y, Z = unidentified ninhydrin-reacting spots; – = not visible on chromatogram; tr = trace.

THE OXYGEN ENVIRONMENT WITHIN THE LICHEN THALLUS

This aspect is of particular relevance to the nitrogen metabolism of those lichens with a blue-green phycobiont, since the nitrogen-fixing process is essentially anaerobic although various mechanisms that enable it to proceed rapidly under aerobic conditions have been developed by micro-organisms. The case of the blue-green algae is of special interest as they are photosynthetic, with water as the hydrogen donor, and therefore evolve oxygen. The heterocyst is widely accepted as the site of nitrogenase activity, and these specialized, differentiated cells have a number of features such as a thick cell wall, active respiratory system, and lack of photosystem II which ensure that they are highly reducing, thus allowing an anaerobic process to proceed rapidly in an aerobic external environment.

Investigations of the effect of light on nitrogen fixation in lichens (Kallio *et al.*, 1972; Hitch and Stewart, 1973) have all demonstrated their close correlation. In the field this means that photosynthesis is proceeding at the same time as nitrogen fixation, and there is therefore a possibility of net oxygen evolution by the symbiosis.

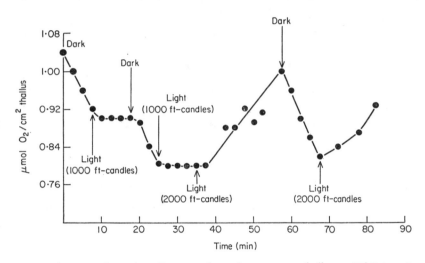

Fig. 3. Evolution and uptake of oxygen by *Peltigera canina* thallus at 25°C in mineral salts medium, pH 7·3, in darkness and at the light intensities shown. (From C. J. B. Hitch, unpublished.)

Using a conventional Clark type oxygen electrode, C. J. B. Hitch (unpublished) has shown with *Peltigera canina* and *Peltigera polydactyla* that a rapid respiratory oxygen uptake occurs in the dark. In light of about 1000 ft candles this was balanced by photosynthetic oxygen evolution by the phycobiont. In brighter light, of about 2000 ft candles, there was a rapid net evolution of oxygen (see Fig. 3). Such light intensities are representative of a north-facing wall and the open shade, respectively, on a sunny day in the north of the British Isles. With an augmented CO_2 supply, known to promote photosynthesis and inhibit photorespiration in blue-green algae (Lex *et al.*, 1972), the rate of photosynthetic oxygen evolution was greatly increased, and the oxygen compensation point achieved in dim light, of approximately 200 ft candles. Nitrogen fixation is also most rapid under conditions of bright light and a high

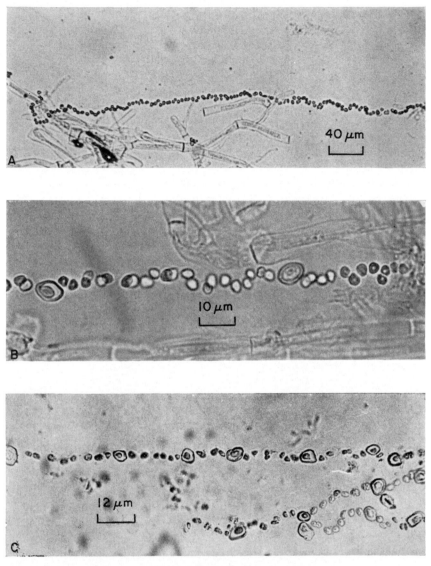

PLATE I

A, B, Filaments of *Nostoc* sp. from *Peltigera canina* after treatment of the thallus with 10% CrO_3 solution for 16 h. **C,** Filaments of *Nostoc* sp. from a cephalodium of *Peltigera aphthosa*, after treatment with 10% CrO_3 solution for 16 h. Note the heterocysts and the length of the convoluted filaments. (From C. J. B. Hitch, unpublished.)

P_{CO_2}, when the rate of oxygen evolution is maximal. In those lichens with phycobiont filaments containing a large proportion of heterocysts this state of affairs does not evoke any surprise. However, in *Peltigera canina*, and other species where the proportion of heterocysts is exceptionally low (Table II), it would seem that either the nitrogenase activity per heterocyst is unusually high or some vegetative cell fixation may be taking place.

Preliminary investigations of the internal P_{O_2} of the thallus of various lichens using oxygen micro-electrodes, have indicated that in *Peltigera polydactyla* the P_{O_2} of the algal zone of the cortex is appreciably below that of the atmosphere being 0·16–0·18 (air = 0·21) and is affected by the intensity of illumination, though without a very rapid response. The P_{O_2} within the cephalodia of *Peltigera aphthosa* when illuminated at about 1500–2000 ft candles approximated to that of air. They proved, however, to be extremely sensitive to variation in illumination, and with a very rapid response, the P_{O_2} falling to 0·04 within about 30 s of darkening, and rising well above atmospheric in about the same time if brightly illuminated (6000 ft candles). This species is thus exceptionally sensitive to changes in its immediate environment.

DISCUSSION

It is suggested that the rates of nitrogen fixation in lichens are governed by physical factors such as P_{O_2}, light intensity, temperature and moisture content. The high throughput of nitrogen is perhaps a manifestation of a permanently existing "lag" or "stationary" phase due to mineral or other nutritional deficiencies imposed on the phycobiont (see Jones and Stewart, 1969). Since there is no lack of nitrogen, oxygen and carbon in the case of blue-green algae, the mycobiont could perhaps be imposing some chemical constraint on, for example, protein synthesis. Thus, both the nitrogen and carbon metabolism of the blue-green phycobiont, apparently never very closely controlled, would be to some extent deranged. This could cause large losses and consequent advantage to the fungus. Alternatively, and perhaps additionally, the mycobiont may have developed a mechanism for absorbing surplus algal metabolic products extremely rapidly and completely, thus promoting a rapid rate of fixation and throughput of nitrogen.

The finding of low proportions of heterocysts in many lichens, allied with rapid rates of nitrogen fixation, in aerobic conditions, remains enigmatic.

REFERENCES

BOND, G. and SCOTT, G. D. (1955). An examination of some symbiotic systems for fixation of nitrogen. *Ann. Bot.* **19**, 67–77.

DREW, E. A. and SMITH, D. C. (1967). Studies in the physiology of lichens VII. The physiology of the *Nostoc* symbiont of *Peltigera polydactyla* compared with cultured and free living forms. *New Phytol.* **66**, 379–388.

FOGG, G. E. and STEWART, W. D. P. (1968). *In situ* determinations of biological nitrogen fixation in Antarctica. *Bull. Br. Antarct. Surv.* **15**, 39–46.

FOGG, G. E., STEWART, W. D. P., FAY, P. and WALSBY, A. E. (1973). "The Blue-Green Algae." Academic Press, London and New York.

GRIFFITHS, H. B., GREENWOOD, A. D. and MILLBANK, J. W. (1972). The frequency of heterocysts in the *Nostoc* phycobiont of the lichen *Peltigera canina*. *New Phytol.* **17**, 11–13.

HENRIKSSON, E. and SIMU, B. (1971). Nitrogen fixation by lichens. *Oikos* **22**, 119–121.

HILL, D. J. and WOOLHOUSE, H. W. (1966). Aspects of the autecology of *Xanthoria parietina* agg. *Lichenologist* **3**, 202–214.

HITCH, C. J. B. (1971). "A Study of some Environmental Factors affecting Nitrogenase Activity in Lichens." M.Sc thesis, University of Dundee.

HITCH, C. J. B. and MILLBANK, J. W. (1975a). Nitrogen metabolism in lichens, VI. The blue-green phycobiont content, heterocyst frequency and nitrogenase activity in *Peltigera* spp. *New Phytol.* **74**, 473–476.

HITCH, C. J. B. and MILLBANK, J. W. (1975b). Nitrogen metabolism in lichens, VII. Nitrogenase activity and heterocyst frequency in lichens with blue green phycobionts. *New Phytol.* **75**, 239–244.

HITCH, C. J. B. and STEWART, W. D. P. (1973). Nitrogen fixation by lichens in Scotland. *New Phytol.* **72**, 509–524.

JONES, K. and STEWART, W. D. P. (1969). Nitrogen turnover in marine and brackish habitats, III. The production of extracellular nitrogen by *Calothrix scopulorum*. *J. mar. biol. Ass. U.K.* **49**, 475–488.

KALLIO, P., SUHONEN, S. and KALLIO, H. (1972). The ecology of nitrogen fixation in *Nephroma arcticum* and *Solorina crocea*. *Rept Kevo Subarctic Res. St.* **9**, 7–14.

KERSHAW, K. A. and MILLBANK, J. W. (1970). Nitrogen metabolism in lichens II. The partition of cephalodia fixed nitrogen between the mycobiont and phycobionts of *Peltigera aphthosa*. *New Phytol.* **69**, 75–79.

KULASOORIYA, S. A., LANG, N. J. and FAY, P. (1972). The heterocysts of blue-green algae, III. Differentiation and nitrogenase activity. *Proc. R. Soc. Lond.*, B, **181**, 199–209.

LEX, M., SILVESTER, W. B. and STEWART, W. D. P. (1972). Photorespiration and nitrogenase activity in the blue-green alga, *Anabaena cylindrica*. *Proc. R. Soc. Lond.*. B, **180**, 87–102.

MILLBANK, J. W. (1972). Nitrogen metabolism in lichens IV. The nitrogenase activity of the *Nostoc* phycobiont in *Peltigera canina*. *New Phytol.* **71**, 1–10.

MILLBANK, J. W. (1974). Nitrogen metabolism in lichens V. The forms of nitrogen released by the blue-green phycobiont in *Peltigera* spp. *New Phytol.* **73**, 1171–1181.

MILLBANK, J. W. and KERSHAW, K. A. (1969). Nitrogen metabolism in lichens I. Nitrogen fixation in the cephalodia of *Peltigera aphthosa*. *New Phytol.* **68**, 721–729.

MILLBANK, J. W. and KERSHAW, K. A. (1970). Nitrogen metabolism in lichens III. Nitrogen fixation by internal cephalodia of *Lobaria pulmonaria*. *New Phytol.* **69**, 595–597.

POSTGATE, J. R. (1972). The acetylene reduction test for nitrogen fixation. *In* "Methods in Microbiology" (J. R. Norris and D. W. Ribbons, eds), **6B**, 343–356. Academic Press, London and New York.

ROGERS, R. W., LANGE, R. T. and NICHOLAS, D. J. D. (1966). Nitrogen fixation by lichens of arid soil crusts. *Nature, Lond.* **209**, 96–97.

SCOTT, G. D. (1956). Further investigation of some lichens for fixation of nitrogen. *New Phytol.* **55**, 111–116.

SMITH, D. C., MUSCATINE, L. and LEWIS, D. H. (1969). Carbohydrate movement from autotrophs to heterotrophs in parasitic and mutualistic symbiosis. *Biol. Rev.* **44**, 17–90.

STEWART, W. D. P. (1967). Nitrogen turnover in marine and brackish habitats II. The use of $^{15}N_2$ in measuring nitrogen fixation in the field. *Ann. Bot.* **31**, 385–406.

STEWART, W. D. P. (1970). Algal fixation of atmospheric nitrogen. *Pl. Soil* **32**, 555–588.

WATANABE, A. and KIYOHARA, T. (1963). Symbiotic blue-green algae of lichens, liverworts and cycads. *In* "Studies in Micro-algae and Photosynthetic Bacteria" (Japanese Society of Plant Physiologists, eds), pp. 189–196. University of Tokyo Press, Tokyo.

19 | The Physiology of Lichen Symbiosis

D. J. HILL

Department of Extra-mural Studies, University of Bristol, England

Abstract: Symbiosis in lichens appears to be based on the movement of photosynthetic products from the alga to the fungus. The path of carbon is followed from carbon dioxide in the air to the growing points in the fungus. Emphasis is placed on the release of carbohydrate by the alga as this is the feature characteristic of the symbiosis. The methods used in the study of this movement are discussed. The metabolism of blue-green algae is different from that of green algae; blue-green phycobionts release glucose while green ones release a polyol. The release is affected by light, DCMU*, pH and external carbohydrate. The amount of carbon moving between the symbionts is substantial but it is not clear whether it is the only carbon obtained by the fungus. Current views on how the release of carbohydrate occurs include diffusion, active pump and hydrolysis of extracellular products. The fungus may possibly stimulate release by a substance or factor, altered algal membrane potentials, pH or control of cell division. A section deals with the use of electron microscopy and of growth of the symbionts in culture in studying the physiology of the symbiosis.

INTRODUCTION

One of the principal features of symbiosis is that two dissimilar organisms living juxtaposed are physiologically co-ordinated in such a way that they behave as a single biological unit. In most symbiotic systems there appears to be at least one major nutritional link between the two organisms (Scott, 1969; Smith *et al.*, 1969). For example, in legume root nodules the symbiotic bacteria fix atmospheric nitrogen which is passed to the host, the host being unable to obtain as much nitrogen from other sources. The host makes a unique structure (the nodule) and the bacterium *Rhizobium* makes specialized cells (bacteroids). The bacteroids depend on the host for a supply of carbohydrate. In ectotrophic

---

Systematics Association Special Volume No. 8, "Lichenology: Progress and Problems", edited by D. H. Brown, D. L. Hawksworth and R. H. Bailey, 1976, pp. 457–496. Academic Press, London and New York.

* DCMU = 3-(3,4-dichlorophenyl)-1,1-dimethyl urea.

mycorrhizas, the fungal partner utilizes carbohydrate obtained from the host but can supply the host with more phosphate from poor soils than can be obtained by non-mycorrhizal roots. Just as the transfer of nutrients is one of the major physiological features of other symbioses, it seems to be the basis for the symbiosis in lichens. In lichens, the most obvious nutrient transfer is the movement of carbohydrate from the alga to the fungus. The fungus in the lichen habitat is likely to be unable to obtain carbohydrate from other sources. One other important process directly related to the symbiosis in many lichens is the transfer of the nitrogen compounds to the fungus from nitrogen fixing blue-green* phycobionts. This aspect of the physiology of lichens is dealt with in Chapter 18. While nutrition of the fungus appears to be the main link between the symbionts in lichens, other physiological processes may be influencing each of the partners. Because the whole lichen behaves as a single organism, features of the growth of the two symbionts must be closely synchronized so that the partners form the organized structure of the thallus. As it appears that a high degree of specificity can exist in the association of fungi and algae in lichens, recognition of the right alga by a fungus, or vice versa, may well occur. These features, other than carbohydrate movement, could be based on the interchange of unknown substances (activating genes or affecting existing physiological processes) and vitamins or even immunological reactions.

The main objective of this paper is first to evaluate what is known about the transfer of carbohydrate from the alga to the fungus and secondly to bring together current views on how the fungus may possibly cause the alga to release carbohydrate.

THE TRANSFER OF CARBOHYDRATE FROM THE ALGA TO THE FUNGUS

This process was first studied by Smith (1961), who exposed *Peltigera polydactyla* to radioactive bicarbonate. The photosynthetic incorporation of $^{14}CO_2$ led to the appearance of ^{14}C in the medulla which contained no algae, demonstrating for the first time transfer of photosynthetic products from the alga to the fungus (Fig. 1). Since then much more has been learned about the transfer. The following account will be based on the path of carbon from the air (CO_2) to a point of utilization (growth) in the fungus. This path is represented diagrammatically in Fig. 2 and the numerals for the most part refer to the sections which follow.

1. The Absorption of Carbon Dioxide
There is little information on the effect of the fungus, especially the fungal

---

* Note: the term "blue-green algae" is used rather than the more recent "blue-green bacteria".

cortex, on the absorption of carbon dioxide by the alga. When the cortex is fully saturated with a film of water over the cells rather than just less than fully saturated, the permeability of the cortex to carbon dioxide could be lowered, leading to a lower photosynthetic rate. Coupled with this there are likely to be

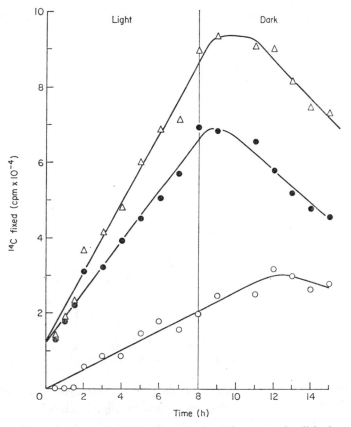

Fig. 1. Distribution of ^{14}C fixed in photosynthesis between the "algal zone" and medulla of *Peltigera*. △, whole thallus; ●, algal layer; ○, medulla. (Adapted from Smith, 1961.)

changes in the light transmission of the cortex (Ertl, 1951) which may increase the light available to the alga. In some lichens the presence of cyphellae or pseudocyphellae may function in the facilitation of gaseous exchange by reducing the diffusion resistance of the cortex. Carbon dioxide generated endogenously by fungal respiration would also be available for photosynthesis.

2. The Carboxylation Reaction

In *Xanthoria aureola* (algae = green) Bednar and Smith (1966) found that, after 30 s photosynthesis, phosphoglyceric acid (with hexose monophosphates) was more heavily labelled than other substances. The proportion of label then declined rapidly as other compounds, especially mannitol, became labelled. This pattern of ^{14}C incorporation is consistent with the view that ribulose

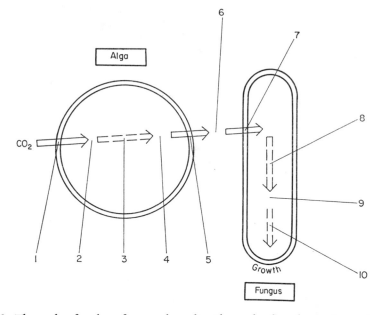

FIG. 2. The path of carbon from carbon dioxide to the fungal growing points. The numerals are referred to in the text by the marginal headings.

diphosphate is the acceptor of carbon dioxide in *Trebouxia*. In *P. polydactyla* (algae = blue-green), Drew (1966) found similarly high levels of ^{14}C in phosphoglyceric acid after 30 s, indicating the operation of the same carboxylation reaction, as did Feige (1969) in *Cora pavonia* (algae = blue-green). Work on other free-living species suggests that this mechanism is usual in blue-green algae (Pelroy and Bassham, 1972).

3. Carbon Metabolism within the Algal Cell

For the purpose of this section, the blue-green and green algae will be discussed separately.

(a) Blue-green phycobionts. There is little information about metabolism in the blue-green algal symbiont in lichens. Information about metabolism in

free-living blue-green algae (Carr and Whitton, 1973) is of interest here as it seems directly related to what we know about carbohydrate movement (Fig. 3). In the non-symbiotic species *Anacystis nidulans*, carbohydrate is stored mainly as glycogen. In the dark glycogen is broken down to glucose-6-phosphate and oxidized by the pentosephosphate cycle. Glycolysis is of limited importance in carbohydrate breakdown and the tricarboxylic acid cycle cannot oxidize

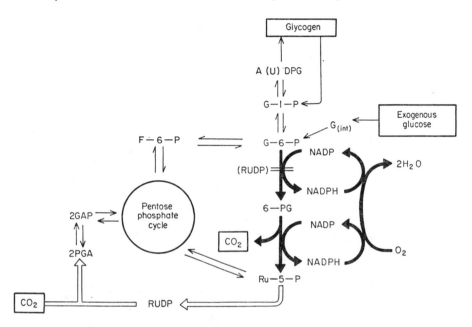

FIG. 3. A simplified metabolic map to show the pathways of primary carbon metabolism in unicellular blue-green algae. Substrates and end-products are enclosed in square boxes. Reactions specific to photosynthetic CO_2 assimilation are indicated by heavy white arrows; reactions specific to respiratory metabolism, by heavy black arrows. The site of allosteric control by ribulose-1,5-diphosphate is indicated by a double line across the arrow connecting glucose-6-phosphate and 6-phosphogluconate (Adapted from Stanier in Carr and Whitton, 1973.).

Abbreviations: A(U)DP-glucose, A(U)DPG; glucose-1-phosphate, G-1-P; glucose-6-phosphate, G-6-P; intracellular glucose pool, G(int); fructose-6-phosphate, F-6-P; 6-phosphogluconate, 6-PG; ribulose-5-phosphate, Ru-5-P; ribulose-1,5-diphosphate, RuDP; 3-phosphoglycerate, PGA; glyceraldehyde-3-phosphate, GAP.

pyruvate because α-ketoglutarate dehydrogenase is absent. Glucose breakdown, primarily by the pentose phosphate cycle, is reduced in the light owing to allosteric inhibition of glucose-6-phosphate dehydrogenase by ribulose diphosphate.

By exposing *P. polydactyla* to $^{14}C_2$, Hill (1972) showed that ^{14}C in an insoluble substance, a glucose polymer, was formed and broken down during the process of [^{14}C]-carbohydrate transfer to the fungus (Fig. 4). It was shown later (Peveling and Hill, 1974) by high-resolution autoradiography that the glucose polymer

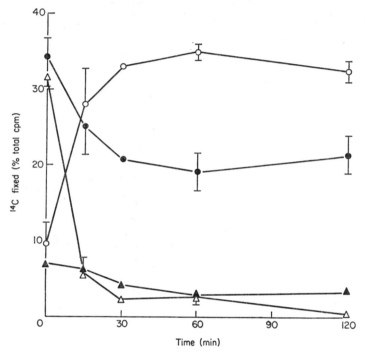

Fig. 4. The redistribution of incorporated ^{14}C during a "chase" period after a pulse of $H^{14}CO_3^-$ in the light. The vertical lines represent the range of duplicate values. ○, mannitol (in the fungus) ($\times \frac{1}{2}$); ▲, glucose; ●, 80% ethanol insoluble fraction (glycogen in alga); △, unknown substances (including sugar phosphates). (From Hill, 1972.)

was located exclusively within the algae (Plate I). Unpublished data have shown that as this substance can be hydrolysed by α- and β-amylase it is therefore chemically similar to glycogen. The pattern of labelling in the glucan can be revealed during hydrolysis with β-amylase (which removes successive maltose units from the non-reducing end) and the release of [^{14}C]-glucose by this enzyme indicated that terminal residues were labelled with ^{14}C immediately after a pulse of $^{14}CO_2$ (Table I). After a chase period with $^{12}CO_2$ (during which half the photosynthate had moved to the fungus), only internal residues were apparently labelled. As the total amount of label in the glucan has decreased during the chase period (Fig. 4), it is assumed that this pattern of labelling in the glucan resulted from breakdown as well as synthesis in the chase period. Cell-free extracts of the lichen can apparently hydrolyse the [^{14}C]-glucan from the

TABLE I. Release of ^{14}C from [^{14}C]-insoluble fraction by β-amylase

Length of enzyme incubation (min)	^{14}C released into solution (% of initial c.p.m. in suspension)	
	after 10 min "pulse" of H^{14}CO$_3^-$ only	after "pulse" then 60 min "chase"
0	9·8	10·7
5	40·0	11·1
10	46·1	22·4

Discs cut from the thalli of *P. polydactyla* were exposed to ^{14}CO$_2$ for 10 min at 10 000 lx at 20°C (pulse) and then extracted with hot 80% ethanol and water. Others were given a pulse and then placed in buffer (0·02 M MES pH 6·5) alone for 60 min (chase) before extraction. After extraction, the discs were ground to a fine suspension with a glass homogenizer.

lichen yielding some [^{14}C]-glucose (Table II). Hill (1970) found that the effect of pH was to increase the amount of ^{14}C in the insoluble fraction (glucan) at lower pH values and this was associated with a decrease in the amount of [^{14}C]-glucose released (see p. 477). These data suggested that there may be a relationship between glucose release and glucan formation. Indeed, the amount of [^{14}C]-mannitol accumulated in the fungus during the transfer of carbohydrate could only be accounted for if it was assumed that ^{14}C in the glucan contributed to ^{14}C in mannitol (Hill, 1972). The [^{14}C]-glucan must have been converted into glucose (or more likely glucose-6-phosphate) before being released by the alga. Although it seems very probable that the α-1-4-glucan is in some way involved in the carbohydrate movement in *P. polydactyla* it can be seen that its precise rôle is not clear. It is also probable that this glucan is the same as that observed by Boissière (1972) who demonstrated the presence of polyglucoside granules in symbiotic as well as free-living *Nostoc*, a finding which contrasts with a previous observation by Peat (1968) who found α-granules (= polyglucoside, see Lang, 1968) absent in *P. polydactyla*. The rôle of glycogen in blue-green algae may be similar to that of starch in green algae (see below).

(*b*) *Green phycobionts*. In green algae the compartmentalization of the chloroplast adds further complexities to the difficult problem of understanding the metabolism of the alga in the lichen. Photosynthetic carbon dioxide fixation is carried out in the chloroplast, but apart from that fact little is known of what happens in the chloroplast as opposed to the cytoplasm, although we know from electron microscope studies that starch is present only in the chloroplast. However, the starch content varies considerably and although it has been suggested that

TABLE II. Effect of cell-free lichen homogenate on release of ^{14}C from [^{14}C]-insoluble fraction

Reaction mixture	^{14}C solubilized (% of total c.p.m.)	[^{14}C]-glucose formed (% of total c.p.m.)
homogenate+insoluble fraction from sample after "pulse"	66	8
homogenate+insoluble fraction from sample after "pulse" then "chase"	15	0
boiled homogenate+insoluble fraction from sample after "pulse"	8	0

[^{14}C]-insoluble fraction was prepared as in Table I.

starch is only present in moist lichen thalli and absent from dry ones (Brown and Wilson, 1968; Jacobs and Ahmadjian, 1971), other evidence is contrary to this finding (Peveling, 1970; Harris and Kershaw, 1971). Clearly more evidence is required before the rôle of starch is fully understood in the symbiont although it may well be a means of storing excess carbohydrate that is not transported to the fungus (Harris and Kershaw, 1971). Lipid droplets are a well-known feature of the phycobiont chloroplast but nothing is known of lipid metabolism in lichens. Of direct interest in the study of carbohydrate movement is the problem of how the carbohydrate moves out of the chloroplast. In higher plants carbon probably leaves the chloroplast as dihydroxyacetone phosphate (Heber, 1974). Sucrose is probably synthesized in the cytoplasm rather than the chloroplast (Bird *et al.*, 1974) although in *Acetabularia* sucrose synthesis may occur inside the chloroplast (Sheppard and Bidwell, 1973). In higher plants, the inner membrane of the double chloroplast envelope appears to be the site of selective permeability control and transport (Heldt and Sauer, 1971). The permeability characteristics of the membrane of lichen algae are not known and may differ from those in higher plants. There is the possibility that the fungus may be able to affect algal metabolism at any point in its physiology and the chloroplast envelope could be one such target. The permeability of the chloroplast envelope in the alga within the lichen may be very difficult to examine and this is an example of the problems facing investigators of the physiology of the lichen symbiosis. Nevertheless, before we can understand the rôle of the fungus effectively we must know more about the metabolism of the alga in the lichen.

4. Precursor of the Released Carbohydrate

The form of the carbohydrate immediately before it crosses the plasmalemma is uncertain. In *P. polydactyla* glucose may not be an abundant intermediate in the alga as it is only present in trace amounts in the thallus (Hill, 1972). In *Coccocarpia* sp. (alga *Scytonema*), however, the thallus contained measurable amounts of glucose which could have been in the alga (Hill, 1970). It is possible that the glucose may be phosphorylated until it is released from the cell. In *Trebouxia* there is evidence that there is a substantial intracellular pool of ribitol (Richardson and Smith, 1968b; Green, 1970). It is clearly possible in these last two cases that the carbohydrate may be accumulated in the alga before it leaves the cell.

5. Carbohydrate Release and its Mechanism

(a) *The type of carbohydrate released.* Table III summarizes information available about the identity of the carbohydrate which is transferred from alga to fungus

TABLE III. The transferred carbohydrate of lichens (and phycobionts) which have been investigated

Alga	Lichen	Carbohydrate transferred between the symbionts		Reference
GREEN ALGAE				
Trebouxia	*Cladonia cristatella*	ribitol	$\begin{array}{c} H \\ HCOH \end{array}$	2, 8
	Hypogymnia physodes	ribitol	HCOH	7
	Lecanora conizaeoides	ribitol		4
	Lepraria chlorina	ribitol	HCOH	4
	L. incana	ribitol	HCOH	4
	Parmelia saxatilis	ribitol	HCOH	4
	Pseudevernia furfuracea	ribitol	H	4
	Ramalina siliquosa var. crassa	ribitol		6
	R. subbreviuscula	ribitol		6
	Sphaerophorus globosus	ribitol		3
	Umbilicaria pustulata	ribitol		4
	Xanthoria aureola	ribitol		9
Myrmecia	*Dermatocarpon hepaticum*	ribitol		4
	Lobaria amplissima	ribitol		5
	L. laetevirens	ribitol		4, 5
	L. pulmonaria	ribitol		4, 5

TABLE III.—continued

Alga	Lichen	Carbohydrate transferred between the symbionts	Reference
Coccomyxa	Botrydina vulgaris	ribitol	2
	Peltigera aphthosa	ribitol	1, 4
	Solorina saccata	ribitol	4
Hyalococcus	Dermatocarpon fluviatile	sorbitol	4
	D. miniatum	sorbitol	1, 2, 4
Stichococcus	Endocarpon pusillum	sorbitol	2
	Staurothele clopima	sorbitol	2
Trentepholia	Dirina stenhammarii	erythritol	4
	Gyalecta jenensis	erythritol	4
	Porina lectissima	erythritol	3
	Roccella fuciformis	erythritol	4
	R. montagnei	erythritol	3
	R. phycopsis	erythritol	4

BLUE–GREEN ALGAE

Alga	Lichen	Carbohydrate transferred between the symbionts	Reference
Nostoc	Collema auriculatum	glucose	4
	Leptogium sp.	glucose	4
	Lobaria amplissima (cephalodia)	glucose	5
	L. scrobiculata	glucose	4, 5
	Peltigera canina	glucose	1, 4
	P. horizontalis	glucose	4
	P. polydactyla	glucose	9
	Sticta fuliginosa	glucose	4, 5
	Sticta sp.	glucose	4
Calothrix	Lichina pygmaea	glucose (? and glucan)	4
Scytonema	Coccocarpia sp.	glucose	3, 4

Carbohydrate structures (Fischer projections shown alongside):

ribitol / sorbitol:
```
    H
  HCOH
  HOCH
  HCOH
  HCOH
  HCOH
   H
```

erythritol:
```
    H
  HCOH
  HCOH
  HCOH
  HCOH
   H
```

glucose:
```
  HCO
  HOCH
  HCOH
  HCOH
  HCOH
  HCOH
   H
```

References: 1, Green (1970); 2, Hill and Ahmadjian (1972); 3, Hill and Smith (1972); 4, Richardson, Hill and Smith (1968); 5, Richardson, Smith and Lewis (1967); 6, Komiya and Shibata (1971); 7, Farrar (1973b); 8, Maruo et al. (1965); and 9, references too numerous to mention individually (see text).

Notes: (a) This table does not imply a critical assessment of the experimental evidence.

(b) The algal symbiont are according to Ahmadjian (1967).

(c) Hill (1970) reported sorbitol movement in Verrucaria hydrela and suggested it in a Verrucaria sp. and in Polyblastia henscheliana. The identity of the algal symbionts is not known but it may possibly be Heterococcus (Xanthophyceae) in V. hydrela.

in the lichens (or phycobionts) which have been examined. The following conclusions may be drawn: (i) the alga releases a simple low molecular weight carbohydrate; (ii) the type of carbohydrate depends on the kind of alga; (iii) blue-green algae release glucose and green algae a polyol; (iv) the last three carbon atoms of all the carbohydrates are identical.

The evidence for the identification of the carbohydrate released by the algae was based on a number of techniques (especially the "inhibition" technique and isolation of the phycobiont from thallus homogenates—see below) but in some cases the evidence is stronger than in others. It will be interesting to know whether the pattern of Table III would change if more lichens (especially those with other genera of algae) were examined in detail.

What conclusions and inferences can we draw from these facts? The fact that a single carbohydrate is released suggests the existence of a single definite mechanism for the release of carbohydrate rather than a general release of soluble substances that results, for example, from general cell lysis. The fact that the nature of the carbohydrates released depends on the alga complicates the issue of finding a single mechanism for all algae, but because there is a similarity in the structure of the carbohydrates it is possible that there may be a similarity in the mechanisms involved. The reason for green algae all releasing a polyol is not clear as polyols do not, as far as is known at present, seem to have any particular rôle in the metabolism of microalgae. It is possible that they are concerned, in the free-living forms of lichen algae, with an aspect apparently unrelated to carbohydrate movement such as desiccation resistance. That the blue-green algae release glucose is less enigmatic, as glucose when phosphory-lated has a well known rôle in the metabolism of cells.

(b) *Studying the release mechanism.* The problems of studying the release mechanism of carbohydrate by the alga partly stem from the difficulty in designing experiments which do not affect the release process. To exemplify this statement let us examine the two major lines of experimental approach.

(i) *The so-called "inhibition" technique; allusion, elusion and illusion.* In their classic experiments on *P. polydactyla* Drew and Smith (1967b) found that when they incubated thalli in a glucose solution containing $H^{14}CO_3^-$ in the light, transfer of ^{14}C from the alga to the fungal medulla did not take place. During the experiment, the solution acquired acid-stable radioactivity and the amount could be accounted for by a decline in the amount of ^{14}C in the lichen. The acid stable ^{14}C in the glucose solution after incubation was found, by chromatography, to be exclusively [^{14}C]-glucose. When Drew and Smith (1967a) found that freshly isolated algal cells from the lichen released glucose, it became clear that with the intact lichen glucose in the medium was probably interfering with the movement of [^{14}C]-glucose in the intracellular space between the alga and the fungus, resulting in the diffusion of the [^{14}C]-glucose into the medium. The scheme in Fig. 5 was developed by Richardson *et al.* (1968) to explain the results of this

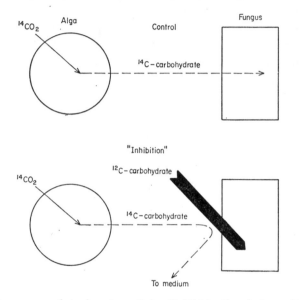

FIG. 5. The first proposed explanation of the "inhibition" technique. (Adapted from Richardson *et al.*, 1968.)

type of experiment. The [^{12}C]-carbohydrate in the medium is envisaged as successfully competing against the [^{14}C]-carbohydrate (released by the alga) for uptake by the fungus. As a consequence the [^{14}C]-carbohydrate has to diffuse into the medium. The scheme results from an allusion to competitive inhibition of enzymes and uptake processes (e.g. 2-deoxy-glucose inhibition of glucose uptake). Drew and Smith (1967b) studied the effects of other carbohydrates (fructose, galactose, mannose and mannitol) on the movement of glucose but they did not inhibit the transfer of [^{14}C]-carbohydrate to the fungus. The only carbohydrates to have an effect comparable to that of glucose were the glucose analogues, 2-deoxy-glucose and 3-methyl-O-glucose. In order to find out if the carbohydrate in their medium was inhibiting the uptake of [^{14}C]-glucose by the fungus, Drew and Smith incubated the lichen in a low concentration of [^{14}C]-glucose and observed the effect of higher concentrations (0·1 M) of other carbohydrates on the uptake of the [^{14}C]-glucose. The effect was broadly similar to the effect of the carbohydrates on [^{14}C]-glucose transfer during inhibition experiments. Assuming that all uptake was by the fungus, this indicated that the carbohydrate in the medium was indeed preventing [^{14}C]-glucose uptake during carbohydrate transfer.

The fact that the [^{12}C]-carbohydrate in the medium has to be the same as or similar to the one moving between the symbionts in order to cause the appearance of [^{14}C]-carbohydrate in the medium, has proved useful in identifying the carbohydrate moving in other lichens. For example, Richardson and Smith (1968a) used this fact in building the evidence that ribitol moves in *X. aureola* (Table IV) and Richardson *et al.* (1968) also used it to show that sorbitol moved in *Dermatocarpon miniatum* (Table V).

TABLE IV. Effect of carbohydrates on the release of ^{14}C from *Xanthoria aureola* (from Richardson and Smith, 1968a)

Carbohydrate in medium	^{14}C released into medium (% of total fixed)
ribitol	23·6
arabitol	20·2
xylitol	10·5
mannitol	3·6
glucose	1·8
trehalose	1·7
sucrose	1·4
ribulose	0·7
control (water)	0·5

TABLE V. Effect of carbohydrates on the release of ^{14}C from *Dermatocarpon miniatum* (from Richardson *et al.*, 1968)

Carbohydrate in medium	^{14}C released into medium (% of total fixed)
sorbitol	56·6[a]
arabitol	27·2[a]
mannitol	7·3[a]
glucose	5·4
ribitol	1·9
erythritol	1·3
control (water)	1·1

[a] Radioactivity shown to be as [^{14}C]-sorbitol by chromatography.

Q

Thus there is an experimental method which appears to prevent labelled carbohydrate from entering the fungus, allowing it to be released by the alga directly into the medium. Therefore, it should be possible to observe the release of carbohydrate by the alga without disrupting the thallus. There are, however, some problems which make the technique somewhat elusive. The presence of

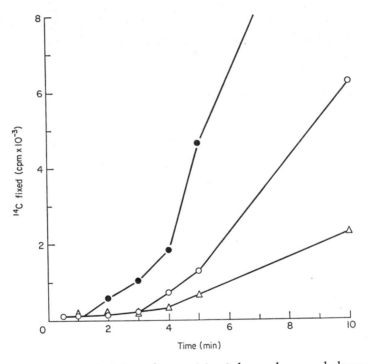

FIG. 6. The appearance of ^{14}C in media containing 2-deoxy-glucose and glucose and of ^{14}C in mannitol (water medium). ●, mannitol ($\times \frac{1}{2}$); ○, 2-deoxy-glucose medium; △, glucose medium. (Adapted from Drew and Smith, 1967b.)

a high concentration of the "inhibiting" carbohydrate in the medium may well affect the kinetics of the release by the alga. Indeed, there is evidence that this is the case. In *P. polydactyla* glucose causes substantially slower release of ^{14}C]-glucose into the medium compared with the rapid formation of ^{14}C]-mannitol in the fungus in the absence of glucose. This was shown by Drew and Smith (1967b) (Fig. 6) who measured the initial rate of release ^{14}C]-glucose into the medium in the presence of glucose and found that it was slower than the rate of ^{14}C]-mannitol formation in the absence of glucose. Similar evidence was reported by Hill and Smith (1972) who found that the rate of release, calculated graphically (Fig. 7), of [¹⁴C]-glucose in the presence of glucose (after a pulse of ¹⁴CO₂) was apparently of two types. A rapid release occurred immediately after a pulse of ¹⁴CO₂ followed by a slower release. After 24 h pretreatment with glucose, all the [¹⁴C]-glucose was released at the slower rate. The rate of [¹⁴C]-mannitol formation (in the absence of glucose in the medium) was, on the other

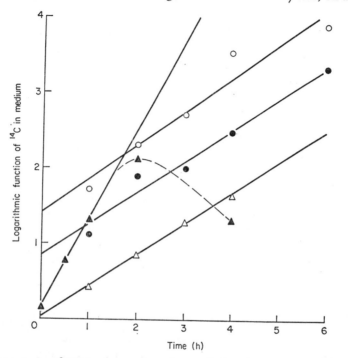

Fig. 7. Appearance of ¹⁴C in the medium during "inhibition" and in mannitol in the fungus without inhibition after a pulse of H¹⁴CO₃⁻ in the light. ○, no previous treatment with glucose; ●, 3 h pretreatment; △, 24 h pretreatment; ▲, mannitol. N.B. the *rate* of appearance of ¹⁴C is given by the gradient. (Adapted from Hill and Smith, 1972.)

hand, much more rapid than the slow glucose release. These data indicated that the initial rapid release of [¹⁴C]-glucose (without pretreatment with glucose) was probably closer to the actual release of [¹⁴C]-glucose by the alga, in the absence of glucose in the medium, and that the slower rate was a result of the effect of the glucose in the medium. Hill (unpublished) found that if the concentration of glucose in a medium containing 0·02 M 2-deoxy-glucose is raised from 0·01 M to 0·10 M and to 1·00 M the amount of [¹⁴C]-glucose released is drastically reduced (Fig. 8). If, on the other hand, the concentration of 2-deoxy-glucose in media containing 0·02 M glucose is raised similarly, the amount of [14C]-glucose released at 0·01 M and 0·10 M is not reduced and the reduction of 1·00 M is much less than that with glucose. On the basis of this experiment and of that of Drew and Smith, 2-deoxy-glucose appears to be a "better" inhibitor of [14C]-glucose movement than glucose as it affects the rate of [14C]-glucose

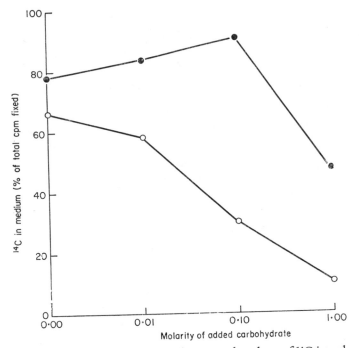

FIG. 8. The effect of glucose and 2-deoxy-glucose on the release of 14C into the medium during "inhibition" experiments. ●, release of 14C into 2-deoxy-glucose medium (containing 0·02 M glucose); ○, release of 14C into glucose medium (containing 0·02 M 2-deoxy-glucose).

release much less than does glucose. However, Komor and Tanner (1974a) have found that 2-deoxy-glucose can be taken up by *Chlorella* via a hexose transport system and that it can be phosphorylated. Clearly, 2-deoxy-glucose-6-phosphate, if it were shown to be formed in the phycobiont, may also affect the release of [14C]-glucose.

Hill and Smith (1972) found that high concentrations (above 0·1 M) of ribitol caused a reduction in 14C released by *X. aureola* suggesting again that the [14C]-ribitol release process may be affected by the presence of ribitol in the medium. Sorbitol release by *D. miniatum* was not apparently affected by high concentrations of sorbitol but, as the final amount of [14C]-sorbitol in the medium rather than the rate of ^{14}C release was measured, it is possible that changes in the rate could have been overlooked. It would be interesting to know whether analogues of polyols can inhibit fungal uptake in a manner comparable to 2-deoxy-glucose inhibition of glucose uptake.

It is felt that as more is now known about the effect of external carbohydrates on the movement of [^{14}C]-carbohydrate from the alga to the fungus, a slight modification of the original scheme proposed for the "inhibition" technique would be of value (Fig. 9) in order to avoid the illusion that the "inhibition"

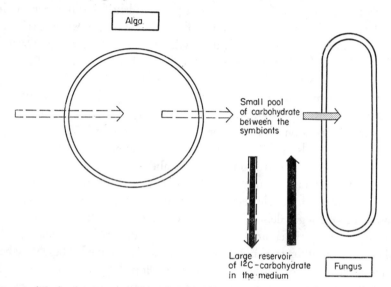

Fig. 9. Modified scheme to explain the so-called "inhibition" technique. Continuous arrows represent ^{12}C and broken arrows ^{14}C.

involves any inhibition in the biochemical sense. When the external carbohydrate in the medium is identical with, but not when it differs from, the mobile carbohydrate, a lowering of the specific activity of the [^{14}C]-carbohydrate between the symbionts occurs as the carbohydrate from the medium diffuses in and out of this space. When the amount of carbohydrate between the symbionts is raised to a sufficiently high level, the specific activity of the carbohydrates is reduced to make the amount of ^{14}C absorbed by the fungus small. Radioactive molecules then have time to diffuse out of the tissues into the medium before being taken up, giving rise to the observed release of ^{14}C into the medium with the "inhibition" technique. Unfortunately, in raising the amount of carbohydrate between the symbionts, a rise in the concentration is inevitable and may affect the release of the [^{14}C]-carbohydrate by the alga at the site of [^{14}C]-carbohydrate release through the algal membrane and/or internally after being absorbed by the alga. Further, much needed, study of the effect of exogenous carbohydrates on the release of carbohydrate by the alga may yield useful information about the release mechanism. When the inhibiting carbohydrate is not the same as that moving between the symbionts, it is assumed that fungal uptake would be competitively inhibited in a similar way to uptake by single organisms.

(ii) *Rapid isolation of algae from thallus homogenates.* Identifying the carbohydrate released by algae isolated within minutes from thallus homogenates has been invaluable in finding out the type of carbohydrate released by the alga (Drew and Smith, 1967a; Richardson and Smith, 1968b; Richardson *et al.*, 1968; Green, 1970). It must be stressed however, that, as with the "inhibition" technique, there are problems with this approach when it is used in the study of the mechanism of carbohydrate release. The most important is that release of the carbohydrate stops soon after isolation (in a matter of hours). The physiological state of the algal cells probably changes rapidly immediately they are isolated. Furthermore, Green (1970) found that the ability of the algae to release the carbohydrate was rather unreliable. He obtained conflicting results for the amount of carbohydrate released and suggested that this may have been due to the presence of fungal debris. There are other considerations as well. When collecting algae by differential centrifugation of thallus homogenates, only a small proportion of the total number of algal cells is obtained. These may be the less fragile cells and/or those less tightly enveloped by fungal hyphae. It is possible that they may represent one part of a range of states of algae in a heterogeneous population. It could be argued that isolates which did not release much carbohydrate might have been composed of cells which were releasing relatively little when in the thallus. Lichens differ in their ability to yield good algal isolates. Green found that *P. aphthosa* provided excellent, relatively pure suspensions of *Coccomyxa* cells and these represented a high proportion of the thallus population. With *Roccella*, on the other hand, it was not possible to obtain a suspension with more algae than fungal fragments. Examination of the thallus suggested that the algae of *P. aphthosa* are relatively loosely enveloped with fungal hyphae while the *Trentepohlia* cells in *Roccella* are extremely intricately enveloped. Ideally, in order to minimize problems of how representative the isolate is, a lichen such as *P. aphthosa*, which yields a high proportion of the cells, would seem to be the best choice for this type of work. It appears that the information available at present is only the tip of an iceberg of data awaiting collection.

(c) *Factors affecting the release of carbohydrate*

(i) *The effect of dark and DCMU.* Using the "inhibition" technique with *P. polydactyla*, Hill and Smith (1972) found that darkness reduced the release of [^{14}C]-glucose after labelling with a pulse of ^{14}C (Fig. 10). When the lichen was returned to the light, release resumed. Release of [^{14}C]-ribitol by the alga in *X*.

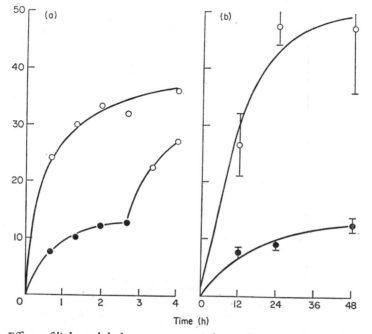

FIG. 10. Effects of light and dark on percentage release of ^{14}C into the medium during "inhibition" after pulse of $H^{14}CO_3^-$ in the light. (a) *Peltigera polydactyla*; (b) *Xanthoria aureola.* ○, samples in the light; ●, samples in dark. (Adapted from Hill and Smith, 1972.)

aureola was also reduced in the dark. The interpretation of the effect of dark is not obvious. One explanation is that the cessation of photosynthetic incorporation of carbon dioxide could stop the "chasing" out of the label. However, as the algae in *X. aureola* contain relatively large amounts of ribitol (Green, 1970), one would have expected release to continue at its expense during the dark period. Green has pointed out, however, that there may be more than one pool of ribitol in the alga. An alternative explanation is that light energy is required to supply ATP (from photophosphorylation) so that carbohydrate is, in some way, actively secreted from the cells. To test this possibility the effect of DCMU, which stops CO_2 fixation but apparently not ATP formation, was studied (C. Swindlehurst, B. Jordan and D. J. Hill, unpublished). The effect of

DCMU was similar to that of darkness (Figs 11 and 12) in both *Peltigera polydactyla* and *Parmelia saxatilis*. Indeed, autoradiographs of chromatograms of the labelled intermediates after treatment of *Peltigera polydactyla* with DCMU apparently show it to inhibit non-cyclic photophosphorylation and NADPH formation but not cyclic photophosphorylation. Cyclic ATP-synthesis maybe

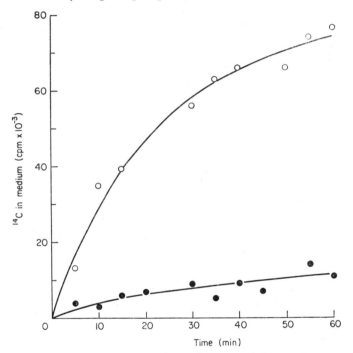

Fig. 11. The effect of DCMU on the release of ^{14}C into 2-deoxy-glucose medium by *Peltigera polydactyla*. ●, + DCMU; ○, − DCMU. (C. Swindlehurst and D. J. Hill, unpublished.)

ubiquitous in photosynthetic organisms but there is no direct evidence for its occurrence in lichen algae. However, there is evidence that it occurs in *Anacystis* (Bornefeld and Simonis, 1974) and *Chlorella* (Klob *et al.*, 1973), where it may drive glucose uptake, and these free-living algae might perhaps be considered as physiological relatives of the blue-green and green lichen algae, respectively.

(ii) *The effect of pH.* Green (1970) studied the effect of pH on ribitol release by *Trebouxia* cells isolated from *X. aureola* and glucose release by *Nostoc* cells from *Peltigera canina*. Ribitol release by *Trebouxia* was greatest at low pH (4·0) but there was no optimum value (see p. 487). Glucose released by *Nostoc* was optimal at pH 6 but rose again at pH 8 and pH 9. Using the intact lichen and the "inhibition" technique, Hill and Smith (1972) found that [14C]-ribitol release by *Trebouxia* in *X. aureola* increased slightly but not significantly at low pH (pH 5) while [14C]-glucose release by *Nostoc* in *P. polydactyla* fell below pH 5; there was no optimum in either case. At present it is difficult to draw any

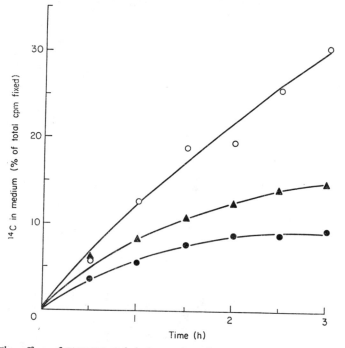

FIG. 12. The effect of DCMU and dark on the release of 14C into ribitol medium by *Parmelia saxatilis*, ●, +DCMU (light); ▲, −DCMU (dark); ○, −DCMU (light). (B. Jordan and D. J. Hill, unpublished.)

definite conclusions from these data, although glucose release seemed to be greatest under neutral and alkaline conditions whilst ribitol release was greatest at acid ones. In conjunction with the effect of pH on the release of [14C]-glucose by *P. polydactyla*, Hill (1970) found that there was an increase in the amount of 14C in the insoluble fraction when the pH was reduced (Fig. 13). It was possible that the reduction of glucose release at lower pH was associated with an increase in the insoluble compound (see p. 462).

6. The Carbohydrate Between the Symbionts

The mobile carbohydrate, after being released by the alga, presumably exists as a solution between the symbionts and unbounded by membranes before it is taken up by the fungus. This presumption is based on structure of the lichen and the fact that the "inhibition" technique works. As the carbohydrate is unbounded by membranes, why are there no epiphytes (e.g. bacteria) taking advantage of this free supply of carbohydrate? The gap between the symbionts could be readily invaded by micro-organisms occurring in or on the thallus. Indeed,

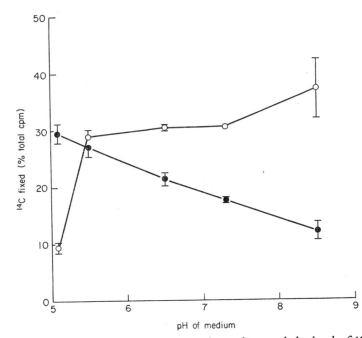

FIG. 13. Effect of pH on the release of ^{14}C into the medium and the level of ^{14}C in the insoluble fraction during "inhibition" experiment. ○, release of ^{14}C into medium; ●, ^{14}C in insoluble fraction. Samples exposed to pulse of [14]CO$_2$ before placing in media of different pHs. (Adapted from Hill, 1970.)

Jacobs and Ahmadjian (1971) located bacteria in the fibrillar sheath of the fungal hyphae in the medulla but no bacteria have ever been found next to the alga (as far as can be ascertained from published reports). Using large volumes of medium, Hill (1972) did not find that much glucose would be leached out of the thallus of *P. polydactyla* suggesting that release and uptake processes are very closely integrated. Nevertheless, at present, there do not seem to be any data available to suggest whether or not epiphytes are prevented from scavenging because they are unable to compete with the fungal uptake mechanism. A suggestion which might find acceptance is that lichen substances inhibit epiphyte growth by their antibiotic properties.

7. Fungal Uptake

There is no reason to suppose from the evidence available at present that the uptake of the carbohydrate released by the alga differs in any way from the uptake of exogenous carbohydrates by free-living fungi. Clearly the uptake mechanism is highly efficient as very little of the carbohydrate diffuses away into the environment (Hill, 1972). There are no figures on the relative efficiency of carbohydrate uptake by the fungus, by the alga which released the carbohydrate and by other organisms. It is probable that the carbohydrate is phosphorylated as soon as it enters the fungal cytoplasm.

8. Utilization of Absorbed Carbohydrate in the Fungus

In *P. polydactyla*, absorbed glucose is primarily recovered as mannitol, as has been found during the photosynthetic incorporation of $^{14}CO_2$ (Fig. 14) (Drew and Smith, 1967b), resulting in the transfer of [^{14}C]-glucose from the alga to the fungus, and during the incorporation of exogenously supplied [^{14}C]-glucose (Smith, 1963; Drew, 1966). Mannitol can be assumed to be present only in the fungus on the basis that (*a*) it is a common fungal carbohydrate, (*b*) it is not known in *Nostoc*, (*c*) it occurs in the fungal medulla (where there are no algal cells) and (*d*) during photosynthesis with $^{14}CO_2$ inhibition of fungal uptake of [^{14}C]-glucose results in no [^{14}C]-mannitol being formed.

In the absence of inhibitors, [^{14}C]-mannitol is formed within 5 min of the onset of photosynthesis with $^{14}CO_2$, indicating that it is an early product in the fungus (Drew and Smith, 1967b). Smith (1963) and Drew (1966) exposed *P. polydactyla* to uniformly labelled [^{14}C]-glucose and found that it was converted into mannitol without detectable amounts of glucose being accumulated.

When the specific activity was low (Smith, 1963) a larger proportion of glucose was converted into mannitol than when the specific activity was higher (Drew, 1966). Incubation of *P. polydactyla* in glucose solution led to an increased mannitol level in the lichen (Hill, 1970; Smith and Molesworth, 1973). There is no information on the path by which glucose is converted into mannitol. In non-lichenized fungi, mannitol may be formed either from fructose or from fructose-6-phosphate via mannitol phosphate (Lewis and Smith, 1967); but it is not known, at present, whether one of these pathways could be involved in lichens.

The respiration of the fungus cells in the algal zone, probably the most physiologically active fungal tissue, may be fed directly from the absorbed glucose. A substantial proportion (*c.* 20%) of the ^{14}C from absorbed exogenous [^{14}C]-glucose is released as carbon dioxide within 24 h (Smith, 1963). It is not known from which metabolic path this carbon dioxide is evolved. It would be very interesting to understand more about the respiratory pathways in lichen fungi but, as has been pointed out by Burnett (1968), there are great technical problems in investigating these pathways in fungi.

In lichens in which the mobile carbohydrate is ribitol, as in most species of lichens, the ribitol is converted into mannitol and arabitol with little or no

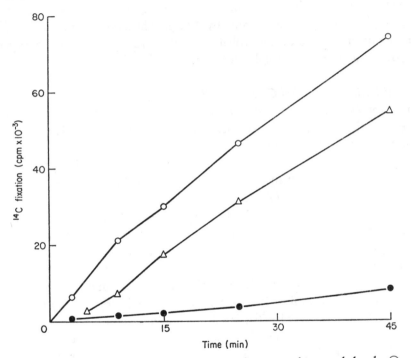

FIG. 14. Progress of ^{14}C fixation during photosynthesis in *Peltigera polydactyla*. $\bigcirc$, net light fixation into ethanol soluble fraction; $\triangle$, mannitol and $\bullet$, total dark fixation. (Adapted from Drew and Smith, 1967b.)

ribitol being accumulated in the fungus (Hill, 1970). The paths of conversion of ribitol into mannitol, arabitol, polysaccharide and carbon dioxide are a matter of speculation at present, though the enzymes of the pentose phosphate cycle would seem likely to be involved, ribitol having a C-5 skeleton. Lewis and Smith (1967) have reviewed the metabolism of polyols and there does not seem to be much in the literature since then on polyols in lichens (but see Chapter 15).

9. Storage of Carbohydrate in the Fungus

The accumulation of ^{14}C in polyols in the fungus during photosynthesis in the presence of $^{14}CO_2$ (Fig. 14) and the high level of these carbohydrates in the thallus (up to several per cent dry weight) (Lewis and Smith, 1967) leads one to suspect that they are storage products of the fungus. That the large pools of polyols are acting principally as carbohydrate reserves and not in some other capacity is not, however, so obvious. The evidence seems to be as follows. In the dark the mannitol content of *X. aureola* declined but arabitol declined faster and reached zero before a marked decline in mannitol was observed (Fig. 15)

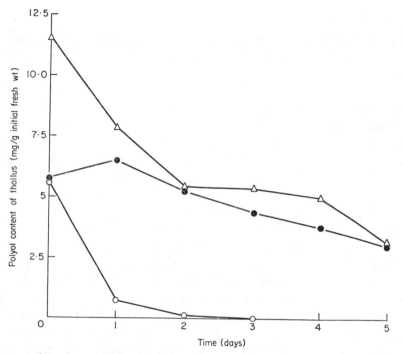

FIG. 15. Effect of starvation in the dark on the sugar alcohol content of *X. aureola*. Total sugar alcohol (△); mannitol (●) and total pentitols (○). (Adapted from Richardson and Smith, 1966.)

(Richardson and Smith, 1966). Farrar (1973b) found that wetting/drying cycles caused loss of polyols from *Hypogymnia physodes* with arabitol falling the most. In *P. polydactyla*, Smith and Molesworth (1973) found that wetting and drying cycles caused a decline in mannitol content. This could have been due to the utilization of this carbohydrate for the increased levels of respiration observed on the rewetting of the dry thalli. It would seem then that both mannitol and arabitol (when present) can act as carbohydrate reserves. Arabitol in *X. aureola* seems to act as a relatively short-term reserve compared with mannitol (see Chapter 15).

Hill (1970) reported that up to 60–70% of the absorbed ribitol (when uptake was saturated) found its way into an 80% ethanol-insoluble fraction (possibly a polysaccharide) in *X. aureola* and Smith (1963) found that 20% of the absorbed [^{14}C]-glucose was found in an insoluble glucose polymer in *P. polydactyla*. These insoluble substances may well occur in the fungus (?glycogen) but as the intact thallus was used, it is impossible to rule out their occurrence in the alga.

10. Translocation

There are two lines of approach which suggest that translocation of substances derived from photosynthesis may occur but only to a limited extent.

First, the work of Smith (1961) on the incorporation of $^{14}CO_2$ into the medulla shows that translocation of photosynthetic products has occurred in the fungus from the algal layer to the medulla. This represents vertical translocation. The time taken for the transport (less than 1 mm distance) was about 2–4 h (Fig. 1) compared with the 2–3 min required to transfer carbohydrate from the alga to the fungus as detected by mannitol formation (Fig. 14).

Secondly, an initial phase has been reported in the linear radial growth of lichens in which the rate of growth is slower than in larger thalli (Beschel, 1958; Phillips, 1963)*. The evidence, however, is not conclusive (Hale, 1974). It is possible that an increasing growth rate (radial increment) occurs in establishing thalli and this would be expected if the translocation of substances occurs from light-absorbing regions in the centre of the thallus to the growing point at the edge (e.g. 0–5 mm). However, radial growth is found to be constant in medium to large colony sizes. This suggests that areas in the centre of the thallus are no longer feeding the growth, implying no effective horizontal translocation over this distance. Clearly, experimental investigation is needed on the possibility of horizontal translocation in the thallus. It would be interesting to know whether fruticose species such as *Usnea* with elongated hyphae in the central core are more efficient at translocation than species without this type of tissue. While carbohydrate produced at the edge of foliose and crustose thalli would probably be used for radial growth, that produced in the centre may well be used for reproductive structures and for growth in thickness of the thallus.

11. The Quantity of Carbohydrate Moving Between the Symbionts

The question which must be answered is whether the carbohydrate detected by using ^{14}C methods is sufficient for all the requirements of the fungus or does the

---

* See also Chapter 12.

fungus obtain organic carbon from the alga in addition to this carbohydrate? There are observations that algal cells die and decompose in the thallus of *Lecanora muralis* (Peveling, 1968), *Squamarina crassa* and *Aspicilia* sp. (Galun *et al.*, 1970c), *Parmelia sulcata* (Webber and Webber, 1970) and *Ochrolechia tartarea* (R. Hill, unpublished). Galun *et al.* (1970a) found that in *S. crassa* the fungus penetrated senescent and decaying cells. In both *S. crassa* and *Aspicilia* sp. they found that the cell wall remnants were pushed into the upper cortex. In *O. tartarea* Hill found that the dead algal cell walls formed a layer below the active algal layer. There is therefore a suggestion that the fungus may obtain substances from "harvesting" dead algal cells from the population of algae in the thallus. However, the number of dying algal cells observed by Galun *et al.* (1970a) was small in proportion to the total number of algal cells present.

Hill and Smith (1972) found that the movement of glucose from the alga to the fungus in *P. polydactyla* was substantial and amounted to 0·7–1·1 mg per g dry weight thallus per hour. They also found that 30–70% of the total CO_2 fixed was released by the alga. Smyth (1934) obtained values of about 1·5 mg $CO_2/g/h$ for photosynthesis in *P. polydactyla* under conditions similar to those used by Hill and Smith (1972) (10 000 lx 20°C). This agrees well with the figures obtained by Hill and Smith for amounts of glucose released and the proportion of the photosynthate released by using the "inhibition" technique. There is also the question of the movement of nitrogenous compounds. Whether this release is the total amount of organic carbon is hard to say. Presumably more data on the destruction of algal cells in the older and younger parts of the thallus would be of value. If the number of algal cells in the older parts does not increase, then one may ask what happens to the carbohydrate retained by the alga if it amounts to as much as half the photosynthate. Some of it will be used by the alga for respiration, while the rest may eventually find its way to the fungus as the products of cell lysis when the algae die. Algal cells could be replaced by cell division.

THEORIES OF THE RELEASE MECHANISM AND HOW THE FUNGUS CONTROLS IT

It can be seen from the discussions above that the central feature which appears to be exclusive to the biological phenomenon of symbiosis is the release of carbohydrate by the alga; the other physiological processes are apparently closely related to those occurring in non-symbiotic organisms. For this reason it is thought important to discuss the problems of possible mechanisms of carbohydrate release by the alga, how the fungus might cause the release and whether control of the growth of the alga is important.

1. Possible Mechanisms of Carbohydrate Release by the Alga

Rather than search for the nature of the message which the fungus transmits to the alga, we can study processes in the alga to see what the fungus could be modifying.

(a) *Diffusion through the membrane.* Most carbohydrates cross membrane barriers by some kind of diffusion process. Glucose uptake by yeast and most mammalian cells depends on the phosphorylation of the absorbed glucose to create a concentration gradient and hence a flux of glucose. The glucose-6-phosphate formed (by the action of hexokinase or glucokinase) cannot diffuse out of the cell as the outer membrane is impermeable to sugar phosphates. The result is a net uptake of glucose. Glucose release into the blood by liver cells involves the action of a specific enzyme, glucose-6-phosphatase (bound on to the endoplasmic reticulum), on glucose-6-phosphate, yielding glucose which then diffuses into the blood. In micro-organisms, the uptake mechanism of glucose proposed by Roseman (Kaback, 1970) involves a complex of enzymes bound to the membrane at a specific site. The glucose is attached onto a protein, and is then phosphorylated by phosphoenolpyruvate yielding glucose-6-phosphate. As glucose-6-phosphate appears to be the first product of glucose absorption by blue-green algae and as it is a convenient end-product of the Calvin cycle, it is possible that glucose release could occur by the action of a phosphate on glucose-6-phosphate leading to the diffusion of glucose out of the cell down a concentration gradient maintained by fungal uptake. There are two possible ways in which this process of release could be controlled: by the inception of glucose-6-phosphatase activity in the algal cell or by inhibiting of glucose utilization in the alga (see p. 496).

In *Trebouxia*, ribitol exists in substantial quantities in the algal cell (Green, 1970) and so a concentration gradient, down which ribitol could move, may well occur in the lichen. Green found that ribitol release was increased at lower algal densities (Fig. 16) and in this feature it was similar to the release of extra-cellular products by phytoplankton (Watt, 1966). This suggests that ribitol release may have been inhibited by a build-up of ribitol in the medium. Green suggested that ribitol release was a feature of the membrane permeability characteristics, which did not include permeability to sucrose. At low pH, ribitol (but not sucrose) release was increased and it is possible that a reduction in pH may cause a specific permeability to ribitol. As Green was working with freshly isolated algae (which were probably undergoing rapid physiological change), there is the possibility that the isolates may not have reflected precisely the state of the algae in the lichen thallus. Follmann (1960) found that the permeability of *Trebouxia* algae to glucose was greater in algae immediately after isolation from *Cladonia furcata* than from algae grown for 1 month in culture. This indicates that the alga in the lichen may have an unselectively increased permeability to carbohydrates.

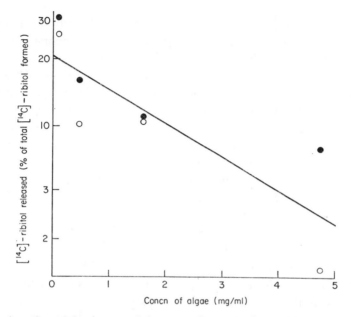

FIG. 16. The effect of the density of alga in medium on release of [¹⁴C]-ribitol of *Trebouxia*. ●, immediately after isolation; ○, 24 h after isolation. (Line is a linear regression *r* = 0·83.) (Adapted from Green, 1970.)

(*b*) *Hydrolysis of extracellular polysaccharide produced by the alga.* Drew and Smith (1967a, b) suggested that glucose released by the alga in *P. polydactyla* could have resulted from a modification of the synthesis of cell wall complex polysaccharide. Richardson *et al.* (1968) suggested a similar possibility for glucose release by *Calothrix* in *Lichina* where the glucose released was accompanied by short chain glucose polymers. In "pulse-chase" experiments, Hill (1972) showed that a substantial amount of ¹⁴C in *P. polydactyla* passed through an insoluble glucose polymer during transfer to the fungus. The pattern of labelling and the results of experiments using [¹⁴C]-glucose suggested that the insoluble glucose polymer could have been such a polysaccharide released by the alga. However, Peveling and Hill (1974) have found, using high-resolution autoradiography, that the insoluble polysaccharide was inside, rather than outside the alga (Plate I), a finding which raises doubts about the validity of the suggestion that glucose was formed as a result of the hydrolysis of a polysaccharide released by the alga. It has now been established that the polysaccharide is an α-1-4 glucan (Hill, unpublished), a known storage product of free-living blue-green algae (Carr and Whitton, 1973) (see pp. 462–3).

(c) *Active pump through the algal membrane.* There are relatively few instances in which carbohydrates are actively transported across membranes against a concentration gradient without being metabolized. Most transport mechanisms involve phosphorylation of the absorbed carbohydrate (see above). However, in inducible lactose uptake by micro-organisms and in glucose uptake by mammalian gut endothelium, active pumps do occur and in micro-organisms lactose uptake is linked to sodium transport. Clearly, one of the essential requirements for active transport is the expenditure of energy. However, at present, there is no evidence that carbohydrate release by algae in lichens requires energy, although positive evidence may be obtained. The evidence is as follows. In *P. polydactyla*, Hill and Smith (1972) found that glucose release was inhibited by darkness (Fig. 10). The requirement for light could have been due to a specific requirement of energy for the release process. In blue-green algae as in other plants, it is possible to inhibit the formation of NADPH by photosynthetic electron transport without affecting the level of ATP with the inhibitor DCMU. However, as already mentioned (pp. 475–6), DCMU inhibits the release of carbohydrate to the same extent as darkness, suggesting that ATP alone cannot drive the carbohydrate out of the cell.

2. *Possible Means by which the Fungus might control Carbohydrate Release by the Alga*

(a) *"Substance" or "factor" produced by the fungus.* Using algae freshly isolated from *X. aureola* and a cultured strain of *Trebouxia*, Green (1970) found that lichen homogenates and thallus washings did not stimulate the algal cells to release ribitol. It is of interest that factors which stimulate the release of substances by symbiotic algae have been found in animals (see Chapter 20). Green tried many other substances such as lichen acids, ATP and kinetin, without success. He attributed the finding by Follmann and Villagran (1965) that lichen acids increased cell permeability to the lowering of the pH of the medium (see p. 487). Although the search for a "factor" was fruitless, one should not rule out the possibility that a substance produced by the fungus modifies the alga in some way. All one can say is that if there is a substance which causes that alga to release carbohydrate, it is not easy to find. Hill and Ahmadjian (1972) found that when the alga and fungus of *Cladonia cristatella* were grown together in the same culture, the pattern of carbon incorporation in the alga was much more in favour of ribitol synthesis than when the alga was in liquid culture; but the difference could be attributed to the contrasting cultural conditions rather than to the existence of a "factor" produced by the fungus (see p. 491).

(b) *The reversal of membrane potentials.* It may be possible that the exit of carbohydrate from the algal cell can be driven by the entry of protons into the cell resulting in a reversal of the membrane potential (D. C. Smith, 1974). Komor and Tanner (1974b), working with *Chlorella*, have found that the efflux of 6-deoxy-glucose from the cells after they were transferred from a 6-deoxy-glucose solution was associated with a proton movement and that the efflux could be used to drive glucose uptake. The efflux of 6-deoxy-glucose was down a concentration gradient and it is possible that the release of carbohydrate by the alga in lichens may occur down such a gradient (see p. 484).

(c) *pH.* Green (1970), exposed *Trebouxia* cells isolated from X. *aureola* to $^{14}CO_2$ and found that the proportion of the $[^{14}C]$-ribitol which was released was greatly increased by reducing the pH below 5. At pH 4, 67% of the $[^{14}C]$-ribitol was released in 3 h from the freshly isolated alga; after the alga had been cultured for 24 h, 43% of the ribitol was released at pH 4·2, although at pH 6 only 6–9% of the ribitol was released by these alga. From these experiments there appeared to be enhanced release at low pH but there is the possibility that pH also affected the rate of CO_2 fixation, making detailed interpretation difficult. Could the fungal hyphae maintain the pH of the environment of the algal cells at around 4 or less (e.g. by lichen acids)? It would be interesting to know whether or not membrane potentials could be reversed at this pH. It is, however, hard to imagine that low pH could be of overriding importance as there are many lichens which are endolithic in calcium carbonate rocks.

(d) *Control of algal cell division.* When the alga is functioning as a carbohydrate donor, the endogenous carbohydrate requirement of the algal cell itself might have to be reduced so that extra carbohydrate is available for the fungus. If one assumes that photosynthesis is not working at a grossly sub-maximal rate, it seems to follow that, if the sink for release is greater than that for endogenous utilization, algal growth could be restricted by carbohydrate starvation. It would also follow that reduction in algal growth may be a requirement for the release of carbohydrate, at least in the substantial amounts observed.

Drew and Smith (1967a) and Green (1970) both showed how the symbiotic alga, once it is isolated from the fungus, stops releasing carbohydrate and uses nearly all the carbon from photosynthesis internally. In the alga the carbon was diverted from soluble carbohydrate to 80% ethanol insoluble substances (e.g. proteins, polysaccharides and nucleic acids). This change in metabolism, resulting from releasing the alga from the fungus, could be due to the use of all the photosynthate for algal growth. How could the fungus control the growth of the alga? Is it conceivable that the fungus envelops the algal cells so tightly that they cannot divide? Could the fungus deprive the alga of nutrient essential for cell growth? Are there specific substances ("factors") that inhibit cell division? The stage in the cell cycle at which the fungus would be affecting algal cell growth could be an important consideration since the pattern of utilization of photosynthetic products varied during the cell cycle (Kanazawa *et al.*, 1970). In the stages of cell growth before nucleic acid synthesis (which immediately precedes nuclear division) there is a peak in the synthesis of carbohydrate from CO_2. Perhaps halting the cell cycle at this stage could lead to the formation of carbohydrate in excess of the endogenous needs of the algal cell (see also p. 490).

PHYSIOLOGY OF THE LICHEN SYMBIOSIS FROM THE POINT OF VIEW OF FINE STRUCTURE AND THE CULTURING OF THE SYMBIONTS

1. Structural Studies

In the last 10 years the use of the electron microscope has opened up a new avenue in the study of lichens. The following discussion is aimed at selecting some of the information that has accumulated to see whether anything can be learned which is of interest to the study of the physiology of the lichen symbiosis.

Haustoria are a classical feature of biotrophic plant pathogens. They are functionally specialized fungal cells which enter the host cell wall and apparently absorb nutrients from the host. Usually they do not break the host plasmalemma. Bracker and Littlefield (1973) have written a detailed review discussing and classifying the ultrastructure of the contact between pathogens (and symbionts) and their hosts. The presence of haustoria in lichens is obviously of interest to lichenologists. However, because haustoria are not a common feature of lichens, these structures cannot be considered an essential feature for the lichen symbiosis in general. Nevertheless it may be worthwhile to consider their occurrence briefly.

Plessl (1963) found with the light microscope that primitive members of the Lecanoraceae had intracellular haustoria and advanced members had haustoria penetrating the cell wall only. Using electron microscopy, Galun *et al.* (1970a) confirmed Plessl's findings. They made an additional observation that in the advanced members haustoria did penetrate decaying cells (see section 11). Their observations were extended to the Lecideaceae (Galun *et al.*, 1971) where they found a similar situation. It would appear that the advanced lichens, such as those used in most physiological experiments, may have evolved from the ones with a host–parasite relationship like the biotrophic fungi of higher plants. Therefore, the advanced lichen symbiosis may have arisen from biotrophic parasitism rather than directly from a synoecious association.

Ben-Shaul *et al.* (1969) and Galun *et al.* (1970b) found that in *Caloplaca aurantia* and in *Lecanora radiosa* (respectively) the extent of fungal penetration was related to environmental conditions of the lichen. In xeric conditions, haustoria penetrated the algal cell wall and in *L. radiosa* were fully intracellular. In more moderate conditions the fungus did not form haustoria. The penetration of haustoria has been ascribed by Jacobs and Ahmadjian (1971) to "an enzymic process" to account for the way in which the algal cell wall appears to dissolve during the penetration by the fungus. Is it conceivable that, if enzymes are released by the fungus at the site of contact and if the lichen is under xeric conditions the enzyme would remain concentrated at the point of contact leading to a localized penetration of the algal cell wall? Under moister conditions, the enzyme might well diffuse over a wider area of the algal cell surface and not lead to the weakening of the cell wall at any particular point. This tentative discussion is aimed at probing the possible ways in which desiccation, a frequent and possibly essential feature of lichens in their natural habitat (Farrar, 1973a), may be related to the symbiosis. The application of electron microscopy in the physiology of the symbiosis may be of great value. For example, there seem to be very few histochemical studies (see p. 496 for acid phosphatase). High activities of this enzyme (along with many others) are associated with breakdown of lysosomes when the cell begins to lyse. In orchid mycorrhizas, peloton "digestion" is associated with high activities of acid phosphatase, and "digestion" may be considered a type of lysis (Williamson, 1973).

Peveling (1969) has found, in the algal layer of lichens, invaginations or peg-like intrusions in the fungal plasmalemma. Similar structures have been found in the fungus of an endotrophic mycorrhizal fungus (Hadley *et al.*, 1971) and similar structures are characteristic of "transfer cells" in higher plants (Pate and Gunning, 1972). Peveling has also observed membrane bodies between the two symbionts (see Chapter 2). It is possible that these invaginations increase the surface area of the fungal plasmalemma thus aiding a more efficient uptake of carbohydrate from the alga.

2. Culture Studies

One of the most frequent questions asked by non-lichenologists about lichens is whether the symbionts can be grown in culture and then recombined to form the lichen thallus anew. This aspect of lichenology has occupied many researchers over the last century since Schwendener proposed the dual nature of lichens. Because the physiology of the symbionts in the lichen is to a large extent concerned with growth and development the results of this research are particularly relevant to the subject of this chapter.

(a) *The growth of the symbiont in culture.* There are two major features which the symbionts in pure culture share: firstly their slow growth and secondly the lack of any cellular organization that gives a clue as to the control of morphogenesis in the lichen. As carbohydrate release by the alga is a central issue in this paper the following discussion is restricted to the physiology of the alga in culture.

Fox (1967) found that *Trebouxia* could fix $^{14}CO_2$ almost as rapidly as a strain of free-living *Chlorella*, and the difference was attributed to different chlorophyll contents. However, *Trebouxia* grew at a very slow rate compared with *Chlorella*. Assuming that the carbohydrate production of photosynthesis was not the cause of the slow growth, perhaps the utilization of the carbon was the limiting factor. Indeed, when glucose was included in the medium as well as minerals, the rate of growth was increased many fold. Above 0·5%, glucose had a declining effect which could be reversed by the addition of casamino acids. These results indicated that slow growth on an unsupplemented mineral medium may, in part, be due to an inefficient supply of suitable carbohydrates and amino acids from photosynthesis. Here is a paradox: the alga in the lichen supplies the fungus with carbohydrate as well as itself and yet in culture it seems unable to use the expected extra carbohydrate for more rapid growth. Obviously, many more experiments are required in this field which appears to be potentially so rewarding. For example, we should know more about the metabolism of polyols in lichen algae in culture.

Hill and Ahmadjian (1972) found that all the green algae isolated from lichens which they examined were capable of forming in culture the carbohydrate which they released in the lichen. Therefore the formation of the mobile polyol by green phycobionts does not appear to be a manifestation of the symbiosis. It is more probable that the symbiotic habit in these algae results from their ability to form polyols, because polyols have mainly been reported from those genera of the Chlorophyta which occur in lichens (Lewis and Smith, 1967).

Drew and Smith (1967a) investigated the incorporation of $^{14}CO_2$ into freshly isolated *Nostoc* from *Peltigera polydactyla*, the cultured phycobiont and a free-living species of Nostoc. They found that the cultured phycobiont did not release carbohydrate (glucose), while the freshly isolated alga did. Both the freshly isolated and the cultured phycobiont incorporated less ^{14}C into insoluble substances than the free-living species, with most of the radioactivity remaining in the soluble fraction in the cultured phycobiont. Green (1970) found that the culturing of *Trebouxia* after isolation from the lichen led to a change in the pattern of utilization of photosynthetic products from the accumulation of ribitol to synthesis of insoluble substances and sucrose. Similarly, Hill and Ahmadjian (1972) also found that the cultured alga of *Cladonia cristatella* (*Trebouxia erici*) incorporated less of its photosynthetic products into ribitol than the alga in the lichen. The proportion of photosynthate that was used in ribitol synthesis could be greatly enhanced by growing the alga with the fungus on an agar medium (as opposed to mineral liquid medium). However, it was not clear whether the increased ribitol formation was due to the presence of the fungus or to the different culture technique. The understanding of how the symbiotic alga differs from the cultured phycobiont would be greatly helped if novel methods of culturing the alga were developed which would simulate to some extent the physical conditions that exist in the lichen. Indeed, some of the differences already observed may be due to the cultural conditions rather than to the symbiotic state.

It is worth mentioning another type of difference that has been noticed between the alga in the lichen and a cultured *Nostoc* species (Hill, unpublished). In carrying out the procedure of isolating the alga from *P. polydactyla* (Drew and Smith, 1967a) excessive grinding of the lichen in the pestle and mortar can lead to the destruction of many of the algal cells. Continuous grinding of cultured *Nostoc* does not yield any noticeable breakage of cells. It is a possibility that investigation of the cell wall complex of lichen algae would give rise to some interesting results.

(*b*) *Dual growth.* Many early authors, but more recently Ahmadjian (see Ahmadjian, 1967), have attempted to resynthesize lichen thalli by inoculating the two cultured symbionts on to media capable (and incapable) of supporting their independent growth. While on the whole they met with less success than one would have hoped, many of their observations have turned out to be most interesting. Most of the following account is abstracted from Ahmadjian (1967).

(1) When the fungus and the alga met, the fungus produced hyphal branches which envelop the algal cells. After several weeks of culturing, the algal cells

enveloped by hyphae remained green and "healthy" and survived much longer than cells not enveloped. The enveloped cells became smaller. These observations indicated a pattern of cellular development in the symbionts arising from the contact of the two organisms. To what extent these responses are thigmotrophic or physiological is not understood.

(2) The formation of distinct fungal tissues (an "algal layer" and "cortex"), which were not present in the pure fungal culture, occurred in the dual culture. In the longer term, it may be possible to investigate the almost axiomatic assumption that the alga has a special rôle in the morphogenesis of the lichen thallus. For a discussion of the current evidence on this aspect of the lichen symbiosis see Chapter 3.

(3) The integration of the two symbionts was favoured by a low level of nutrient and a drying out of the culture. The possibility of nutrients, especially phosphate, being involved in the symbiosis by controlling the growth of the alga has already been mentioned. Further discussion of this aspect can be found in Chapter 15. The fact that drying out favours the establishment of the symbiosis is of great interest as lichens are characteristically plants of habitats exposed to rapid changes of wetting and drying. Although there is no obvious explanation, there is the possibility that frequent drying tends to cause some substance or substances to accumulate in the thallus either from the environment or secreted from within. However, some lichens live under very constant conditions and are intolerant of drying out. Interestingly, Harris and Kershaw (1971) found that drying and wetting were not the only cyclic factor required in the environment. *Hypogymnia physodes* and *Parmelia sulcata* would only continue to grow when they were given a 12 h photoperiod, as well as watering on alternate days (r.h. 90%).

(4) The culture of the two symbionts together could give insight into the control the fungus has over algal metabolism, carbohydrate release or growth (see p. 486).

(5) The only obvious answer to the problem of the extent of specificity of the alga for the fungus, and vice versa, is to carry out synthesis experiments exploring various combinations of alga and fungus. Ahmadjian and Heikkilä (1970) tried this by swapping the phycobionts of *Endocarpon pusillum* and *Staurothele clopima* but thalli failed to develop although the algae were both species of *Stichococcus* and the fungi were closely related. This indicated that fairly narrow specificity is possible. Ahmadjian (1967) suggested that there was a wide specificity involved for species of *Trebouxia* and *Myrmecia biatorellae*, as dissimilar fungi were symbiotic with very similar strains of *Trebouxia* as judged by the cultural characteristics of the alga. Wang-Yang (Ahmadjian, 1970) reported

that many distinct strains of *Trebouxia* were isolated from *Cladonia rangiferina* and *Parmelia caperata* collected from different parts of the world.

Endocarpon pusillum is one of the few lichens that have been successfully grown in culture from spore to spore, i.e. through the complete life cycle (Bertsch and Butin, 1967; Ahmadjian and Heikkilä, 1970). As this species is a terricolous species, lichens growing on soil may be especially suitable for synthesis experiments. This line of research may provide some of the fundamental facts of the lichen symbiosis in the future.

REFERENCES

AHMADJIAN, V. (1967). "The Lichen Symbiosis." Blaisdell, Waltham, Massachusetts.
AHMADJIAN, V. (1970). The lichen symbiosis: its origin and evolution. *In* "Evolutionary Biology" (T. Dobzhansky, M. K. Hecht and W. C. Steere, eds), Vol. 4, pp. 163–184. Meredith Corporation, New York.
AHMADJIAN, V. and HEIKKILÄ, H. (1970). The culture and synthesis of *Endocarpon pusillum* and *Staurothele clopima. Lichenologist* **4**, 259–267.
BEDNAR, T. W. and SMITH, D. C. (1966). Studies in the physiology of lichens VI. Preliminary studies of photosynthesis and carbohydrate metabolism of the lichen *Xanthoria aureola. New Phytol.* **65**, 211–220.
BEN-SHAUL, Y., PARAN, N. and GALUN, M. (1969). The ultrastructure of the association between phycobiont and mycobiont in three ecotypes of the lichen *Caloplaca aurantia* var. *aurantia. J. Microscopie* **8**, 415–422.
BERTSCH, A. and BUTIN, H. (1967). Die Kultur der Erdflechte *Endocarpon pusillum* im Labor. *Planta* **72**, 29–42.
BESCHEL, R. (1958). Flechtenvereine der Städte, Stadtflechten und ihr Wachstum. *Ber. Naturw.-med. Ver. Innsbruck* **52**, 1–158.
BIRD, I. F., CORNELIUS, M. J., KEYS, A. J. and WHITTINGHAM, C. P. (1974). Intracellular site of sucrose synthesis in leaves. *Phytochemistry* **13**, 59–64.
BOISSIÈRE, J. C. (1972). Mise en évidence cytochimique en microscopie électronique de polyglucosides de réserve chez des *Nostoc* libres et lichénisés. *C. r. hebd. Séanc. Acad. Sci., Paris, Sér. D* **274**, 2643–2646.
BORNEFELD, T. and SIMONIS, W. (1974). Effects of light, temperature, pH and inhibitors on the ATP level of the blue-green alga *Anacystis nidulans. Planta* **115**, 309–318.
BRACKER, C. E. and LITTLEFIELD, L. J. (1973). Structural concepts of host–pathogen interfaces. *In* "Fungal Pathogenicity and the Plant's Response" (R. J. W. Byrde and C. V. Cutting, eds), pp. 159–318. Academic Press, London and New York.
BROWN, R. M. and WILSON, R. (1968). Electron microscopy of the lichen *Physcia aipolia* (Ehrh.) Nyl. *J. Phycol.* **4**, 230–240.
BURNETT, J. H. (1968). "Fundamentals of Mycology." Arnold, London.
CARR, N. G. and WHITTON, B. A., eds. (1973). "The Biology of Blue-green Algae." Blackwell, Oxford.
DREW, E. A. (1966). "Some Aspects of the Carbohydrate Metabolism of Lichens." D.Phil. thesis, University of Oxford.

DREW, E. A. and SMITH, D. C. (1967a). Studies in the physiology of lichens VII. The physiology of the *Nostoc* symbiont of *Peltigera polydactyla* compared with cultured and free-living forms. *New Phytol.* **66**, 379–388.

DREW, E. A. and SMITH, D. C. (1967b). Studies in the physiology of lichens VIII. Movement of glucose from alga to fungus during photosynthesis in the thallus of *Peltigera polydactyla*. *New Phytol.* **66**, 389–400.

ERTL, L. (1951). Über die Lichtverhältnisse in Laubflechten. *Planta* **39**, 245–270.

FARRAR, J. F. (1973a). Lichen physiology: progress and pitfalls. *In* "Air Pollution and Lichens" (B. W. Ferry, M. S. Baddeley and D. L. Hawksworth, eds), pp. 238–282. University of London, Athlone Press, London.

FARRAR, J. F. (1973b). "Physiological Lichen Ecology." D.Phil thesis, University of Oxford.

FEIGE, B. (1969). Stoffwechselphysiologische Untersuchungen an der tropischen Basidiolichene *Cora pavonia* (Sw.) Fr. *Flora, Jena* **160A**, 169–180.

FOLLMANN, G. (1960). Die Durchlässig Keitseigenschaften der Photoplasten von phycobionten aus *Cladonia furcata* (Huds.) Schrad. *Naturwissenschaften* **47**, 405–406.

FOLLMANN, G. and VILLAGRAN, V. (1965). Flechtenstoffe und Zellpermeabilitat. *Z. Naturf.* **20b**, 723.

Fox, C. H. (1967). Studies of the cultured physiology of the lichen alga *Trebouxia*. *Physiologia Pl.* **20**, 251–262.

GALUN, M., BEN-SHAUL, Y. and PARAN, N. (1971). The fungus–alga association in the Lecideaceae: an ultrastructural study. *New Phytol.* **70**, 483–485.

GALUN, M., PARAN, N. and BEN-SHAUL, Y. (1970a). The fungus–alga association in the Lecanoraceae: an ultrastructural study. *New Phytol.* **69**, 599–603.

GALUN, M., PARAN, N. and BEN-SHAUL, Y. (1970b). An ultrastructural study of the fungus alga association in *Lecanora radiosa* growing under different environmental conditions. *J. Microscopie* **9**, 801–806.

GALUN, M., PARAN, N. and BEN-SHAUL, Y. (1970c). Structural modifications of the phycobiont in the lichen thallus. *Protoplasma* **69**, 85–96.

GREEN, T. G. A. (1970). "The Biology of Lichen Symbionts." D.Phil. thesis, University of Oxford.

HADLEY, G., JOHNSON, R. P. C. and JOHN, D. A. (1971). Fine structure of the host–fungus interface in orchid mycorrhiza. *Planta* **100**, 191–199.

HALE, M. E. (1974) ["1973"]. Growth. *In* "The Lichens" (V. Ahmadjian and M. E. Hale, eds), pp. 473–492. Academic Press, New York and London.

HARRIS, G. P. and KERSHAW, K. A. (1971). Thallus growth and the distribution of stored metabolites in the phycobionts of the lichens *Parmelia sulcata* and *P. physodes*. *Can. J. Bot.* **49**, 1367–1372.

HEBER, U. (1974). Metabolite exchange between chloroplasts and cytoplasm. *A. Rev. Pl. Physiol.* **25**, 393–421.

HELDT, H. W. and SAUER, F. (1971). The inner membrane of the chloroplast envelope as the site of specific metabolite transport. *Biochim. biophys. Acta* **234**, 83–91.

HILL, D. J. (1970). "The Carbohydrate Movement between the Symbionts of Lichens." D.Phil. thesis, University of Oxford.

HILL, D. J. (1972). The movement of carbohydrate from the alga to the fungus in the lichen *Peltigera polydactyla*. *New Phytol.* **71**, 31–39.

HILL, D. J. and AHMADJIAN, V. (1972). Relationship between carbohydrate movement and the symbiosis in lichens with green algae. *Planta* **103**, 267–277.

HILL, D. J. and SMITH, D. C. (1972). Lichen physiology XII. The inhibition technique. *New Phytol.* **71**, 15–30.

JACOBS, J. B. and AHMADJIAN, V. (1971). The ultrastructure of lichens. II. *Cladonia cristatella*: the lichen and its isolated symbionts. *J. Phycol.* **7**, 71–82.

KABACK, H. R. (1970). Transport. *A. Rev. Biochem.* **39**, 561–598.

KANAZAWA, T., KANAZAWA, K., KIRK, M. R. and BASSHAM, J. A. (1970). Regulation of photosynthetic carbon metabolism in synchronously growing *Chlorella pyrenoidosa*. *Pl. Cell Physiol.* **11**, 149–160.

KLOB, W., KANDLER, O. and TANNER, W. (1973). The role of cyclic phosphorylation *in vivo*. *Plant Physiol.* **51**, 825–827.

KOMIYA, T. and SHIBATA, S. (1971). Polyols produced by the cultured phyco- and mycobionts of some *Ramalina* species. *Phytochemistry* **10**, 695–699.

KOMOR, E. and TANNER, W. (1974a). The nature of the energy metabolite responsible for sugar accumulation in *Chlorella vulgaris*. *Z. Pflanzenphysiol.* **71**, 115–128.

KOMOR, E. and TANNER, W. (1974b). Can energy generated by sugar efflux be used for ATP synthesis in *Chlorella? Nature, Lond.* **248**, 511–512.

LANG, N. J. (1968). The fine structure of blue-green algae. *A. Rev. Microbiol.* **22**, 15–70.

LEWIS, D. H. and SMITH, D. C. (1967). Sugar alcohols (polyols) in fungi and green plants. I. Distribution, physiology and metabolism. *New Phytol.* **66**, 143–184.

MARUO, B., HATTORI, T. and TAKAHASHI, H. (1965). Excretion of ribitol and sucrose by green algae into the culture medium. *Agric. biol. Chem.* **29**, 1084–1089.

PATE, J. S. and GUNNING, B. E. S. (1972). Transfer cells. *A. Rev. Pl. Physiol.* **23**, 173–196.

PEAT, A. (1968). Fine structure of the vegetative thallus of the lichen *Peltigera polydactyla*. *Arch. Mikrobiol.* **61**, 212–222.

PELROY, R. A. and BASSHAM, J. A. (1972). Photosynthetic and dark carbon metabolism in unicellular blue-green algae. *Arch. Mikrobiol.* **86**, 25–38.

PEVELING, E. (1968). Electronenoptische Untersuchungen an Flechten I. Strukturveränderungen der Algenzellen von *Lecanora muralis* (Schreber) Rabenh. (– *Placodium saxicolum* (Nyl.) sec. Klem.) beim Eindringen von Pilzhyphen. *Z. Pflanzenphysiol.* **59**, 172–183.

PEVELING, E. (1969). Electronenoptische Untersuchungen an Flechten III. Cytologische Differenzierungen der Pilzzellen im Zusammenhang mit ihrer symbiontishcen Lebensweise. *Z. Pflanzenphysiol.* **61**, 151–164.

PEVELING, E. (1970). Das Vorkommen von Stärke in *Chlorphyceen*-phycobionten. *Planta* **93**, 82–85.

PEVELING, E. and HILL, D. J. (1974). The localisation of an insoluble intermediate in glucose production in the lichen *Peltigera polydactyla*. *New Phytol.* **73**, 769–771.

PHILLIPS, H. C. (1963). Growth Rate of *Parmelia isidiosa* (Mull. Arg.) Hale. *J. Tenn. Acad. Sci.* **38**, 95–96.

PLESSL, A. (1963). Über die Deziehungen von Haustorientypus und Organisationshohe bei Flechten. *Öst. bot. Z.* **110**, 194–269.

RICHARDSON, D. H. S. and SMITH, D. C. (1966). The physiology of the symbiosis in *Xanthoria aureola* (Ach.) Erichs. *Lichenologist* **3**, 202–206.

RICHARDSON, D. H. S. and SMITH, D. C. (1968a). Lichen physiology IX. Carbohydrate movement from the *Trebouxia* symbiont of *Xanthoria aureola* to the fungus. *New Phytol.* **67**, 61–68.

RICHARDSON, D. H. S. and SMITH, D. C. (1968b). Lichen physiology. X. The isolated algal and fungal symbionts of *Xanthoria aureola*. *New Phytol.* **67**, 69–77.

RICHARDSON, D. H. S., HILL, D. J. and SMITH, D. C. (1968). Lichen physiology XI. The role of the alga in determining the pattern of carbohydrate movement between lichen symbionts. *New Phytol.* **67**, 469–486.

RICHARDSON, D. H. S., SMITH, D. C. and LEWIS, D. H. (1967). Carbohydrate movement between the symbionts of lichens. *Nature, Lond.* **214**, 879–882.

SCOTT, G. D. (1969). "Plant Symbiosis." Edward Arnold, London.

SHEPPARD, D. C. and BIDWELL, R. G. S. (1973). Photosynthesis and carbon metabolism in a chloroplast preparation from *Acetabularia*. *Protoplasma* **76**, 289–307.

SMITH, D. C. (1961). The physiology of *Peltigera polydactyla* (Neck.) Hoffm. *Lichenologist* **1**, 209–226.

SMITH, D. C. (1963). Studies in the physiology of lichens IV. Carbohydrates in *Peltigera polydactyla* and the utilization of absorbed glucose. *New Phytol.* **62**, 205–216.

SMITH, D. C. (1974). Transport from symbiotic algae and symbiotic chloroplasts to host cells. *Symp. Soc. exp. Biol.* **28**, 485–520.

SMITH, D. C. and MOLESWORTH, S. (1973). Lichen physiology XIII. Effects of rewetting dry lichens. *New Phytol.* **72**, 525–533.

SMITH, D. C., MUSCATINE, L. and LEWIS, D. H. (1969). Carbohydrate movement from autotrophs to heterotrophs in parasitic and mutualistic symbiosis. *Biol. Rev.* **44**, 17–90.

SMYTH, E. S. (1934). A contribution to the physiology and ecology of *Peltigera canina* and *Peltigera polydactyla*. *Ann. Bot.* **48**, 781–818.

WATT, W. D. (1966). Release of dissolved organic material from the cells of phytoplankton populations. *Proc. R. Soc.*, B **164**, 521–551.

WEBBER, M. M. and WEBBER, P. J. (1970). Ultrastructure of lichen haustoria: symbiosis in *Parmelia sulcata*. *Can. J. Bot.* **48**, 1521–1524.

WILLIAMSON, B. (1973). Acid phosphatase and esterase activity in orchid mycorrhiza. *Planta* **112**, 149–158.

Notes added in proof
1. The literature survey for the above review was up to June 1974.
2. Smith (*Symp. Soc. exp. Biol.* **29**, 373–405, 1975) reports that treatment of thalli of *Peltigera polydactyla* with digitonin severely affected the fungal membrane with little effect on the alga. Fungal glucose uptake was abolished but algal glucose release continued at a high rate, showing that release and uptake are not linked. The release, insensitive to many inhibitors, was characteristic of facilitated diffusion; but how it is induced in symbiosis remains obscure. When air-dry thalli are immersed in water, membrane permeability barriers are not re-established for 1–2 min., so that substances might also move passively between symbionts in this period.
3. Boissière (*C. r. hebd. Séanc. Acad. Sci., Paris, Sér. D* **277**, 1649–1651, 1973) found neutral phosphatase activity near the outer membrane in *Nostoc* in *Peltigera canina* but not in free-living *Nostoc*.

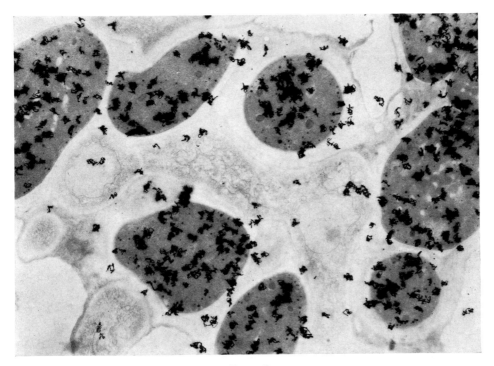

PLATE I

High resolution autoradiograph of [^{14}C]-insoluble substance in *Peltigera polydactyla*. The black silver grains indicate the presence of the ^{14}C within the algal cells. (Reproduced with permission from Peveling and Hill, 1974.)

20 | A Comparison between the Lichen Symbiosis and other Symbioses

D. C. SMITH

Department of Botany, The University, Bristol, England

Abstract: A framework for describing the features of a symbiosis is outlined. It has six main characteristics: (i) primary exploitation (i.e. the original physiological properties which the symbionts exploited in each other, and the mechanisms involved); (ii) ancillary modifications to facilitate nutrient transfer; (iii) balanced growth between the symbionts; (iv) structural changes; (v) dependence for complex metabolites; and (vi) specificity. Within this framework, lichens are compared to other associations. Many points of similarity are demonstrated, and it is shown that lichens are particularly well suited to the study of some of the general problems of symbiosis.

INTRODUCTION

This paper compares the main features of the lichen symbiosis with those of other associations, and assesses the extent to which investigations of lichens are valuable in understanding some of the more general problems of the phenomenon of symbiosis.

The types of association with which lichens will be compared are those described as *biotrophic*. In biotrophic associations, nutrients move between the living cells of the symbionts, and there is usually little or no destruction of one symbiont by the other. This is distinct from necrotrophy, in which one organism first destroys part or all of the other before absorbing nutrients from dead tissues. Biotrophy has the advantage that one symbiont continuously exploits a biochemical property possessed by the other over a long period of time, but the success of biotrophy necessitates the development of mechanisms for transferring nutrients between the living cells of the different symbionts.

In recent years, there have been experimental investigations into a wide

Systematics Association Special Volume No. 8, "Lichenology: Progress and Problems", edited by D. H. Brown, D. L. Hawksworth and R. H. Bailey, 1976, pp. 497–513. Academic Press, London and New York.

range of biotrophic associations. Mycorrhizal and other types of association between fungi and higher plants have been extensively reviewed (e.g. Harley, 1969; Lewis, 1973) as have legume nodules (e.g. Nutman, 1963; Bergersen, 1971) and associations between non-leguminous plants and nitrogen-fixing prokaryotes (e.g. Bond, 1963, 1967). About 10% of all insects are probably symbiotic: sometimes with yeasts, protozoa and other micro-organisms in particular regions of the digestive tract (cf. Brooks, 1963), but many have special organs separate from the digestive tract called *mycetomes*. These contain unidentified prokaryotes (*not* yeasts or other fungi) which are best termed Blochman Bodies (cf. Lanham, 1968). Many animals besides insects have digestive tract symbionts, sometimes in distinct structures such as the enormous rumen of ruminants.

Of particular interest to lichenologists are the variety of associations between algae and other heterotrophic organisms, especially invertebrates (cf. Smith, 1973). In tropical oceans, almost all coelenterates contain large numbers of zooxanthellae (symbiotic dinoflagellates) which typically pack each cell of the gastrodermal layer (Yonge, 1968). Outstanding among these are the reef-building corals, where the zooxanthellae probably account for the very high productivity of the reef ecosystem as well as being essential for the massive extent of calcium carbonate deposition on the reef. Zooxanthellae are also very abundant in the mantle of giant clams. There have been elegant studies of the symbiosis between *Chlorella* spp. and green *Hydra* (cf. Muscatine, 1975a, b), and *Paramecium bursaria* (cf. Karakashian, 1975). The flatworm *Convoluta roscoffensis* is associated with the green alga *Platymonas convolutae* and was the subject of a classic study by Keeble (1910), and has again been the subject of more recent study (Muscatine et al., 1974; Holligan and Gooday, 1975). Finally, there is the remarkable association between chloroplasts of siphonaceous seaweeds such as *Codium* and *Caulerpa* and some of the sacoglossan molluscs which feed on them; chloroplasts continue photosynthesis for 1–2 months after entry into animal cells lining the digestive tract (Hinde and Smith, 1972; Trench et al., 1973a, b). The great majority of algal symbionts of invertebrates are intracellular; sometimes the animals retain their original holozoic feeding mechanisms (as in *Hydra* and corals) but sometimes they cease to feed (e.g. *Convoluta* and some zoanthids).

A FRAMEWORK FOR COMPARING BIOTROPHIC ASSOCIATIONS

Before discussing different biotrophic associations, it is first essential to devise some descriptive framework within which to make broad comparisons. Such a framework can be constructed from an essentially physiological and structural standpoint based upon the assumption that a key feature in the evolution of a biotrophic association is the development of mechanisms of nutrient transfer between the symbionts. At the same time, ancillary modifications to improve nutrient transfer will presumably evolve, accompanied by other physiological and structural changes which improve conditions for the donor symbiont, enabling it to function more efficiently. In describing this framework, it will be convenient to term symbionts "donor" and "recipient" with respect to a specific nutrient, and to consider the recipient as "exploiting" the donor, rather than "receiving benefit" from it. Viewed in this way, six main characteristics of a biotrophic association may be recognized.

1. Primary Exploitation

This describes the physiological properties of the donor originally exploited by the recipient. Typically, it concerns major nutrients present in the environment which are either in a form unavailable to the recipient or in a form which the donor can acquire much more efficiently. Examples of primary exploitation are shown in Table I.

Donors may be exploited for more than one kind of nutrient (e.g. algal endosymbionts of animals may be exploited both for photosynthetically fixed carbon and for their ability to convert nitrogenous waste back to usable amino acids). In environments where nutrients are principally in a form unavailable to recipients there will be strong selective pressure to develop associations with appropriate donor symbionts. Hence the tendency for symbiotic associations to become common in certain types of habitat, such as lichens on bare rock, mycorrhizas in nutrient-poor soils and symbiotic coelenterates in the nutrient-poor waters of the tropics. Finally, in mutualistic associations, each symbiont may exploit the other so that there is a two-way movement of nutrients.

2. Ancillary Modifications to Facilitate Nutrient Transfer

These are of two principal types: (a) improvement or enlargement of the area of contact for transfer (e.g. haustoria, convoluted plasmalemma of "transfer cells", etc.); (b) biochemical or physiological changes which increase the amount of nutrients in the donor available for transfer.

3. Balanced Growth Between the Symbionts

Achievement of a degree of balanced growth between the symbionts has the following advantageous consequences: (a) exploitation of one symbiont by the other is prevented from becoming excessive; (b) nutrient transfer can then occur over very prolonged periods; and (*c*) the growth and differentiation of special structures composed of both symbionts is facilitated.

TABLE I. Principal types of primary exploitation in biotrophic associations

Nutrient	Environmental form	Examples of donors
Carbon	CO_2	Autotrophs (e.g. lichen algae, hosts of mycorrhizal fungi)
	Polysaccharides non-utilizable by recipient	Rumen and digestive tract micro-organisms
Nitrogen	N_2	Nitrogen fixers (e.g. *Rhizobium*)
	NO_3^-	Organisms with NO_3^- reductase (?some algal endosymbionts of animals)
	$NH_4^=$	Organisms converting $NH_4^=$ to amino acids (e.g. zooxanthellae)
	Organic N which recipients cannot utilize such as animal waste products	Organisms converting them to usable amino acids (e.g. algal symbiont of *Convoluta*)
Sulphur	$SO_4^=$	Organisms converting inorganic S to organic (e.g. ?aphid symbionts ?zoochlorellae)
Minerals	Soil minerals	Flowering plant hosts of mistletoes (i.e. recipient not in contact with soil) Mycorrhizal fungi (i.e. more efficient uptake than recipient)

4. Structural Changes

Frequently, both symbionts participate in the formation of a structure whose functions may be to improve the environmental conditions of the donor, enabling it to perform efficiently (e.g. legume nodule).

5. Dependence for Complex Metabolites

This refers to situations where it is presumed a symbiont that originally possessed a physiological property, but subsequently lost it on entering the association and now relies upon the donor. Such a tendency to eliminate unnecessary duplication of metabolic pathways will presumably occur in associations which have developed a degree of intimacy such that the relevant metabolites can move freely between the symbionts. This may explain the absolute requirements for complex compounds such as vitamins shown by some symbionts in isolated culture, or indeed the complete unculturability of others (e.g. the Blochmann Bodies in the mycetomes of insects).

6. Specificity

In all biotrophic associations, the symbionts show some degree of specificity for each other. In general, recipients show limited or no ability to associate with non-symbiotic donors possessing the same primary physiological properties as the natural donor. In many cases, the level of specificity is such that recipients cannot form associations with organisms which donate the required nutrients to other, but different recipients.

A COMPARISON BETWEEN THE LICHEN AND OTHER ASSOCIATIONS

Lichens will now be compared to other biotrophic associations with respect to each of the six main characteristics outlined above.

1. Primary Exploitation

(a) *Carbon.* The movement of photosynthetically fixed carbon from autotrophs to heterotrophs has been studied in many associations (Smith *et al.*, 1969). Lichens have proved particularly suitable for study because of the ease with which the symbionts can be separated and because the fungus converts the fixed carbon it receives mainly to compounds not present in the alga: transfer can therefore be studied without breaking up the symbiosis.

The principal characteristics of movement in those symbioses involving algae are summarized in Table II. It is remarkable that despite the wide variety of associations, they almost always show the following four main features. (1) Transport is substantial—usually about 40% or more of the carbon fixed in photosynthesis moves from the alga to its heterotrophic partner. (2) This movement is selective in that most of the carbon moves as only one or a few types of molecule. (3) The molecules most commonly involved in transport are glucose, polyols or neutral amino acids. (4) The algae usually cease to release large amounts of carbon very soon after isolation from the symbiosis; the loss may occur immediately (as with symbiotic chloroplasts or the alga of *Convoluta roscoffensis*) or over a period of several hours (as with zooxanthellae and lichen algae).

In all cases, the presence of the heterotroph is necessary to induce the massive efflux of fixed carbon from the symbiotic algae. The principal problem is how

R

TABLE II. Summary of the characteristics of transport of photosynthate from symbiotic algae and symbiotic chloroplasts to host cells (Based upon Smith, 1974, modified by later, unpublished results)

	LICHENS		Hydra viridis	Dinoflagellate zooxanthellae	Convoluta roscoffensis	Symbiotic chloroplasts in molluscs
	Blue-green symbionts	Green symbionts				
Principal form in which photosynthetically fixed ^{14}C moves to heterotroph — Glucose	+++	−	+++ (maltose and glucose)	−	−	+++
Polyols	−	+++	−	+++ (glycerol)	−	−
Neutral amino acids	?	−	−	++ (esp. alanine)	+++ (glutamine)	+ (esp. alanine)
Proportion of total fixed ^{14}C moving to heterotroph	25–70%		30–40%	25–50%	~40%	36–40%
Stimulus required to cause autotroph to release photosynthate	?Physical contact?		acid pH	Water-soluble, thermolabile "factors" in host tissues		
Loss of release after isolation of autotroph from symbiosis	Rapid (within 6–12 h)		No loss	Rapid (less than 10 h)	Immediate	Immediate

different types of heterotrophs stimulate efflux by similar types of transport system. In many symbiotic marine invertebrates the animal tissues contain water-soluble, thermolabile "factors" which stimulate efflux from their symbiotic algae (or chloroplasts); "factors" from one animal can affect the algal symbionts of another, indicating that they are not highly specific (Muscatine *et al.*, 1972; Smith, 1974). Gallop (1974) suggested that such "factors" may be naturally present in animal cells as part of the mechanism for regulating function of organelle membranes, and that their concentrations become markedly increased in the presence of a symbiont.

The widespread detection of such "factors" in symbiotic invertebrates led to a search for similar compounds in lichens. In particular, Green (1970) made exhaustive studies of *Xanthoria aureola* but failed to obtain any thallus extract which had a consistent effect on the symbiotic *Trebouxia*; lichen acids had no effect on permeability other than that normally due to acid pH. For various reasons, Smith (1974) concluded that the fungus may not affect the alga by extracellular release of chemical compounds, but by some aspect of physical contact between the symbionts. One possibility he suggested was physical modification by the fungus of the transmembrane potential of the alga, so that the normal direction of transport of certain molecules becomes reversed to produce an efflux from the alga.

Although lichens differ in this respect from the other examples of algal symbiosis shown in Table II, it may well be because they are the only example of extracellular symbiosis in the Table. It is intriguing to speculate that transport between the lichen symbionts may be generally similar to that between many biotrophic fungi and their higher plant hosts; indeed, there could even be similarities with the normal process of cell-to-cell transport in plant tissues. The morphological simplicity of lichens, coupled with the fact that fixed carbon accumulates in different compounds in the two symbionts, has made the lichen one of the best systems in which to study intercellular transport.

(*b*) *Nitrogen.* In those lichens with blue-green algae which have been appropriately investigated, nitrogen is actively fixed and released to the fungus. This subject has already been reviewed by Millbank and Kershaw (1974) and Millbank (Chapter 18), and therefore will not be described further. The extent to which transport between the symbionts resembles that in other nitrogen-fixing symbioses is less easy to evaluate since nitrogen transport is much more difficult to study than carbon transport. According to the general principles implicit in Table II it might be expected that in lichens nitrogen moves primarily as neutral amino acids—but I have found no evidence of this, and Millbank believes that peptides are involved; if so, lichens would appear to be unlike other nitrogen-fixing symbioses. Given the high level of nitrogen fixation and transfer found by Millbank, it can be calculated that substantial amounts of carbon would have to move with the nitrogen if either amino acids or peptides were involved in transfer. Movement of such carbon has not been detected in photosynthetic ^{14}C fixation experiments, although there could be two reasons why it might occur without detection. (1) Photosynthetic ^{14}C fixation experiments are carried out under relatively high light intensities where the rate and extent of oxygen production might reduce nitrogenase activity. (2) The carbon used for nitrogen transport might be derived from preformed insoluble reserves which do not become labelled during the relatively short term of photosynthesis experiments.

(c) *Minerals: to what extent is there movement from fungus to alga?* It is part of the traditional dogma about the lichen symbiosis that "the fungus supplies the alga with minerals". It must be recognized that this is not yet proven by any experiment. Since the alga is not located within the fungus and lies external to it, there is no absolute necessity for any or all minerals absorbed by the alga to have first passed through the fungal cytoplasm; minerals could enter the alga from external solutions by passing through inert cell walls. Such penetration can be rapid, and it has been calculated that in short-term photosynthesis experiments with *Peltigera polydactyla*, $H^{14}CO_3^-$ passes from the external solution to the photosynthetic centres of the algae in 15 s.

An association in which there is undoubted and substantial movement of phosphate, and probably other minerals, from a fungus to its autotrophic symbiont is ectotrophic mycorrhiza (e.g. Harley, 1969). However, here the autotrophic symbiont is a large, actively growing tree with a high demand for nutrients. In the lichen, the autotrophic symbiont is small in volume compared to the fungus, grows very slowly, and presumably therefore has a very low demand for nutrients. It may be that any mineral transport is too small to be easily detected in experiments. Further, even after prolonged uptake of substantial amounts of $^{32}PO_4^=$, lichens show little or no tendency to release absorbed phosphate (Farrar, 1973).

On the other hand, it would be foolish to ignore the fact that lichen fungi have remarkably efficient mechanisms of absorption and that absorbed minerals do not in some way pass to the alga. A possible mechanism for this is suggested by two observations: (a) the great majority of lichens pass through regular cycles of wetting and drying and indeed seem to require such cycles for healthy growth (cf. Fletcher, Chapter 14; Farrar, Chapter 15); and (*b*) when water is added to air-dry lichen thalli, there may be substantial loss of both organic and inorganic substances during the first minute before membrane permeability is re-established (Farrar, 1973, and Chapter 15; H. Outred, unpublished). In nature, lichens are infrequently wetted by a sudden torrent of water and so are unlikely to show the kinds of losses to the external environment found in laboratory experiments. Nevertheless, in the localized area of an air-dry thallus suddenly saturated by a rain drop, one might imagine a process of instant release followed by reabsorption of substances occurring in the affected area: if it included the algal layer, then one could visualize some passive movement of substances from fungus to alga during the early part of the rewetting period. If this mechanism of transfer between symbionts were to prove important, then it would be one respect in which lichens differ from other types of symbiosis.

2. Ancillary Modifications to Facilitate Nutrient Transfer

(*a*) *Improved contact between symbionts.* In all biotrophic associations, there are modifications such that the type of contact between the symbionts is very different from what would be expected in casual or accidental associations. The area of contact is usually much increased, with the extreme situation presented by endosymbionts which are completely enveloped by their host. In other cases—especially those involving fungi, one symbiont may penetrate the other to a greater or lesser extent by haustoria, or one symbiont may simply be closely appressed to the other over an appreciable area. Lichens fall somewhere between the latter two situations: haustorial penetration of the alga has been observed in some species, but not in others. Unfortunately, the great majority of those reporting haustoria in lichens give no indication of the proportion of algal cells so penetrated. For some lichens, the occurrence of haustoria depends partly on the environment (Ben-Shaul *et al.*, 1969; Galun *et al.*, 1970c), the presumed phylogenetic position (Galun *et al.*, 1970a, 1971) and the stage of maturity (Galun *et al.*, 1970b). The rather scanty literature leaves the impression that in the great majority of species, haustoria are either absent or not very common· Where haustoria have been reported, they penetrate the algal cell wall but the algal cell membrane is only rarely broken.

In a comprehensive and detailed review of the ultrastructure of interfaces between symbionts, Bracker and Littlefield (1973) showed that a large number of types exist. Their classification of interface types does not include presence or absence of haustoria, so it is convenient to consider the place of lichens in their scheme (which is primarily based on the presence or absence of the walls of one

R*

or both symbionts, the occurrence of an intervening matrix of wall appositions or other material). Although lichens are difficult material for study by electron microscopy, such evidence as is available shows that they probably belong to one of the common "interface types" of Bracker and Littlefield, with either symbiont walls in contact (type 8) or with some form of intervening matrix (types 13 or 24). Indeed, Bracker and Littlefield's survey shows that in the very great majority of biotrophic associations, the symbiont membranes are kept apart so that mobile substances have to traverse an appreciable distance between them; the reason for this is not clear.

The various wall and matrix modifications seen in the interface in lichens disappear when the symbionts are grown in isolated culture, showing that the modifications such as reduced wall thickness are due to repression rather than loss of any ability. The mechanism of repression is not known. It has been suggested that reduction in algal wall formation may be due to a modification of wall synthesis mechanisms so as to produce a release of carbohydrates for the host (Hill, 1972) though later evidence casts doubt on this (cf. Smith, 1974). Suppression of wall formation also occurs in a wide range of endosymbiotic algae (Smith *et al.*, 1969). It is of interest that, as in certain lichen algae, formation of flagellae is inhibited, although basal bodies are still recognizable in electron micrographs of zooxanthellae (e.g. R. K. Trench, unpublished).

(*b*) *Biochemical and physiological changes.* Parallel to the morphological changes outlined above, there may also be biochemical and physiological changes which ultimately result in improved nutrient transport from donor to recipient. An outstanding example is the way in which the translocation stream in higher plants may become diverted towards the site of association with fungal biotrophs (Smith *et al.*, 1969). Another case is illustrated by the legume nodule, which provides the essential environmental conditions for efficient nitrogen fixation by the symbiotic bacteria—a good supply of energy and carbon skeletons, coupled with a low free oxygen tension maintained by the leg-haemoglobin.

Similar modifications presumably occur in lichens. An obvious example is the very low incorporation of photosynthetically fixed carbon into insoluble compounds in the alga, enabling a substantial surplus of the mobile carbohydrate to develop. Relevant to this may be the observations of Peveling (1974) that, unlike free-living algae, no starch sheath is formed around the pyrenoid of *Trebouxia* symbionts of lichens; similarly, Peat (1968) noted that polyglucoside granules are not formed by the *Nostoc* from *Peltigera* when in symbiosis.

3. Balanced Growth between Symbionts

In mutualistic associations, there is balanced growth between the symbionts. Sometimes it may be very precise, as in the protozoan *Paulinella* which always contains two *Synechococcus* cells. In this organism, synchrony of division between the symbionts is exact, though out of phase; the animal first divides, each daughter cell receiving one alga, then the algae divide, restoring the number of algae per animal to two. In green *Hydra*, the number of zoochlorellae per gastrodermal cell remains remarkably constant (Pardy and Muscatine, 1973). In *Paramecium bursaria* there are elegant experiments showing how the algal populations per animal varies with environmental conditions such as light and level of animal nutrients (Pado, 1965, 1967).

Lichens also show excellent balance of growth—particularly when it is remembered that they persist for very much longer than any other associations: for example, some foliose *Parmelia* sp. may last for 15–80 years, while some crustaceous lichens may live well over 100 years (e.g. cf. Hale, 1974). The exact proportion of fungus to alga has been measured in very few lichens: nearly all *Peltigera* spp. (Drew and Smith, 1967; Millbank, 1972; N. Morris, unpublished), where the fungus comprises 90–98% of the weight of the thallus. Although the amount of algae may thus seem low, it is clearly adequate to maintain the extremely slow growth rate of lichens; in any case, thalli show no sign of carbon shortage and are rich in soluble carbohydrates. Casual field observations of *Peltigera* spp. indicates that the thickness of the fungal medulla varies with the moistness of the habitat, but more investigations are needed to verify this.

While all mutualistic associations normally show balanced growth, the mechanism of control remains unknown. One symbiont—probably the larger one—can presumably in some way set an upper limit to the growth of the other. This could be through various combinations of physical constraint, restriction in the supply of nutrients, or interference with mitosis or other aspect of development. In some associations, the balance may break down under certain abnormal environmental conditions. In both *Paramecium bursaria* and corals, prolonged incubation in the dark results in loss of algal endosymbionts (cf. Pado, 1965; Muscatine, 1975a): in *Paramecium* because animal growth in the dark far outstrips algal, and in corals because shrinkage of tissues under starvation results in expulsion of algae. In the case of lichens, two factors which can lead to the breakdown of the symbiosis are an excess of nutrients (Tobler, 1925) or incubation for prolonged periods in moist conditions (see Fletcher, Chapter 14). Although this suggests that both periodic drying and restricted nutrient supply (perhaps fungus to alga) may be important factors in maintaining balanced growth, much further work is needed on this topic. Finally, cephalodiate lichens show that one fungus can establish a balanced relationship with two quite different algae.

4. Structural Changes
In virtually all associations involving extracellular symbionts, and in many of those involving intracellular ones, the host shows some general modifications which presumably result in increased advantage to the association, or at least in improved functioning of the other symbiont. Sometimes these modifications disappear on isolation (e.g. lichens) and sometimes they do not (e.g. mycetomes of insects).

In the case of lichens it is important to remember that too little is known of thallus biology to be able to state the definite functions of different regions of the thallus. The following are some of the *suggestions* that have been made (cf. Smith, 1962), which deserve further investigation. (*a*) The medulla may function as a primitive reservoir of moisture for the algal layer; (*b*) within the algal layer air pockets are maintained even in fully saturated thalli (through crystals of non-wettable lichen substances) and this improves gaseous diffusion to and from the algae; (*c*) lichen substances function as antibiotics to protect the association from microbial attack. There is additionally the frequently stated but vague proposition that in some way the fungus "protects" the alga. Clear evidence for the nature of the protection is still lacking; certainly, the widespread assumption that algae in the thallus need protection from "insolation" in those particular environmental conditions requires investigation—especially since free-living algae grow in most lichen habitats.

5. Dependence for Complex Metabolites
Most lichen symbionts do not manifest complex requirements when grown in isolated culture with the exception of the general requirement of the fungi for biotin and thiamine (cf. Ahmadjian, 1967; Hale, 1967). Relevant to this may be the solitary observation (Bednar, 1963) that the isolated lichen alga *Coccomyxa* produces 17 times more biotin than free-living *Chlorella*. However, it is not known whether the level of biotin excretion by *Coccomyxa* is unusually high for an alga and whether, as with other aspects of excretion, there are marked changes after entry into symbiosis. In this respect it might be useful to study biotin excretion by lichen algae immediately after isolation from the symbiosis. Apart from this, there is little to suggest that dependence for complex metabolites has developed on an extensive scale in lichen symbionts. In so far as other types of symbiosis are concerned, obvious examples are mostly shown only by those

symbionts which are intracellular, such as the Blochman Bodies of insects. Where the association is largely extracellular, as in ectotrophic mycorrhizas, there is little evidence.

6. Specificity

In all types of biotrophic association, the symbionts show a measure of specificity for each other. Sometimes the specificity is of a high degree as in the case of the Florida strain 61 of *Hydra viridis* which will associate with only two out of a number of symbiotic *Chlorella* (Pardy and Muscatine, 1973); the mycorrhizal fungus *Boletus elegans* associates only with larch. In the case of lichens, the problem is complicated by the fact that the taxonomy of lichen algae is poorly understood. Although several different strains of algae may occur in the same thallus (Ahmadjian, 1967), the specificity of any one algal taxon for different lichen fungi is not known.

During the early stages of lichen synthesis, specificity is lax, but increases as synthesis proceeds (Scott, 1964). This is unlike the situation in *Hydra viridis*, where the host animal appears to be able to "recognize" its own symbiont almost instantly on contact (Pardy and Muscatine, 1973). Both in *Hydra* and in lichens specificity presumably begins, at least in part, when the walls of the symbionts first come into contact. Pardy and Muscatine speculate that the "recognition" of symbionts for each other must involve interaction of molecules at cell surfaces, which may well have definite "recognition sites"; a great deal more information about the structure of the cell walls of symbionts in general is thus highly desirable. On the other hand, in some associations, such as that between *Rhizobia* and legumes, secretion of substances by one or both symbionts seems to be involved in determining specificity. Although secretions have not yet been demonstrated in any association involving algae, further investigation is required.

Poor knowledge of symbiont taxonomy as an obstacle to understanding specificity is a problem unfortunately not restricted to lichens. Despite the ubiquity and ecological importance of coelenterate–zooxanthella associations, especially in tropical oceans, very little information exists about the types of dinoflagellate involved and whether they are common components of the phytoplankton. So far, only three species have been isolated and described: *Gymnodinium adriaticum*, *Amphidinium chattonii* and *A. klebsii* (Taylor, 1973); yet it is difficult to believe that only these three species occur in the bewildering variety of different hosts and different habitats throughout the world. In legumes, where specificity relationships are quite well understood, it is evident that morphologically identical Rhizobia can show marked differences in host specificity, and similar phenomena are well known in the black stem rust of wheat, *Puccinia graminis*, and many other plant-pathogenic fungi.

A related and potentially important problem was highlighted by James and Henssen (see Chapter 3) when they showed that the same lichen fungus adopted a strikingly different morphology when associating with green or blue-green algae—even though the algae comprised but a small fraction of the mass of the thallus; this led to the question of whether different strains of *Trebouxia* symbionts could "evoke" differences in thallus morphology, and so be a complicating factor in lichen taxonomy (which is based upon fungal characteristics). It is not impossible that similar problems may be found in other types of symbiotic association, especially those presenting taxonomic difficulties, such as zoanthids and corals.

CONCLUSIONS

This paper has shown that many features of the lichen symbiosis are typical of other types of biotrophic associations. Most of the characteristics that have caused lichens to be considered "odd" or "unusual" relate to the way in which they are adapted to existence in their habitats.

As experimental material for investigating general problems of symbiosis, lichens have proved exceptionally useful in the study of carbon transfer. Indeed, it is striking that the pattern first established for lichens was subsequently shown to be closely similar to that in many other autotroph–heterotroph associations. Although technically more difficult to study, it is also possible that lichens may yet prove particularly apt for the study of nitrogen transfer.

Lichens are less satisfactory for studying ultrastructural aspects of symbiosis since they prove difficult material in fixing and sectioning for transmission electron microscopy. Nevertheless, studies such as those of Peveling (Chapter 2) and Galun *et al.* (1970a, b, c, 1971) show that careful techniques can still give quite clear pictures of the nature of the contact between the symbionts. Furthermore, the toughness of the thallus has made lichens particularly good objects for studying by scanning electron microscopy (Hale, Chapter 1); if improved techniques can prevent the collapse and disruption of algal cells, then it may be possible to get an unusually clear and good three-dimensional picture of inter-relationships in a symbiosis.

Aspects of symbiosis for which lichens do not represent good experimental material are primarily those in which slow growth and inadequate algal taxonomy put them at a disadvantage. However, one topic in which slow growth may be an advantage is in studies of specificity mechanisms: this is suggested by the fact that one of the major difficulties of studying *Hydra* is that "recognition" of algae occurs far too rapidly to permit satisfactory study of the mechanisms involved. The large investment of time and effort which has been devoted to developing techniques for the laboratory synthesis of lichens might be put to good purpose in studying mechanisms of specificity in the initial stages of lichenization.

ACKNOWLEDGEMENTS

I am very grateful for helpful advice and discussion with Dr H. Outred and Dr L. J. Littlefield during the preparation of this article.

REFERENCES

AHMADJIAN, V. (1967). "The Lichen Symbiosis." Blaisdell, Waltham, Massachusetts.

BEDNAR, T. W. (1963). "Physiological Studies on the Isolated Components of the Lichen *Peltigera aphthosa*." Ph.D. thesis, University of Wisconsin, Madison.

BEN-SHAUL, Y., PARAN, N. and GALUN, M. (1969). The ultrastructure of the association between phycobiont and mycobiont in three ecotypes of the lichen *Caloplaca aurantia* var. *aurantia. J. Microscopie* **8**, 415–422.

BERGERSEN, F. J. (1971). Biochemistry of symbiotic nitrogen fixation. *A. Rev. Pl. Physiol.* **22**, 121–140.

BOND, G. (1963). The root nodules of non-leguminous angiosperms. *Symp. Soc. gen. Microbiol.* **13**, 72–91.

BOND, G. (1967). Fixation of nitrogen by higher plants other than legumes. *A. Rev. Pl. Physiol.* **18**, 107–126.

BRACKER, C. E. and LITTLEFIELD, L. J. (1973). Structural concepts of host-pathogen interfaces. *In* "Fungal Pathogenicity and the Plant's Response" (R. J. W. Byrde and C. V. Cutting, eds), pp. 159–318. Academic Press, London and New York.

BROOKS, M. A. (1963). Symbiosis and aposymbiosis in arthropods. *Symp. Soc. gen. Microbiol.* **13**, 200–231.

DREW, E. A. and SMITH, D. C. (1967). Studies in the physiology of lichens. VII. The physiology of the *Nostoc* symbiont of *Peltigera polydactyla* compared with cultured and free-living forms. *New Phytol.* **66**, 379–388.

FARRAR, J. F. (1973). "Physiological Lichen Ecology." D.Phil. thesis, University of Oxford.

GALLOP, A. M. (1974). Control of photosynthate release from the symbiotic chloroplasts in *Elysia viridis. New Phytol.* **73**, 1111–1117.

GALUN, M., BEN-SHAUL, Y. and PARAN, N. (1971). The fungus–alga association in the Lecideaceae: an ultrastructural study. *New Phytol.* **70**, 483–585.

GALUN, M., PARAN, N. and BEN-SHAUL, Y. (1970a). Structural modifications of the phycobiont in the lichen thallus. *Protoplasma* **69**, 85–96.

GALUN, M., PARAN, N. and BEN-SHAUL, Y. (1970b). The fungus–alga association in the Lecanoraceae: An ultrastructural study. *New Phytol.* **69**, 599–603.

GALUN, M., PARAN, N. and BEN-SHAUL, Y. (1970c). An ultrastructural study of the fungus–alga association in *Lecanora radiosa* growing under different environmental conditions. *J. Microscopie* **9**, 801–806.

GREEN, T. G. A. (1970). "The Biology of the Lichen Symbionts." D.Phil. thesis, University of Oxford.

HALE, M. E. (1967). "The Biology of Lichens." Arnold, London.

HALE, M. E. (1974) ["1973"]. Growth. *In* "The Lichens" (V. Ahmadjian and M. E. Hale, eds), pp. 473–492. Academic Press, New York and London.

HARLEY, J. L. (1969). "The Biology of Mycorrhiza" (2nd edition). Leonard Hill, London.

HILL, D. J. (1972). The movement of carbohydrate from the alga to the fungus in the lichen *Peltigera polydactyla*. *New Phytol.* **71**, 31–39.

HINDE, R. T. and SMITH, D. C. (1972). Persistence of functional chloroplasts in *Elysia viridis* (Opisthobranchia, Sacoglossa). *Nature, New Biol.* **239**, 30–31.

HOLLIGAN, P. M. and GOODAY, G. W. (1975). Symbiosis in *Convoluta roscoffensis*. *Symp. Soc. exp. Biol.* **29** (in press).

KARAKASHIAN, M. J. (1975). Symbiosis in *Paramecium bursaria*. *Symp. Soc. exp. Biol.* **29** (in press).

KEEBLE, F. (1910). "Plant-Animals: A Study in Symbiosis." Cambridge University Press, Cambridge.

LANHAM, U. N. (1968). The Blochmann Bodies: Hereditary intracellular symbionts of insects. *Biol. Rev.* **42**, 269–286.

LEWIS, D. H. (1973). Concepts in fungal nutrition and the origin of biotrophy. *Biol. Rev.* **48**, 261–278.

MILLBANK, J. W. (1972). Nitrogen metabolism in lichens. IV. The nitrogenase activity of the *Nostoc* phycobiont in *Peltigera canina*. *New Phytol.* **71**, 1–10.

MILLBANK, J. W. and KERSHAW, K. A. (1974) ["1973"]. Nitrogen metabolism. *In* "The Lichens" (V. Ahmadjian and M. E. Hale, eds), pp. 289–307. Academic Press, New York and London.

MUSCATINE, L. (1975a). *Chlorella* symbiosis in *Hydra*. *Symp. Soc. exp. Biol.* **29** (in press).

MUSCATINE, L. (1975b). Endosymbiosis of cnidarians and algae. *In* "Recent Perspectives in Coelenterate Biology." Academic Press, New York and London.

MUSCATINE, L., BOYLE, J. E. and SMITH, D. C. (1974). Symbiosis of the acoel flatworm *Convoluta roscoffensis* with the alga *Platymonas convolutae*. *Proc. R. Soc. Lond.* B **187**, 221–234.

MUSCATINE, L., POOL, R. R. and CERNICHIARI, E. (1972). Some factors influencing selective release of soluble organic material by zooxanthellae from reef corals. *Mar. Biol.* **13**, 298–308.

NUTMAN, P. S. (1963). Factors influencing the balance of mutual advantage in legume symbiosis. *Symp. Soc. gen. Microbiol.* **13**, 51–71.

PADO, R. (1965). Mutual relations of protozoans and symbiotic algae in *Paramecium bursaria*. I. The influence of light on the growth of symbionts. *Folia biol., Praha* **13**, 173–182.

PADO, R. (1967). Mutual relations of protozoans and symbiotic algae in *Paramecium bursaria*. II. Photosynthesis. *Acta Soc. Bot. Pol.* **36**, 97–108.

PARDY, R. L. and MUSCATINE, L. (1973). Recognition of symbiotic algae by *Hydra viridis*. A quantitative study of the uptake of living algae by aposymbiotic *H. viridis*. *Biol. Bull. mar. biol. Lab. Woods Hole* **145**, 565–579.

PEAT, A. (1968). Fine structure of the vegetative thallus of the lichen *Peltigera polydactyla*. *Arch. Mikrobiol.* **61**, 212–222.

PEVELING, E. (1974) ["1973"]. Fine structure. *In* "The Lichens" (V. Ahmadjian and M. E. Hale, eds), pp. 147–182. Academic Press, New York and London.

SCOTT, G. D. (1964). The lichen symbiosis. *Advmt Sci., Lond.* **20**, 244–248.

SMITH, D. C. (1962). The biology of lichen thalli. *Biol. Rev.* **37**, 537–570.

SMITH, D. C. (1973). "Symbiosis of Algae with Invertebrates." [Oxford Biology Readers No. 43.] Oxford University Press, London.

SMITH, D. C. (1974). Transport from symbiotic algae and symbiotic chloroplasts to host cells. *Symp. Soc. exp. Biol.* **28**, 485–520.

SMITH, D. C., MUSCATINE, L. and LEWIS, D. H. (1969). Carbohydrate movement from autotrophs to heterotrophs in parasitic and mutualistic symbiosis. *Biol. Rev.* **44**, 17–90.

TAYLOR, D. L. (1973). Algal symbionts of invertebrates. *A. Rev. Microbiol.* **27**, 171–187.

TOBLER, F. (1925). "Biologie der Flechten." Borntraeger, Berlin.

TRENCH, R. K., BOYLE, J. E. and SMITH, D. C. (1973a). The association between chloroplasts of *Codium fragile* and the mollusc *Elysia viridis*. I. Characteristics of isolated *Codium* chloroplasts. *Proc. R. Soc. Lond.* B **184**, 51–61.

TRENCH, R. K., BOYLE, J. E. and SMITH, D. C. (1973b). The association between chloroplasts of *Codium fragile* and the mollusc *Elysia viridis*. II. Chloroplast ultrastructure and photosynthetic carbon fixation in *E. viridis*. *Proc. R. Soc. Lond.* B **184**, 63–81.

YONGE, C. M. (1968). Living corals. *Proc. R. Soc. Lond.* B **169**, 329–344.

Author Index

The numbers in *italic* indicate the pages on which names are mentioned in the reference lists

Taxonomic Index

Chemical Index

Metabolites and metabolically active compounds only are included

A

acetyl-CoA, 151, 153
N-acetyl glucosamine, 416, 427
alanine, 449, 451, 502
alectoronic acid, 12, 145, 160, 172
aluminium, 434
amides, 427–8
amino acids, 197, 397, 449–51, 490, 500–1, 504
ammonia, 61, 395, 397, 450, 500
α-amylase, 462
β-amylase, 462–3
anthraquinones, 152–3, 185, 188
anziaic acid, 160
arabitol, 390, 392, 396, 469, 480–1
arginine, 451
aspartic acid, 427, 449–51
ATP, 475–6, 486
atranorin, 10, 12, 62, 160–2, 167, 169, 186, 193–4, 196, 202–3, 207–10, 328
A(U)DP-glucose, 461
A(U)DP-glucose-1-phosphate, 461

B

baeomycesic acid, 163, 170
barbatic acid, 163–4, 170
bellidiflorin, 63
biotin, 398, 508

C

caesium, 420–1
calcium, 365, 367–8, 375–82, 420, 429, 434
calcium carbonate, 487
calcium oxalate, 11, 40
calcium pectate, 367
calycin, 10

caperatic acid, 158, 160
carbohydrates, see Subject Index
carbon dioxide, see Subject Index
carotenoids, 18, 173
casamino acids, 490
chitin, 416, 427–8
chloride, see Subject Index
chlorophyll, 199, 323, 408, 490
citric acid, 434
citrulline, 451
cobalt, 421–6
α-collatolic acid, 160
congyrophoric acid, 61–2
connorstictic acid, 172
constictic acid, 63, 172
copper, 366, 421–6, 428
cryptochlorophaeic acid, 162, 186, 196, 200–1
cutin, 8

D

DCMU, 3-(3,4-dichlorophenyl)-1, dimethyl urea, 457, 475–7, 486
decarboxythamnolic acid, 164
6-deoxy-glucose, 487
2-deoxy-glucose, 468, 470–3
2-deoxy-glucose-6-phosphate, 472
depsides, 61, 143, 146, 151–3, 158, 161, 165, 169, 195–6, 434, see also lichen substances in Subject Index
β-depsides, 152, see also lichen substances in Subject Index
depsidones, 61, 143, 146, 151–3, 158, 165, 169, 197, 434, see also lichen substances in Subject Index
β-depsidones, 123, 152, see also lichen substances in Subject Index

Subject Index

The Systematics Association Publications

LONDON. Published by the Association

Systematics Association Special Volumes

*Published by Academic Press for the Systematics Association